Emerging Trends in Cybersecurity Applications

Kevin Daimi • Abeer Alsadoon • Cathryn Peoples •
Nour El Madhoun
Editors

Emerging Trends in Cybersecurity Applications

Springer

Editors
Kevin Daimi
University of Detroit Mercy
Detroit, MI, USA

Abeer Alsadoon
Kent Institute Australia
Sydney, NSW, Australia

Cathryn Peoples
Ulster University
Ulster, UK

Nour El Madhoun
EPITA Engineering School
Paris, France

ISBN 978-3-031-09639-6 ISBN 978-3-031-09640-2 (eBook)
https://doi.org/10.1007/978-3-031-09640-2

This Springer imprint is published by the registered company Springer Nature Switzerland AG
The registered company address is: Gewerbestrasse 11, 6330 Cham, Switzerland

Preface

With the constantly increasing reliance of organizations on applications that span business, banking, insurance, education, marketing, healthcare, engineering design, manufacturing military, government, and communication sectors, protecting these applications is vital for their survival and continuity.

Cybersecurity applications describes how the security measures, techniques, methodologies, approaches, concepts, and procedures are specifically implemented in these applications to make them immune against various attacks and to eliminate or minimize security vulnerabilities, and mitigating risks and threats. By doing so, data or code within these applications will be protected against theft or hijacking. The implementation of these application-protecting techniques and methodologies involves both software and hardware.

Emerging Trends in Cybersecurity Applications provides an essential compilation of relevant and cutting-edge academic and industry work on key cybersecurity applications topics. Furthermore, it introduces cybersecurity applications to the public at large to develop their cybersecurity applications knowledge and awareness. The book can be a valuable resource to applied cybersecurity experts towards their professional development efforts and to students as a supplement to their cybersecurity courses.

This book concentrates on a wide range of advances related to cybersecurity applications which include, among others, applications in the areas of data science, Internet of Things, artificial intelligence, the Web, high-tech systems, cyber-physical systems, mobile devices, cloud computing, distributed systems, vehicles, software, energy, and education. It introduces the concepts, techniques, methods, approaches, and trends needed by cybersecurity applications specialists and educators to keep current their cybersecurity applications knowledge. Further, it provides a glimpse of future directions where cybersecurity applications are headed. It is a rich collection of carefully selected and reviewed manuscripts written by diverse cybersecurity applications experts in the listed fields and edited by prominent cybersecurity applications researchers and specialists.

Emerging Trends in Cybersecurity Applications has several important features. It provides an excellent professional development resource for educators and practitioners on state-of-the-art cybersecurity applications materials, contributes towards the enhancement of the community outreach and engagement component of cybersecurity applications, and introduces various techniques, methods, and approaches adopted by cybersecurity applications experts. In addition, it provides detailed explanation of the cybersecurity applications concepts that are pertinently reinforced by practical examples, a road map of future trends that are suitable for innovative cybersecurity applications training, and a rich collection of manuscripts in highly regarded cybersecurity applications topics that have not been creatively compiled before. It is written by cybersecurity professors and industry security professionals with long experience in the field of cybersecurity applications.

The book is organized into seven parts: the first part deals with cybersecurity applications in Internet of Things; the second concentrates on cybersecurity applications in the Internet, networking, and the cloud; the third involves cybersecurity applications in vehicles; the fourth covers cybersecurity applications in mobile computing; the fifth spans cybersecurity applications in energy systems; the sixth introduces cybersecurity applications in cyber-physical systems, artificial intelligence, and software; and the seventh includes miscellaneous cybersecurity applications. A brief overview of the chapters is introduced below.

The Internet of things (IoT) consists of various physical objects equipped with sensors and actuators, computing capabilities, and other hardware and software technologies that communicate to exchange data with other objects through the Internet or other communication media. To demonstrate how such systems could be protected, chapters on Elliptic Curve cryptography, transfer learning model for intrusion detection systems are introduced to safeguard data exchange, and end-to-end security are used. These chapters are enhanced with a third chapter on application of intrusion detection system.

To demonstrate how cybersecurity is applied to the Internet, network, and the cloud, several authors contributed their chapters. These included guiding users towards less revealing Internet browsers, analyzing the threat landscape inside the dark web, securing the 5G network slices auction broker, applying zero trust architecture and probability-based authentication to preserve security and privacy of data in the cloud, and using data mining for prevention of cross-site scripting (XSS).

The security of vehicles is critical, especially after fully autonomous vehicles are introduced. To cover this area, authors presented two chapters. The first treats cyberattacks and risks in V2X-assisted autonomous highway merging, and the second applies machine learning framework to detect various intrusions in vehicle communications.

Nowadays, everyone owns at least one type of small electronic equipment that connects to the Internet, such as mobile phones and tablets. As a result of this connection, these devices must be protected. To this extent, two chapters are introduced: implementation of uncertainty models for fraud detection on mobile advertising, and improving Android applications quality through extendable, automated security testing.

Deployment of smart energy minimizes global warming pollution, improves public health, and cuts costs on traditional fuels. Attacks on such energy sources could be fatal. Hence, smart energy demands protection. To determine possible protection against attacks, chapters on provably secure data sharing scheme for smart gas grid in fog computing environment, countering cybersecurity threats in smart grid systems using machine learning, and preserving the privacy and cybersecurity of home energy data are presented.

Cyber-physical Systems, artificial intelligence, and software applications are essential for many fields. They are being used extensively in business, health, and government. Chapters on non-stationary watermark-based attack detection to protect cyber-physical control systems, data-driven software vulnerability assessment and management, and application of homomorphic encryption in machine learning reveal examples of some approaches to protect them.

Further chapters deal with design of ethical service-level agreements to protect ethical cyber attacks and victims, and defend against adversarial attack on knowledge graph embedding.

The editors of *Emerging Trends in Cybersecurity Applications* aim to deliver a book that is valuable for faculty, researchers, security professionals, students, and the society at large to understand how cybersecurity could be implemented in various fields.

Detroit, MI, USA — Kevin Daimi
Sydney, NSW, Australia — Abeer Alsadoon
Ulster, UK — Cathryn Peoples
Paris, France — Nour El Madhoun

Acknowledgments

This book would not have been possible without the effort, contribution, and support of many people. We would like to thank all authors who submitted their chapters and congratulate those whose chapters were accepted. It gives us great pleasure to thank our reviewers, listed below, who spent long time in reviewing the chapters and selecting only quality chapters that are suitable for this book, and providing excellent comments to authors to improve their chapters. Finally, we would like to express our appreciation to Mary James, Zoe Kennedy, and Brian Halm at Springer for their helpfulness, courteousness, and professionalism.

Laila Ahddar, IDEMIA, France
Abeer Alsadoon, Charles Sturt University (CSU), Australia
Allen Ashourian, ZRD Technology, USA
Richard Bean, University of Queensland, Australia
Biodoumoye George Bokolo, Sam Houston State University, USA
Agathe Blaise, Thales Group, France
Rachid Boudour, Badji Mokhtar-Annaba University, Algeria
Samia Bouzefrane, CNAM University, France
Oliver Buckley, University of East Anglia, UK
Jinli Cao, La Trobe University, Australia
Khalil Challita, Notre Dame University-Louaize, Lebanon
Kevin Daimi, University of Detroit Mercy, USA
Ahmed Dawoud, Western Sydney University, Australia
Mehrdad Dianati, University of Warwick, UK
George Dimitoglou, Hood College, USA
Ioanna Dionysiou, University of Nicosia, Cyprus
Badis Hammi, EPITA, France
Dinh Thai Hoang, University Technology of Sydney, Australia
Mary Ann Hoppa, Norfolk State University, USA
Rida Khatoun, Telecom-ParisTech, France
Irene Kopaliani, Princeton University, USA
Konstantinos Koufos, University of Warwick, UK

Nour El Madhoun, EPITA Engineering School, France
Christophe Maudoux, CNAM University, France
Douglas Millward, University of Essex, UK
Esmiralda Moradian, Stockholm University, Sweden
Mais Nijim, Texas A&M University-Kingsville, USA
Nkaepe Olaniyi, Kaplan Open Learning, UK
Pierre Parrend, EPITA, France
Rachana Patil, Pimpri Chinchwad College of Engineering, India
Cathryn Peoples, Ulster University, UK
Avi Purkayastha, National Renewable Energy Laboratory, USA
Karpoor Shashidhar, Sam Houston State University, USA
Muhammad Siddiqi, Kent Institute Australia, Australia
Gaurav Somani, University of Rajasthan, India
Jenny Torres, Escuela Politécnica Nacional (EPN), Ecuador
Israr Ullah, Ulster University, UK
Syed Muhammad Unsub Zia, Ulster University, UK

Contents

About the Editors

Kevin Daimi received his PhD from the University of Cranfield, England. He has a long academic and industry experience. His research interests include computer and network security with emphasis on vehicle network security, software engineering, data science, and computer science and software engineering education. He has published a number of papers on vehicle security. He is the editor of three books in cybersecurity: *Computer and Network Security Essentials*, *Innovation in Cybersecurity Education*, and *Advances in Cybersecurity Management*, which were published by Springer. He has been chairing the annual International Conference on Security and Management (SAM) since 2012. He is also program chair of the 2022 International Conference on Innovations in Computing Research (ICR'22), Athens, Greece. Kevin is a senior member of the Association for Computing Machinery (ACM), a senior member of the Institute of Electrical and Electronic Engineers (IEEE), and a fellow of the British Computer Society (BCS). He is the receipient of the Outstanding Achievement Award from the 2010 World Congress in Computer Science, Computer Engineering, and Applied Computing (WORLDCOMP'10) in recognition and appreciation of his leadership, service, and research contributions to the field of network security. He is currently Professor Emeritus of Computer Science and Software Engineering at the University of Detroit Mercy.

Abeer Alsadoon received her PhD from the University of Technology, Baghdad. She has published more than 100 papers in A- and B-ranking journals and has more than 100 conference papers published in different IEEE conferences. Abeer received 20 awards for teaching and research excellence from different Australian Universities. She had been recognized nationally as one out of five finalists for the prestigious 2019 Australian Women's Agenda Leadership Awards in the category of Emerging Female Leader in the Government and Public Sector. She chairs of the 2022 International Conference on Health Informatics and Medical System (HIMS'22), Las Vegas, USA, and is program chair of the 2022 International Conference on Innovations in Computing Research (ICR'22), Athens, Greece. Abeer has been chairing the program and technical committees of the Annual IEEE International Conference on Innovative Technologies in Intelligent Systems & Industrial Application (CITISIA'20), Sydney, since 2020. She has more than 25 years of academic and industry experience. Abeer is currently a dean at Kent Institute Australia. As a leader, she oversees research, scholarly activities, and professional development for all disciplines at that institute.

Cathryn Peoples received her BA degree (with honors and diploma in industrial studies) in business studies with computing, MSc degree in telecommunications and Internet systems, and PhD degree in networking from Ulster University, UK, in 2004, 2005, and 2009, respectively. She is currently employed by The Open University in the School of Computing and Communications as an associate lecturer. Cathryn is also employed on a part-time basis as a research associate at Ulster University working on Internet of Things (IoT) management. She became the editor-in-chief of the EAI-endorsed *Transactions on Cloud Systems* in January 2020. Her research interests include smart cities, green IT, network management, and secure software development.

Nour El Madhoun received her master's degree in networks/computer science from Sorbonne Université/Télécom ParisTech in 2014, and her PhD degree in cybersecurity/computer science from Sorbonne Université in 2018. In 2019, she joined ISEP – Engineering School, Paris, as Associate Professor of Cybersecurity in addition to overseeing the engineering cycle – Digital Security and Networks. In 2018, Nour gained industry experience through work as a postdoctoral researcher at Orange Labs. At Sorbonne Université, she became an ATER in 2017. From 2020 to 2022, Nour joined, EPITA, an engineering school in Paris, as Associate Professor of Cybersecurity and Blockchain. Her current research focuses on network security, cryptographic protocols, EMV payment, NFC technology, and blockchain and smart-contracts technologies. Nour is currently Associate Professor of Computer Science, Cybersecurity, and Blockchain at ISEP in Paris. She is also an associate researcher at Sorbonne Université (LIP6-PHARE Team).

Part I
Internet of Things Applications Security

Ephemeral Elliptic Curve Diffie-Hellman to Secure Data Exchange in Internet of Medical Things

Osman Salem and Ahmed Mehaoua

1 Introduction

With the advances in information and communication technologies, the Internet of Medical Things (IoMT) becomes a promising solution for remote healthcare monitoring, where a set of wearable biosensors are used to collect the physiological data from the monitored patient, and to transmit the acquired measurements to a Local Processing Unit (LPU—such as Smartphone or tablet) for processing and alerting the healthcare professionals when an emergency is detected. Such monitoring systems are able to assist the healthcare professionals by analyzing the acquired physiological data in the edge of the network, and raising an alarm when an anomaly is detected by highlighting abnormal changes in monitored parameters. The use of IoMT for remote monitoring, and for the detection of chronic diseases gives impetus to the development and implementation of enriched and ubiquitous health services.

The use of IoMT devices provides a tool to improve the Quality of Life (QoL) by allowing the monitored patient to continue their Activity of Daily Living (ADL) while being monitored and followed-up. Their fast deployment has an impact on reducing the number of beds occupied by patients kept under monitoring. The COVID-19 pandemic has driven an exponential rise in IoMT, with quarantine and stay-at-home orders, which accelerated trends in telemedicine and telehealth.

However, the medical data involves stringent security requirements which are not available in sensors with restricted resources [1]. The collected sensitive medical data is transmitted to the LPU for processing using wireless technologies, and an attacker in vicinity can eavesdrop or modify the intercepted data [2] leading to false

O. Salem (✉) · A. Mehaoua
Borelli Research Center, University of Paris, Paris, France
e-mail: osman.salem@u-paris.fr; ahmed.mehaoua@u-paris.fr

K. Daimi et al. (eds.), *Emerging Trends in Cybersecurity Applications*,
https://doi.org/10.1007/978-3-031-09640-2_1

alarms, or can conduct a black hole attack by preventing information from being transmitted to the LPU, in order to prevent the system from raising alarms. The attacker may also exploit the vulnerabilities [3] in the software of IoMT device to increase the transmission rate and deplete the energy of sensors or to flood the LPU. Therefore, a security framework is required to ensure the integrity of the exchanged data.

Several mechanisms have been proposed and tested in the literature for securing the exchanged data between the sensors and the LPU [4]. The Bluetooth Low Energy (BLE) is widely implemented today in IoMT to transmit data from sensors to the LPU. The IoMT object requires a short range communication, low bandwidth, low delay, and reduced energy consumption. BLE exchanges less data than normal Bluetooth to reduce energy, and devices can stay in "sleep mode" until the next interaction. These advantages have led to this wireless technology being widely deployed in IoMT for remote monitoring of patients during long periods of time (months and even years) without charging or changing the battery.

Devices in BLE are classified into two types: central and peripheral. The central device (e.g., smartphone) has higher computational power and storage than peripherals and sends commands and collects data from peripherals. Conversely, the peripheral or the slave cannot initiate a connection and can only connect to a single master. It only executes received orders and sends packages to advertise its presence. The peripherals stop sending advertising packets when they receive a specific packet, indicating that they are connected to a central device. Peripherals are sensors that collect and send data to the central device for processing, such as the collection of blood pressure, SpO2 and body temperature, and other physiological parameters by sensors, as well as their transmission to a central processing unit (smartphone or tablet).

BLE operates using radio frequency on 2.4–2.8 GHz band within a distance of 10 m. It operates with 40 physical channels, against 80 for legacy Bluetooth, for frequency and time multiplexing thanks to the L2CAP layer. The difference between two channels is found to be 2 MHz. The devices in advertising mode send packets of 31 bytes at regular intervals. This task is conducted only on 3 of the channels: 37, 38, and 39. The other channels are reserved for data exchange between devices [5].

To establish a connection, the central device alternates between scanning for pairing requests and sending advertising packets. It scans to check if it can find a peripheral to begin the exchange with it. The scanning process is expensive, so the scan usually does not run indefinitely. The BLE devices exchange their services, their capabilities, their inputs (such as the presence of keyboard or not) and output resources, their names and their manufacturers' information, authentication method, etc. during the first phase of pairing, which is not encrypted. However, the second phase is for key exchange and needs to be secured.

In the second phase of pairing, one of the devices generates a Temporary Key (TK) which will be known from both devices. Confirmation of the key is made through the exchange of random values, encrypted and then decrypted. With the TK and random values, a Short-Term Key (STK) is derived by devices without traveling in the network. The connection will be encrypted with this key at the link layer level. Eventually, a Long Term Key (LTK) can be exchanged for bonding.

Four pairing models are supported by BLE: "Just Works," Out of Band, Numerical Comparison, and Passkey. The BLE secures the communication using the Advanced Encryption Standard (AES) algorithm with a key length of 128 bits. However, when the object does not have I/O capabilities, the BLE "Just Works" pairing mode does not provide any protection against MitM (Man in the Middle) or eavesdropping. As the IoMT device does not have display or keyboard, the default value of pairing code $0x00$ is used as value for TK ($TK = 0$), which in turn is used to derive the STK and the LTK.

In other words, we can connect to any BLE device that uses the "Just Works" pairing mode and access the exchanged medical data. In fact, this pairing mode is deployed in several healthcare devices available in the market, and it does not provide any protection against MitM and must not be used in healthcare monitoring services. In the real world, sensors do not have I/O interfaces and this mode is currently deployed in healthcare products available in the market. The illegal access to medical data causes a huge violation to the privacy of the monitored patient, and the injection of faulty measurements may threaten the life of patient with a decision based on faulty measurements.

In this chapter, we implement the ECDHE with key renewal process to secure the communications and prevent MitM attacks while using the same security mechanisms in BLE for confidentiality and integrity. We use the Elliptic Curve Cryptography (ECC) with pre-distributed public keys used to derive the encryption key. The ECC has a small key size compared to RSA, where a 384 bits key is equivalent to 3072 bits in RSA [6]. Elliptic curve is more convenient for IoMT devices with constrained resources, where its usage is limited to derive a shared key using ECDH. The AES-CCM implemented in the BLE standard is used in our approach to provide encryption and integrity, and to prevent the MitM from conducting eavesdropping or injection attacks.

The IoMT devices are susceptible to various exploits and an attacker can easily change the behavior of compromised devices to increase the transmission rate and flood the LPU. Such change increases the energy consumption of the compromised devices and the LPU and threatens the functioning of the network. There is a need of a suitable system to detect such intrusion and to alert the user. We applied the sequential change point detection algorithm PELT [7] and the box-and-whisker plot on the number of received packets by the LPU to detect such changes and raise a network alert for user.

The rest of this chapter is organized as follows. In Sect. 2, we review recent related work. Section 3 presents our proposed approach for securing the communication link between the devices and to detect anomaly in the physiological parameters and in the number of received packets. In Sect. 4, we present our experimental results from the application of our proposed framework on real physiological data. Finally, Sect. 5 concludes the chapter and presents our future work.

2 Related Work

Despite the security measures adopted in BLE, some attacks are still feasible up to date. They range from simple passive data interception to identity theft and Denial of Service (DoS). Pallavi et al. in [8] review feasible security attacks on IoT devices with BLE transmission technology. Sevier et al. in [9] highlighted BLE vulnerabilities and proved that TK is vulnerable and showed how to sniff and decrypt acquired BLE data. They used Ubertooth dongle to capture BLE packets and to obtain the signal strength of the different channel frequencies. As this dongle is able to capture exchanged packets in the handshake, the TK could be cracked using the Crackle software on the Ubertooth data capture. Therefore, the LTK can be derived from the TK [10], and Wireshark can be used to decrypt the BLE packets when the LTK is provided. As Ubertooth outputs PCAP file, the sniffer Wireshark can read it and decrypt the packets in an automatic manner.

Lounis et al. in [11] confirm the results of Sevier et al. in [9]. Using the "Just Works" pairing mode, they demonstrated its weakness by showing how to generate keys. Moreover, simple technologies have been used for conducting the sniffing attack. Data from smart deadbolt, bike lock, and a lightbulb have been captured and decrypted in their experiments. However, the "Just Works" pairing method is not secure enough to generate a TK.

Cominelli et al. in [12] presented an open-source sniffer based on a Software-Defined Radio framework to capture BLE data packets in a very simple manner. They used the Graphic Processor Unit (GPU) to process the traffic. Even though sniffing can be dangerous for sensitive medical data, the attacker can induce a Denial of Service (DoS) or even spoof a device.

Therefore, the IoMT are vulnerable to various attacks as the data is transmitted using BLE wireless technology from the sensor to the LPU [13]. An adversary can modify, eavesdrop, or delete the data [14]. The impact of such attacks has been highlighted on insulin pumps with over dosage to kill the patient, and on pacemaker [15] to threaten the patient's life.

The work of Lahmadi et al. in [16] demonstrated a MitM attack against BLE and showed the low security features and inherent vulnerabilities. Afterward, they compared two unsupervised learning techniques to detect suspicious data, followed by classification method to tag packets as normal or attack from suspicious measurements. Their work is very near in his spirit to our work, where they combined supervised and unsupervised techniques to detect anomaly. However, the supervised classification requires labelled training data, which is not easy to find or to build. It is interesting to propose a lightweight and reliable sequential and non-parametric approach to prevent passive and active attacks conducted by MitM.

Aghili et al. in [17] proposed a lightweight multi-factor authentication protocol for e-health systems in IoMT. Ayub et al. in [18] proposed a secure authenticated key agreement protocol using the concepts of Physically Unclonable Function (PUF). Other research work focused on authentication, encryption, integrity, and intrusion detection to secure the network of IoMT devices [2]. However, most of the proposed solutions have higher computation complexity which prevents their deployment on the constrained resources in IoMT devices.

Gulen et al. in [19] implemented ECC on the MSP430 micro-controller, which is commonly used in wireless sensor devices to secure wireless transmissions. Their implementation combined number transformation and elliptic curves to reduce the processing complexity. However, the implementation of other elliptic curves with more efficient formulas for key derivation is required to evaluate the complexity of such techniques.

To overcome these problems, Ahmed et al. in [20] proposed an enhanced ECDH for securing the data exchange of IoT applications. Our approach is similar in the spirit to their approach, where we use the Ephemeral ECDH to derive a session key for securing the data exchange of IoMT devices in "Just Works" pairing mode. The use of ephemeral keys allows key renewal in every time period.

On the other hand, the IoMT raises an alarm when a healthcare emergency is detected. Change Point Detection (CPD) algorithms seek to detect abrupt changes in the monitored physiological parameters, such as detecting changes in SpO2 to identify severe hypoxia or patient with COVID-19, or detecting changes in Blood Pressure (BP) to subsequently identify hypertension after vaccine. These changes need to be identified automatically with the large amount of collected data.

Several approaches for identifying changes in monitored data have been proposed in the literature [21]. The most common methods are those based on segmentation. These methods identify one or more points in a dataset where the statistical properties (e.g., mean and variance), change over time, based on the likelihood of the data in the time series. Among the proposed segmentation methods [21–23], window-based change point detection, Binary Segmentation (BS), and Optimal Partitioning (OP).

BS [23] is a sequential approach with a computational complexity $O(nlogn)$ where n is the number of samples in the segment. The principle of this method is to detect a change point in the time series, and to subdivide it into two parts, where the first is before the change and the second is after the change. The operation is repeated on the two resulting parts. BS is fast and seeks to identify the minimum number of change points.

Window-based change point detection is used to perform rapid segmentation of the signal. The algorithm uses two windows that slide along the data stream. The statistical properties of the signals in each window are compared to measure the deviation. Window Segmentation (WS) has low complexity $O(nw)$ where n and w are the number of sample and the size of window, respectively. However, it does not produce optimal segmentations [22]. The OP method has higher computational complexity $O(n^2)$ when compared to the previous two methods (BS & WS) but is able to find the exact global optimum.

Killick et al. in [24] improved the OP by proposing a new approach to search for change points. Their proposed approach is the PELT [7], which is an efficient approximate search method able to detect all change points with respect to the change of the mean or the variance, and regardless of the statistical distribution of the time series. Its basic idea is to divide the time series into several segments where the average of each segment is significantly different from the previous and subsequent segments. The penalty is an adjustable parameter in PELT to control the number of detected change points.

The PELT has several advantages compared to other methods, especially in terms of linear computational complexity $O(n)$ as it uses dynamic programming and pruning [7]. Yeung et al. in [25] used the PELT method to analyze public feelings towards personal masks during the COVID-19 period using Twitter data. Valdez et al. in [26] exploited PELT to identify significant changes in the volume and feeling of tweets to obtain mental health information in the USA during COVID-19 pandemic. The detection of such changes has a significant implication to trigger mitigation efforts.

Several previous work [21, 22] devoted to the search for the most adequate strategy to segment the data and compare many CPD algorithms. Their results proved that PELT provides the best tradeoff between complexity and detection accuracy, where it has the lower complexity and memory requirements when compared to other methods. This is why we will use PELT in our approach for CPD in the measurements to detect healthcare emergency, and in the number of received packets to detect compromised sensors with a high transmission rate, which intends to flood the LPU and deplete the energy.

3 Proposed Approach

Most IoMT devices do not have I/O capabilities and the "Just works" with the default pin code is used. To secure the communication links between devices and the LPU and to prevent attacks conducted by MitM (as shown in Fig. 1), which is able to intercept and alter the data, our proposed approach is based on pre-distributed ECC keys before deployment. These small size pre-distributed keys are used to derive a shared session key to encrypt the communication between devices and LPU using the AES-CCM deployed in BLE.

The creation of asymmetric keys is based on modern public key ECC, which is based on mathematical elliptic curves known to produce a smaller key size than RSA. The reduced key size makes the encryption operation faster and reduces the processing complexity. Let F be a field with N elements, E is an elliptic curve with

Fig. 1 MitM attacks against IoMT

a set of points (x, y), and G is the identity or the neutral element of the curve. E is a function known as the Weierstrass Equation (given in Eq. 1) defined over the field F:

$$(E): \quad y^2 + a_1xy + a_3y = x^3 + a_2x^2 + a_4x + a_6 \tag{1}$$

The coefficients $a_1, a_2, a_3, a_4, a_6 \in F$ have real values. A curve of the Weierstrass equation is said to be smooth if the partial derivatives in x and y of the Eq. 2 do not cancel each other at the same time instant.

$$f(x, y) = y^2 + a_1xy + a_3y - x^3 - a_2x^2 - a_4x - a_6 \tag{2}$$

For their use in cryptography, a simplification of the Eq. 1 is given in Eq. 3:

$$y^2 = x^3 - ax + b \quad \text{with} \quad 4a^4 + 27b^2 \neq 0 \tag{3}$$

To create an asymmetric key pair (P, K), we used openSSL with P-384 (secp384r1) to derive the 384-bit key pair, where P_i is used to denote the public key, which results from ECC point multiplication of G with the private key (η_i):

$$P_i = \eta_i * G \tag{4}$$

The operator "*" is used to denote ECC point multiplication. With a pre-distributed key, the use of ECDH mechanism does not require any exchange between the two devices to derive the shared symmetric encryption key, as shown in Fig. 2 and in Eq. 5. In Fig. 2, P_L denotes the public key of the LPU, P_D denotes the public key of the IoMT device, and S_k denotes the derived shared key.

$$S_k = \eta_i * P_j \quad \text{with} \quad i \neq j \tag{5}$$

where S_k is the secret key used to guarantee the security of exchanged data, and P_j is the public key of the other device. However, the derived secret key is always the same. To renew the key in our approach, the LPU starts by deriving an ephemeral ECC key pair (η_E, P_E) for each IoMT device, and transmits the public key (digitally signed) to the device to derive the same ephemeral secret key (as shown in Fig. 3),

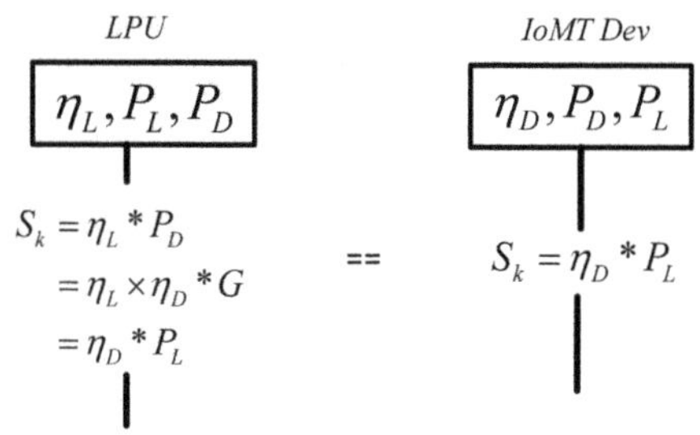

Fig. 2 Elliptic curve Diffie-Hellman (ECDH)

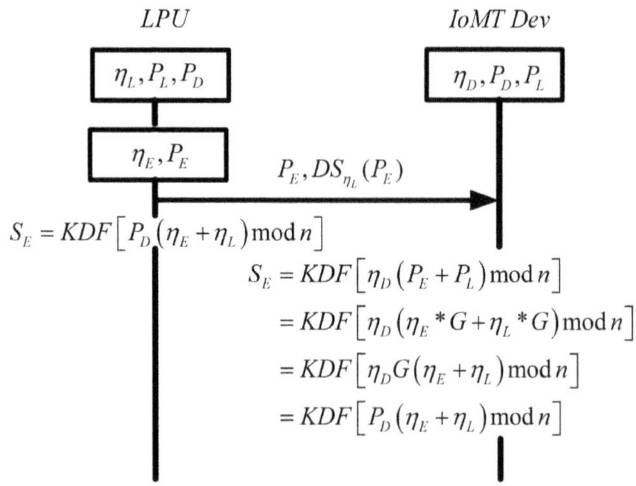

Fig. 3 Ephemeral ECDH

which will change every period of time T_k. In Fig. 3, the Key Derivation Function is denoted by KDF, and the function $DS_{\eta_L}(P_E)$ is used to denote the Digital Signature (DS) of ephemeral key P_E.

The confidentiality and integrity of the exchanged data are provided by AES-CCM to avoid the MitM from accessing the content or modifying the values of measurements. To prevent data suppression by the MitM, the transmission is reliable and must be acknowledged (ACK) in both directions to avoid black hole attack. In the case where the IoMT device does not receive an ACK after 3 retransmissions for k consecutive packets, it raises a local alert (light or sound) to notify user with a network or security problem.

To detect anomaly in acquired vital signs, we start by preprocessing the data over a window of measurements. Let $y_{1:n}$ denote the set of measurements during the period of time T, where $y_{1:n} = (y_1, \ldots, y_n)$ is a set of n physiological measurements with real values. The CPD algorithm is able to identify m changes along with their positions $t_{1:m} = (t_1, \ldots, t_m)$. The position of the change point is an integer between 1 and n. The time series is supposed to be piecewise stationary, which means that some characteristics of the process suddenly change at unknown time instants $t_1 < t_2 < \ldots < t_m$. The data are normalized, and their values are between 0 and 1.

To detect change points, we applied the PELT method that aims to identify the instants of change in $y_{1:n}$. It is based on the OP and pruning method. The OP method aims to minimize cost:

$$\sum_{i=1}^{m+1} \left\{ C(y_{(t_{i-1}+1)}, \ldots, y_{t_i}) + \beta \right\} \tag{6}$$

where C is a cost function for the ith segment, and β is a penalty to prevent over-fitting. Subsequently, PELT uses pruning to increase the efficiency of the OP method while ensuring that the method finds an overall minimum of the cost function. The optimal segmentation is $F(n)$:

$$F(n) = \min_{t} \left\{ \sum_{i=1}^{m+1} \left[C\left(y_{(t_{i-1}+1)}, \ldots, y_t + \beta \right) \right] \right\} \tag{7}$$

The main idea behind the pruning is to remove these values of t which can never be minima of the minimization performed in each iteration. The OP method applies recursive conditioning by starting with a first conditioning on the last change point and calculating the optimal segmentation of the data up to the change point:

$$F(n) = \min_{t} \left\{ \begin{array}{c} \min\limits_{t|t_m} \sum\limits_{i=1}^{m} \left[C(y_{(t_{i-1}+1)}, \ldots, y_{t_i}) + \beta \right] \\ + C(y_{(t_m+1)}, \ldots, y_n) \end{array} \right\} \tag{8}$$

Using Eq. 6 to simplify the previous equation, the internal minimization is equal to $F(t_m)$ and the Eq. 8 can be re-written as:

$$F(n) = \min_{t_m} \left\{ F(t_m) + C(y_{(t_m+1)}, \ldots, y_n) \right\} \tag{9}$$

We applied the PELT on the received measurements and on the number of received packets. The CPD in the received measurements allows to detect emergency and to raise alarms for healthcare professionals, while the CPD in the total number of packets allows to detect compromised sensors with an increased transmission rate. However, the PELT method is sensitive to changes and identify all the change points with several false alarms. To increase the reliability of the system by reducing the False Alarm Rate (FAR), we apply the box-and-whiskers (boxplot) by comparing each identified change point by PELT with robust statistical parameters derived from a window of previous w values in order to confirm its deviation.

Let $Y_i^w = \{y_{t-w,i}, \ldots, y_{t,i}\}$ represents the sliding window of the last w values ($[DPC - w, DPC]$) for the ith monitored attribute. The first quartile Q_1 and the third quartile Q_3 of Y_i^w are used to derive the interquartile range $\hat{\sigma} = IQR = Q_3 - Q_1$. A measurement is considered as abnormal (as shown in Fig. 4) if the following condition is satisfied:

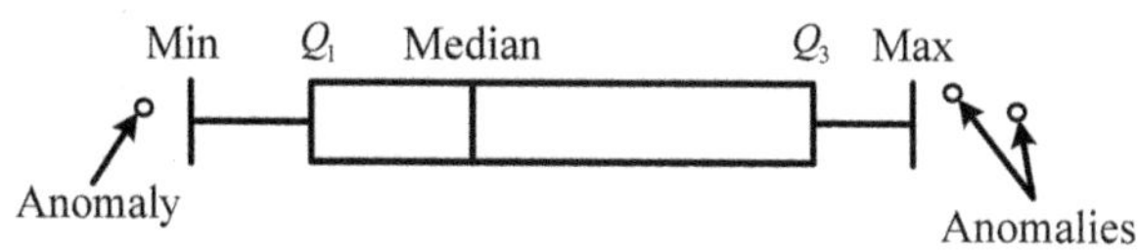

Fig. 4 Box-and-whiskers

$$y_{t,i} \leq Q_1 - 1.5.(Q_3 - Q_1) \vee y_{t,i} \geq Q_3 + 1.5.(Q_3 - Q_1) \tag{10}$$

A medical alarm is raised if the deviation is detected only in the monitored biometric parameters and not in the number of received packets.

4 Experimental Results

To conduct an experiment and analyze the performance of our proposed approach, we used real physiological data collected from a patient with cardiovascular disease. The monitored patient is 68 years old, 1.75 m living independently in his apartment and kept under monitoring. The used dataset is private, collected using other prototype and stored inside a CSV file. We focus only on the chunk with changes in our experiments.

Five vital signs are available in the dataset: ABP Mean (Ambulatory BP), Heart Rate (HR), Pulse, SpO2, and Respiration Rate (RR). The measurement units are: mmHg for BP, beat per minute (bpm) for HR and Pulse, respiration per minute (rpm) for RR and % for SpO2. A value of SpO2 lower than 95% is symptomatic of asphyxia and requires ventilator and assistance. To simulate a real life scenario in Fig. 1, we used two Raspberry Pi 4B, with 8 GB of RAM and BLE as IoMT devices that read data from the CSV file and transmit records to the LPU (Android tablet) for processing. The first device transmits SpO2 and Pulse, while the second is used to transmit BP, HR, and RR.

We start our experiments by using AdaFruit USB stick (presented in Fig. 5) as BLE sniffer and Wireshark to prove the ability of MitM to access the data in the BLE pairing mode. The captured data by Wireshark sniffer in "Just works" mode is shown in Fig. 12, where the clear text value of the HR is 96 bpm. We refer to [16] and several tutorials available online to conduct such an attack using kali Linux [27].

To prevent security attacks and leakage of sensitive data, we start by implementing our approach for ephemeral key derivation from ECDH, which is used to encrypt the data. We also configure the two devices to renew the key every 10 minutes to

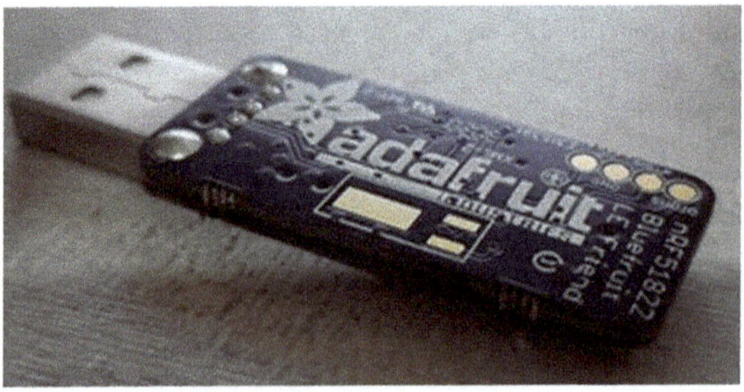

Fig. 5 Sniffer BlueFruit

prevent off-line password guessing. The anomaly detection is implemented in the LPU and aims to identify changes in physiological and total number of received packets. The received data on the LPU from the two Raspberry devices are decrypted before processing.

The Continuous Noninvasive Atrial BP measurement (CNAP) is used to measure the BP continuously in real-time. Several CNAP monitors based on PhotoPlethysmoGraphy (PPG) are available in the market [28]. The variations of ABP Mean (denoted by BP) measurements are presented in Fig. 6, where the heavy change is visible around the time instant 18,000 sec and lasts until the end. The ABP Mean is derived from Systolic Blood Pressure (SBP) and Diastolic Blood Pressure (DBP) as given in Eq. 11:

$$ABPMean = \frac{1}{3}SBP + \frac{2}{3}DBP \tag{11}$$

Similarly, the variations of the HR and PULSE are shown in Figs. 7 and 8 where correlated changes occur at the same instant as the BP. The variations of the RR and SpO2 are presented in Figs. 9 and 10, respectively. The SpO2 falls down and becomes lower than 90% (asphyxia) at the same time instant 18,000 sec, and this explains the simultaneous increase in the number of RR and in the measurements of BP, HR, and PULSE. The patient tries to get more oxygen by increasing his respiration and making more effort. In fact, the patient needs oxygen assistant in this chunk of data.

The variations of whole physiological parameters (BP, HR, Pulse, Respiration, SpO2) are presented in Fig. 11, where we can identify a correlated change point around 18,000 sec for approximately whole parameters. The HR and PULSE superpose as they measure the same information (Fig. 12).

Fig. 6 Blood pressure

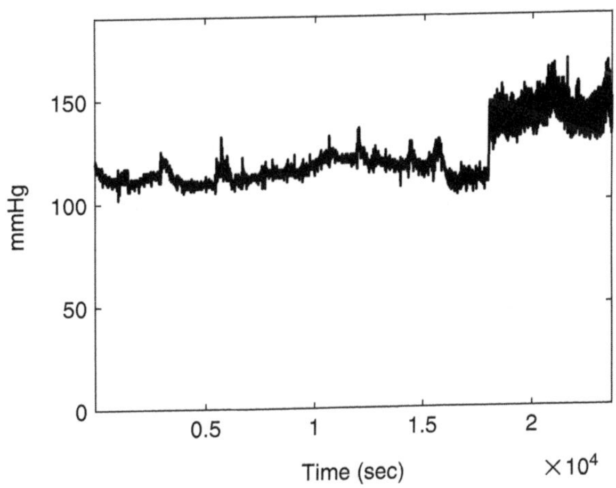

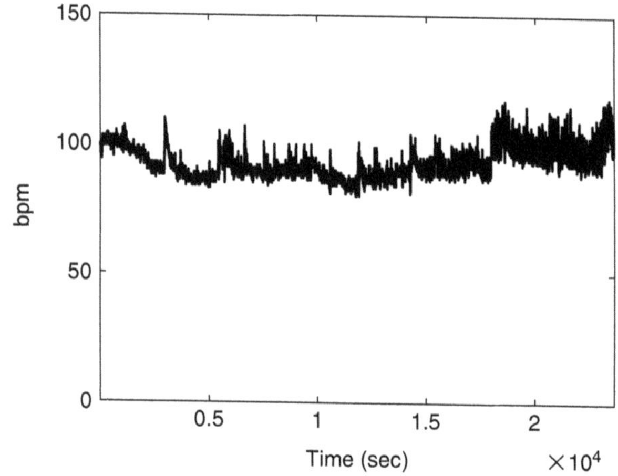

Fig. 7 Heart rate

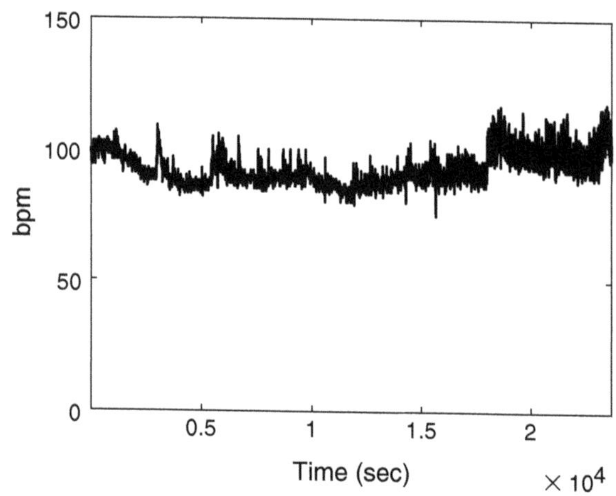

Fig. 8 PULSE

In the second set of experiments, we start by conducting a MitM attack to capture and verify the encryption of the data. A screenshot of the captured data with Wireshark is presented in Fig. 13, where we can notice that encrypted data cannot be decoded by the sniffer. Afterward, we test the security of our approach by assuming the worst case scenario to simulate MitM attack, where an attacker successfully compromises both IoMT devices by exploiting software vulnerability. We start by increasing the transmission rate and the value of measurements for only one device in the beginning, followed by simultaneous increase in the rate of the second device (as shown in Fig. 14a) to deplete the energy of compromised sensors, and to flood the LPU with packets containing modified values. The measurements of HR in the beginning of attack can be distinguished from the Pulse as shown in Fig. 14b, where the variations are surrounded by an ellipse.

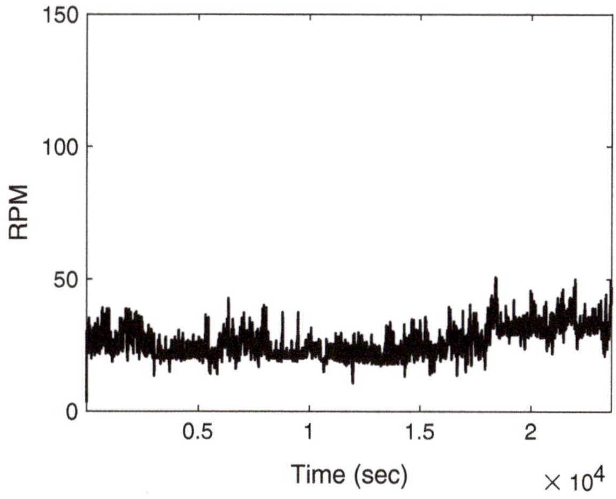

Fig. 9 Respiration rate

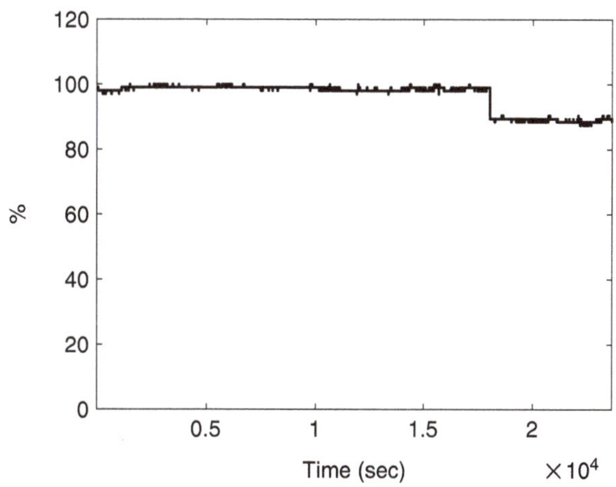

Fig. 10 SpO2

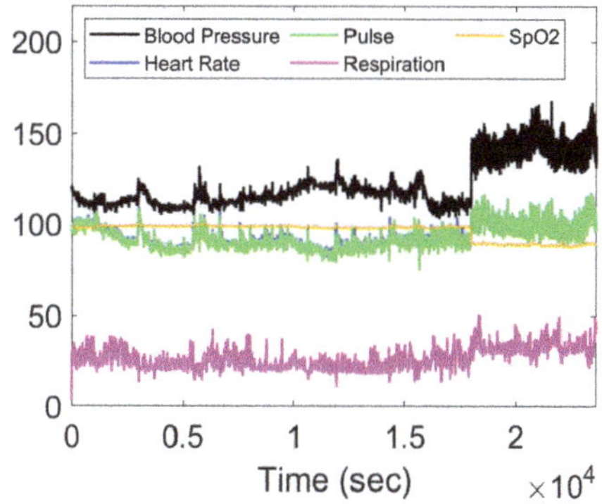

Fig. 11 All parameters

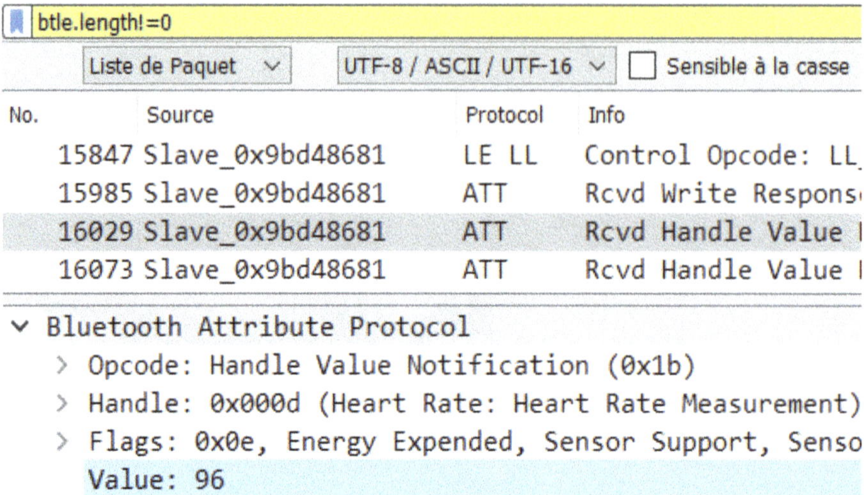

Fig. 12 MitM: Wireshark with the value of HR

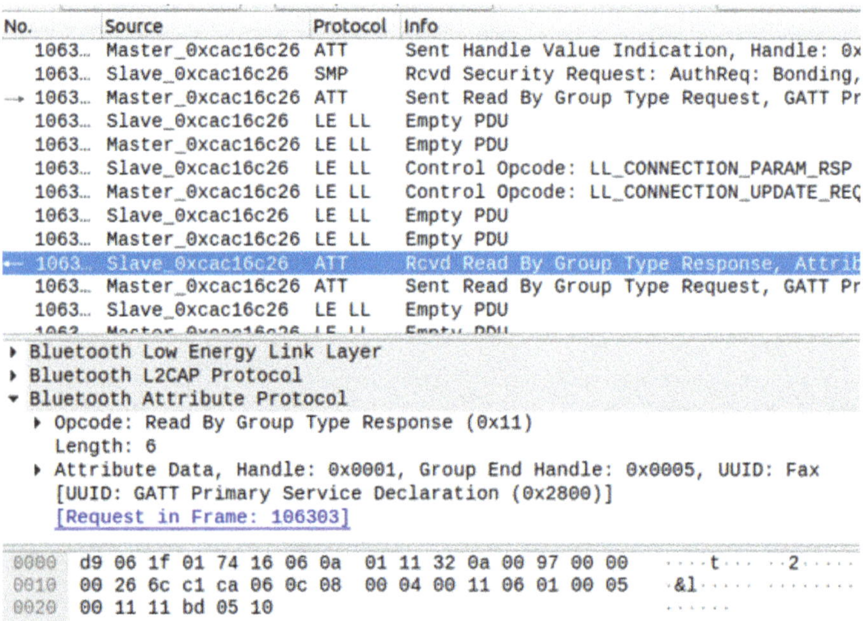

Fig. 13 MitM: encrypted data

The average of received measurements in each second was derived and used in Fig. 14b. Our approach detects a change in the number of received packets for these variations and raises a local alert for user as a network connection alert. In such

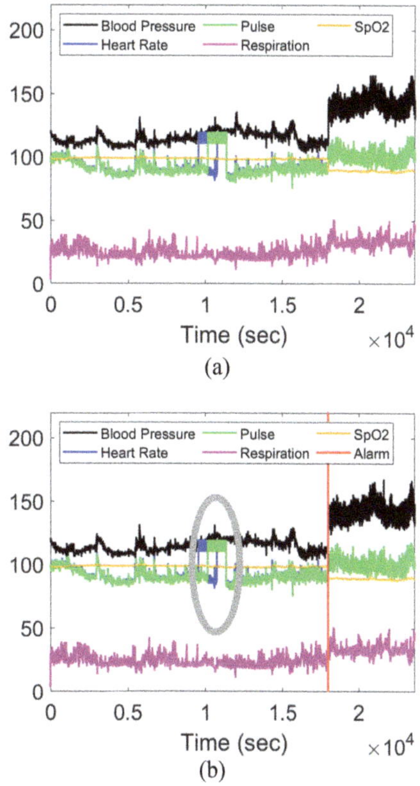

Fig. 14 Injected measurements. (**a**) Injected values. (**b**) Normal and alarm

situation, the user must re-initialize the system to force the change of the encryption key.

The raised medical alert is represented by vertical red line in Fig. 14b and triggered only if there is no change point in the number of received packets.

In the third set of experiments, we conduct a performance comparison using the Receiver Operating Characteristic (ROC) to study the impact of the threshold on the accuracy of the system in terms of True Positive Rate (TPR) and False Alarm Rate (FAR). The TPR and FAR are given in the following equations:

$$\text{TPR} = \frac{TP}{TP + FN} \tag{12}$$

$$\text{FAR} = \frac{FP}{FP + TN} \tag{13}$$

where TP is the number of True Positives, FP is the number of False Positives, FN is the number of False Negatives, and TN is the number of True Negatives. The ROC represents the variation of TPR with respect to FAR when changing the value of the score. A value of TPR closer to 100% indicates a high detection accuracy, while a lower value of FAR is desirable to achieve to enhance the reliability of the system. However, increasing the value of TPR induces an increase of FAR, and decreasing

Fig. 15 ROC

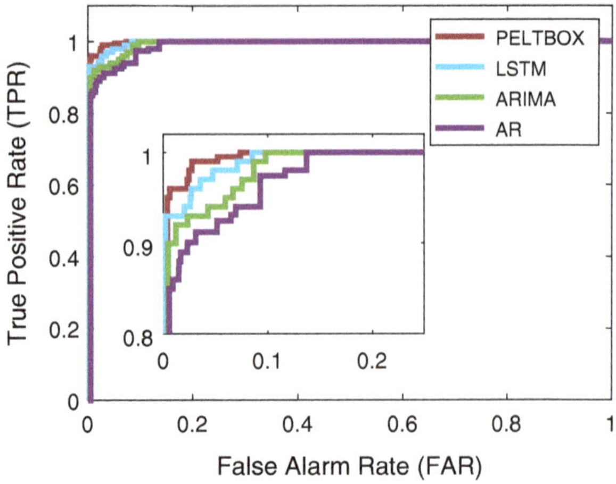

the FAR induces a reduction in TPR. Therefore, a tradeoff between TPR and FAR is required by changing the value of the decision threshold.

The ROC curve presented in Fig. 15 shows the relationship between the TPR and FAR for our proposed approach. To prove the effectiveness of our approach, we also conduct a performance comparison with existing works [29] which are based on the difference between predicted and measured values to identify changes in time series. The prediction of the current measurement was achieved using Long Short-Term Memory (LSTM), AutoRegressive Integrated Moving Average $ARIMA(p, d, q)$, and Auto Regressive $AR(p)$, with $p = 4$, $d = 1$, and $q = 2$.

The obtained ROC is presented in Fig. 15 where for a TPR of 99%, our approach has a FAR of 6%, followed by LSTM with 8%, ARIMA with 9% and AR with 12%. In fact, the use of our approach slightly outperforms the LSTM in term of FAR. On the other hand, even if the four methods have a linear computational complexity $O(n)$, our method has less execution time for processing one record than LSTM, where the decision delay of our method is 25.56 sec while the delay for LSTM is 39.63 sec, followed by ARIMA with 20.61 sec and AR with 18.48 sec.

5 Conclusion

In this chapter, we proposed a framework to secure the exchange of medical data in IoMT and to detect anomaly in the number of received packets and in the acquired vital signs from monitored patient. We used the ECDHE to exchange the session key in "Just Works" pairing mode, while keeping the same mechanisms used in BLE to ensure confidentiality and integrity. To detect healthcare emergency, we applied the PELT algorithm followed by boxplot to detect changes in the monitored physiological parameters with reduced FAR and low computational complexity. Furthermore, to detect attacks aiming to deplete the energy of sensors or to flood LPU, we applied the same change point detection algorithm on the number of received packets in LPU to raise network alarms.

We conducted several experiments on data from different subjects for performance analysis and we compare the performance of our approach with previous works. Our experimental results on real physiological data showed that our approach is effective and able to achieve a good detection accuracy with a FAR of 6%. The comparison results showed that our system slightly outperforms LSTM and regression based systems. Our future work will focus on anomaly detection in the amount of energy consumed by compromised IoMT device.

References

1. J. Fiaidhi, S. Mohammed, Security and vulnerability of extreme automation systems: the IoMT and IoA case studies. IT Professional **21**(4), 48–55 (2019)
2. G. Thamilarasu, A. Odesile, A. Hoang, An intrusion detection system for internet of medical things. IEEE Access **8**, 181560–181576 (2020)
3. G. Hatzivasilis, O. Soultatos, S. Ioannidis, C. Verikoukis, G. Demetriou, C. Tsatsoulis, Review of security and privacy for the internet of medical things (IoMT), in *15th International Conference on Distributed Computing in Sensor Systems (DCOSS)* (2019), pp. 457–464
4. D. Koutras, G. Stergiopoulos, T. Dasaklis, P. Kotzanikolaou, D. Glynos, C. Douligeris, Security in IoMT communications: a survey. Sensors **20**(17), 4828 (2020)
5. Bluetooth SIG. Bluetooth Radio Versions. https://www.bluetooth.com/learn-about-bluetooth/radio-versions/, Last visited: February 2022
6. Australian Government Autralian Cyber Security Center. Information Security Manual. https://www.cyber.gov.au/sites/default/files/2022-03/22.%20ISM%20-%20Guidelines%20for%20Cryptography%20%28March%202022%29.pdf, March 2022
7. R. Killick, I. Eckley, changepoint: an R package for changepoint analysis. J. Statist. Softw. **58**(3), 1–19 (2014)
8. S. Pallavi, V.A. Narayanan, An overview of practical attacks on BLE based IOT devices and their security, in *5th International Conference on Advanced Computing Communication Systems (ICACCS'19)* (2019), pp. 694–698
9. S. Sevier, A. Tekeoglu, Analyzing the security of bluetooth low energy, in *International Conference on Electronics, Information, and Communication (ICEIC'19)* (2019), pp. 1–5
10. K. Ren, Bluetooth Pairing Part 3 – Low Energy Legacy Pairing Passkey Entry (2016). https://www.bluetooth.com/blog/bluetooth-pairing-passkey-entry/
11. K. Lounis, M. Zulkernine, Bluetooth low energy makes "Just Works" Not Work, in *3rd Cyber Security in Networking Conference (CSNet'19)* (2019), pp. 99–106
12. M. Cominelli, P. Patras, F. Gringoli, One GPU to snoop them all: a full-band bluetooth low energy sniffer, in *Mediterranean Communication and Computer Networking Conference (MedComNet'20)* (2020), pp. 1–4
13. Wencheng Sun, Zhiping Cai, Yangyang Li, Fang Liu, Shengqun Fang, Guoyan Wang, "Security and Privacy in the Medical Internet of Things: A Review", Security and Communication Networks, vol. 2018, Article ID 5978636, 9 pages, 2018. https://doi.org/10.1155/2018/5978636
14. T. Yaqoob, H. Abbas, M. Atiquzzaman, Security vulnerabilities, attacks, countermeasures, and regulations of networked medical devices – a review. IEEE Commun. Surv. Tutor. **21**(4), 3723–3768 (2019)
15. H.A.M. Puat, N.A. Abd Rahman, IoMT: a review of pacemaker vulnerabilities and security strategy. J. Phys. Conf. Ser. **1712**(1), 012009 (2020)

16. A. Lahmadi, A. Duque, N. Heraief, J. Francq, MitM attack detection in BLE networks using reconstruction and classification machine learning techniques, in *2nd Workshop on Machine Learning for Cybersecurity (MLCS'20)* (2020), pp. 1–16
17. S.F. Aghili, H. Mala, M. Shojafar, P. Peris-Lopez, LACO: lightweight three-factor authentication, access control and ownership transfer scheme for e-health systems in IoT. Future Gener. Comput. Syst. **96**, 410–424 (2019)
18. M.F. Ayub, M.A. Saleem, I. Altaf, K. Mahmood, S. Kumari, Fuzzy extraction and PUF based three party authentication protocol using USB as mass storage device. J. Inf. Secur. Appl. **55**, 102585 (2020)
19. U. Gulen, S. Baktir, Elliptic curve cryptography for wireless sensor networks using the number theoretic transform. Sensors **20**(5), 1507 (2020)
20. M.I. Ahmed, G. Kannan, Secure end to end communications and data analytics in IoT integrated application using IBM Watson IoT platform. Wirel. Personal Commun. **120**, 1–16 (2021)
21. C. Truong, L. Oudre, N. Vayatis, Selective review of offline change point detection methods. Signal Process. **167**, 107299 (2020)
22. G.J.J. van den Burg, C.K.I. Williams, An evaluation of change point detection algorithms. arXiv, abs/2003.06222 (2020)
23. S. Kovács, H. Li, P. Bühlmann, A. Munk, Seeded binary segmentation: A general methodology for fast and optimal change point detection (2020). Preprint arXiv:2002.06633
24. R. Killick, P. Fearnhead, I.A. Eckley, Optimal detection of changepoints with a linear computational cost. J. Amer. Statist. Assoc. **107**(500), 1590–1598 (2012)
25. N. Yeung, J. Lai, J. Luo, Face off: Polarized public opinions on personal face mask usage during the covid-19 pandemic, in *IEEE International Conference on Big Data (Big Data)* (2020), pp. 4802–4810
26. D. Valdez, M. Ten Thij, K. Bathina, L.A. Rutter, J. Bollen, et al., Social media insights into us mental health during the covid-19 pandemic: longitudinal analysis of twitter data. J. Med. Int. Res. **22**(12), e21418 (2020)
27. B. Hills, Machine in the Middle (MitM) BLE Attack (2020). https://www.blackhillsinfosec.com/machine-in-the-middle-mitm-ble-attack/
28. A. Paviglianiti, V. Randazzo, S. Villata, et al. A Comparison of Deep Learning Techniques for Arterial Blood Pressure Prediction. Cognitive computation (2021). https://doi.org/10.1007/s12559-021-09910-0, DOI: 10.1007/s12559-021-09910-0, (EPUB). https://link.springer.com/content/pdf/10.1007/s12559-021-09910-0.pdf Open access paper.
29. A. Khamparia, R.H. Mondal, P. Podder, B. Bhushan, V.H.C. de Albuquerque, S. Kumar, *Computational Intelligence for Managing Pandemics*, vol. 5. (Walter de Gruyter GmbH & Co KG, Berlin, 2021)

End-to-End Security for IoT Communications: A Practical Implementation

Sairath Bhattacharjya and Hossein Saiedian

1 Introduction

Internet-of-Things (IoT) is rapidly gaining momentum in recent years. From the inception of the term in 1999 [4], IoT has emerged as the solution to many application areas. Such solutions are making a mark in every domain, including agriculture, transportation, health care, supply chain, smart homes, smart cities, and many more. Emerging technologies like cloud computing, fog computing, machine learning have further enhanced the usability of these devices. Along with IoT, there has been a growth in machine-to-machine (M2M) interactions. Gartner has predicted that the number of interconnected devices would grow to 20.4 billion by the end of 2022. The number of M2M connections is expected to grow up to 27 billion by 2024. Looking at these estimates, it becomes clear that IoT would be the next cornerstone for the ever-growing digital economy. The enterprise IoT platform market will grow to $7.6 billion in 2024 [26]. The reason for this huge spike is the usability of these devices. From the monitoring of a patient's health at all times to efficient consumption of electricity to continuous monitoring of goods in transportation. With just an internet-connected mobile phone in our hand, all these pieces of information are available in seconds [3]. IoT has made life easy for everyday users and industries alike.

The prominence of IoT in the market has attracted the attention of both security researchers and adversaries. To meet the growing need, manufacturers are focusing on time-to-market, giving less consideration to the security issues. Numerous experiments have proven that these devices get hacked with equipment that is

S. Bhattacharjya
Abbott, Kansas City, MO, USA

H. Saiedian (✉)
The Institute for Information Sciences and Electrical Engineering & Computer Science, The University of Kansas, Lawrence, KS, USA
e-mail: saiedian@ku.edu

K. Daimi et al. (eds.), *Emerging Trends in Cybersecurity Applications*,
https://doi.org/10.1007/978-3-031-09640-2_2

readily available [30]. Many manufacturers delegate the responsibility of securing the devices to the user. Devices get configured with default credentials, and the user is expected to change the password to make the device secure. In some implementations, the credential change is not enforced, and the device continues to operate with the default user privileges. From a survey conducted, 57% of the consumers believe that the responsibility of securing these devices should be on the vendor, and ironically 48% admitted they did not know these devices could be used to conduct cyberattacks [11, 18].

The negligence of security consideration from the manufacturers exposed sensitive and personal information about the consumers to the adversaries. Poorly architected devices allow the execution of arbitrary code, allowing a malicious user to use them as an entry point into someone's private network [6, 9]. Smaller manufacturers neglect to provide options to patch these vulnerabilities after being discovered. Malware like Mirai, Reaper, Hajime, and EchoBot take advantage of the situation by converting these devices into bots. Perpetrators used such botnets to cause massive DDoS attacks [6]. Mirai caused a 1.1 Tbps attack using 148,000 IoT devices. The number of infected endpoints has doubled after the Mirai source code was made public. Major DDoS attacks like the GitHub attack in 2012 and Dyn Inc. DNS servers attack in 2016 were notable using botnets.

With the exponential growth of IoT and its usage, device security must be considered from the design phase [12]. The heterogeneous nature of the IoT ecosystem poses a challenge to the researchers to define a unified solution for the end-to-end security for IoT. Lack of standards and regulations gives manufacturers the freedom to build custom solutions. These architectures pose a challenge as each vendor implements its custom solutions without considering all risks. The challenge is further enhanced by the resource limitations of these devices, in terms of energy, storage, and computation power. Such limits prevent the application of standard security solutions implemented in traditional network devices like routers.

In this chapter, we defined end-to-end security architecture for IoT devices. We start by understanding the security issues in IoT in Sect. 2. Then we deep dive into the most common cloud-based architecture in Sect. 3. With the understanding of security issues and architecture, in Sects. 4 and 5, we discuss in detail how the plug-pair-play (P3) model could be used to provide end-to-end security in communication and firmware updates. In Sect. 6 we cross-check data security using the model with the security issues defined in Sect. 2. We also evaluate the memory utilization of the device. In Sect. 7 we conclude and state the future works in IoT security.

2 Security Issues in IoT Devices

Researchers [27] scrutinized approximately 1.2 Gigabytes of data and correlated it with the Shodan and MaxMind databases to find a distribution of exploited IoT devices. Of the 19,629 devices uniquely probed, they identified that China hosts the most vulnerable devices (3,345), followed by Brazil (1,326) and Indonesia (1,191).

Out of the vulnerable devices, Internet Service Providers hosted around 25%. These statistics raise the question as to what are the security risks associated with IoT devices.

We can divide IoT applications into four layers, namely sensing layer, network layer, middleware layer, and application layer [10, 17]. Each layer presents its threat and issues. The sensing layer represents the physical access to the sensors and actuators. The network layer deals with problems associated with transmitting data from the devices to a larger computational unit for processing. The middleware provides an abstraction between the network layer and the application layer. It provides the API for the application layer to function. The application layer is responsible for providing the service requested by the user.

Summarizing the security issues in the different layers of an IoT application:

- **Lack of physical security:** Many of these devices operate autonomously in unattended environments [25]. A perpetrator gaining access to these devices can replace the node with a vulnerable one, or physically damage them. The attacker can also inject malicious code into these devices gaining access to the entire network. Such issues result in node capturing, side-channel attack, eavesdropping, or even sleep deprivation attack.
- **Resource limitation:** Constrained resources, especially for the sensors and actuators are a major concern raised by all researchers. Lack of available resources limits the implementation of security solutions as present in a traditional network device. Thus, making these devices a source of interest for the attackers. A simple DoS/DDoS attack or a flooding attack can make the device unavailable, inducing operational disruption.
- **Insufficient user authentication:** As established earlier, manufacturers provide minimal security solutions using default credentials and not enforcing strict rules to change them. An attacker uses the lack of proper authentication to spoof requests to the device and gain total control over them. They can also use these devices and turn them into a botnet. This can be a step in a chain of attacks to use these botnets to perform a DDoS attack on a larger target [3].
- **Inadequate encryption:** Encryption effectively defuses the data and makes it unreadable to any prying eyes. Cryptography depends on the randomness of the algorithm and key size to effectively morph the data. Due to limited storage, it becomes difficult to store large keys. Similarly, limitation in computation power makes it difficult to run complex algorithms. A hacker takes advantage of it by performing a brute force attack to break the encryption. An attacker can perform a sniffing attack to expose critical data resulting in a breach of confidentiality [9, 33].
- **Inefficient access control:** Manufacturers enable the functionality of these devices with users having higher privilege. Once an attacker gains control of the device, they can perform any operation with that access. Using root privilege the attacker can disrupt the entire network on which the device is connected. The attacker can monitor all traffic on that network to gain unauthorized access to systems that they would not have otherwise.

- **Improper patch management:** While most bigger players in the market provide some means to provide security patches to their devices, there are numerous incidents where the smaller players do not provide such options. With a known vulnerability, the attackers can easily target these devices and gain total control. Others provide an option to patch, but the patches are not checked for integrity, making them susceptible to code injection attacks.

Other issues like weak programming practices and insufficient audit logging are common to both traditional software development and IoT applications. IoT applications suffer from some of the same issues that are seen commonly in web and mobile development. Limited energy and computational power enhance the problem further. It is difficult to implement traditional solutions on IoT devices.

3 Cloud-Based IoT Architecture

As described in Sect. 2, IoT applications face similar challenges as traditional web and mobile applications. However, they do not possess the computational power and storage to implement solutions like anti-virus, intrusion detection systems, and a firewall. To counter these issues, the research community proposed diverse architectural solutions. Due to the heterogeneous nature of the IoT ecosystem, the diversity in technology has increased. Communication standards like Z-Wave, Zigbee, 6LOWPAN, NFC, RFID, and others help interact with the devices. To connect them to the cloud backend standards like MQTT, SMQTT, JavaScript IoT are used.

Bluetooth Low Energy (LE) is used in most smartphone devices to connect with other peripherals. Researchers have demonstrated the use of IPv6 over Bluetooth LE [28]. Bluetooth LE Link-layer security protects wireless communication using Cipher Block Chaining Message Authentication Code (CCM). The OpenConnect project proposed the integration of these devices in a cloud-based architecture [29]. REST API endpoints integrate the devices with the central command center. The integration service maintains the security of the architecture. An arithmetic computer-based information hiding technique was developed to provide features like IP watermark, digital fingerprinting, and lightweight encryption ensuring energy efficiency to resource-constrained devices [16].

Many proposals embark on the authentication issue for resource-constrained devices. A certificate-based authentication technique address the default password problem [3]. The solution proposes to provide a certificate to every entity in the system by a trusted certification authority. Another solution uses a one-time password (OTP) scheme secured using elliptic curve cryptography. It uses the Lamport algorithm to generate the OTP. One research used the physical properties of a device for authentication them in a smart home environment [20]. A random set of challenges along with symmetric key cryptography secure the device.

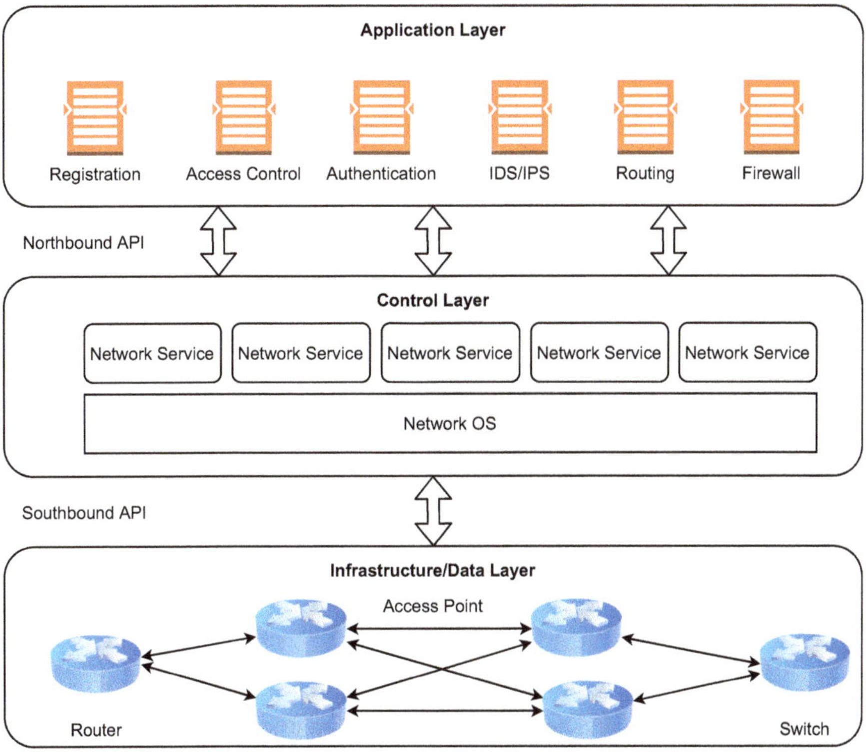

Fig. 1 Software-defined network (SDN)

Software-Defined Network (SDN) is gaining a lot of momentum in recent years. Software-Defined Networking (SDN) is an approach to networking that uses software-based controllers or application programming interfaces (APIs) to communicate with underlying hardware infrastructure and direct traffic on a network [24]. SDN controllers consist of three layers, namely infrastructure layer, control layer, and application layer [21] as shown in Fig. 1. The infrastructure layer interacts with the device through open interfaces like OpenFlow. The application layer interacts with all third-party libraries to cater to users' needs. The control plane is a collection of network APIs. It helps the interaction between the infrastructure layer and the application layer.

IoT architectures can take advantage of the layered approach in SDN architecture. IoT on an SD network avoids the complexity associated with generating, storing, and forwarding data through a traditional network. Security mechanisms like authentication, access control are maintained in the application layer. This separation of concern between layers helps the infrastructure layer focus on data forwarding and routing without concern about data privacy and security. Focusing on providing an end-to-end security model, we would use a generalized cloud-based architecture. The

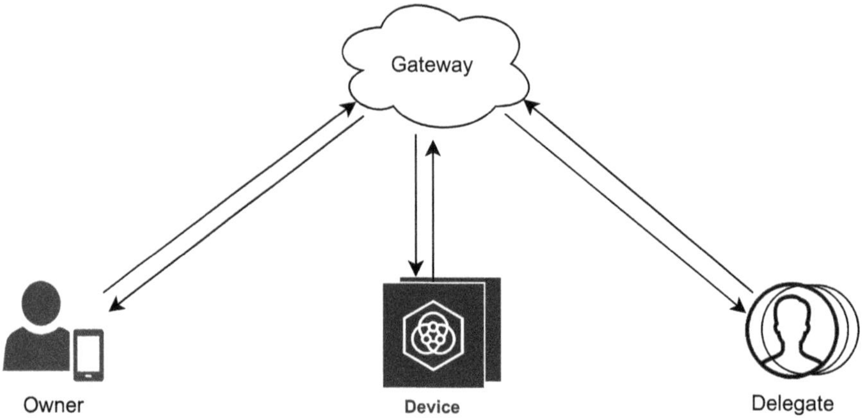

Fig. 2 Proposed architecture for IoT ecosystem

model presented in this chapter would implement data protection in the application layer following the SDN paradigm. End-to-end security solutions are proposed for IoT devices on SDN network [23]. The proposed architecture uses policies to secure the devices, network components, and services. The separation of the control plane and data plane is used to implement these policies.

An cloud-based IoT architecture consists of three primary entities as shown in Fig. 2:

- **Device** represents the IoT endpoint that performs a specific operation. It represents the sensors and actuators that can operate autonomously and generate data passed to a cloud system for further processing and analysis. Some devices like a motion detection camera or a light bulb are designed only to fulfill operations requested by the user. On such occasions, data analysis is not necessary.
- **User** requests information from the device. In this architecture, we have categorized the user group as owners and delegates. Each device is associated with only one owner. The owner has total authority over the machine. The delegates represent all other people or systems that interact with the appliance. Delegates can interact with the device after the owner approves the pairing. The owner grants access to a delegate to perform specific operations on a device. Throughout this chapter, we have addressed the owner and delegate separately when needed and collectively referred to them as users on concepts that apply to both.
- **Gateway** consists of API endpoints that are hosted in the cloud. They help coordinate the communications between users and devices. It also holds information about users, devices, registrations, and transaction logs. Each new device that is manufactured is recorded in the device database. The gateway holds the identity and public key of the device. It takes the computation and memory-intensive operations like data analytics and forensics away from the device. During user registration, a record gets added to the user database.

4 The Plug-Pair-Play (P3) Model to Establish a Secure Communication Channel

With the security issues presented in Sect. 2, having an end-to-end security solution is essential. Researchers have provided multiple different solutions to provide overall security. There have been proposals for using machine learning and deep learning techniques to detect malware and DDoS attacks [2, 19]. Some research used geospatial attributes of the devices to secure them in the transportation industry. Sensors are used within the smart city to identify the vehicles. End-to-end solutions are proposed for smart home systems as well [22]. One of the big challenges of providing an end-to-end solution for IoT devices is the heterogeneous nature of the devices. The proposed solutions either focus on specific sectors or do not describe the full spectrum from initiation of the devices to secure communication. SDN-dependent security solutions [23] are still in the research phase and cannot get easily integrated into the current network architecture.

A step-by-step approach to secure communications with IoT devices that can be easily integrated into the current network architecture is described by [7]. Internet-connected devices are rapidly replacing regular electronic devices. Users are very used to the plug-and-play model used widely in the industry. For a remote control car, we connect the battery, and it instantly starts operating. The section describes a similar plug-pair-play (P3) model for IoT devices. The model depends on a cloud architecture, where there is a gateway available to manage the registration of the users and devices. The model can operate in both traditional network architecture and SDN architecture. Since SDN is still under research, we would use traditional networking solutions like TLS/SSL to protect data in transit.

The P3 connection model is effective in multiple scenarios to maintain a secure channel of communication with the devices. It relies on a personal area network (PAN) channel for establishing communication. In all the below architectural implementation of the model, Bluetooth LE is taken as an example of PAN. Other technologies like Zigbee, Near-Field Communication (NFC), and InfraRed (IR) can also be considered to provide the same outcome [13]. Once the initial handshaking is completed, the model uses Wi-Fi for every other request and response with the device.

4.1 Secure Communication Between User and Gateway

As described in Sect. 3, the user provides commands and instructions to the device to perform specific operations. The framework relies on a few prebuilt security mechanisms to secure transactions. The model dictates that all communications over the internet between user and device go through the gateway. It provides a few key benefits:

- The gateway can keep track of all communications for forensics.
- For device data that require analytics, the gateway can provide the computing power to perform the operation.
- The device need to verify that the request is coming from the gateway and reject every other request.

The user registers with the gateway before using the device. Mobile apps are common nowadays to perform such operations. The app provides the interface for the user to register and connect to the gateway. After logging in, an authentication header accompanies any request to the gateway. JSON Web Token (JWT) [31] is a common industry standard for authenticating requests and identifying users. The token consists of three sections, namely header, payload, and signature. The header section contains information about the algorithm used for the signature. The payload holds user data, commonly called user claims. The signature section has the digital signature of the encoded payload. The signature verifies the integrity of the data and ensures that the payload is not tampered with.

Transport Layer Security (TLS) protects the requests and responses in transit. TLS encrypts the header and body before it is put in the wire. This prevents any wiretapping or man-in-the-middle attack. Figure 3 shows the Wireshark output showing the encrypted communication to the gateway. The transaction use port 443.

4.2 Secure Communication Between Device and Gateway

A public-private key pair provides authentication when the device and gateway communicate. The manufacturing process generates a unique key pair for each device. The gateway keeps the public key in a secure data store while the device holds the private key. Similarly, the gateway has its public-private key pair. The device stores the public key of the gateway in EEPROM. We recommend using Elliptic Curve Cryptography (ECC) for asymmetric encryption. ECC turns out to be the preferred choice of public-key cryptography for resource-constrained devices. The

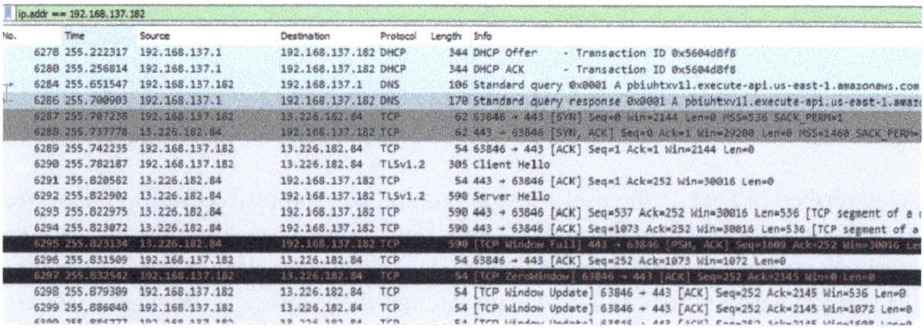

Fig. 3 Wireshark logs showing use of TLS

smaller key size and efficient algorithm make it more appealing for IoT devices. The operational time for signing and verification is comparable between ECC and RSA. A 256 bits ECC key can provide the same level of security as the 2048 bits RSA key.

Along with the keys, each device also has a unique `device_id`. The identifier helps the user and gateway to recognize the device. The below formula shows the approach to provide authentication and integrity checks for every request between the device and gateway.

<device_id, current_timestamp, raw_data> →data
<data, Enc{H(data), PrivKey$_{device}$}> → package

The `raw_data` represents the dataset that the device is trying to send to the user. We would look deeper into how to secure `raw_data` in Sect. 4.5. The `raw_data` combined with the `device_id` and the `current_timestamp` forms the payload. When a request goes from the device to the gateway, the `device_id` identifies the device. The `current_timestamp` ensures that the quest is not a replay attack. There are various ways to use the timestamp to determine if a request is stale. One of the common techniques is to save the timestamp in a database. If the new request has a timestamp less than or equal to the saved value, then the request is stale. The security operations center (SOC) can look into these requests for validation.

The payload hash is signed using the private key of the device. The hashed signed value can act as a digital signature for the device. The hashed value provides the integrity check for the `payload` and the digital signature provides authentication and non-repudiation.

The gateway uses the `device_id` to extract the correct public key. The public key helps verify the digital signature. Then the hash is used to validate the integrity of the `payload`. As mentioned above, the `current_timestamp` ensures the request is not old. After all the validations and verifications, the gateway can safely transmit the `raw_data` back to the user. Figure 4 shows the validation workflow for a request from the device to the gateway. A similar technique can protect the communication back to the device.

4.3 Setting Up Shared Key for Owner

The security communication between user and device cannot be preset. In this section, we describe the steps to set up a security key between the owner and the device. The model defines that every device has only one owner. Any other user would be considered a delegate and require the owner's permission to interact with the device. Section 4.4 describes the steps to set up the keys for a delegate. Since these keys are auto-generated, a symmetric key algorithm would best serve the purpose. The random unique key avoids password-based authentication or setting up default credentials in the device. The step is performed only once during the device initiation.

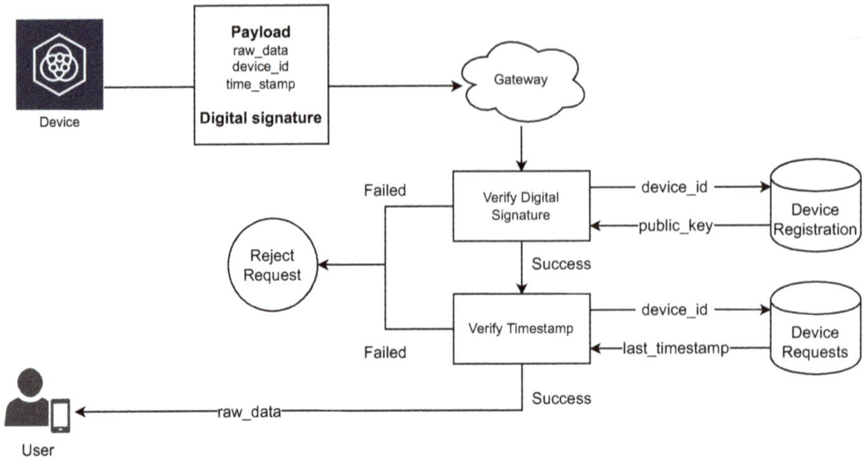

Fig. 4 Verification of a request from the device to the gateway

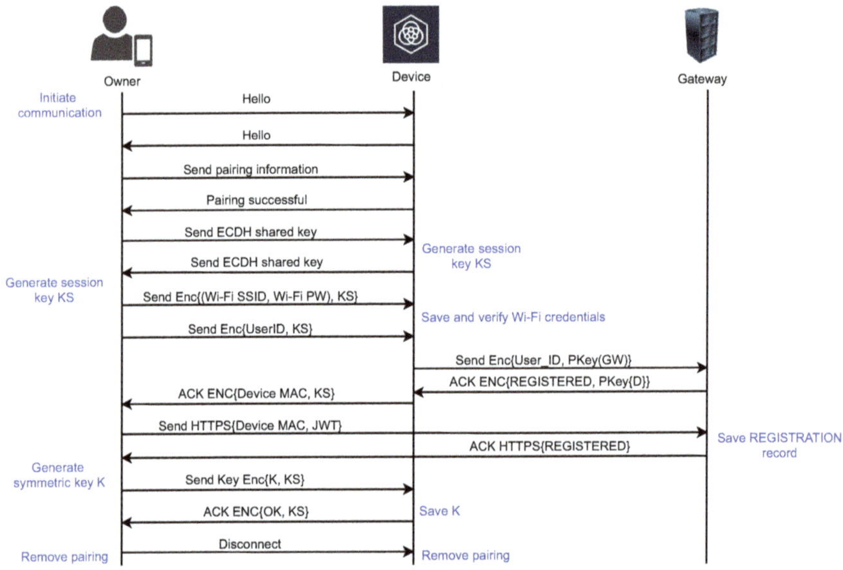

Fig. 5 P3 connection between owner and device

Figure 5 shows the steps to set up the shared key between the device and its owner. The workflow assumes that the user is already registered to the gateway and is logged in to the mobile app as described in Sect. 4.1. A PA network initiates communication with the device. In our demonstration, we use Bluetooth LE. Bluetooth LE is similar to classic Bluetooth and effective for ultra-low power applications [28]. It is present in all modern mobile phones and connected devices. The attack vector is small for a PAN network and is an efficient choice to set up keys.

- **Pairing:** The initial pairing creates that handshaking between the owner and the device to perform the remaining steps. The owner uses his mobile app to search for the available device to pair. Each device has a unique identifier that the app locates to establish the pairing. In this process, the owner's phone acts as a master, and the device acts as a slave.
- **Generate session key:** In the next steps of the workflow, we would share passwords and keys between the owner and device. We need to secure all these communications to avoid an eavesdropping attack. Curve25519 is an elliptic curve algorithm using 128 bits of key and designed for Elliptic Curve Diffie-Hellman (ECDH) key exchange. Here, both the owner and device generate a key and exchange the public part. On receiving the public part of the key they both generate the session key K_S using Diffie-Hellman to protect the next steps in the workflow.
- **Connect Wi-Fi:** The owner encrypts the Wi-Fi SSID and password with the session key Enc{<WiFi SSID, WiFi password>, K_S} and sends it. The device uses the credential to connect to the internet and ensures a successful connection with the gateway. The Wi-Fi information gets stored in memory for the remaining workflow. It returns a "success" to the owner.
- **User verification:** The device needs to verify the identity of the owner. The owner sends the `user_id` encrypted Enc{`user_id`, K_S} using the session key. The device forwards this identifier to the gateway along with the device's digital signature for verification. The gateway validates the identity of the device. Then the `user_id` is matched against the user database. On successful verification, a partial registration record gets created.
- **Device verification:** On a successful response from the gateway, the device returns an encrypted `device_mac` to the owner Enc{`device_mac`, K_S}. The owner forwards the mac and the authentication header to the gateway for device verification. The gateway verifies the user identity and then compares the `device_mac` against the newly created partial registration record. Once all the validations pass, the registration record is marked complete, and success gets returned to the user.
- **Generate and share the symmetric key:** Both the device and owner have validated the identity of each other. The owner creates the symmetric key by generating 256 bits key and a 128 bits initialization vector. The owner stores the keys and shares them with the device Enc {K, K_S}. The device saves the keys and the `user_id`. The Wi-Fi credentials are also stored. Then an acknowledgment is sent back to the user.
- **Disconnect:** The workflow established a secure key between the device and the owner. The Bluetooth interface is no longer needed to be active. The owner sends a termination request to the device, and the device complies.

The key generated in this session can secure communication between the user and device. After the initiation, all future communications can flow through the public internet, and the key can be used to maintain the confidentiality of the data. As established earlier, all communications flow through the gateway. In scenarios where data analytics is not required, the key can reside only with the user and device. The gateway does not need to interpret the message. The registration record can ensure that unauthorized requests to the device are terminated at the gateway.

Recycling a symmetric key is essential to avoid key detection. A similar process can help generate a new key. After the handshaking and establishing a session key, the user can generate a new key and send it to the device. The device uses the `user_id` to locate the previous key and replace it with the new one. A general industry recommendation is to recycle keys every 90–180 days.

4.4 Setting Up Shared Key for Delegate

The previous section described how an owner connects to a device without prior registration. Here we will discuss the situation where the device has an owner, and another user is trying to establish a connection. As described earlier, a delegate refers to people and appliances that want to connect to the device and is not the owner. When a delegate wants to pair with the device, they need approval from the owner. This workflow allows the owner to provide access control on the IoT device. It prevents unauthorized access and prevents perpetrators from misusing internet-connected systems.

Figure 6 shows the steps for a delegate to connect to the device. A similar workflow is followed for delegates as described in Sect. 4.3. The delegate initiates the pairing, similar to the owner. After pairing and setting up the session key, the delegate sends its `user_id` to the device. The device is already connected to the Wi-Fi, so no credentials are further required. The device sends the user's identifier to the gateway for verification.

The user verification process adds an extra step of approval from the owner. The gateway verifies the registration records and finds the device has an existing owner. The gateway sends a notification to the owner requesting permission to create the partial registration record. If the owner approves, the gateway creates the partial record and responds to the device. The device sends the encrypted `mac_address` to the delegate. The delegate verifies the identity of the device with the gateway and the registration record is completed. Then a secret is generated by the delegate and sent to the device. The device stores it in EEPROM along with the `user_id`. Following that, the delegate sends a termination request. All further communications between the delegate and the device go through the public internet. The secret maintains confidentiality when communicating between the delegate and the device.

If the owner rejects, the transaction gets terminated. The device returns a failed verification code to the delegate. This process provides an option to the owner to decide who gets access to the device. In this article, we concentrated on the security of data communication. The steps shown in this workflow provided equal authorization for all delegates. Role-based access control (RBAC) can be effective

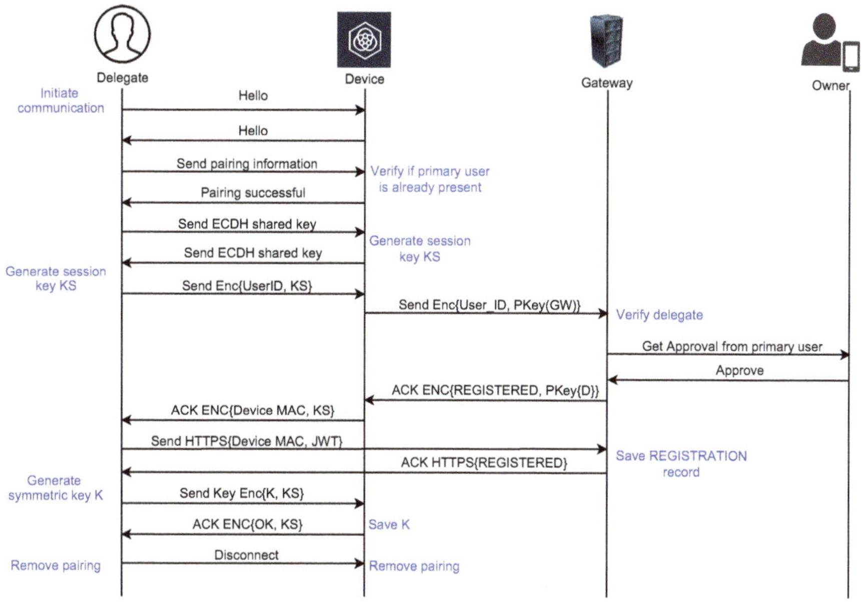

Fig. 6 P3 connection between delegate and device

on devices that perform multiple operations. That would give more control to the owner and they can define what operations can be performed by each delegate.

The secret key generated in the P3 connection model identifies each pair of users and devices. As described above, there are no predefined secret or default credentials in this workflow. A dynamic symmetric key is generated dynamically that protects all communications over the public internet. This approach is particularly effective for a novice user. This process works in the background, and the user does not have to configure or remember any additional details to enable security. It also plays well with the plug-and-play paradigm that general users are well accustomed to.

4.5 Secure Communication Between User and Device

When data flows over the internet, it goes through a chain of network devices. It is practically impossible to secure every one of them from being wiretapped. The shared secret K generated in the P3 connection model can be used to maintain the confidentiality of information flowing between user and device.

Researchers have proposed multiple solutions for a user to communicate with the device. The heartbeat approach is popular for network devices. The device sends out a pulse at a regular interval, indicating the device is active and functioning. The same technique could be effective for IoT devices. The device can inform the gateway that

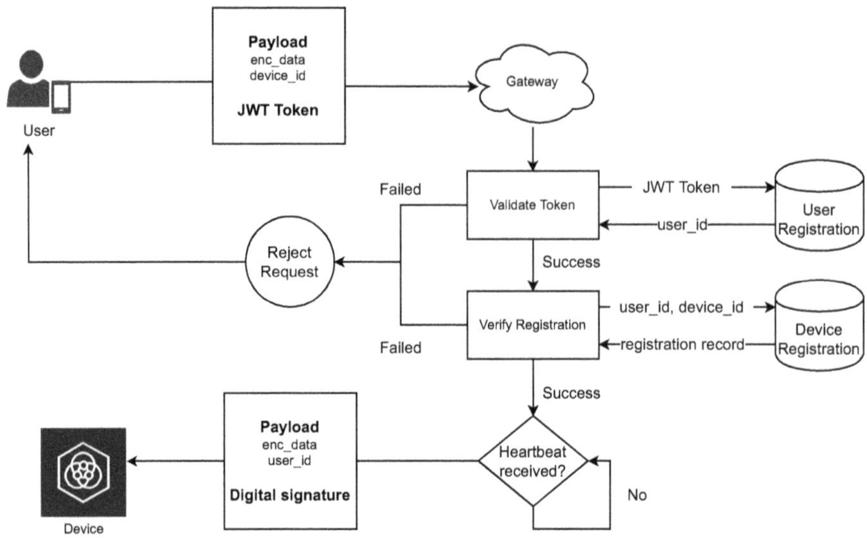

Fig. 7 Communication send from user to device via gateway

it is operating. The gateway could use the pulse response forwarding requests from the user.

```
<command, user_id> → data
Enc(data, K) → enc_data
```

Figure 7 shows the data flowing from the user to the device via the gateway. Before sending the data to the gateway, the user uses the key `K` to encrypt the `command` and `user_id` to generate `enc_data`. The user sends the `enc_data` to the gateway along with `device_id` and the JWT token in the auth header. The gateway on receiving the request verifies the user's identity using the JWT token. Next, the gateway extracts the device registration using the `user_id` and `device_id`. After all the validation, the gateway waits for the device's heartbeat. On receiving a pulse, the gateway forwards the user's request along with the `user_id` and the device's digital signature.

```
Dec(enc_data, K) → <command, user_id>
```

On receiving a user request in the pulse response, the device validates the digital signature to ensure the response is from the gateway. The `user_id` is used to extract the correct P3 key from the EEPROM. The device uses the key to extract out the command and the `user_id` send from the user from `enc_data`. The device matches both the user identifiers for validation.

The `command` is used to perform the specific operation requested by the user. The generated payload is again encrypted back using `K` before sending back to the gateway. Figure 8 shows the workflow back from the device to the user via the

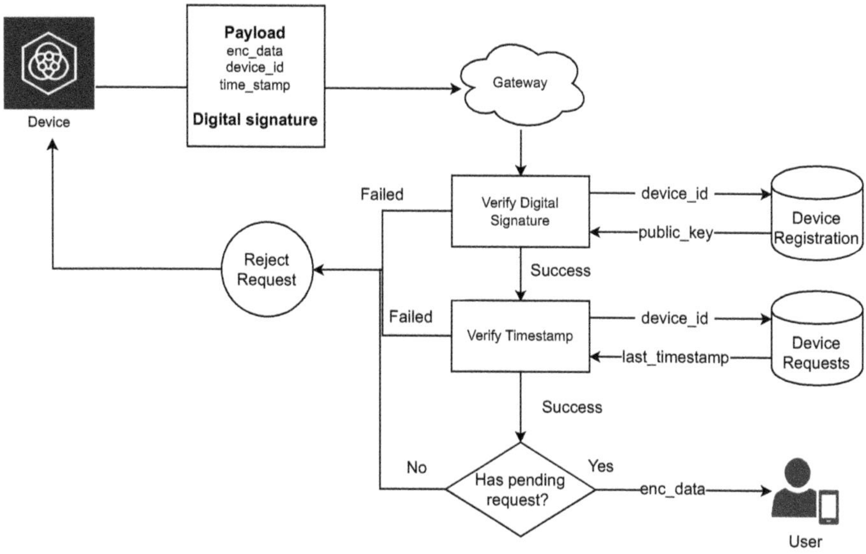

Fig. 8 Communication send back from device to user via gateway

gateway. On receiving the encrypted request from the device the gateway verifies the digital signature to ensure the integrity of the request. The `time_stamp` is validated to rule out replay attacks. Then the gateway checks if there are any pending requests from the user for this device. If yes, the enc_data is forwarded to the user.

To summarize the whole workflow,

- The user sends out a command to the device encrypted using K and the authentication header to identify itself to the gateway.
- The gateway verifies the user and the registration record.
- The request is forwarded to the device along with the user's identifier as a pulse response.
- The device verifies the gateway's digital signature and then extracts the key K using the user's identifier. The device uses K to decrypt the command. Then it performs the required operation to formulates the response and encrypts it with K.
- The encrypted response is sent to the gateway, which gets forwarded to the user.
- The user decrypts the response using K and completes the cycle.

Thus, end-to-end confidentiality, integrity, and authentication are maintained when communicating from a user to a device. One thing to note, all communications are encrypted in transit using TLS. So if there is wiretapping somewhere in the middle, the information flowing in the network is protected.

Here the request from the user and response from the device is encrypted. Only the pair holding the secret can decrypt this conversation. Auto-generation of the key can be used to integrate multiple users and devices. The model makes security work in the background and becomes user-friendly for general users.

5 Using P3 Connection Model to Update Device Firmware

The attack surface is evolving, and zero-day vulnerabilities are coming out more frequently. Having an option for IoT devices to get firmware updates is equally important as any desktop or mobile solution. The research community has provided multiple solutions to safely update the firmware [5, 14, 34]. The IETF SUIT working group is working on a standard framework update process for IoT devices. The group defines that the manifest should include the information about the firmware and a security wrapper to protect the metadata end-to-end.

The software in an IoT device consists of two parts, namely the bootloader and firmware. The bootloader helps load the firmware in memory for the device to function. The developer makes changes and recompiles the code to generate a new firmware. A content delivery network (CDN) hosts the updated firmware from where the device can access it. The device downloads the new firmware in one o0f the available slot, and point the bootloader to this new location. Then the device is restarted and operates with the new code. The old firmware is kept for rollback and eventually discarded.

Here we describe how we can use the P3 key to perform the firmware update. The P3 key generated in Sects. 4.3 and 4.4, provides an end-to-end secure communication between the user and device. We can use this channel to pass the metadata information securely from the user to the device. This metadata includes the CDN URL and the hash of the download. Figure 9 describes the steps by which the device can safely download and update the firmware.

- The first step is for the gateway to notify the owner that the device needs an update. As mentioned previously, the owner has total authority over the device and thus is responsible for informing the device of the update.
- The owner requests the gateway to provide the metadata information about the new update. The request contains the auth header to identify the owner to the gateway. The gateway validates the identity and sends the manifest back containing the CDN URL and the hash of the download. This separation of the hash from the firmware download maintains integrity. If an attacker somehow gets access to the CDN and alters the firmware, the hash would prevent the device from using it.
- The owner encrypts the metadata using the P3 key K and forwards the same to the device via the gateway. Section 4.5 explains the steps for secure communication from the owner to the device.
- The device on receiving the package validates the identity of the gateway. Then the package is decrypted using the same key K.

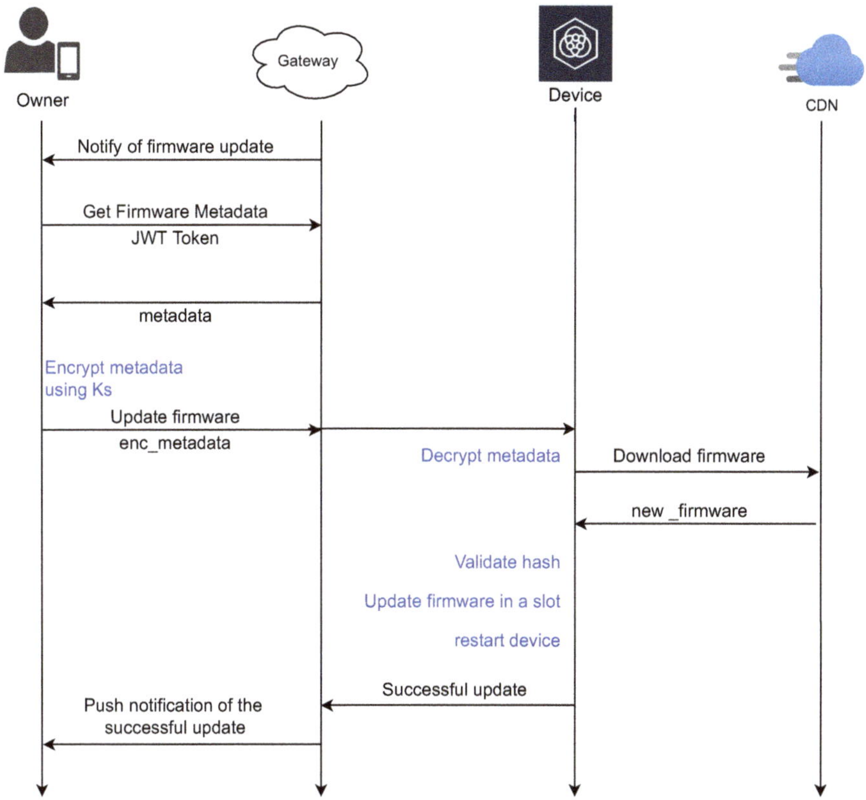

Fig. 9 Firmware update workflow using the P3 key

- the device extracts the CDN URL from the metadata and requests to download the updated firmware. A hash is generated of the downloaded and compared against the provided hash in the metadata from the owner. If there is a mismatch, the device returns an error to the user via the gateway. Otherwise, a successful download message is returned.
- The firmware is kept in one of the available slots, and the bootloader points to this new slot. The device restarts to implement the change. If everything looks good, the gateway gets notified by a success message. If the update fails, a rollback operation takes effect, and the bootloader points back to the old slot. Gateway is notified of the failure.
- The gateway informs the owner using a push notification of the update status.

As seen above, the automated symmetric key generated in the P3 connection model can help perform device updates. Updating the device is of utmost importance to improve the security and performance of the device. Software keep improving, and so should the devices. Here we have considered the scenario where there is a full update of the firmware. There are situations, where only a part of the firmware is updated. We would need further research in this area. There are also scenarios where the bootloader needs an update. Those are complex situations and need further research as well.

6 Model Evaluation

To test the P3 connection model, we implemented a humidity and temperature sensor. The following components make up the IoT device:

- The NodeMCU v3 ESP8266 microcontroller acts as the computing unit.
- HC-05 Wireless RF transceiver acted as a PAN endpoint operating on Bluetooth LE.
- The DTH-22 sensor recorded the reading of the environment.
- UCTRONICS 0.96 inch OLED module for the device display.

The ESP8266 microcontroller has 512 KB of EEPROM storage, 64 KB of instructional RAM, and 96 KB of data RAM.

The gateway was developed on AWS API Gateway using lambda functions to support the REST calls. The user registration was stored using the AWS Cognito service. DynamoDB acted as a data warehouse for the gateway to store the device and registration records.

We developed a React Native app on an Android platform to act as the user.

To validate the model's performance, we evaluated two aspects, data security, and device memory utilization.

6.1 Data Security

Throughout the chapter, we focused on data security. The framework implemented end-to-end security for communications between the user and the device. Looking back at the security issues pointed out in Sect. 2, let us evaluate how the model can address each of the problems:

- **Lack of physical security:** We did not go into the depth of physical security in this chapter. We focused mainly on data privacy. However, the heartbeat pulse can be used to maintain physical security. A heartbeat pulse is sent from the device to the gateway at a regular interval. If the gateway fails to receive the pulse, it can send a push notification to the owner notifying that the device is offline. This is a more reactive approach to physical security and does not cover the fact that the device can be physically manipulated by an attacker. This area needs more research.

- **Resource limitation:** In our implementation we used both symmetric and asymmetric encryption and were able to achieve the desired result with a limited resource. One thing to note here, the number of users that can connect to a device depends on the EEPROM storage. In our experiment, we could connect a maximum of four with the 96 KB of data RAM. Section 6.2 details the memory utilization of the device in an end-to-end secure communication.
- **Insufficient user authentication:** P3 model enforces authentication at all levels. We tried to implement a zero-trust ecosystem, where every request is authentication. The gateway and device use digital signatures amongst them. The user identifies itself using the authentication header. And we generated a symmetric key for communication between the user and device. Every request between the entities are authenticated before any operation is performed.
- **Inadequate encryption:** We used elliptic curve cryptography (ECC) for asymmetric and AES 256 for symmetric encryption. Both are suited well for constrained devices. We were able to successfully perform an end-to-end communication from the user to the device.
- **Inefficient access control:** We distinguished the users into two categories, owner and delegates. The owner has total control over the device and gets notifications for firmware updates and critical situations. The delegates are like general users and can interact with the device with approval from the owner. In our workflow, we did not distinguish amongst the delegates, but that can be easily achieved using the same model.
- **Improper patch management:** Sect. 5 detailed the steps on how we can achieve a secure firmware update. Following the IETF SUIT group's guidelines, we separated the metadata and provided end-to-end security in sending them to the device from the owner. The hash gets passed in the metadata which validates the new firmware's integrity.

6.2 Memory Utilization

The end-to-end secure communication was the most memory-intensive cryptographic operation. The device verified the digital signature of the gateway to ensure the sender's authenticity. Post that, it extracts the encrypted message from the user and then decrypts it using AES 256. After performing the given command, the device encrypts the response using AES 256 and attaches its digital before sending a response to the gateway.

The memory utilization for the device was tracked using an inbuilt ESP library. We added a wrapper around the `ESP.getFreeHeap()` function to print the available memory on the Arduino console.

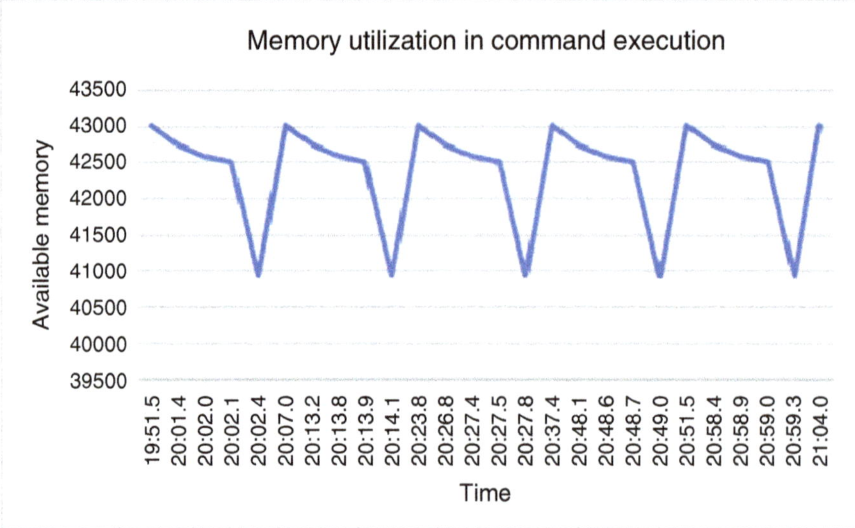

Fig. 10 Memory utilization during command execution

```
// Function to print the current memory usage
void availableMemory(){
    Serial.print("Memory available: ");
    Serial.println(ESP.getFreeHeap());
}
```

Figure 10 shows the memory utilization for the five consecutive command execution. We must note that the heartbeat continues to function as a command gets executed, so the memory start and end times are different from the two operations. We noticed that the available memory at the start of each cycle for command execution is around 42 MB. Each execution cycle took around 2080 bytes. We repeated the experiment five times, and we got the same results.

7 Conclusion

The P3 connection model, described in this chapter, provides the groundwork for end-to-end secured communication with IoT devices. The model integrates millions of users and devices seamlessly. The framework relies on the principle of zero trust. More research in the area of zero-interaction authentication (ZIA) can provide the required solution to protect the privacy of data [15]. In the workflow, we used Bluetooth LE for the initial pairing. Other technologies like cellular IoT, near-field communication (NFC), and long-term evolution (LTE) could be potential alternatives to Bluetooth [32]. LPWAN [1] could be used to avoid the dependency of home routers. This is an area of research for efficient pairing of devices within proximity.

The threat to the IoT devices is genuine, and with the growing number of internet-connected devices, the attack vector is growing [8]. Eliminating trust from the security framework could help establishing trust among the users. The P3 connection model securely set up a secret key for users and devices. Verification of the parties eliminates the threat of unauthorized access. In our implementation we built a device with minimal storage and computing power to represent a resource-constrained device. We were able to prove that we can implement cryptographic solutions over the limited resource. The technique used to generate the key can also be used to refresh it at a regular interval. Recycling of the keys avoid side-channel attacks. We also demonstrated how the shared key could help update the device firmware securely.

The technique described here helps maintain the security triad of integrity, confidentiality, and authentication. This process establishes the foundation of a zero-trust principle for IoT communication, "never trust, always verify." Every request and response is validated before the operation is performed. Another advantage of the model is that it can operate on a traditional network architecture as well as software-defined network (SDN). We have implemented the security layer on top of the application layer. The P3 connection helps create a private channel between every user and device on top of TLS.

IoT is the next breakthrough in the digital world. These devices are specialized in performing single operations and are efficient in doing it. They are gradually turning out to be an important part of our day-to-day lives. With home automation systems and home assistants like Google Nest and Amazon Alexa on the rise, we are using natural language to communicate them. These devices are recording our personal information and transacting with our financial data. However, this is just the tip of the iceberg for the potential of these widgets. We need a secure infrastructure to communicate with them. To ensure security and privacy P3 approach provides a zero-trust architecture to maintain privacy of the users.

References

1. T. Adame, A. Bel, B. Bellalta, Increasing LPWAN scalability by means of concurrent multiband iot technologies: an industry 4.0 use case. IEEE Access **7**, 46990–47010 (2019)
2. M.A. Al-Garadi, A. Mohamed, A.K. Al-Ali, X. Du, I. Ali, M. Guizani, A survey of machine and deep learning methods for internet of things (IoT) security. IEEE Commun. Surv. Tutor. **22**(3), 1646–1685 (2020)
3. H. Almuhimedi, F. Schaub, N. Sadeh, I. Adjerid, A. Acquisti, J. Gluck, L.F. Cranor, Y. Agarwal, Your location has been shared 5398 times! a field study on mobile app privacy nudging, in *CHI '15: Proceedings of the 33rd Annual ACM Conference on Human Factors in Computing Systems* (ACM, New York, 2015), pp. 787–796 https://doi.org/10.1145/2702123.2702210
4. K. Ashton, That "internet of things" thing: In the real world things matter more than ideas. RFID J. **22**, 97–114 (2009)

5. N. Asokan, T. Nyman, N. Rattanavipanon, A.-R. Sadeghi, G. Tsudik, Assured: architecture for secure software update of realistic embedded devices. IEEE Trans. Comput. Aided Design Integr. Circuits Syst. **37**(11), 2290–2300 (2018)
6. E. Bertino, N. Islam, Botnets and internet of things security. Computer **50**(2), 76–79 (2017) https://doi.org/10.1109/MC.2017.62
7. S. Bhattacharjya, H. Saiedian, Establishing and validating secured keys for IoT devices: using p3 connection model on a cloud-based architecture. Int. J. Inf. Secur. **21**, 1–10 (2021). https://doi.org/10.1007/s10207-021-00562-7
8. S. Bhattarai, Y. Wang, End-to-end trust and security for internet of things applications. Computer **51**(4), 20–27 (2018)
9. B. Bryant, H. Saiedian, Improving SIEM alert metadata aggregation with a novel kill-chain based classification model. Comput. Secur. **94**, 101817 (2020)
10. B. Bryant, H. Saiedian, An evaluation of videogame network architecture, performance, and security. Comput. Netw. **192**, 108128 (2021)
11. Canonical Ltd. Who should bear the cost of IoT security: consumers or vendors? (2017). https://tinyurl.com/bdbwze24
12. Congress.gov. H.R.1668 - 116th Congress (2019–2020): IoT Cybersecurity Improvement Act of 2020, December 4, 2020. https://www.congress.gov/bill/116th-congress/house-bill/1668
13. S. Cotton, W. Scanlon, Characterization and modeling of the indoor radio channel at 868 MHz for a mobile bodyworn wireless personal area network. IEEE Antennas Wirel. Propag. Lett. **6**, 51–55 (2007)
14. B. Cyr, J. Mahmod, U. Guin, Low-cost and secure firmware obfuscation method for protecting electronic systems from cloning. IEEE Int. Things J. **6**(2), 3700–3711 (2019)
15. M. Fomichev, M. Maass, L. Almon, A. Molina, M. Hollick, Perils of zero-interaction security in the internet of things. Proc. ACM Interact. Mob. Wearable Ubiquitous Technol. **3**(1), 1–38 (2019)
16. M. Gao, Q. Wang, M.T. Arafin, Y. Lyu, G. Qu, Approximate computing for low power and security in the internet of things. IEEE Comput. **50**(6), 27–34 (2017)
17. V. Hassija, V. Chamola, V. Saxena, D. Jain, P. Goyal, B. Sikdar, A survey on IoT security: application areas, security threats, and solution architectures. IEEE Access **7**, 82721–82743 (2019). https://doi.org/10.1109/ACCESS.2019.2924045
18. C. Horan, H. Saiedian, Cyber crime investigation: landscape, challenges, and future research directions. J. Cybersecur. Privacy **1**(4), 580–596 (2021)
19. F. Hussain, R. Hussain, S. Hassan, E. Hossain, Machine learning in IoT security: current solutions and future challenges. IEEE Commun. Surv. Tutor. **22**(3), 1686–1721 (2020). https://doi.org/10.1109/COMST.2020.2986444
20. C. Huth, J. Zibuschka, P. Duplys, T. Guneysu, Securing systems on the internet of things via physical properties of devices and communications, in *2015 Annual IEEE Systems Conference (SysCon) Proceedings* (2015), pp. 8–13. https://doi.org/10.1109/SYSCON.2015.7116721
21. W. Iqbal, H. Abbas, M. Daneshmand, B. Rauf, Y.A. Bangash, An in-depth analysis of IoT security requirements, challenges, and their countermeasures via software-defined security. IEEE Int. Things J. **7**(10), 10250–10276 (2020). https://doi.org/10.1109/JIOT.2020.2997651
22. N. Karie, N. Sahri, W. Yang, C. Valli, V. Kebande, A review of security standards and frameworks for IoT-based smart environments. IEEE Access **9**, 121975–121995 (2021)
23. K. Karmakar, V. Varadharajan, S. Nepal, U. Tupakula, SDN-enabled secure IoT architecture. IEEE Int. Things J. **8**(8), 6549–6564 (2021). https://doi.org/10.1109/JIOT.2020.3043740
24. D. Kreutz, F. Ramos, P. Verissimo, C.E. Rothenberg, S. Azodolmolky, S. Uhlig, Software-defined networking: a comprehensive survey. Proc. IEEE **103**(1), 14–76 (2015). https://doi.org/10.1109/JPROC.2014.2371999
25. R. Mahmoud, T. Yousuf, F. Aloul, I. Zualkernan, Internet of things (IoT) security: Current status, challenges and prospective measures, in *2015 10th International Conference for Internet Technology and Secured Transactions (ICITST)* (2015), pp. 336–341. https://doi.org/10.1109/ICITST.2015.7412116

26. P. Middleton, A. Velosa, F. Biscotti, Forecast analysis: enterprise IoT platforms, worldwide (2020). gartner.com/en/documents/3983783/forecast-analysis-enterprise-iot-platforms-worldwide
27. N. Neshenko, E. Bou-Harb, J. Crichigno, G. Kaddoum, N. Ghani, Demystifying IoT security: an exhaustive survey on IoT vulnerabilities and a first empirical look on internet-scale IoT exploitations. IEEE Commun. Surv. Tutor. **21**(3), 2702–2733 (2019). https://doi.org/10.1109/COMST.2019.2910750
28. J. Nieminen, C. Gomez, M. Isomaki, T. Savolainen, B. Patil, Z. Shelby, M. Xi, J. Oller, Networking solutions for connecting Bluetooth low energy enabled machines to the internet of things. IEEE Netw. **28**(6), 83–90 (2014)
29. N. Pazos, M. Muller, M. Aeberli, N. Ouerhani, ConnectOpen - automatic integration of IoT devices, in *2015 IEEE 2nd World Forum on Internet of Things (WF-IoT)* (2015), pp. 640–644
30. E. Ronen, A. Shamir, Extended functionality attacks on IoT devices: The case of smart lights, in *2016 IEEE European Symposium on Security and Privacy (EuroS&P)* (2016), pp. 3–12
31. N. Sakimura, M. Jones, J. Bradley, JSON Web Token (JWT) (2015). https://datatracker.ietf.org/doc/html/rfc7519
32. S.K. Sharma, X. Wang, Toward massive machine type communications in ultra-dense cellular IoT networks: current issues and machine learning-assisted solutions. IEEE Commun. Surv. Tutor. **22**(1), 426–471 (2020)
33. S. Swamy, D. Jadhav, N. Kulkarni, Security threats in the application layer in IoT applications, in *2017 International Conference on I-SMAC (IoT in Social, Mobile, Analytics and Cloud)* (2017), pp. 477–480. https://doi.org/10.1109/I-SMAC.2017.8058395
34. K. Zandberg, K. Schleiser, F. Acosta, H. Tschofenig, E. Baccelli, Secure firmware updates for constrained iot devices using open standards: a reality check. IEEE Access **7**, 71907–71920 (2019). https://doi.org/10.1109/ACCESS.2019.2919760

A Novel Transfer Learning Model for Intrusion Detection Systems in IoT Networks

Ly Vu, Quang Uy Nguyen, Dinh Thai Hoang, Diep N. Nguyen, and Eryk Dutkiewicz

1 Introduction

Internet of Things, or IoT, describes objects and sensors that are integrated in electronic devices, buildings, and vehicles to perform transmission functions in a wireless environment. With the strong development of the Internet today, IoT devices bring more and more benefits to human lives [1] (e.g., in healthcare, smart city, smart transportation, smart building). However, this fast growth has also led to a sharp increase in cyberattacks aimed at harming Internet and IoT users. According to a study in [2], network attacks have caused damage valued at more than US$500,000, including lost revenue, customers, opportunities, and so on.

With expanding IoT network, attackers are exploiting undefended gaps in the security of network environments [2]. Un-patched and un-monitored IoT devices are easily attacked to destroy the operation of IoT networks. Meanwhile, IoT attacks are developing along with the IoT network becoming more automated. Moreover, the resources (power and computation) of IoT devices are often limited, making it difficult for them to deploy effective protection methods [3, 4]. Consequently,

L. Vu
University Technology of Sydney, Ultimo, NSW, Australia

Le Quy Don Technical University, Hanoi, Vietnam
e-mail: vu.ly@lqdtu.edu.vn; Ly.T.Vu@student.uts.edu.au

Q. U. Nguyen
Le Quy Don Technical University, Hanoi, Vietnam
e-mail: uynq@lqdtu.edu.vn

D. T. Hoang (✉) · D. N. Nguyen · E. Dutkiewicz
University Technology of Sydney, Ultimo, NSW, Australia
e-mail: hoang.dinh@uts.edu.au; diep.nguyen@uts.edu.au; eryk.dutkiewicz@uts.edu.au

K. Daimi et al. (eds.), *Emerging Trends in Cybersecurity Applications*,
https://doi.org/10.1007/978-3-031-09640-2_3

development of IoT attack detection systems to protect IoT devices from IoT attacks is a crucial task to expand the applications of IoT [5–7].

An IoT attack detection system performs network traffic data analysis to detect anomalous behavior in IoT network. Basically, three popular approaches are often used for analyzing network traffic to detect network attacks [8], i.e., knowledge-based methods, statistic-based methods, and machine learning-based methods. First, in order to detect IoT attacks, knowledge-based methods generate network attack rules or signatures to match network behaviors. The popular knowledge-based method is an expert system that extracts features from training data to build the rules to classify new traffic data. Knowledge-based methods can detect attacks robustly in a short time. However, they need high-quality prior knowledge about cyberattacks. Moreover, these methods are often unable to detect zero-day attacks.

Second, statistic-based methods consider network traffic activity as normal traffic. In the sequel, an anomaly score is calculated by some statistical methods on the currently observed network traffic data. If the score is more significant than a certain threshold, it will raise the alarm for this network traffic [8]. There are several statistical methods, such as information gain, information entropy, and conditional entropy [9]. These methods explore the network traffic distribution by capturing the essential features of network traffic. Then, the distribution is compared with the predefined distribution of normal traffic to detect anomalous behaviors.

Third, machine learning-based methods to detect network attack are of paramount interest recently because they provide highly efficient attack detection models [4, 10, 11]. The main idea of machine learning methods is to build a model for detecting attacks automatically based on training datasets. As a result, the attack detection model can identify the attack traffic based on the general attributes of network traffic behaviors.

Although machine learning, especially deep learning, has achieved great success in network attack detection, there are still some issues that can affect the accuracy of IoT attack detection systems. One of these issues is that the machine learning-based methods are usually based on the assumption that we can collect labelled data of both normal and attack classes. However, the labelling process is usually performed manually by humans, which is time-consuming and expensive. Thus, in some problem domains, e.g., IoT environment, due to the quick evolution of network attacks, it is often unable to label for all samples when they are collected from different IoT devices. In other words, it is desirable to develop detection models that can detect attacks to various IoT devices without labelled information.

To handle the above issue, this chapter introduces a novel deep transfer learning (DTL) model based on AutoEncoders (AEs). The new model is called Multi-Maximum Mean Discrepancy AE (M2DA) that includes two AEs. M2DA is trained using both labelled dataset (source domain) and unlabelled dataset (target domain). Here, source domain and target domain are network traffic data collected from two different IoT data. By this way, the knowledge (label information) is transferred from the source domain to the target domain, thereby improving accuracy of IoT attack detection on the target domain. Specifically, the first AE in M2DA (AE_1) inputs data and label from the source domain while the second AE (AE_2) inputs only data from the target domain. In the training phase, the latent representation of AE_2 is learnt to be close to the latent representation of AE_1. Consequently, the latent representation of AE_2 can classify data from the target domain more correctly.

The major contributions of this chapter include:

- We introduce a novel DTL model, namely M2DA, to use the knowledge of label information of a related data domain. This model aims to alleviate the problem of "lack label information" in IoT network traffic.
- We propose to use the Maximum Mean Discrepancy (M2D) metric to make the latent representation of AE_1 and those of AE_2 to be as close as possible. This metric is also very helpful to enhance the efficiency of the transferring knowledge between AE layers.
- We conduct a large number of experiments using nine IoT attack datasets to investigate the effectiveness of M2DA. The results are compared with the canonical AutoEncoder and the recently proposed DTL models [12, 13].

2 Background

2.1 *Transfer Learning*

A model that can use the knowledge learned from a source problem for a learning task on a target problem is a Transfer Learning (TL) technique. Usually, the target problem is different from the source problem but they are related data distributions. Here, we define two IoT datasets as the source problem and the target problem that present two related data distributions. More specifically, we can consider two IoT data distributions that are different due to being collected from different IoT devices but related data distributions due to presenting normal and IoT attacks.

We assume that X is the input data space and Y is its label space. D_S and D_T are the domain distributions of a source problem and a target problem, respectively. The source domain has data and label as $D_S = (X_S, Y_S) = (x_S^i, y_S^i)_{i=1}^{n_S}$ and the target domain does not have a label set, i.e., $D_T = (X_T) = (x_T^i)_{i=1}^{n_T}$. n_S and n_T present the numbers of data samples of the source domain and the target domain, respectively. The learning task is transferring label information from the source domain to the target domain. As a result, the TL model can classify data in the target domain. By minimizing the distance between the source domain and the target domain at the latent representation space, the latent representation of the target domain is easy to be classified as the source domain. In the TL field, researchers usually use Maximum mean discrepancy (M2D) [14] or Kullback–Leibler divergence (KL) [13] due to their effectiveness to minimize distance between the two data distributions. Moreover, TL that is based on deep neural networks is called Deep Transfer Learning (DTL). In this chapter, we present a DTL method based on a neural network, namely AutoEncoder (AE), to enhance the accuracy in IoT attack detection. The next subsection will present the fundamental background of the M2D metric and AE.

2.2 Maximum Mean Discrepancy (M2D)

Similar to Kullback–Leibler divergence (KL) [13], M2D is used to quantify the distance between two distributions. Comparing to KL, M2D is more flexible due to the ability of estimating the nonparametric distance [14]. Moreover, M2D can avoid computing the intermediate density of distributions. The calculation of M2D is presented as Eq. (1).

$$\text{M2D}(X_S, X_T) = |\frac{1}{n_S}\sum_{i=1}^{n_S}\Gamma(x_S^i) - \frac{1}{n_T}\sum_{i=1}^{n_T}\Gamma(x_T^i)|, \quad (1)$$

where n_S and n_T are the numbers of samples in the source and target problems, respectively, Γ present the representation of the original data x_S^i or x_T^i.

2.3 AutoEncoder

An AutoEncoder (AE) is a neural network that has two parts: an encoder and a decoder. The aim of this neural network is reconstructing its input at its output [15–17]. As shown in Fig. 1, the encoder inputs the data $x = x_1, x_2, \ldots, x_d$ and outputs the latent representation of x, i.e., z where d is the number of dimension of the input data x. The decoder receives z as input and its output is $\hat{x}$. The objective function of AE aims to minimize the distance between x and $\hat{x}$. Thus, x and $\hat{x}$ have the same dimension as d.

We denote $\psi = \{W, b\}$ and $\omega = \{W', b'\}$ to be the parameter sets of the encoder and decoder where W, W', b, and b' are the weights and the biases, respectively. Let $X = x^1, x^2, \ldots, x^n$ be the training dataset. The q_ψ projects the input x^i to the latent representation z^i. The decoder p_ω aims to map the latent representation z^i to output $\hat{x}_i$. The output $\hat{x}_i$ has a similar structure as the input x_i. The calculations of the encoder and decoder are described in Eqs. (2) and (3), respectively.

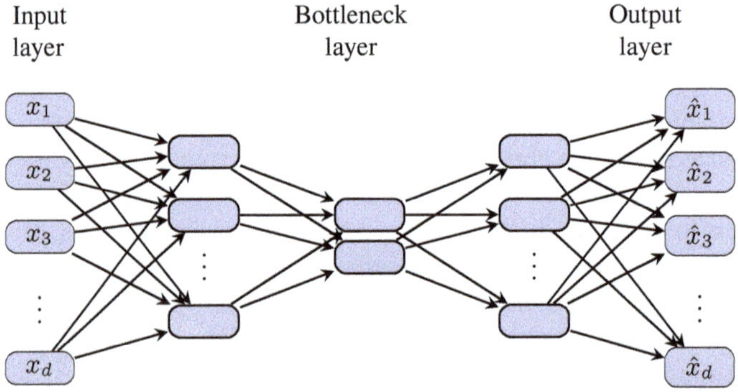

Fig. 1 Architecture of AutoEncoder (AE)

$$z = q_{\psi}(x) = a_f(Wx + b), \tag{2}$$

$$\hat{x} = p_{\omega}(z) = a_g(W'z + b'), \tag{3}$$

where a_f and a_g represent the activation functions of the encoder and the decoder, respectively. Figure 1 illustrates an AE structure where the input dimension, the number of layers, and the bottleneck layer size are d, 5, and 2, respectively.

Specifically, the objective of AE is to reconstruct the input layer at the output layer. Thus, the training process of AE aims to minimize the Reconstruction error (RE). RE is the loss function of AE described in Eq. (4). This is the expected negative log-likelihood of the i-th data sample. The expectation is taken over the representation with respect to the encoder's distribution. Minimizing this loss function motivates the decoder to reconstruct the input data. If the decoder cannot reconstruct the input data well, it will incur a cost in this loss function.

$$\delta_{AE}\left(x^i, \psi, \omega\right) = \mathbf{E}_{q_{\psi}(z|x^i)}\left[logp_{\omega}(x^i|z)\right]. \tag{4}$$

The training process is minimizing the summation of the loss function, $\epsilon\left(x^i, \hat{x}^i\right)$,[1] over all training data samples. This loss function is the difference between the input x^i and the output $\hat{x}^i$. The mean squared error (MSE) and cross-entropy loss are commonly used to represent the difference between input and output data. Thus, the loss function of AE in Eq. (4) is rewritten as Eq. (5).

$$\delta_{AE}\left(x^i, \psi, \omega\right) = \frac{1}{n}\sum_{i=0}^{n}\epsilon\left(x^i, \hat{x}^i\right). \tag{5}$$

3 Related Work

Using machine learning to detect network attacks has been widely used in recent studies. [4, 17–19]. A real-time unsupervised network anomaly detection algorithm that uses a discrete-time sliding window to continuously update the feature space and clustering to detect anomalies has been proposed by Juliette et al. [18]. Ibidunmoye et al. [19] proposed a TL technique that is used to estimate a temporal property of the stream. They used statistically robust control charts to recognize deviations. However, this technique highly depends on the selected threshold value. Another work [17] tried to represent network traffic data into a latent representation space to improve the classification tasks. However, general machine learning algorithms need to assume that the data to be predicted in the future has a similar distribution to the

[1] $\epsilon\left(x, \hat{x}\right) = \frac{1}{n}\sqrt{\sum_{i=1}^{n}\left(x^i - \hat{x}^i\right)^2}$.

data used to train the model. However, there are many data analysis applications that do not guarantee this assumption [20, 21].

Therefore, TL technologies including Deep Transfer Learning (DTL) are being widely used in data analysis applications today. Based on using deep neural networks, DTL can be categorized into four groups, such as instance based DTL, mapping based DTL, network based DTL, and adversarial based DTL [21–23].

The instance based DTL method selects data samples from the source domain to complement the training set in the target domain. These data samples are assigned appropriate weights in the training set in the target domain. Wan et al. [24] proposed the bi-weighting domain adaptation that can map the feature spaces of both source and target domains to the common coordinate system. In the new representation space of features, samples in the source domain are assigned to appropriate weights. To learn sample weights, [25] introduced a DTL metric framework together with learning a distance between two domains. This framework makes knowledge transfer between domains more effective. An ensemble DTL introduced in [26] can leverage samples from the source domain to train on the target domain.

The mapping based DTL approach aims to map samples from both the source domain and target domain into a new feature space. Thus, in the new representation space, the source domain representation and the target domain representation are similar and are considered as a training set of a deep neural network. The work in [27] introduced an adaptation layer and an additional domain confusion loss based on M2D to learn a new representation space. In this new representation space, the source domain and target domain are invariant. To measure the distance of the joint distributions, joint M2D was introduced in [28] to transfer the data distribution between different domains.

The adversarial based DTL approach refers to the use of adversarial networks inspired by generative adversarial nets (GAN) [29] to get a representation space. This new representation space is suitable to both the source domain and target domain. Tzeng et al. [30] introduced a new loss function of GAN combining a discriminative model with a DTL method. A randomized multi-linear adversarial network was introduced in [31] to find the multiple feature layers.

The network based DTL approach reuses the network structure and parameters trained in the source domain for training the target domain. Oquab et al. [32] used some pretrain weights of front-layers of Convolutional Neural Network (CNN) trained on the ImageNet dataset to map images of other datasets to intermediate image representation. It helps to transfer knowledge learned from a big dataset, e.g., ImageNet, to other object recognition tasks with a small training set size. Long et al. [33] introduced a method to learn adaptive classifiers with a residual function. Several DTL approaches based on AEs were introduced in [12, 13, 34]. They used different AE-based models such as AE [13], denoising AE [34], and sparse AE [12] for some specific TL applications. In previous work based on AE, the transferring process is executed only on the bottleneck layer. They may use the Kullback–Leibler divergence (KL) metric [13] or M2D metric [12] to minimize the representation data at the bottleneck layers of the source domain and the target domain.

To take the advantages of deep neural networks for learning network traffic representation, we develop a network based DTL method. In the IoT environment, we are unable to label data collected from all IoT devices. To handle this issue, we only label IoT traffic data from several IoT devices and transfer the label information to other devices to enhance accuracy in the IoT attack detection problem. More specifically, we can transfer knowledge at the higher level of features from the source domain with labels to the target domain without labels. This approach improves the accuracy of learning tasks on the target domain with limited samples and no labels.

In this chapter, we introduce a DTL model, i.e., M2DA that uses a non-linear mapping of a neural network (i.e., AE) to enhance the accuracy of IoT attack detection on some datasets collected IoT devices without label information. The key idea of the M2DA model (compared with previous AE-based DTL methods [12, 13]) is that the transferring task is executed on *every encoding layer* of AE while it is only executed in the bottleneck layer (latent representation) of AE in previous work. It helps to greatly encourage the latent representation of the target domain to be similar to the latent representation of the source domain. The results illustrate that using M2DA can enhance the accuracy of the classification task at the target domain for IoT attack detection problems.

4 Proposed Deep Transfer Learning Model

In this section, we first describe the system architecture of the IoT attack detection system. Then, we present the M2DA model in detail.

4.1 System Structure

Figure 2 illustrates the architecture of IoT attack detection model that uses a DTL model. The system include two phases. In the first phase, we gather data from the IoT devices using a data collection function. This data can be labelled or not depending on the cost of labelling. We can select to label data from some IoT devices. The labelling process usually includes the following two steps [35]. First, the Tcptrace tool is used to extracted data samples from the captured network traffic [36]. Second, the data samples are labelled manually by analyzing the network flow by a tool, e.g., the Wireshark tool [37]. Due to the expensive cost of the second step and the large number of IoT devices in the system, the number of IoT devices that has labelled data is often smaller than those with unlabelled data.

In the second phase, the data samples in the form of feature vectors are used for training the M2DA model. During the training, the labelled information is used to adjust the latent representation of the first AE in M2DA. Moreover, the latent representation of the second AE is also trained to mimic that of the first AE. This is achieved by minimizing the distance between the latent representations of the

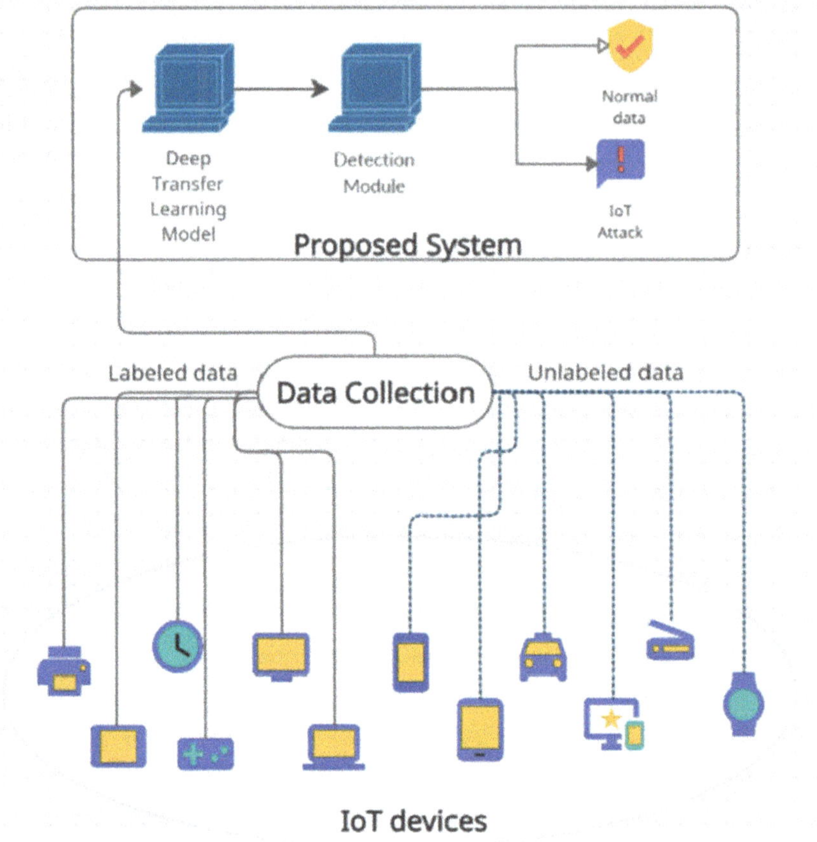

Fig. 2 Proposed system structure

first and the second AE. When the training is completed, the DTL model is used to classify the incoming traffic from the IoT devices into normal or attack. We will describe the DTL model in detail in the next subsection.

4.2 *Multi-Maximum Mean Discrepancy AutoEncoder*

This subsection introduces the Multi-Maximum Mean Discrepancy AutoEncoder (M2DA). M2DA (Fig. 3) consists of two AEs (i.e., AE_1 and AE_2) that are trained together. The first AE, i.e., AE_1 is trained on the data samples of the source problem x_S while the second AE, i.e., AE_2, is trained on the data samples of the target problem x_T. The loss function of M2DA has three components: The first component is the

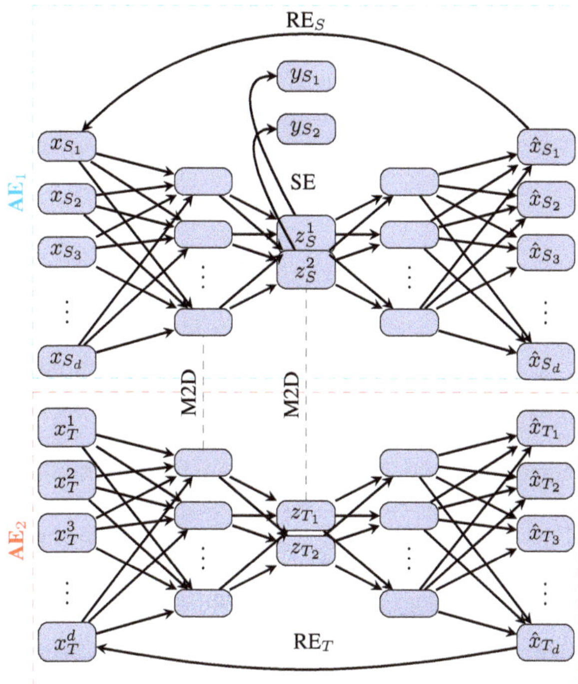

Fig. 3 Architecture of M2DA

reconstruction error, (δ_{RE}), the second component is the supervised error, (δ_{SE}), and the third component is the Multi-Maximum Mean Discrepancy (δ_{M2D}).

Define ψ_S, ω_S and ψ_T, ω_T be the weight set of the encoder and decoder of AE$_1$, AE$_2$, respectively. The reconstruction errorδ_{RE} is the summation of RE of AE$_1$, i.e., RE$_S$ and RE of AE$_2$, i.e., RE$_T$ in Fig. 3. This term aims to force the output of AE$_1$ and AE$_2$ closely to their input. Specifically, the RE$_S$ attempts to reconstruct x_S at $\hat{x}_S$ and RE$_T$ tries to reconstruct x_T at $\hat{x}_S$. Since the output of the AEs is constructed from the latent vector, the RE term helps the AEs to encode the useful information of the input data at the latent vector. Thus, the latent representation can be used as a new representation for the input data. In other words, we can do the classification task on the latent vector instead of the original data. Based on above analysis, the δ_{RE} term is computed as follows:

$$\delta_{\text{RE}}(x_S^i, \psi_S, \omega_S, x_T^i, \psi_T, \omega_T) = \epsilon(x_S^i, \hat{x}_S^i) + \epsilon(x_T^i, \hat{x}_T^i), \quad (6)$$

where $\epsilon()$ is the mean square error function [17], x_S^i, $\hat{x}_S^i$ are input and output of the source domain, respectively. x_T^i and $\hat{x}_T^i$ are the input and the output of the target domain, respectively.

The second component in the M2DA loss δ_{SE} aims to classify the input sample AE$_1$ into normal or attack using the labelled information. Specifically, this term tries to map the latent representation of AE_1, i.e., z_S, to the label y_S of the input sample. We use a softmax function [16] to learn to map z_S to y_S. To calculate the softmax function, z_S and y_S must have the same dimension. Therefore, the size of the latent vector must equal to the number of the class label. This loss term helps to distinguish the latent representation space into separated class, e.g., normal and attack. Thus, it can be defined as the following equation:

$$\delta_{\mathrm{SE}}(x_S^i, y_S^i, \psi_S.\omega_S) = -\sum_{j=1}^{C} y_S^{i,j} \log(z_S^{i,j}), \tag{7}$$

where z_S^i and y_S^i are the latent vector and label of the i-th data sample in the source problem x_S^i. $y_S^{i,j}$ and $z_S^{i,j}$ represent the j-th element of the vector y_S^i and z_S^i, respectively.

The third component of the M2DA loss, δ_{M2D}, aims to transfer the knowledge from the first AE to the second AE. This is achieved by minimizing the M2D distance between every layer in the encoder of AE_1 and the corresponding layer in the encoder of AE_2. Thus, this loss forces the latent vector of the AE_2 (the target problem) closely to the latent vector of the AE_1 (the target problem). The δ_{M2D} loss is calculated in Eq. (8):

$$\delta_{\mathrm{M2D}}(x_S^i, \psi_S, \omega_S, x_T^i, \psi_T, \omega_T) = \sum_{k=1}^{K} \mathrm{M2D}(\Gamma_S^k(x_S^i), \Gamma_T^k(x_T^i)), \tag{8}$$

where K is the number of layers of the encoders in the AEs. $\Gamma_S^k(x_S^i)$ and $\Gamma_T^k(x_T^i)$ are the k-th layer of the encoders of AE_1 and AE_2, respectively, M2D(,) is the Multi-Maximum Mean Discrepancy distance in Eq. (8). Finally, the loss of M2DA consists of the loss components in Eqs. (6), (7), and (8) are rewritten in Eq. (9).

$$\delta = \delta_{\mathrm{SE}} + \delta_{\mathrm{RE}} + \delta_{\mathrm{M2D}}. \tag{9}$$

The key idea of M2DA compared with the previous transfer learning model [12, 13] is that the transferring task is executed in multiple encoding layers. The previous models [12, 13] transfer only the bottleneck layer from the first to the second AE. Therefore, M2DA allows more knowledge to be transferred from the source problem (AE1) to the target problem (AE2). One possible drawback of M2DA is that its training process is more time consuming due to the multiple calculations of the distance between the layers in the encoders of AE_1 and AE_2. However, in the inferring process, only AE_2 is used to detect IoT attacks in the target problem. Thus, the M2DA model does not increase the detection time compared to the other models [12, 13].

4.3 *Training and Predicting Process Using M2DA*

4.3.1 Training Process

Algorithm 1 presents the steps for training the M2DA model. The first step is to normalize data samples from the source domain and the target domain. The second step is fitting the data x_S and label y_S to input of AE_1 and fitting the data x_T to input of AE_2. The final step is the training process that is executed by minimizing the loss term Eq. (9).

After training, we have the trained M2DA model that includes two AEs.

Algorithm 1 Training process of M2DA

INPUT:
Data sample set x_S and corresponding label set y_S of the source domain
Data sample set x_T of the target domain
OUTPUT: Trained M2DA model.
BEGIN:
Step 1. Normalizing x_S and x_T
Step 2. Fitting x_S and y_S to AE_1, fitting x_T is to AE_2.
Step 3. Training M2DE by minimizing the loss function Eq. (9).
return Trained M2DE model including two parts: AE_1 and AE_2.
END.

4.3.2 Predicting Process

After training, the weights of AE_2 are used to predict a new data sample of the target domain in four steps as in Algorithm 2. First, a network traffic data sample $x_{i,T}$ is extracted as the feature vector and normalized. Second, we put it to the input of AE_2. Third, we take $z_{i,T}$ as the latent representation of $x_{i,T}$ by AE_2. Finally, we calculate the label $y_{i,T}$ by the softmax function. The predicting process is calculated based on only one AE, i.e., AE_2. Consequently, the predicting process time is similar to an AE-based model.

5 Experiment Description

5.1 *Bot-IoT Datasets (IoT Datasets)*

In the simulations, we consider 9 IoT attack datasets [4] that were collected from 9 IoT devices under 2 common IoT botnet types, These datasets were collected from nine commercial IoT devices with two most popular IoT botnet families, Mirai and BASHLITE (Gafgyt). Each botnet type includes 5 dissimilar IoT attacks. Here, it is noted that 3 datasets IoT3, IoT6, and IoT7 contain only one type of botnet type. Thus,

Algorithm 2 Predicting process of the M2DA model

INPUT:
$x_{i,T}$: A testing sample of the target domain.
INPUT:Trained weights of AE_2 model, a data sample $x_{i,T}$
OUTPUT: Label $y_{i,T}$
BEGIN:
Step 1. Normalizing $x_{i,T}$
Step 2: Putting normalized $x_{i,T}$ vector to the input of the trained AE_2
Step 3: Calculating $z_{i,T}$ is the latent representation of $x_{i,T}$ at the bottleneck layer of AE_2
Step 4. Calculating $y_{i,T}$ = softmax $(z_{i,T})$
return $y_{i,T}$
END.

each of these datasets includes five types of IoT attacks generated from one type of botnet family. All other datasets contain both Mirai and BASHLITE attacks. Thus, one of these datasets includes ten types of IoT attacks. The features are designed for over a sliding window of 100 connections as in Table 1.

We encode the categorical feature by one-hot encoding and remove some identify features, such as source IP address (saddr), destination IP address (daddr). Then, each data sample has 115 features. These features can be categorized into three groups: stream aggregation, time-frame, and the statistics attributes. We divide each IoT dataset into a training set and a testing set. The testing set contains the different IoT attacks compared with attacks in the training set. The details of nine IoT datasets are presented in Table 2.

5.2 *Evaluation Metric*

We use the AUC score to evaluate the effectiveness of our proposed model. AUC stands for Area Under Receiver Operating Characteristics (ROC) Curve [38] that is calculated by plotting the True Positive Rate (*TPR*) against the False Positive Rate (*FPR*) at various threshold settings of a classifier. We assume that the classifier need to classify positive data samples and negative data samples. The *TPR* score evaluates the number of actual positive data samples that predicted correctly (Eq. (10)). The *FPR* score is the ratio of the real negative data samples that are incorrectly predicted (Eq. (11)).

$$TPR = \frac{TP}{TP} + FN, \tag{10}$$

$$FPR = \frac{FP}{TN} + FP, \tag{11}$$

Table 1 Feature description of IoT datasets

Feature name	Feature description
pkSeqID	Identifier of row
Stime	Start time to collect sample
Flgs	State of flow flags in transactions
Flgs number	Representation of flags in number
Proto	Representation of transaction protocols present in network flow
Proto number	Representation of protocols in number
Saddr	Source IP address
Sport	Source port number
Daddr	Destination IP address
Dport	Destination port number
Pkts	Total packets in transaction
Bytes	Number of bytes in transaction
State	State of transaction
State number	Representation of feature state in number
Ltime	Last time of collected data sample
Seq	Sequence number
Dur	Total duration of collected data sample
Mean	Average duration of aggregated data samples
Stddev	Standard deviation of aggregated data samples
Sum	Total duration of aggregated data samples
Min	Minimum duration of aggregated data samples
Max	Maximum duration of aggregated data samples
Spkts	Number of Source-to-destination packets
Dpkts	Number of Destination-to-source packets
Sbytes	Number of Source-to-destination bytes
Dbytes	Number of Destination-to-source bytes
Rate	Total packets per second in transaction
Srate	Number of Source-to-destination packets per second
Drate	Number of Destination-to-source packets per second
TnBPSrcIP	Total bytes per source IP
TnBPDstIP	Total bytes per destination IP
TnP_PSrcIP	Total packets per source IP
TnP_PDstIP	Total packets per destination IP
TnP_PerProto	Total packets per protocol
TnP_Per_Dport	Total packets per destination port
AR_P_Proto_P_SrcIP	Mean of rate per protocol per source IP (calculated by packets per duration)
AR_P_Proto_P_DstIP	Mean of rate per protocol per destination IP
N_IN_Conn_P_SrcIP	Number of inbound connections per source IP
N_IN_Conn_P_DstIP	Number of inbound connections per destination IP
AR_P_Proto_P_Srcport	Mean of rate per protocol per source port
AR_P_Proto_P_Dstport	Mean of rate per protocol per destination port
Pkts_P_State_P_Protocol_P_DestIP	Number of packets grouped by state of flows and protocols per destination IP
Pkts_P_State_P_Protocol_P_SrcIP	Number of packets grouped by state of flows and protocols per source IP

Table 2 9 IoT datasets

Dataset	Device name	Training attacks	Number of training samples	Number of testing samples
IoT1	Danmini Doorbell	combo, ack	239,488	778,810
IoT2	Ecobee Thermostat	combo, ack	59,568	245,406
IoT3	Ennio Doorbell	combo, tcp	174,100	181,400
IoT4	Philips B120N10 Baby Monitor	TCP, syn	298,329	800,348
IoT5	Provision PT 737E Security Camera	combo, ack	153,011	675,249
IoT6	Provision PT 838 Security Camera	ack, udp	265,862	261,989
IoT7	Samsung SNH 1011 N Webcam	combo, tcp	182,527	192,695
IoT8	SimpleHome XCS7 1002 WHT Security Camera	combo, ack	189,055	674,001
IoT9	SimpleHome XCS7 1003 WHT Security Camera	combo, ack	176,349	674,477

where *TP* and *FP* are the number of positive samples predicted correctly and incorrectly, respectively. *TN* and *FN* are the number of negative samples predicted correctly and incorrectly, respectively.

The space under the ROC curve represents the AUC score. It measures the average quality of the classification model at different thresholds. With a perfect classifier, the ROC curve will plot in the top left corner ($FPR = 0$, $TPR = 100\%$). Whereas the ROC curve of a worst classifier will score in the bottom right corner ($FPR = 100\%$, $TPR = 0$). The AUC score of a random classifier is 0.5, and the AUC score for a perfect classifier is 1.0.

5.3 *Experimental Setting*

5.3.1 Hyper-Parameter Setting

We compares the performance of the M2DA model with previous AE-based models. We set the same hyper-parameter set for all the AE-based models. These hyper-parameters are common for AE-based models training with network traffic dataset. The value of these hyper-parameters are presented in Table 3. This experiment setting is based on the AE-based models for detecting network attacks in the literature [3, 4, 17]. Because we have the δ_{SE} loss term of M2DA, the number of neurons in the bottleneck layer is the same as the number of classes in the IoT dataset. Thus, the bottleneck layer has two neurons to represent two classes, i.e., normal and attack. The reason is that we need to classify into two classes in this bottleneck layer.

Table 3 Hyper-parameter setting for the DTL models

Hyper-parameter Hyper-parameter	Number of layers	Neuron number of bottleneck layer	Optimization Optimization	Activation function
Value	5	2	Adam	ReLu

The number of layers of each AE is 5. This follows the previous work for network attack detection [17]. Here, we note that the ADAM algorithm [39] has been widely adopted in the literature to optimize weights of trained models in training processes. For our proposed model, the Sigmoid function is used at the last layers (for the encoder and decoder), while the active function ReLu is used at all the AE's layers.

5.3.2 Experimental Set

We have carried out three sets of experiments in this chapter. The first set is to evaluate the effectiveness of M2DA in the transferring task. Here, we show the differences of the M2D distances between the bottleneck layer of AE_1 trained on the source domain and that of AE_2 trained on the target domain after training the three scenarios. These transfer learning processes are performed in one, two, and three encoding layers. It can be observed that the transfer learning process will be more effective if the M2D distance is low [40].

The second experiment set is to compare the AUC scores of M2DA with that of (1) the conventional AE learning model, (2) the proposed transfer learning model using KL metric (i.e., KLA) [13] and (3) the proposed transfer learning model using the M2D metric (i.e., MDA) [12]. We choose two these DTL models for comparison due to two reasons. First, these models are based on the AE's architecture and the AE-based models are the most effective with network traffic analysis as proved in previous work [3, 4, 17] and (2) these DTL models are of the similar type of transfer learning as the M2DA model. Specifically, we focus on the DTL models where the source domain has data and label information and the target domain has data without label information. All models are trained on the training sets of both source domain and target domain and evaluated on the testing sets of the target domain.

The third set is to measure the complexity and processing time of the experimental models. To measure the complexity of a neural network, we report the model size to measure the complexity of evaluated models. The model size is calculated by the number of the trainable parameters. The detailed results of the three experimental sets are presented in the next section.

6 Result and Discussion

6.1 Effectiveness of Transferring Task

M2DA implements the transferring task between every layer in the encoder of AE_1 and AE_2. This transferring task aims to push the latent vector of AE_1 closer to the latent vector of AE_2. To evaluate if this transferring task is effective, we have conducted an experiment in which IoT1 and IoT2 are used as source and target datasets, respectively. We measure the M2D distance between the latent vectors of AE_1 and AE_2 in the three scenarios. We train the M2DA model where the M2D loss as in Eq. (8) is calculated on one, two, and three layers of the encoder of M2DA. The lower distance value represents more knowledge transferred from the source to the target dataset.

The value of the M2D distance between the latent vectors of AE_1 and AE_2 is shown in Fig. 4. This figure shows that if more layers are deployed in the transfer learning tasks, the value of the M2D distance is smaller. It means that more knowledge is transferred from the source to the target problem. This proves that the more layers of the encoders in M2DA are deployed transferring tasks, the more effectively the learning model can perform.

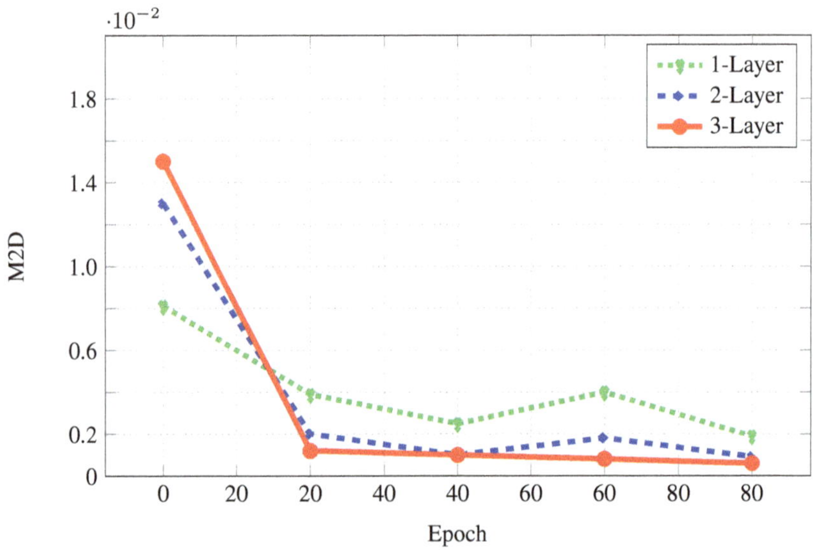

Fig. 4 M2D value that is transferred (i.e., the transferring task is executed by minimizing the M2D distance between corresponding encoding layers of the two AEs in 1-Layer, 2-Layer, and 3-Layer models) from the bottleneck layer of AE_1 to that of AE_2 when training on the source domain (IoT1) and the target domain (IoT2)

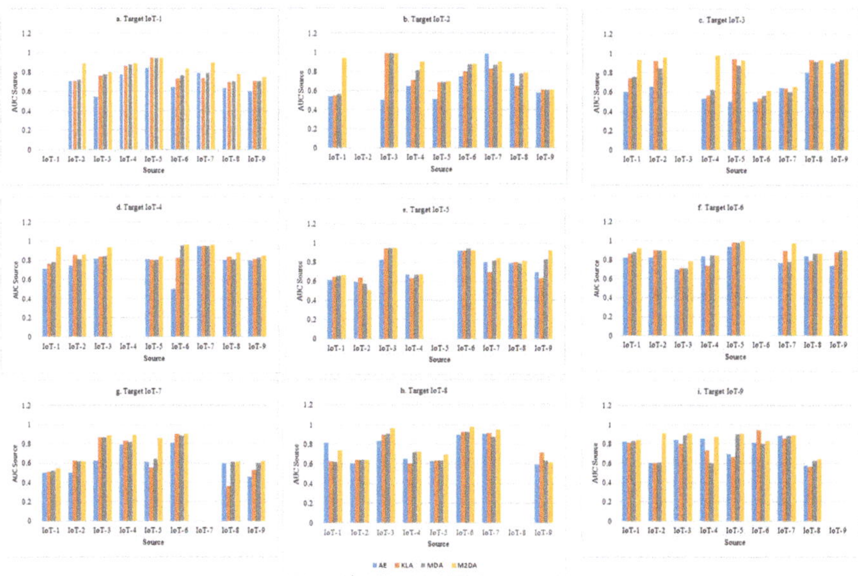

Fig. 5 AUC score on testing data of target datasets

6.2 Accuracy Comparison

Figure 5 presents the AUC scores of the tested models on the testing sets of the target datasets, i.e., (a) IoT1, (b) IoT2, (c) IoT3, (d) IoT4, (e) IoT5, (f) IoT6, (g) IoT7, (h) IoT8, and (i) IoT9. In each subfigure, the horizontal axis shows source datasets and the vertical axis shows the AUC score of a target dataset. At each source dataset in the horizontal axis, we present the results of the tested methods such as AE, KLA, MDA, M2DA. Moreover, the results of M2DA in Fig. 5 are shown in the yellow bars. It can be seen that AE provides the lowest accuracy among the four methods. This is not surprising since the AE model is trained and then tested on different datasets (e.g., trained on IoT1 and then tested on IoT2 datasets). Moreover, transfer learning is not implemented in the AE model, thus the performance of the AE is not satisfactory.

As a result, it can be observed that the performance (in term of the prediction accuracy) obtained by the AE model is much lower than those of the proposed learning models. For example, if we take IoT3 dataset as the source domain and IoT2 dataset as the target domain, then the accuracies (in terms of AUC) obtained by the KLA and MDA, respectively, will be increased to 0.921 and 0.848, (vs. 0.658 obtained by the AE). These results clearly show that in the cases when IoT devices have data without label information, our proposed learning model can significantly improve the performance of the conventional AE model, especially in detecting IoT attacks.

Among the transfer learning models, M2DA often achieves the best AUC score in almost all datasets. To give an example, when the source and target datasets are IoT2 and IoT3, respectively, the AUC score of M2DA is 0.956 compared to 0.659, 0.922, 0.849 of AE, KLA, and MDA, respectively. We also observe similar results on the other datasets. These results shows that the M2DA model transfers more effectively the knowledge from the source to the target dataset. Consequently, M2DA usually enhances accuracy compared to AE, KLA, and MDA for IoT attack detection in the target problem without label information.

6.3 Complexity

In this section, we mainly focus on discussing the complexity of the proposed framework in terms of computing time. In particular, we first select the IoT2 data as the source domain for training and IoT1 data as the target domain for testing. Then, we preform the experiments and record time for training and testing data. The results are shown in Fig. 6a, b in which the training time is represented by the grey bar with the unit of hours (Fig. 6a) and the testing time is represented by the green bar with the unit of seconds (Fig. 6b). Here, it can be observed that the training time for the proposed learning models is much higher than that of the conventional AE model. The main reason is that, unlike conventional AE learning models, the DTL models (including KLA, M2DA, and MDA) require to calculate distances between encoding layers of the encoders of the first and second AE in every training iteration to execute the transferring task. Moreover, the time for training M2DA is also much longer than those of KLA and MDA. This is because M2DA executes the transfer learning in every encoding layer of the encoders whereas this task is only done in one layer (the bottleneck layer) of KLA and MDA.

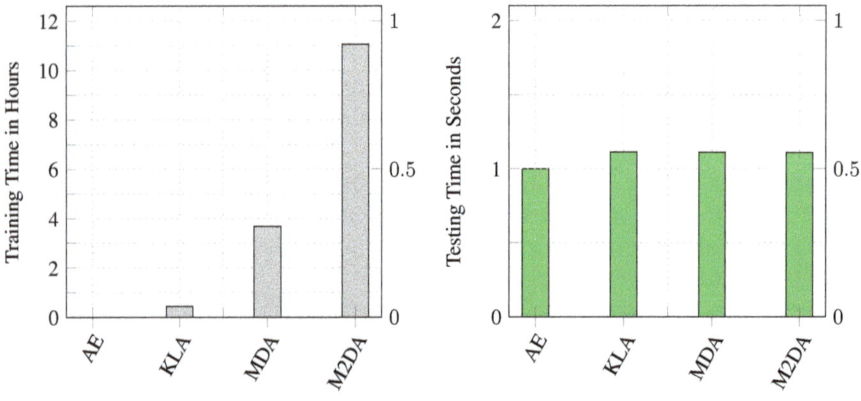

(a) Training time (Hours). (b) Testing time (Seconds).

Fig. 6 Training time (**a**) and Average Testing time for one data sample (**b**) of evaluated models (AE, KLA, MDA, and M2DA) on the source domain as IoT1 dataset and the target domain as IoT2 dataset

Table 4 Complexity of DTL models

Models	AE [16]	KLA [13]	MDA [12]	M2DA
No. parameters	25,117	150,702	150,702	150,702

Although, the training times of the transfer learning models are much higher compared to that of the AE, the inferring time of all the methods are mostly equal. The reason is that in the testing process, the data sample is only forwarded to one AE without the need for the distance calculation. For instance, the inferring times of AE, KLA, MDA, and M2DA on the IoT1 dataset are 1.001, 1.112, 1.110, and 1.108 second, respectively.

Table 4 shows the complexity (the number of parameters) of the evaluated models. This table shows that while the number of the parameters of the AE model is much smaller, the number of the parameters of the three transfer learning models is mostly equal. This is due to the fact that the three models use a similar structure of neural networks. Moreover, in the inferring phase, the data sample is forwarded to only one AE (the second AE) of KLA, MDA, and M2DA. Thus, the times for detecting IoT attacks of all the four tested models are roughly equal as in Fig. 6.

7 Conclusion

In this chapter, we resolve "the lack of label information" in the IoT attack detection problem. In some situations, we are unable to collect network traffic data with its label information. For example, we cannot label all collected IoT traffic data due to the cost of labelling tasks. Moreover, data distributions of data samples collected from different IoT devices are not the same. Thus, we develop a DTL technique, namely M2DA, that can transfer the knowledge of label information from a domain (i.e., data collected from one IoT device) to a related domain (i.e., data collected from a different IoT device) without label information. The results show that the M2DA model can help to identify IoT attacks more accurately.

However, the technique is subject to some limitations. M2DA is more complex due to the transferring process executed in multiple layers. It leads to time-consuming tasks for training M2DA. However, the predicting process of M2DA is mostly similar to that of the other AE-based models because it only uses one AE architecture for the predicting task.

In future work, to enhance the privacy of IoT data, we plan to use federated learning to train the M2DA model. In that case, we do not need to send IoT traffic data to the processing center. Besides that, we will test the effectiveness of the DTL model with network traffic from several different network environments. Last but not least, we will attempt to expand this model to other neural networks such as Convolutional Neural Network, Deep Belief Network.

References

1. N.C. Luong, D.T. Hoang, P. Wang, D. Niyato, D.I. Kim, Z. Han, Data collection and wireless communication in internet of things (IoT) using economic analysis and pricing models: a survey. IEEE Commun. Surv. Tutor. **18**(4), 2546–2590 (2016)
2. 2018 annual cybersecurity report: the evolution of malware and rise of artificial intelligence. (2018). https://www.cisco.com/c/en_in/products/security/security-reports.html#~about-the-series. Accessed 10 May 2019
3. Y. Meidan, M. Bohadana, A. Shabtai, M. Ochoa, N.O. Tippenhauer, J.D. Guarnizo, Y. Elovici, Detection of unauthorized IoT devices using machine learning techniques (2017). Preprint arXiv:1709.04647
4. Y. Meidan, M. Bohadana, Y. Mathov, Y. Mirsky, A. Shabtai, D. Breitenbacher, Y. Elovici, N-BaIoT—network-based detection of IoT botnet attacks using deep autoencoders. IEEE Pervas. Comput. **17**(3), 12–22 (2018)
5. N. Vlajic, D. Zhou, Iot as a land of opportunity for DDoS hackers. Computer **51**(7), 26–34 (2018)
6. C. Kolias, G. Kambourakis, A. Stavrou, J. Voas, DDoS in the IoT: Mirai and other botnets. Computer **50**(7), 80–84 (2017)
7. R. Gow, F.A. Rabhi, S. Venugopal, Anomaly detection in complex real world application systems. IEEE Trans. Netw. Serv. Manag. **15**(1), 83–96 (2018)
8. X. Jing, Z. Yan, W. Pedrycz, Security data collection and data analytics in the internet: a survey. IEEE Commun. Surv. Tutor. **21**(1), 586–618 (2018)
9. W. Lee, D. Xiang, Information-theoretic measures for anomaly detection, in *Proceedings 2001 IEEE Symposium on Security and Privacy. S&P 2001* (IEEE, Piscataway, 2001), pp. 130–143
10. S. Khattak, N.R. Ramay, K.R. Khan, A.A. Syed, S.A. Khayam, A taxonomy of botnet behavior, detection, and defense. IEEE Commun. Surv. Tutor. **16**(2), 898–924 (2014)
11. S. Nomm, H. Bahsi, Unsupervised anomaly based botnet detection in IoT networks, in *2018 17th IEEE International Conference on Machine Learning and Applications (ICMLA)* (2018), pp. 1048–1053
12. L. Wen, L. Gao, X. Li, A new deep transfer learning based on sparse auto-encoder for fault diagnosis. IEEE Trans. Syst. Man Cyber. Syst. **49**(1), 136–144 (2017)
13. F. Zhuang, X. Cheng, P. Luo, S.J. Pan, Q. He, Supervised representation learning: Transfer learning with deep autoencoders, in *Twenty-Fourth International Joint Conference on Artificial Intelligence* (2015)
14. A. Gretton, K. Borgwardt, M. Rasch, B. Schölkopf, A.J. Smola, A kernel method for the two-sample-problem, in *Advances in Neural Information Processing Systems* (2007), pp. 513–520
15. Y. Bengio, P. Lamblin, D. Popovici, H. Larochelle, Greedy layer-wise training of deep networks, in *Advances in Neural Information Processing Systems* (2007), pp. 153–160
16. I. Goodfellow, Y. Bengio, A. Courville, *Deep Learning* (MIT Press, Cambridge, 2016). http://www.deeplearningbook.org
17. V.L. Cao, M. Nicolau, J. McDermott, Learning neural representations for network anomaly detection. IEEE Trans. Cyber. **49**(8), 3074–3087 (2019)
18. J. Dromard, G. Roudière, P. Owezarski, Online and scalable unsupervised network anomaly detection method. IEEE Trans. Netw. Serv. Manag. **14**(1), 34–47 (2017)
19. O. Ibidunmoye, A. Rezaie, E. Elmroth, Adaptive anomaly detection in performance metric streams. IEEE Trans. Netw. Serv. Manag. **15**(1), 217–231 (2018)
20. J. Lu, V. Behbood, P. Hao, H. Zuo, S. Xue, G. Zhang, Transfer learning using computational intelligence: a survey. Knowl. Based Syst. **80**, 14–23 (2015)
21. S.J. Pan, Q. Yang, A survey on transfer learning. IEEE Trans. Knowl. Data Eng. **22**(10), 1345–1359 (2009)
22. C. Tan, F. Sun, T. Kong, W. Zhang, C. Yang, C. Liu, A survey on deep transfer learning, in *International Conference on Artificial Neural Networks* (Springer, Berlin, 2018), pp. 270–279
23. K. Weiss, T.M. Khoshgoftaar, D. Wang, A survey of transfer learning. J. Big Data **3**(1), 9 (2016)

24. C. Wan, R. Pan, J. Li, Bi-weighting domain adaptation for cross-language text classification, in *Twenty-Second International Joint Conference on Artificial Intelligence* (2011)
25. Y. Xu, S.J. Pan, H. Xiong, Q. Wu, R. Luo, H. Min, H. Song, A unified framework for metric transfer learning. IEEE Trans. Knowl. Data Eng. **29**(6), 1158–1171 (2017)
26. X. Liu, Z. Liu, G. Wang, Z. Cai, H. Zhang, Ensemble transfer learning algorithm. IEEE Access **6**, 2389–2396 (2018)
27. E. Tzeng, J. Hoffman, N. Zhang, K. Saenko, T. Darrell, Deep domain confusion: Maximizing for domain invariance (2014). Preprint arXiv:1412.3474
28. M. Long, H. Zhu, J. Wang, M.I. Jordan, Deep transfer learning with joint adaptation networks, in *Proceedings of the 34th International Conference on Machine Learning*, vol. 70. JMLR. org (2017), pp. 2208–2217
29. I. Goodfellow, J. Pouget-Abadie, M. Mirza, B. Xu, D. Warde-Farley, S. Ozair, A. Courville, Y. Bengio, Generative adversarial nets, in *Advances in Neural Information Processing Systems* (2014), pp. 2672–2680
30. E. Tzeng, J. Hoffman, K. Saenko, T. Darrell, Adversarial discriminative domain adaptation, in *Proceedings of the IEEE Conference on Computer Vision and Pattern Recognition* (2017), pp. 7167–7176
31. M. Long, Z. Cao, J. Wang, M.I. Jordan, Domain adaptation with randomized multilinear adversarial networks (2017). Preprint arXiv:1705.10667
32. M. Oquab, L. Bottou, I. Laptev, J. Sivic, Learning and transferring mid-level image representations using convolutional neural networks, in *Proceedings of the IEEE Conference on Computer Vision and Pattern Recognition* (2014), pp. 1717–1724
33. M. Long, H. Zhu, J. Wang, M.I. Jordan, Unsupervised domain adaptation with residual transfer networks, in *Advances in Neural Information Processing Systems* (2016), pp. 136–144
34. C. Kandaswamy, L.M. Silva, L.A. Alexandre, R. Sousa, J.M. Santos, J.M. de Sá, Improving transfer learning accuracy by reusing stacked denoising autoencoders, in *2014 IEEE International Conference on Systems, Man, and Cybernetics (SMC)* (IEEE, Piscataway, 2014), pp. 1380–1387
35. S. García, A. Zunino, M. Campo, Botnet behavior detection using network synchronism, in *Privacy, Intrusion Detection and Response: Technologies for Protecting Networks* (IGI Global, Pennsylvania, 2012), pp. 122–144
36. TCPtrace Tool, http://www.tcptrace.org/. Accessed: 30 Nov 2019
37. Wireshark Tool, https://www.wireshark.org/. Accessed: 30 Nov 2019
38. D. Powers, Evaluation: From precision, recall and F-measure to roc, informedness, markedness and correlation. J. Mach. Learn. Technol. **2**, 37–63 (2007)
39. D.P. Kingma, J. Ba, Adam: A method for stochastic optimization (2014). Preprint arXiv:1412.6980
40. J. Yang, R. Yan, A.G. Hauptmann, Cross-domain video concept detection using adaptive SVMs, in *Proceedings of the 15th ACM International Conference on Multimedia* (2007), pp. 188–197

Part II
Internet, Network and Cloud Applications Security

An Approach to Guide Users Towards Less Revealing Internet Browsers

Fadi Mohsen, Adel Shtayyeh, Marten Struijk, Riham Naser, and Lena Mohammad

1 Introduction

Web browsers are programs that allow users to search for and view the content of the World Wide Web [9]. Though, these browsers do more than just simply rendering HTML (Hypertext Markup Language) pages and displaying the results. They enable users to use search engines, make online purchases, communicate with each other using social media sites, and much more [26]. However, there are issues related to maintaining the privacy of users and the security of their devices while surfing the web using these programs. These issues can possibly result in compromising user's devices and access their personal data such as browser history and auto-fill information [2, 18]. For instance, vulnerable browsers could give attackers the opportunity to exploit their security gaps to steal information, delete files, and other malicious activities [30]. Though, executing such attacks is normally proceeded by collecting detailed information about the target. For example, the version of the operating system, the version of the hardware, and browsing software type. In fact, each browser has its own distinctive User-Agent request header [12]. The User-Agent request header exposes information about the software being used, the operating system, and the installation of certain plugins [19]. This information can also be leveraged to track user activities on the web via a process called fingerprinting [10, 17]. Fingerprinting is the process where the combination of fields exposed by the browser leads to an (almost) unique combination [5]. Although most people are aware of the role of cookies in fingerprinting and tracking users across the Internet,

F. Mohsen (✉) · M. Struijk
Information Systems Group, Bernoulli Institute for Mathematics, Computer Science and Artificial Intelligence, University of Groningen, Groningen, The Netherlands
e-mail: f.f.m.mohsen@rug.nl

A. Shtayyeh · R. Naser · L. Mohammad
An-Najah National University, Nablus, Palestine

K. Daimi et al. (eds.), *Emerging Trends in Cybersecurity Applications*,
https://doi.org/10.1007/978-3-031-09640-2_4

the use of the User-Agent request header is relatively unknown. Users do not get prompted to share such information. There are numerous privacy and security risks related to the User-Agent request header because of the amount and the sensitivity of the information it exposes. Yet, browsers differ tremendously with regard to the amount of information they provide in their User-Agent strings. In this work, we propose an approach to aid users to select less revealing browsers. The approach relies on assigning an exposure score to each browser based on the content of its User-Agent string and the number and severity of its security vulnerabilities. In effect, the vendors of these browsers will be forced into reducing the exposure by limiting the amount of information that they include in the User-Agent request headers and reducing the number of vulnerabilities. Thus, in this chapter, we start with a trivial exposure score that is merely based on the information items that exist in the User-Agent request header. We then show how such method can be improved by incorporating the Common Vulnerabilities and Exposures (CVE) of a browser.[1] We provide a full implementation for our method as a web tool. We follow that by conducting a user study to gather feedback from real users about our method. The feedback is then employed to further improve the look and the feel of this web tool, which is deployed here [28]. The source code and data set are publicly available here [20].

The rest of the paper is organized as follows. In Sect. 2, we start with the preliminaries. In Sect. 3 we go over the methodology including the initial exposure score, the user study, and then the final exposure score. In Sect. 4 we discuss the data set that we used in implementing our approach. In Sect. 5 we discuss the implementation details. In Sect. 6 we go over the related works. Finally, we conclude the paper in Sect. 7.

2 Preliminaries

In this section we give a brief overview to some relevant topics such as the Hypertext Transfer Protocol (HTTP), User-Agent request header, the Common Vulnerabilities and Exposures (CVE), the National Vulnerability Database (NVD), and the Common Vulnerability Scoring System (CVSS).

2.1 HTTP

The Hypertext Transfer Protocol is the foundation of the World Wide Web's data communication. The World Wide Web provides access to resources and documents by pressing on or navigates their Uniform Resource Locators (URLs) using a

[1] CVE is a list of publicly disclosed computer security flaws.

client/browser. When a user presses on a URL or types it into the browser's address bar, the browser/client and the hosting web-server exchange a series of HTTP requests and response messages [29].

2.2 *User-Agent Request Header*

The User-Agent request header is simply a feature string that allows the communicating parties on the web to identify the software, operating system, and version of the browser [21]. It is also known as User-Agent string. The User-Agent in this context is the software that sets on the user-end and submits requests to remote servers. For example, a bot scraping web pages, a download manager, or a browser. User-Agents have a self-identifying User-Agent HTTP header called a User-Agent (UA) string that is included with each request they make to a server. This string is often used by web servers to tailor their response/contents accordingly. For example, some web servers provide custom layouts across multiple devices. On the other hand, the UA string can be misused by attackers and trackers.

2.3 *CVE*

The Common Vulnerabilities and Exposures (CVE) List was launched by MITRE[2] as a community effort in 1999 [4]. The CVE List contains records of publicly known cyber security vulnerabilities. Each record contains unique number, a description, and at least one public reference. The CVE Records are used in numerous cyber security research, solutions, and services from around the world such as the National Vulnerability Database (NVD).

2.4 *NVD*

The National Vulnerability Database (NVD) is populated and kept up to date using the CVE List. The NVD enables automation of vulnerabilities detection, classification, and quantification. The NVD contains information such as software names, vulnerabilities, and misconfigurations, and impact metrics. Each record in the NVD database is given an impact or severity score using the Common Vulnerability

[2] The Mitre Corporation is an American not-for-profit organization.

Scoring System (CVSS) [23].[3] There are three versions of CVSS, the NVD supports both v2.0 and v3.X standards [23].

2.5 CVSS

The CVSS score of a particular vulnerability depends on how easy it is to exploit it, and the consequences of exploiting it. A score ranges from 0 to 10, with 0 being the least severe. The CVSS has three score types: Base, Temporal, and Environment. The Base score is the most widely used among the three. In this chapter, we also rely on the Base score.

3 Methodology

As a first step, we devise a formula to calculate a relative exposure score for web browsers based on the content of their User-Agent strings in relative to other browsers. Next, we implement the formula into a website to conduct a usability study of the approach. Based on the feedback that we got from the participants of the user study, we decide to extend the formula by incorporating the vulnerability records of browsers. The content and the look of the website is also improved based on this feedback. In the following sections, we first start with the relative score, then, we will discuss the user study, and finally, we cover the final exposure score.

3.1 The Relative Score

The purpose of the initial formula is to assign a relative exposure score for browsers. The score is based on the amount of information that are revealed by the browser's User-Agent string in relative to the other browsers in the data set. Equation 1 shows the exposure score of browser i based on all j's, the attributes that were contained in the User-Agent string. For each element j, the sensitivity and visibility scores and constants are calculated based on other equations, which we do not show here. For further information about this equation, we refer you to [19].

$$EXP(i) = \sum_j EXP(i, j) = \sum_j (S(j) + S_j) \times (V(i, j) + V_j) \quad (1)$$

[3] CVSS is a free and open industry standard for assessing the severity of computer system security vulnerabilities.

3.2 User Study

The aim of this user study is twofold. First, it aims investigating users' understanding and awareness of their privacy while using Internet browsers. Second, to study whether the implementation of the initial formula/website is appealing to the end users, and if not, obtain the list of new features that need to be added to the it. As such, this study raises three main research questions: (i) Are users aware of the data that is being exposed by their browsers? (ii) How likely are they willing to use such a score-based system? (iii) Are there any vital features missing this system?

In order to answers these questions the following hypothesis were created:

- **H1:** People are likely to use a score system to help them with their privacy based browser choice.
- **H2:** People are generally not aware the data in the user-agent string is being exposed to sites.
- **H3:** People that feel secure are less likely to use the privacy score recommendation.

3.3 Survey Design

The user study is composed of the following sequence of actions: participants will first complete a short questionnaire, then they will get to navigate to our website to view the exposure score of their browsers and the content of their User-Agent strings, finally, participants are asked to provide their feedback. Below is a more detailed view of the user study.

- **P1** Introduction
- **P2** Demographic questions (Appendix A.1)
- **P3** Questions about privacy concerns (Appendix A.2)
- **P4** Questions about browser use and selection (Appendix A.3)
- **P5** Exposure score report & questions about the report (Appendix A.4)
- **P6** Questions about score guides and responsibility (Appendix A.5)
- **P7** Final page, thank participants and show their unique token.

The unique token at the end is connected to their survey answers if they wish to have them removed from the data set.

3.4 Results

A total of 115 participants have successfully completed the user study. The results will be discussed in the following sections. Starting with the demographics then the hypothesis testing.

Demographics About half of the participants who attempted the study decided to complete it. In total 115 responses were recorded and used to create the data set. On average, participants took 8 min to complete the study. Most participants are from the Netherlands (81.7%). The ages ranges mostly from 15–20 (29.6%) to 20–25 (55.8%) as can be seen in Table 2. About 32.2% of the participants was female and 67.0% was male. The education and technical experience of the participants can be noted as relatively high. This is clearly visible in Tables 1 and 3. This shall have an impact on the final results of this study and should be considered when attempting to explain the results of the research.

Table 1 Educational background of participants

Education	Freq
Bachelor	74
High school	19
College	7
Master	6
Associate degree	5
Ph. D.	4

Table 2 Age range of participants

Age range	Freq
0–15	1
15–20	34
20–25	63
25–30	6
30–40	4
40–50	2
50–60	3
60+	2

Table 3 Participants technical experience with browsers

	Freq
Very poor	1
Poor	6
Fair	30
Good	57
Excellent	21

Table 4 Median, mean, and SD of question 18 and 19

	Q18	Q19
Scale	1: Strongly disagree, 7: Strongly agree	1: Extremely unlikely, 7: Very likely
Median	Neutral (4)	Neutral (4)
Mean	3.93	3.83
SD	1.48	1.42

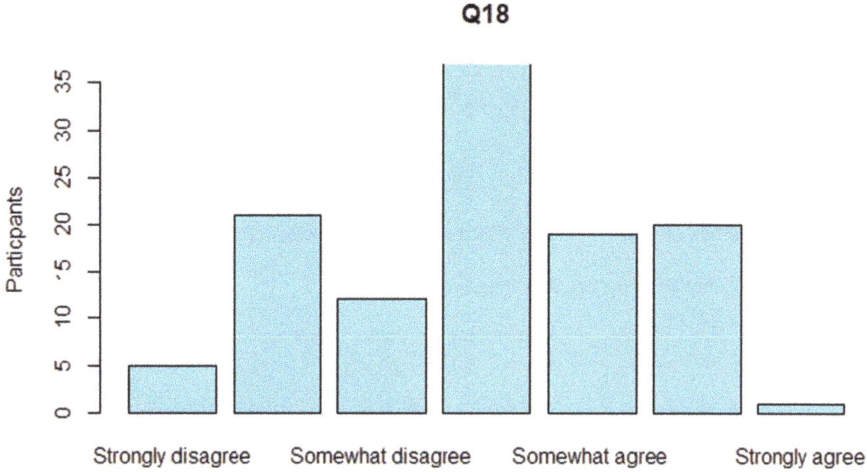

Fig. 1 The distribution of participants' responses to Q18

H1: People Are Likely to Use a Score System to Help Them with Their Privacy Based Browser Choice In order to examine this hypothesis, questions 18 and 19 are of particular interest.

- Q18—I am willing to use one of the browsers shown in the suggestions with a lower privacy score
- Q19—How likely would you be to let a score guide you for browser installation?

The results of these questions have been summarized in Table 4.

For these results, it is also important to look at bar plots of the response data. Figures 1 and 2 do show that the data is somewhat scattered around the neutral options. For question 19 there is a significant drop off after the "somewhat likely" option. Based on the data we can conclude that users are not very likely to use the initial score system. There seems to be no overwhelming majority for either side and users are indifferent. This does not mean that there are no people that would want to use the system, just that the majority will not.

To possibly explain the results of Q18 and Q19, another question/(Q16) was included where participants had to indicate how important certain features were for them regarding choosing a browser. The results can be seen in Fig. 3. To test if the

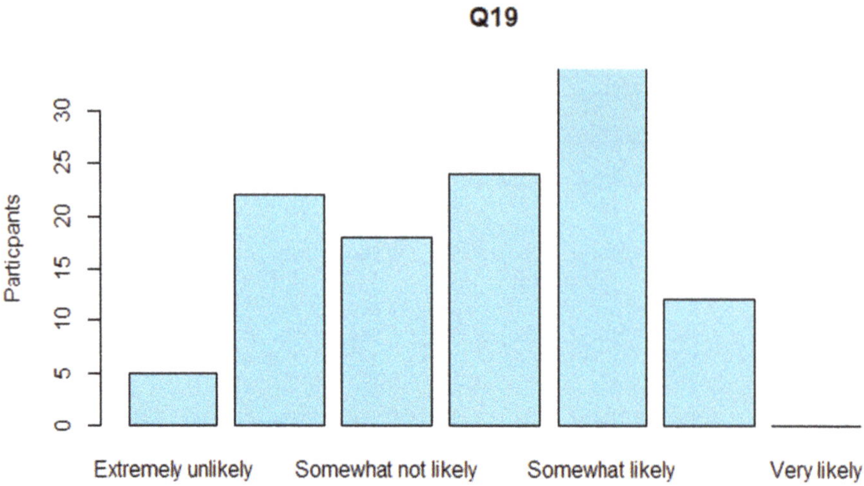

Fig. 2 The distribution of participants' responses to Q19

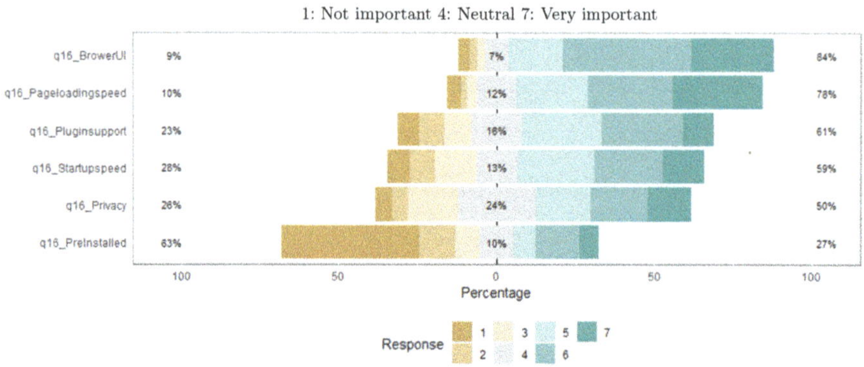

Fig. 3 The importance distribution for each of the six features that were presented to participants.

difference between the groups was statistically significant the Friedman ranked sum test was used ($\alpha = 0.05$). The results can be seen in Table 5.

Based on the results of the test we conclude that browser UI and page loading speed are the most important features for users upon choosing their browsers. After these two comes in the plugin support, startup speed and privacy. The least important feature is if a browser was pre-installed on the machine. Using this data it is possible to explain the results of Q18 and Q19 as follows: although people do consider privacy, it is not the most important feature that people will base their browser choice on. Yielding the neutral response towards the privacy score. Thus, H1 does not hold.

Question 21 allowed participants to elaborate on the reasons they have for using or not using the score system.

Table 5 Results of Friedman ranked sum test

Test	DF	*P* value	Significant ($p < 0.05$)
Browser UI—Page loading speed	1	0.729	No
Browser UI—Page loading speed—Privacy	2	$2.163e - 07$	Yes
Privacy—Start up speed—Plugin support	2	0.8364	No
Privacy—Pre-installed	1	$2.007e - 06$	Yes

Table 6 The summary of participants' responses to questions 8 and 9

Measurement	Q8	Q13
Median	Disagree (2)	Disagree (2)
Mean	2.18	2.86
Mean	1.35	1.55

Participants who chose not to use such a system pointed out some limitations in the exiting implementation. For example, they found the presented data as very technical. Some of them did not see any harm in sharing all the information that are included in the User-Agent strings. On the other side, participants who were in favor of using the system found it useful and solves part of their privacy concerns upon browsing the Internet.

H2: People Are Generally Not Aware That the Data in the User-Agent String Is Being Exposed to Sites Before showing what data was retrieved from participants' User-Agent strings, they were asked the following 2 questions:

- Q8—I am sure my browser does not share data about me and my device.
- Q13—When navigating the Internet I know which data my browser shares with websites.

The results can be found in Table 6. The majority of participants disagreed with the statement that their browser did not share any data about them. However, they indicated not to know what data their browser shared.

After showing the User-Agent strings, participants were asked if they were aware such data was being exposed.

- Q17—I was aware that my browser exposed this range of fields and values to web servers.

We summarized this data in Fig. 4, which shows participants' responses to both questions 5 and 17. The purpose of combining these two questions is because we want to spot any differences between technical people and less technical people.

Based on Fig. 4, the assumption rose that the more technical people might be the more aware they will be of the User-Agent strings, as such they will not be surprised about the exposed data. To test this assumption a Spearman test was conducted. Both values were treated as categorical. This resulted in a p-value of 0.001245 and $rho = 0.2985884$. A moderate correlation between these 2 values is confirmed. The better someone's technical knowledge, the more they were aware of the data being

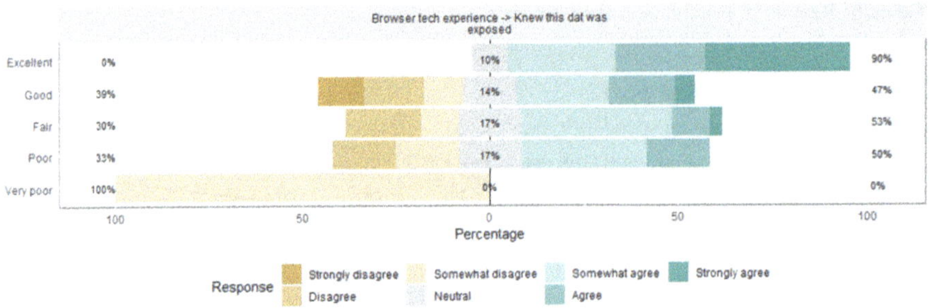

Fig. 4 Participants' responses to questions 5 and 17

Table 7 The results of applying Spearman test on questions 11 and 18, and questions 11 and 19

Test	*P*	*rho*
Q11∼Q18	0.02376	−0.2110193
Q11∼Q19	0.00004809	−0.3720299

exposed. Based on this correlation and the demographic data it was concluded that to answer the hypothesis at hand, one should look at the technical experience of a participant. For the overall data set the median of question 17 is "Somewhat agree." However, to reject the hypothesis, in this case, would be incorrect. Excluding people with excellent knowledge 42.5% of the respondents were not aware this data was being shared ("Strongly disagree," "Disagree," "Somewhat disagree"). To conclude H2. People with high technical knowledge of browsers are not surprised by these values. The lower the technical knowledge though the more users were not aware of the data being shared with websites.

H3: People That Feel Secure Are Less Likely to Use the Privacy Score Recommendation For this hypothesis the following 3 questions are of importance:

- Q11—When navigating the Internet I feel secure.
- Q18—I am willing to use one of the browsers shown in the suggestions with a lower privacy score.
- Q19—How likely would you be to let a score guide you for browser installation?

In order to test for any correlation, a Spearman test was conducted on Q11–Q18 and Q11–Q19. In Table 7 the results can be seen. Both tests have a p-value lower than 0.05, which is significant. Feeling more secure on the Internet has a negative impact on wanting to try one of the suggested browsers and being likely to use a score guide. Thus, H3 holds.

3.5 Discussion

After having viewed the many results it is important to look back at the original research questions. The first research question: Do users/Would users want to use a score-based system to guide them in choosing an Internet browser? And the second research question: Are people aware of the data being exposed? As per the first research question, after having viewed the results from the survey and having connected answers to some questions together it is concluded that people are not likely to use the score guide in its current form. Other features take priority over privacy. In addition, participants do not think the data is sensitive and find the data very technical. Other reasons are that the score calculation is not transparent enough. For the second research question, the answer to this research question depends on the user's technical knowledge. Whereas people with good technical knowledge seem to be aware this data (or at least partly) is being shared with the websites they visit, users with less technical knowledge do not seem to be aware of this. The demographic data of the participants turned out to be young and somewhat well served with technologies like browsers. This resulted in 50+% of the participants being somewhat aware of the data shared. If this survey is to be repeated it is believed this will be lower.

As far as improving the initial approach and the web implementation concern. After studying the results and, in particular, the open questions it becomes clear that some users would use this tool, although not in its current form. In order for the tool to become more attractive a couple of things could be done:

1. Explain why limiting data exposure is important.
2. Show how the score is calculated and explain why.
3. Parse data in a more readable format for users.
4. Recommend the latest version of a browser.

Explaining why limiting exposure could help users show the importance of the tool. In doing so, it is also important to show users how the score is calculated. A score can still be used but highlighting certain key features of why the current user-agent string scores "low" could really help people better understand the process of calculating the score. In addition, the tool should not recommend browsers of an outdated version. Older versions of browsers can contain security flaws that make users vulnerable to cybercrime.

3.6 CVSS Score

In order to expand the initial approach, we looked into ways to incorporate more data when calculating the exposure scores for the Internet browsers. The number of vulnerabilities of a browser and their severity scores would be a significant addition. As such, for each browser we obtain its vulnerabilities from the National

Vulnerability Database (NVD). Attached to each record in the NVD a CVSS score as explained in Sect. 2. The National Vulnerability Database (NVD) is currently supporting CVSS Version 3.x [24] and CVSS Version 2.0 [1, 25]. Thus, a particular vulnerability can have at most two scores. However, since the CVSS 3.0 is not yet complete, many vulnerabilities will only have one score, the CVSS version 2.0. As such, we decided that the CVSS score of a vulnerability is the average of two scores if both existed, otherwise, it is the CVSS 2.0 score. In Eq. 2 we show the formula for calculating the average score of a vulnerability. In Eq. 3 we show formula for calculating the final CVSS score of a particular browser.

$$CVSS\_BaseScore(vuln) = [AVG(CVSS2(vuln), CVSS3(vuln))|(CVSS2(vuln))] \tag{2}$$

$$CVSS\_FinalScore(browser) = (\textstyle\sum_{vuln} CVSS\_BaseScore(vuln))/N \tag{3}$$

3.7 *Final Exposure Score*

In Eq. 4 we show our final formula for calculating the final exposure score for a browser. The final exposure score equals the normalized relative score plus the CVSS score divided by 2. Originally, the relative score has no maximum value. In order to combine it with the CVSS score which are out of 10, we had first to normalize it.

$$FIN\_SCORE(i) = NORM(REL\_SCORE(i) + [AVG(CVSS2, CVSS3)|(CVSS3)]/2 \tag{4}$$

4 Data Set

Participants of the user study indicated that the initial data set contained duplicated records. We, therefore, removed these duplicates. Our final data set contains 52,000 unique browsers. The data set is used testing our approach. Each record in the final data set is composed of 51 columns. Forty-seven columns are for the different attributes that we can possibly retrieve from the User-Agent string. Two columns are for the CVSS and the relative scores. The last two columns are for keeping track of when was the two scores were updated.

4.1 *Summary of the Data Set*

In Table 8 we show the most popular software categories and device types that exist in the final data set. Nearly, 89% of the User-Agent strings are coming from a Browser, the other 11% are divided among other software types such as Application

Table 8 The mostly used browsers, device type vs. count

Software type	Count	Device type	Count
Browser	46,743	Mobile Phone	23,307
Application	3218	Desktop	16,875
Bot/Crawler	1544	Tablet	9907
Email Client	752	unknown	1362
Multimedia Player	78	Mobile Device	668
Offline Browser	35	TV Device	164
Feed Reader	15	Ebook Reader	110

Table 9 The top 10 used platform maker vs. count

Platform maker	Count
Google Inc	27,969
Microsoft Corporation	11,304
Apple Inc	6960
Linux Foundation	1958
unknown	1833
Nokia	564
Oracle	442
Canonical Foundation	346
FreeBSD Foundation	277
Symbian Foundation	245

Table 10 The top 10 most used rendering engines, their makers, and frequency of use. The various pointing approaches existing in the data set

Rendering engine	Maker	Pointing method	Count
Blink	Google Inc.	Touchscreen	22,208
WebKit	Apple Inc.	Touchscreen	7411
Blink	Google Inc.	Mouse	5751
Gecko	Mozilla Foundation	Mouse	5636
WebKit	Apple Inc.	Mouse	2173
Presto	Opera Software ASA	Mouse	1376
Trident	Microsoft Corporation	Mouse	1189
Gecko	Mozilla Foundation	Touchscreen	855
U3	UCWeb Inc.	Touchscreen	818
U2	UCWeb Inc.	Touchscreen	497

6%, Bot/Crawler 3%, Email Client 2%. As per the device types, Table 8 shows their distribution, 45% mobile phones, 33% desktop, 19% tablets, and the other 3% are unknown devices. Table 9 shows the distribution of the platform makers. Regarding the distribution of the platform makers, it comes as no surprise that the top three most used platforms were Google, Microsoft, and Apple, since 45% of the devices are mobile phones. Table 10 shows the distribution of the rendering engines that are used in this browsing software. Finally, the distribution of the used rendering engines shows that Blink is the most used, with 54% share, WebKit 19%, Gecko 13%, Trident 3%, on Presto Opera 3%, and UCWeb 3%.

As per the privacy score and the vulnerability records of the browsing software. Figure 5 shows the average number of vulnerabilities per device type. As can be seen, Desktop devices are among the most vulnerable. Figure 6 shows the distribution of both the exposure and the CVSS scores for all the browsing software that we have in our data set. There is clearly a significant difference between the distributions of both scores. A closer look at both scores as shown in Figs. 7 and 8, the difference

Fig. 5 The average number of vulnerabilities for each device types

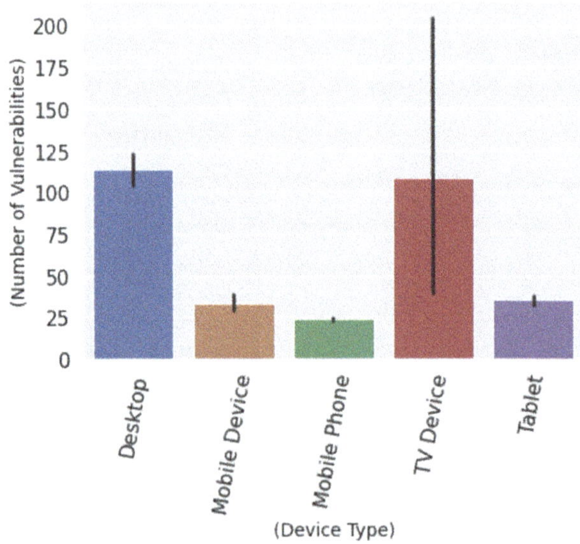

Fig. 6 The difference between relative and CVSS score

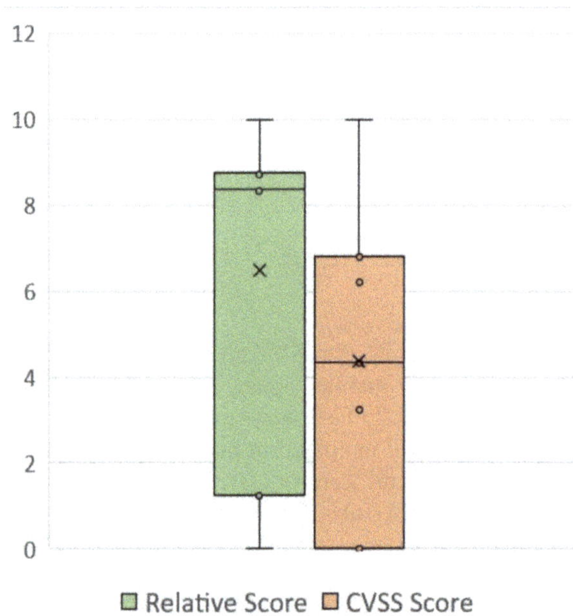

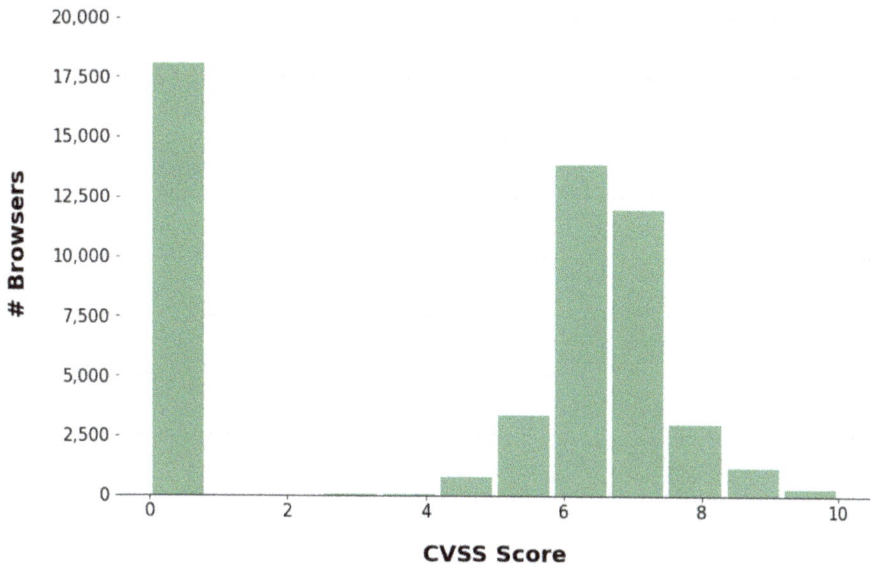

Fig. 7 The distribution of the CVSS score of all browsers in the final data set

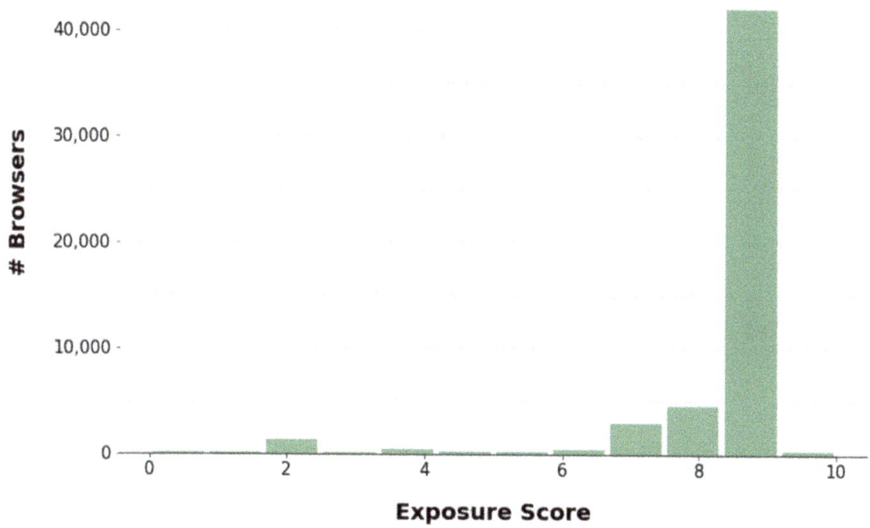

Fig. 8 The distribution of the exposure score of all browsers in the final data set

become more noticeable. Finally, Figs. 9 and 10 shows the distribution of the final combined score for all browsers and grouped by browsing software, respectively.

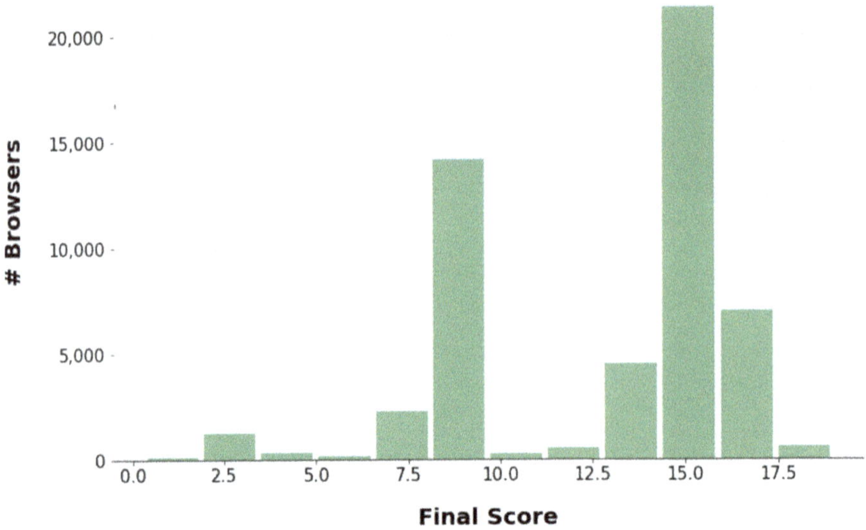

Fig. 9 The distribution of the final privacy score of all browsers in the final data set

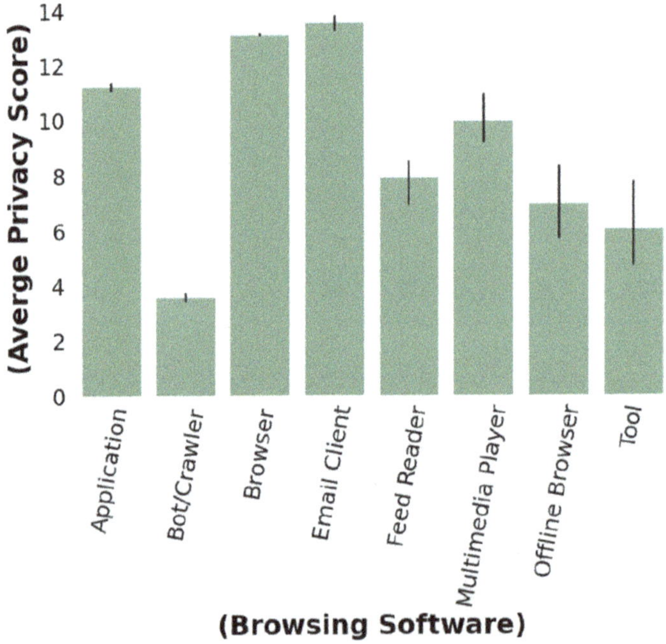

Fig. 10 The final privacy score grouped by browsing software

5 Implementation

As a first step, we calculated the relative scores for all the records in the final data set according to Eq. 1. We then calculated the CVSS scores based on Eq. 4, which took a lot of time because we had to crawl the information from the NVD website. Finally, the final score is calculated by adding up the normalized relative score to the final CVSS score.

The following is an overview of the score-based system that we developed. It uses the aforementioned scores to recommend safer browsing options to the users. The following are the steps that make up the whole procedure:

- Extract and process the User-Agent string.
- Calculate the relative score of user's browser.
- Calculate the CVSS score of user's browser.
- Present the report including the scores and alternative browsers.
- Update the final data set.

We will go through these measures in significant detail.

Extracting and Processing the User-Agent String In this step, the User-Agent string is extracted from the requests coming into our system. Then, the User-Agent string is parsed to extract the 47 features. Next, the final data set is searched to find a match based on the 47 features. The result of this search can be summarized in the following cases:

- Best-case scenario: a match is found, and both scores are present.
- Average-case scenario: a match is found, but one of the scores is missing.
- Worst-case scenario: a match is not found.

Calculating the Relative Score This step is needed in the worst-case scenario. It is also needed in the average_case scenario when the relative score is missing. For the latter case, the relative score is calculated and the browser's record in the final data set is updated accordingly. However, calculating the relative score for a requesting browser that does not exist in the final data set is a quite cumbersome process. This is due to the fact that the relative score of a browser depends on all browsers in the data set. Additionally, calculating a relative score for a new browser entails updating some of the terms and constants in the original equation, Eq. 1 such as n and $|R^j|$. Moreover, it requires updating the relative scores of all browsers in the database. For simplicity and efficiency reasons, the current score system temporarily updates the terms and constants to calculate the relative score of the requesting browser. Though, the relative scores of existing browsers do not get updated instantly.

Calculating the CVSS Score This step is needed in the worst-case scenario. It is also needed in the average_case scenario when the CVSS score is missing. The CVSS score is calculated by first searching the NVD website for related vulnerabilities. The search is conducted based on three keys: the browser name, the browser version, and the platform. The NVD website sends back the results as a JSON file. The

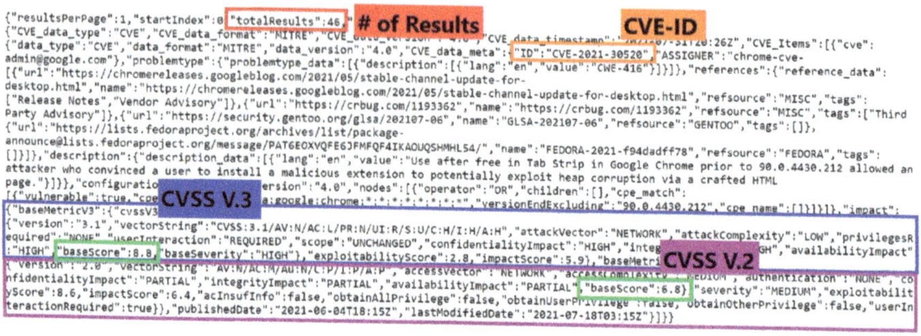

Fig. 11 A snippet of an NVD Json file. It contains the number of vulnerabilities, and the CVE id of each vulnerability and its CVSS score

file contains several information, most notably the number of vulnerabilities, the ID of each vulnerability and the base metric that contains the CVSS score in its two versions. The file shown in Fig. 11 was returned after searching for the following keys: Chrome, 90.0, and Win10. It shows that there are 46 distinct vulnerabilities linked to this particular browser. The CVSS scores for one of these vulnerabilities are highlighted.

The Json file is then parsed into two-dimensional array. For each vulnerability, the final CVSS score is calculated by either averaging both scores or considering one of them if the other is missing. The final CVSS score of a browser would then be the average of all these final CVSS scores.

Calculating the Final Exposure Score The final exposure score for each browsing software in our final data set is then calculated. The maximum possible value for this score is 20. The relative and CVSS scores contributes evenly to this score with 10 each. Browsers with lower scores are better for users because they reveal less information and has less vulnerabilities. In order to understand the relationship between the CVSS score and the relative score of a browser, we calculated the correlation coefficient. It meant to measure the degree of linear association between the two continuous variables. The correlation between the relative score and the CVSS score is 0.18, which is considered weak. Thus, merging these two scores is considered advantageous as it gives us a better representative score. Otherwise, one of these scores would have been enough.

Displaying Scores and Suggestions In this step, a report is displayed to the user pertaining her browsing software. The report contains information such as a relative score, CVSS score, final score, last updated, and alternative browsers. It also shows all the attributes that the current browser reveals. Additionally, it provides a description of the scores and their privacy implications. In Fig. 12, the final exposure score of the user's browser, which is *Chrome 90.0 on Windows 10 64bit*, is 13.97 out of 20, the relative score is 7.37 out of 10, and the CVSS score is 6.6 out of 10. The browsing software reveals numerous attributes such as platform, device type, and device name.

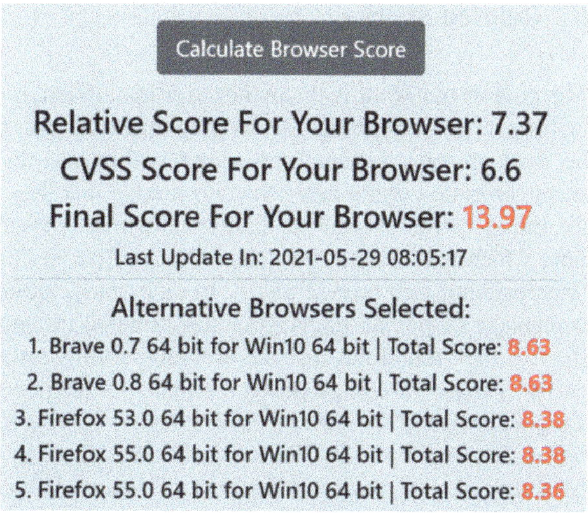

Fig. 12 The current implementation returns a list of alternative browsers with lower final exposure scores that also fit the user's device specifications

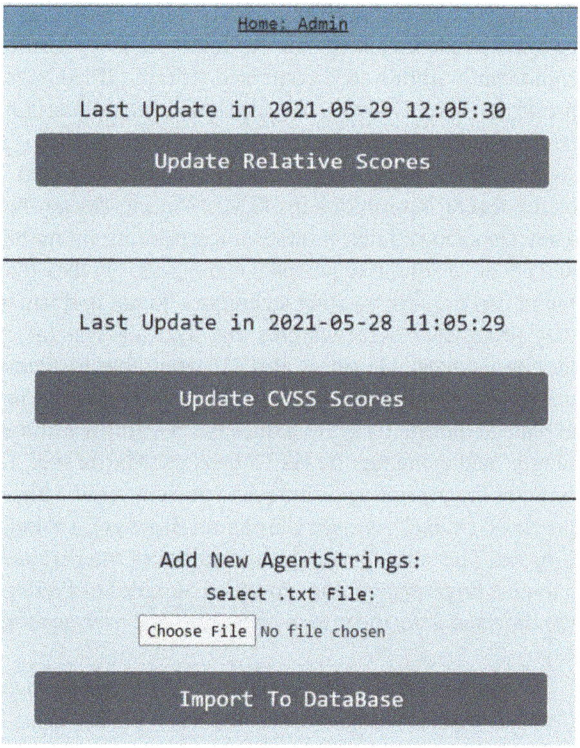

Fig. 13 Admin graphical user interface

Update: Admin Portal The Admin Portal is used to add new User-Agent strings to the data set and update the privacy scores of existing ones, as can be seen in Fig. 13.

6 Related Works

The goal of our work is to counter user identification and tracking through device or browser fingerprinting. Device fingerprinting was first studied by Peter [5]. In his work, modern web browsers were tested in order to determine whether they can be fingerprinted or not using the information that they disseminate while browsing the Internet. One of their key findings was that browsers reveal too much information, which makes them trackable. Gómez-Boix et al. studied the effectiveness of fingerprinting web browsers [10]. In their study, although they found out that the percentage of unique fingerprints has dropped in comparison to previous studies [6, 15], they found out that the non-unique fingerprints are fragile that they can easily change. Takeda proposed a number of techniques to identify the owner of a digital device [27]. One of these techniques was based on analyzing the browsers' fingerprints such as: HTTP Accept Header, Browser Plugins, System Fonts and Screen size and color depth. Yen et al. [31] carried on a large-scale study on month-long anonymized data sets that were collected by the Hotmail web-mail service and the Bing search engine. Their results showed that User-Agent strings can effectively be used to identify hosts on the Internet. The identification accuracy can also be significantly improved if combined with the IP address of the host. Kaur et al. [14] proposed a web browser fingerprinting technique that works despite the security devices and measures that are normally deployed at the corporate network boundary such as VPNs, proxy servers and NAT. Laperdrix et al. [16] demonstrated the effect of the recent innovations in HTML5 on increasing the accuracy of fingerprinting. They also showed that browser fingerprinting on mobile devices is highly possible and effective similar to personal computers. On the other hand, Hupperich et al. [13] found that existing tracking techniques do not perform well on mobile devices; thus, they proposed several features that tracking systems could leverage to fingerprint mobile devices. Martin et al. [22] were able to identify web browsers using the underlying JavaScript engine. As far as the preventive measures, Laperdrix et al. [16] explained different ways to reduce the possibility of fingerprinting, such as removing plugins and using regular HTTP headers. Martin et al. [22] leveraged their proposed browser fingerprinting technique to prevent session hijacking attacks. Baumann et al. proposed DCB (Disguised Chromium Browser), a solution to disguise the Chromium browser. The solution changes the values of the parameters that are used in building a unique fingerprint of the browser. Examples of these parameters are: the language, the time and date, the screen resolution, the user-agent strings, fonts list, and plugins [3].

In addition, there were a number of proposals to counter the privacy threat of browser fingerprinting and tracking users [7, 8, 11]

7 Conclusion and Future Work

In this chapter, we spot light on the privacy risks of the User-Agent strings of Internet browsers. The content of these strings grant advertisement companies and other parties the ability to identify users and track them on the web. Our analysis of thousands of User-Agent strings of numerous browsers revealed that they differ significantly. Thus, we proposed a new approach to quantify the exposure risks of browsers based on the content of their User-Agent strings and their vulnerability records. We also provided a full web implementation for our approach. The implementation was further improved based on the feedback we obtained from conducting a user study. Our validation and performance analysis of our implementation showed that it is accurate and efficient. For instance, the time to get a score takes only 0.85 s if this browser is existing in our database. If this browser is new, it takes 6.16 s to calculate the final score. And updating the entire data set takes 1.82 min. We believe that our approach can be further improved by incorporating the CVSS temporal metrics and the CVSS environmental metrics. The PHP source code, the database schema, and the final data set are made publicly available here [20]. Additionally, the web tool is currently deployed here [28].

Appendix

A Survey Questions

A.1 Demografic Questions

ID	Variable	Question	Type	Options
1	Location	In which country do you live?	Choose 1	All countries
2	Age	How old are you?	Choose 1	0–15, 15–20 ,20–25, 25–30 ,40–50, 50–60, 60+
3	Sex	What is your gender?	Choose 1	Male, Female, Other
4	Education	What is the highest degree of education you finished (or that you are currently following)?	Choose 1	"High school," "Associate degree," "College," "Bachelor," "Master," "Ph.D."
5	Tech	How well are you served with technologies like internet browsers or other technical aspects of the internet?	Choose 1	"Very poor," "Poor," "Fair," "Good," "Excellent"

A.2 Question Set 1

ID	Variable	Question	Type	Options
6	Concern security	In general, how concerned are you about your privacy online?	Choose 1	Not at all concerned A little concerned Somewhat concerned Concerned Very concerned
7	Privacy/Convenience	I prefer convenience over privacy when I browse the internet	Scale 1–7	1: Strongly disagree 7: Strongly agree
8	Awareness	I am sure my browser does not share data about me and my device	Scale 1–7	1: Strongly disagree 7: Strongly agree
9	Shared properties	Present all options: Ask how likely they thing it is being shared: Device name Device type Your last name Language Browser name Operating system	Scale 1–7	1: Not shared 7: Definitely shared
10	Data	When navigating the internet I feel like I know what is happening to my data	Choose 1	Strongly disagree Disagree Somewhat disagree Neutral Somewhat agree Agree Strongly agree
11	Secure feeling	When navigating the internet I feel secure	Choose 1	Strongly disagree Disagree Somewhat disagree Neutral Somewhat agree Agree Strongly agree

(continued)

ID	Variable	Question	Type	Options
12	Tracking	When navigating the internet I feel like no-one follows my browsing actions	Choose 1	Strongly disagree Disagree Somewhat disagree Neutral Somewhat agree Agree Strongly agree
13	Informed Feeling	When navigating the internet I know which data my browser shares with websites	Choose 1	Strongly disagree Disagree Somewhat disagree Neutral Somewhat agree Agree Strongly agree

A.3 Question Set 2

ID	Variable	Question	Type	Options
14	Safe mode	How often do you use safe mode, anonymous mode, incognito mode or any other plugins to increase the privacy of your browser?	Choose 1	Always Very frequently Occasionally Rarely Very rarely Never
15		How many different browsers do you use?	Numeric	
16	Browser choice	What features make you choose a browser? Startup speed Page loading speed Pre-Installed Plugin support Privacy Browser UI	Scale 1–7	1: Not important 2: Very important

A.4 Question Set 3

ID	Variable	Question	Type	Options
17	Awareness	I was aware that my browser exposed this range of fields and values to web-servers.	Choose 1	Strongly disagree Disagree Somewhat disagree Neutral Somewhat agree Agree Strongly agree
18	Willing to try	I am willing to use one of the browsers shown in the suggestions with lower privacy score	Choose 1	Strongly disagree Disagree Somewhat disagree Neutral Somewhat agree Agree Strongly agree

A.5 Question Set 4

ID	Variable	Question	Type	Options
19	Score guide use	How likely would you be to let a score guide you for browser installation?	Choose 1	Extremely unlikely Unlikely Somewhat not likely Neutral Somewhat likely Likely Very likely
20	Responsibility	I think that browser vendors (Google, firefox...) have a responsibility to limit the information exposure to the minimum.	Choose 1	Strongly disagree Disagree Somewhat disagree Neutral Somewhat agree Agree Strongly agree
21	Reasoning	Please explain why or why not you would let a score guide you for browser installation/choice.	Open	

References

1. M.U. Aksu, M.H. Dilek, E. Tatlı, K. Bicakci, H. Dirik, M.U. Demirezen, T. Aykır, A quantitative CVSS-based cyber security risk assessment methodology for it systems, in *2017 International Carnahan Conference on Security Technology (ICCST)*, pp. 1–2 (2017). https://doi.org/10.1109/CCST.2017.8167819
2. R. Barona, E.A.M. Anita, A survey on data breach challenges in cloud computing security: Issues and threats, in *2017 International Conference on Circuit ,Power and Computing Technologies (ICCPCT)*, pp. 1–8 (2017). https://doi.org/10.1109/ICCPCT.2017.8074287
3. P. Baumann, S. Katzenbeisser, M. Stopczynski, E. Tews, Disguised chromium browser: Robust browser, flash and canvas fingerprinting protection, in *Proceedings of the 2016 ACM on Workshop on Privacy in the Electronic Society*, WPES '16 (Association for Computing Machinery, New York, NY, USA, 2016), pp. 37–46. https://doi.org/10.1145/2994620.2994621
4. CVE: Common vulnerabilities and exposures, https://cve.mitre.org/cve/cna.html
5. P. Eckersley, How unique is your web browser? in *International Symposium on Privacy Enhancing Technologies Symposium* (Springer, 2010), pp. 1–18
6. P. Eckersley, How unique is your web browser? in *Proceedings of the 10th International Conference on Privacy Enhancing Technologies*, PETS'10 (Springer, Berlin, Heidelberg, 2010), pp. 1–18
7. A. FaizKhademi, M. Zulkernine, K. Weldemariam, FPGuard: Detection and prevention of browser fingerprinting, pp. 293–308 (2015). https://doi.org/10.1007/978-3-319-20810-7_21
8. U. Fiore, A. Castiglione, A.D. Santis, F. Palmieri, Countering browser fingerprinting techniques: Constructing a fake profile with google chrome, in *2014 17th International Conference on Network-Based Information Systems*, pp. 355–360 (Sep 2014). https://doi.org/10.1109/NBiS.2014.102
9. GCFGlobal: Internet basics - using a web browser, https://edu.gcfglobal.org/en/internetbasics/using-a-web-browser/1/
10. A. Gómez-Boix, P. Laperdrix, B. Baudry, Hiding in the crowd: An analysis of the effectiveness of browser fingerprinting at large scale, in *Proceedings of the 2018 World Wide Web Conference*, WWW '18 (International World Wide Web Conferences Steering Committee, Republic and Canton of Geneva, CHE, 2018), pp. 309–318. https://doi.org/10.1145/3178876.3186097, https://doi.org/10.1145/3178876.3186097
11. A. Gómez-Boix, D. Frey, Y.D. Bromberg, B. Baudry, A collaborative strategy for mitigating tracking through browser fingerprinting, in *MTD 2019 - 6th ACM Workshop on Moving Target Defense*, pp. 1–12, London, United Kingdom (Nov 2019). https://doi.org/10.1145/3338468.3356828, https://hal.inria.fr/hal-02282591
12. C. Hoffman, What is a browser's user agent? https://cutt.ly/DW77C6v
13. T. Hupperich, D. Maiorca, M. Kührer, T. Holz, G. Giacinto, On the robustness of mobile device fingerprinting: Can mobile users escape modern web-tracking mechanisms? in *Proceedings of the 31st Annual Computer Security Applications Conference*, pp. 191–200 (2015)
14. H. Kaur, P. Zavarsky, F. Jaafar, Unauthorised data leakage from corporate networks through web browser fingerprinting vulnerability
15. P. Laperdrix, W. Rudametkin, B. Baudry, Beauty and the beast: Diverting modern web browsers to build unique browser fingerprints, in *2016 IEEE Symposium on Security and Privacy (SP)*, pp. 878–894 (2016). https://doi.org/10.1109/SP.2016.57
16. P. Laperdrix, W. Rudametkin, B. Baudry, Beauty and the beast: Diverting modern web browsers to build unique browser fingerprints, in *2016 IEEE Symposium on Security and Privacy (SP)* (IEEE, 2016), pp. 878–894
17. P. Laperdrix, N. Bielova, B. Baudry, G. Avoine, Browser fingerprinting: A survey. CoRR **abs/1905.01051** (2019). http://arxiv.org/abs/1905.01051
18. S. Matteson, 5 common browser security threats, and how to handle them. https://www.techrepublic.com/article/5-common-browser-security-threats-and-how-to-handle-them/

19. F. Mohsen, M. Shehab, M. Lange, D. Karastoyanova, Quantifying information exposure by web browsers, in *Proceedings of the Future Technologies Conference (FTC)*, vol. 3, 2020, ed. by K. Arai, S. Kapoor, R. Bhatia (Springer International Publishing, 2021), pp. 648–667
20. F. Mohsen, A. Shtayyeh, R. Naser, L. Mohammad, 52k+ user-agent strings and their exposure scores. https://dataverse.nl/dataset.xhtml?persistentId=doi:10.34894/2SVOIE
21. Mozilla, User agent. https://developer.mozilla.org/en-US/docs/Glossary/User_agent
22. M. Mulazzani, P. Reschl, M. Huber, M. Leithner, S. Schrittwieser, E. Weippl, F. Wien, Fast and reliable browser identification with JavaScript engine fingerprinting, in *Web 2.0 Workshop on Security and Privacy (W2SP)*, vol. 5 (Citeseer, 2013)
23. NIST, The common vulnerability scoring system (CVSS). https://nvd.nist.gov/vuln-metrics/cvss
24. NIST, nvd.nist.gov. https://nvd.nist.gov/vuln-metrics/cvss/v3-calculator
25. NIST, nvd.nist.gov. https://nvd.nist.gov/vuln-metrics/cvss/v2-calculator
26. F. Scientific, Introduction to browsing the web. https://www.freedomscientific.com/SurfsUp/Introduction.htm
27. K. Takeda, User identification and tracking with online device fingerprints fusion, pp. 163–167 (2012). https://doi.org/10.1109/CCST.2012.6393552
28. Web tool: What is the exposure score of my browser. https://mybrowserscore.com/
29. w3schools, w3schools.com. https://www.w3schools.com/whatis/whatis_http.asp
30. Wikipedia, Web browser. https://en.wikipedia.org/wiki/Web_browser
31. T.F. Yen, Y. Xie, F. Yu, R.P. Yu, M. Abadi, Host fingerprinting and tracking on the web: Privacy and security implications, in *NDSS*, vol. 62, p. 66 (2012)

Analysing the Threat Landscape Inside the Dark Web

Selahattin Hürol Türen, Rafiqul Islam, and Kenneth Eustace

1 Introduction

The Dark Web can be defined as the encrypted part of the Internet [1]. The deep web includes 95% of the documents on the Internet. In general, it is possible to index and search the surface web, which does not include personal and sensitive data such as credit card information.

Dark Web pages can be accessed from private networks such as TOR (The Onion Routing), I2P (Invisible Internet Project) and Freenet. TOR is widely used by the Dark Web users. Its domain can be defined by a .onion extension and it is free to use. TOR provides anonymous access for the Dark Web users. This anonymity gives them the courage to perform illegal activities like selling fake IDs or distributing child pornography. This kind of activities should be monitored by cybersecurity professionals to avoid illegal activities in the real world. Monitoring those activities is not easy because of the anonymity of the users in the Dark Web. Cybercriminal activities cause billions of dollars of damage across the globe. Since we live with the information superhighway, the cybercriminals took this opportunity to spread their malicious activity rate exponentially. Credit or debit card frauds, DOS (denial-of-service) or DDOS (Distributed Denial of Service) attacks, child pornography and selling weapons are examples of malicious activities. The security agencies and researchers are trying hard to defend against cybercriminal activities; however, there are a lot of challenges.

The new fields of blockchain, cryptocurrency and Internet of Things (IoT) security are among the new targets used by threat actors. The new public and private cloud technologies and IoT small devices such as Raspberry Pi further expand the

S. H. Türen (✉) · R. Islam · K. Eustace
School of Computing, Mathematics and Engineering, Charles Sturt University, Bathurst, NSW, Australia
e-mail: hturen@csu.edu.au; mislam@csu.edu.au; keustace@csu.edu.au

K. Daimi et al. (eds.), *Emerging Trends in Cybersecurity Applications*,
https://doi.org/10.1007/978-3-031-09640-2_5

illegal activities. These technologies and the latest social media world can be abused by hackers easily. These activities are organised in the Dark Web because the Dark Web provides encrypted communication through private networks like TOR.

This study focuses on the tools and technologies which are applied to monitor the illegal activities inside the darknet. This consists of:

- Investigating how criminal activities are happening inside the Dark Web
- Searching for attack patterns, which are made by adversaries in the Dark Web, and the impact of these attacks in the deep web
- Identifying the techniques and tools currently used to protect the Dark Web from these attack patterns

Dark Web users can communicate with each other without using their identities. However, the anonymity feature can encourage many Dark Web users to perform illegal activities. The following topics can be used to group various illegal activities:

- Terror activities: Terror activities can include the following activities:
  - Spreading propaganda documents to gain new members
  - Communication with other terror organisations to plan terror activities
  - Selling illegal weapons
  - Technical information to the members like how to prepare explosive materials [17]
- Drug dealers [18].
- Pornography: Till the middle of the 1990s, we have had only pornographic magazines like Playboy. Illegal pornographic materials like child pornography were not easy to obtain. It is now possible to reach pornographic materials including child pornography. Child pornography is one of the main cybercrime issues of the police in the world [28].
- Cryptocurrency: Since Bitcoin became valid in 2009, there are available APIs in different programming languages to mine the crypto moneys. Cryptocurrencies can be encrypted with a 256-bit key generator and these keys are also desired by the hackers.
- DDOS (distributed denial-of-service) attacks: DDOS attacks are planned to destroy the communication of important systems like power plants, which may damage the infrastructure of a big city. They can also be coordinated from the Dark Web [26].
- Copied credit cards/debit cards [26].
- Password lists of email providers and cloud systems [26].
- Human trafficking: Although we are living in the twenty-first century, we have still slavery in the world, and that is why it is also possible to control human traffic through the Dark Web [36].

Due to the major illegal topics as described above, it is a difficult challenge to verify such illegal activities in real time. The Center for Applied Internet Data Analysis (CAIDA) has developed a real-time lens into dark address space of the Internet to verify the illegal activities in real time [11]. Therefore, it is important to quickly collect data from Dark Web forums and use data science techniques to analyse the dataset for patterns or anomalies. Then machine learning (ML) algorithms can be used to train a model response from analysis of all data collected.

There are a huge number of illegal forum activities happening inside the Dark Web including drug dealers, terrorists, human trafficking, etc. An analysis and survey of the threat landscape of the Dark Web may also extend the effectiveness of previous machine learning algorithms as countermeasures to these threats. Such ML techniques can then be applied to monitor forum activities or gather the web fingerprints of the illegal websites which are visited frequently by Dark Web users, easier and faster. There are three major problems for this project. The first problem is that the detection of Dark Web forum activities should be early as possible before the any user can put their ideas into action. The second problem deals with the accuracy and effectiveness of any countermeasure used to mitigate these attacks on the TOR network. The third problem is ongoing due to the development of technologies (e.g. IoT, blockchain and streaming technologies) that require a constant revision of the algorithms and techniques used to monitor forum activities and any subsequent attacks on the TOR networks.

2 Literature Review

In this literature review section, the authors explain the machine learning and deep learning techniques which are applied to detect and monitor illegal activities in the Dark Web. The deep web and the Dark Web will also be described in this section. It is important to know the difference between the deep web and the Dark Web as well as the types of threats that are common in the Dark Web. There are two types of threat, which are going to be discussed in this literature review. One of these threats are the attacks on the TOR Network from adversaries, whose aim is to deidentify themselves from the target users. The second threat is that current forum activities can trigger cybercrime events in the future, making it vital to monitor all threats and have early detection methods in place. It is important revise and update those threat detection methods (and associated machine learning algorithms) used to monitor forum activities on the TOR Network.

2.1 The Deep Web

The topology of the internet started with simple technology by CERN (Conseil Européen pour la Recherche Nucléaire), as described in CERN [12]. According to this published report, the first World Wide Web software was made public on 30 March 1993. In the beginning, the topology of the Internet was a quite simple model where the HyperText Markup Language (HTML) documents could be shared through the HyperText Transport Protocol (HTTP). However sensitive data was still being kept locally by companies and was not available or published on the Internet. Later, the development of search engines began with services such as Alta Vista and Yahoo, which applied crawler algorithms to search for web documents. The companies have begun to realise the power of the web, and they began to communicate with their customers through the Internet.

The interaction of people all over the world caused another problem. The personal and sensitive data should not be found through search engines, and this data should only be visible for the people, who are involved with the data. That means there was a need to have another layer on the Internet, which would keep the data that is not easily reachable from the other third parties. Today almost 5% of the web can be named as the public net. The common search engines can reach the data on the public web. According to Akyıldız [2], 95% of data on the Internet is kept in the deep web which includes the private data of the people, institutes and companies.

2.2 The Dark Web

The deep web has another layer, the Dark Web. Illegal organisations, cryptocurrency buyers and sellers, money washers, terrorists are all using the deep web for their illegal activities in order not to be caught from the authorities such as FBI and Europol. One famous example was the preparation of terrorist attacks which happened in Paris in November 2015, which was done using the TOR network. Baravalle et al. [5] described the role of organised crime within "the Silk Road" as a hidden service in the TOR network, which was used to trade illegal information and material. It was a kind of black market for the drug dealer, terrorists, etc. The only payment method was Bitcoin. Koch [21] mentioned that the Silk Road was shut down by FBI in 2012, and the Silk Road 2.0 was also shut down later in 2014. Of course, it was not the last black market, which is frequently used in the Dark Web for the exchange of illegal materials.

In 2022 onwards, it is important to monitor the illegal forum activities in the Dark Web, because the users in the Dark Web can communicate with each other and act secretly to breach various resources IoT devices, sensors, social media communications and other important infrastructure resources such as nuclear plants. Their anonymised identities make it difficult to find the real identity of the threat actor(s) behind the scenes. Wang et al. [35] worked on the extraction of the personal attribute data in the Dark Web and the measurement of analysis in darknet. Their paper was also important for the detection of anonymised users.

How Does a TOR Network Work?

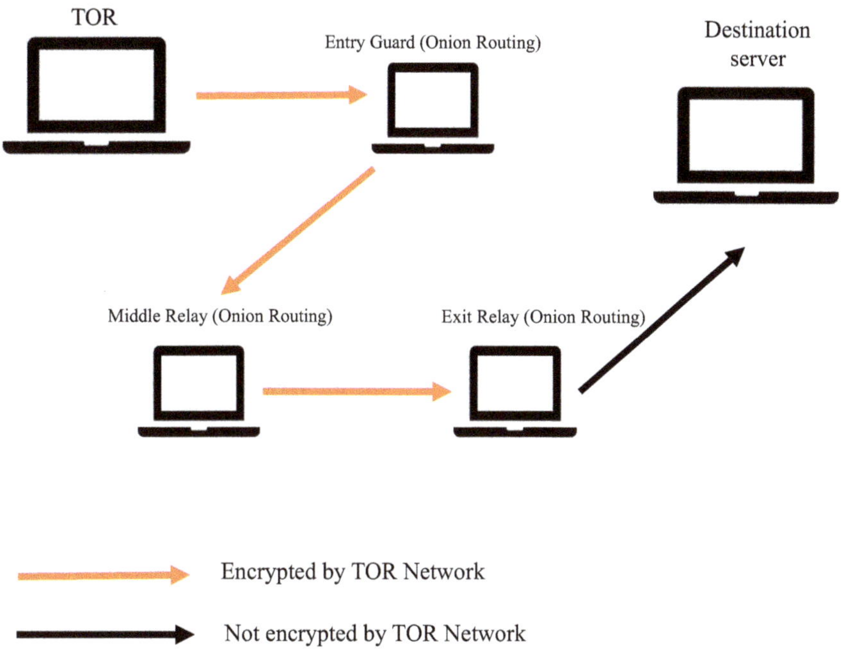

TOR Network is a kind of relay network. This network consists of onion routers, which communicate in a wireless environment.

Each onion router (OR) has its own descriptor which contains the OR's key and address. These descriptors of the ORs are sent to the authority directory in the relay.

How Can a Client Communicate with a Server by Using the TOR Network?

The client connects through an Onion Proxy (OP) which establishes a path to the destination address (server) that creates a circuit. A circuit can include three ORs. The OP connects with an authority directory to choose an entry point for the client which is also called an entry guard. This is the first OR. The entry guard extends this circuit to an exit guard which is the second OR. The exit guard is then connected with the destination OR. The communication of each OR is succeeded through a TLS/TCP channel [6].

2.3 Attacks on TOR

TOR was planned to provide an anonymity service for the user by hiding the location of their servers and to ensure that users are not concerned about the disclosure of their location. On the other hand, TOR has a decentralised architecture which allows opponents to launch distributed hash table attacks (DHT) [32]. A DHT is a distributed hash table, which provides key value pairs of the participating nodes in a network. It is easy to find the value of the any participating node with its given key. The advantage of DHT is the easiness of the adding and removing the key value pairs of the nodes.

There are several types of TOR attacks and some of them are listed in this survey of the literature.

2.3.1 Eclipse Attacks Against TOR Hidden Services

In an eclipse attack the threat actor grabs the DHT, gets the peer information of the onion routers and controls the incoming and outcoming communication of the victims, so that they cannot communicate with the rest of the peers in the network. The hidden services in TOR are kept as their descriptors in the hidden service directories (HSDirs) [32].

What Do the HSDirs Have and How Do They Store the Descriptor Information of the Hidden Services?

If a user tries to reach a hidden service, this user must do the following:

The user must receive a hidden service descriptor. A hidden service contains short messages which are updated in a short period (e.g. every hour). These messages include information about the popular nodes and other information like the descriptor ID. A descriptor ID is a hash code which changes every day, according to O'Cearbhaill [25]. The descriptor of a hidden service will be stored into the HSDir according to the smallest XOR distance between the descriptor IDs and the HSDir fingerprint lists. If the opponent has a good algorithm to find out the descriptor IDs, it is possible for him to grab the HSDir, which can make for him possible to block the user, so that they cannot find the right hidden service in the TOR Network.

What Can Be the Result of an Eclipse Attack on a TOR Network?

- Eclipse attack on TOR can disable the TOR's hidden services for the users.
- It will cost according to the data of the HSDirs.
- It will also have its security implications according to Tan et al. [32].

What Are the Phases of Eclipse Attacks on a TOR Hidden Service?

Phase 1: Computation of the Hidden Service's Descriptor IDs

Each TOR hidden service has descriptors so that the user can connect anonymously with that hidden service. This descriptor obtains information from the hidden service which are the introduction points of hidden services. This information can be easily predicted by the attacker. Some TOR networks upload these descriptors with their HSDir flags. These relays with HSDir flag are called as hidden service directory servers (HSDirs) which has the form of DHT. These servers act like a DNS of TOR Network. The HSDirs are important, because the anonymised user should ask for the information of the onion router (OR) to connect, and this information exists in the HSDirs. That is why it is important for the attackers to grab the HSDirs information [32].

Phase 2: Take Over the Hidden Service Directories and Start an Eclipse Attack The hidden service directories include the descriptor IDs and fingerprints of the onion routers of the site with which the user wants to connect. If an opponent or attacker can compute the descriptor IDs of a HSDir, it is possible for this person to calculate or find the fingerprint of the defined routers. The attacker can have hash table of the given descriptor IDs with their fingerprints, and it is enough to start an eclipse attack on TOR relay.

How Can an Eclipse Attack Be Managed?

The attacker saves the private key in a file which is called as secret ID key. This file is saved in a key directory of a data directory (DataDir) and a TOR instance is restarted with this directory. The TOR instance should run with the fingerprints of the ORs which are expected through the descriptor IDs. This process runs for 96 h and at the end the attacker grabs the HSDir flags. This gives the attacker the chance to take control over the hidden service directories. It is then possible to modify the source code of TOR before running the ORs, and the client receives no data from the targeted hidden data descriptor. It can be considered as a complete DoS. It is also possible to make an eclipse attack persistent. To make an eclipse attack persistent, it is important to mock the HDirs and to have a high secrecy for that mock for a long period of time. It is possible for an attacker to try to deny connection from the clients (users) to the ORs in a probabilistic way without any detection of the attacker. That is why it is important to know how many instances an attacker needs to run and control for a persistent eclipse attack.

According to the preliminary analysis of Tan et al. [32], an attacker needs at least 12 TOR instances to run an eclipse attack 1 day long. That is: 6 HSDirs * 2 (24 h HSDir delay) = 12

It is possible to run 2 TOR instances per each IP address, so that means the attacker needs only 6 static IP addresses for 12 TOR instances.

Biryukov and Weinmann [9] proposed that it is possible to use a shadowing technique to make the eclipse attacks persistent. The attacker should set up the nodes to receive descriptor IDs from the HSDirs. It should take more than 96 h so that all the relays will be able to have the HSDir flags.

There are eight TOR instances to run with static IP. The system builds four groups with two TOR instances for each group. These four groups run with their fingerprints for 96 h to obtain the HSDir flags from the hidden service directory servers. At the end of 4 days, all relays of these 6 static IP addresses will have their HSDir flags, but only 12 of them are going to be visible. The rest of these relays are going to be shadowed. That is why the attacker should make these 12 relays unreachable for the TOR authorities so that the shadow relays appear concurrently. This makes the eclipse attack persistent. That means the attacker needs 48 instances for 4 days:

6 HDirs * 2 (1 day HSDir relay) * 4 days = 48 TOR instances

2.3.2 Website Fingerprinting Attack

There is another type of traffic analysis attack which is called as website fingerprinting attack. A website fingerprinting attack tries to identify the URLs of the visited websites by using the TOR browser. A website fingerprinting attack can be considered as monitoring of the communication to the web pages in order to define the traffic characteristic pattern of the communication. This can be useful to know what kind of users are visiting a specific web page for which purposes, according to Attarian and Hashemi [4]. There are static and dynamic websites in TOR network. The fingerprints of the static websites do not change but the fingerprints of the dynamic websites change over time according to the updated content of the webpage.

Attarian and Hashemi [4] have investigated the usage of stream mining algorithms for the website fingerprinting attacks at the websites with static content. Stream mining algorithms can update their models frequently, and they are able to detect concept drifts, so that is why such a model can also be applied for the website fingerprinting attacks to the websites with dynamic content. This research has applied the stream mining algorithms to the following methods to know which of them is the best method for website fingerprinting attacks:

- *Adaptive Hoeffding tree*: This is a Hoeffding tree with an additional feature for detecting changes, and it gives the stream module the opportunity to adapt itself according to the changes in the data streams [27].
- *Extremely Fast Decision Tree (EFDT)*: EFDT is another modification of a Hoeffding tree, which is the implementation of a Hoeffding Anytime Tree (HATT). Manapragada et al. [22] described the EFDT has a higher predictive sequential accuracy than Very Fast Decision Tree. HATT is actually equal to a Hoeffding tree with one exception: It uses a Hoeffding bound to split the attribute to make sure the quality of the split attribute is better than the non-splitting attribute. The algorithm of HATT is defined in Manapragada et al. [22].
- *OzaBag*: OzaBag is a common bagging and boosting learning method in data streams. This is a part of MOA (Massive Online Analysis) API, which is developed at the University of Waikato in New Zealand. In this case, a Hoeffding tree is used as a data stream.
- *Concept adapting Very Fast Decision Tree (CVFDT)*: CVFDT is based on Very Fast Decision Tree and Hoeffding tree algorithms. The challenge of CVFDT model is to mine dynamic data or content efficiently. This model builds another tree for the old data, and if the updated data is more accurate than the old data, then the model replaces the updated data with the old data in this alternative tree [19].

Attarian and Hashemi [4] have analysed these four Hoeffding tree-based algorithms and have decided that adaptive Hoeffding tree has the better detection rate.

Table 1 Methods which are applied to attack on TOR [4]

Research paper	Method	Accuracy (%)
Feghhi and Leith [14]	Dynamic time warping	68.00
Cai et al. [10]	Damerau-Levenshtein	80.00
Wang and Goldberg [33]	Optimised SVM	88.00
Wang et al. [34]	Weighted KNN	91.00
Bhat et al. [7]	Var-CNN	98.07
Sirinam et al. [30]	CNN	98.60

Adaptive Hoeffding tree updates its model adaptively, and it can detect concept drift so that it can learn from constant content updates. That is why the researchers suggested that adaptive Hoeffding tree has a better recognition rate by detecting the websites which are visited. They also had some previous website fingerprinting attack data and then applied various learning and neural network algorithms and compared the accuracy of each method with these attacks. Here is the table of these attacks with their accuracies (Attarian and Hashemi [4] (Table 1).

These previous website fingerprinting attacks are based on batch learning methods which can be dropped easily over time as soon as the website content is updated. When the content of a website changes, the fingerprint of that website changes too. That is why the accuracy of these website fingerprinting attacks, which are based of batch learning algorithms, drops [4].

Then Attarian and Hashemi [4] tried to find another solution to keep the accuracy of the website fingerprinting attacks stabler. The website fingerprinting attack is divided into the following steps.

1. *Catch traffic traces*: Website traffic traces are captured by some special tools such as tcpdump, Wireshark, etc.
2. *Pre-processing*: Empty and erroneous traces are eliminated.
3. *Extract packet sequence information*: In this phase, data should be prepared for the feature extraction. Data like packet direction, packet orders and inter-packet times are extracted.
4. *Creating website fingerprint*: The fingerprint of a website includes information like the number of incoming and outgoing packets and the total number of incoming packets between two outgoing packets [4].

   Steps 1–4 are applied for both training and test data phases.
5. *Learning from website fingerprints*: Stream mining models are used to learn and classify from the website fingerprints. At the train phase, the model learns from the updated target websites, and that is why it is possible for the model to detect and specify them.
6. *Detecting targeting websites*: In the test phase, it is important to have a class prediction for the test examples. Here, the pre-sequential measure is applied so that the model is tested and trained for each instance (in this case, the detected webpage fingerprints) [8].

Table 2 The results of stream mining algorithms [4]

Methods	Window	Precision %	Recall %	Accuracy %	Kappa %
Adaptive Hoeffding tree	1000	97.97	99.00	98.97	97.80
	5000	98.74	98.99	98.87	98.77
	8000	98.99	99.12	99.20	98.81
EFDT	1000	98.00	98.10	98.20	98.14
	5000	98.10	98.13	98.20	98.13
	8000	98.15	98.14	98.20	98.14
OzaBag	1000	98.07	98.04	98.02	98.04
	5000	98.25	98.40	98.85	97.93
	8000	98.24	98.69	98.88	98.79
CVFDT	1000	98.00	98.14	98.26	98.15
	5000	98.13	98.17	98.24	98.17
	8000	98.11	98.17	98.27	98.16

Attarian and Hashemi [4] have followed these steps and applied the four mentioned stream mining algorithms (adaptive Hoeffding, EFDT, OzaBag, CVFDT). The dataset of Wang et al. [34] is used to test these four algorithms to know which of them has the most accuracy. As it is seen in the tables, the adaptive Hoeffding tree is the most accurate algorithm to make the website fingerprinting attacks (Table 2).

Yang et al. [37] have another approach for active website fingerprinting attacks which attempts to deidentify the TOR client. They have developed a workflow which includes the following steps:

1. Determining the start position
2. Delay of HTTP requests
3. Recording TOR cells
4. Feature extraction
5. Classification of the website

Determining the start position (the first HTTP request): It is essential to determine the HTTP request at the entry point. For that, Yang et al. [37] have analysed the behaviours of TOR protocols to find out the circuit and stream establishment, and data cell transmission, as it is an important issue to know at the entry point. When the first HTTP request is obtained with this information, then it is possible to infer the circuit and stream creation process [37]. This HTTP data can be transmitted between Onion Proxy (OP) and the web service which is remote.

Delay of HTTP requests: The entry node can receive many relay cells. These cells are coming from the OP which browses a website and delivered as encrypted. That is why it is impossible at that moment to identify each HTTP request from these cells. Therefore, Yang et al. [37] decided that they must find out an effective scheme which would find HTTP request positions and delay them from the web objects. To do that, they have developed two algorithms to modify the code of the TOR transmission mechanism, and they could delay the selected cells.

Table 3 Datasets ([37])

Name	Content	Size	Training	Testing
Data-closed	Cell	$100 \times 60 \times 6$	$100 \times 40 \times 6$	$100 \times 20 \times 6$
Data-open	Cell	2000×1	0	2000×1
Data-passive	Cell	100×60	100×40	100×20
Data-passive-link	Packet	100×6	100×40	100×20

Recording TOR cells: Yang et al. [37] have recorded TOR cells at the entry node for their feature extraction.

Feature extraction: It is important to extract features from the recorded TOR cell traces which generate the fingerprint. These features include total bandwidth per direction, positions of outbound cells and others.

Classification of website: According to the extracted features in the previous step, it is important to classify the website. Yang et al. [37] have applied two classifiers and then compared their prediction ability, in order to choose the better one.

Yang et al. [37] had built an experimental environment in PlanetLab (https://www.planet-lab.org/). They deployed three directory servers, one bridge and seven Onion Routers and applied TOR with version 0.2.5.10 on Fedora 8 with Kernel 2.6.32-20. The Onion Proxy is installed on Ubuntu 14.04.4 and Mozilla Firefox is used as web browser. They disabled the browser features to avoid noise traffic. They have used Obfsproxy (https://2019.www.torproject.org/projects/projects) to obfuscate traffic between the bridge and OP.

The data collection of Yang et al. [37] was as follows:

They have selected the most popular websites from Alexa (http://www.alexa.com). There was a problem with those websites. For example, Google has multiple distinct country domains like other most popular websites. They have kept only the most popular domains. They have generated datasets for closed world and open world scenarios which are called data-closed and data-open. Data-closed dataset has six cell delay position schemes. Each scheme includes these top 100 websites with 60 instances of each website. Forty instances of each website were for training and 20 instances were for testing. For the data-open dataset, they have chosen 2000 on monitored websites randomly. Each website had only one instance.

This experiment with these datasets were for the active website fingerprinting attacks. To compare their methods with the previous passive website fingerprinting attacks, Yang et al. [37] have two additional datasets named data-passive and data-passive-link. They have used the same monitor list. The following table includes the information about these four datasets (Table 3).

Yang et al. [37] believed that the entry node can bring more information to the attacker, and this would increase the detection rate. That's why they have implemented the passive website fingerprinting on data-passive and data-passive-link datasets to prove their assumption.

Table 4 Accuracy of passive website fingerprinting attacks ([37])

Scheme	Monitoring location	Dataset	Classifier	Accuracy (%)
Passive 1	Network layer	Data-passive-link	SVM	73.60
Passive 2	Bridge (entry node)	Data-passive	SVM	87.01

Table 5 Delay schemes of different delay positions ([37])

Scheme	Delay position	Delay time (ms)
Passive 1	None	None
Active 1	8, 9, 10, 15, 20	500
Active 2	8, 9, 10, 20	500
Active 3	8, 9, 10, 17, 27	500

Table 6 Detection rates of all schemes with different k values ([37])

Scheme	$k = 1$	$k = 2$	$k = 3$	$k = 4$	$k = 5$	$k = 6$	$k = 7$	$k = 8$	$k = 9$	$k = 10$
Passive 1	0.8595	0.7590	0.6865	0.6215	0.5635	0.5135	0.4690	0.4230	0.3820	0.3580
Active 1	0.9626	0.9429	0.9237	0.9066	0.8869	0.8631	0.8439	0.8222	0.8015	0.7828
Active 2	0.9705	0.9450	0.9297	0.9144	0.9016	0.8822	0.8629	0.8486	0.8328	0.8184
Active 3	0.9754	0.9505	0.9295	0.9075	0.8940	0.8605	0.8445	0.8170	0.8010	0.7845
Active 4	0.9602	0.9454	0.9255	0.9051	0.8903	0.8796	0.8612	0.8510	0.8362	0.8010
Active 5	0.9864	0.9722	0.9626	0.9545	0.9419	0.9222	0.9010	0.8833	0.8742	0.8586
Active 6	0.9737	0.9556	0.9394	0.9202	0.9015	0.8808	0.8621	0.8298	0.7793	0.7455

The following tables show the results of their experiments (Tables 4, 5 and 6).

Yang et al. [37] suggested that the active fingerprinting attacks have more accuracy in the detection phase.

2.3.3 RAPTOR (Routing Attacks on Privacy in TOR)

TOR is an overlay network which anonymised the user login data. TOR protects its user from the analysis of the user's data traffic. That is why millions of users use TOR to communicate. The anonymity of TOR network and the number of users of TOR network make it a target for the adversaries, who want to unleash the users. TOR is unfortunately vulnerable against the traffic attacks of these adversaries. If an adversary can observe both sides of the communication (TOR client – entry relay, exit relay – destination server), it is possible for him to deidentify the user. Border Gateway Protocol (BGP) routing may cause the new RAPTOR attacks, which can deidentify the users and the destination servers.

Sun et al. [31] have tried to counter RAPTOR attacks to find a way to protect the TOR network. They have investigated autonomous systems (AS) adversaries. They have seen that these ASes appear in almost 30% of entry-exit pairs. That is why the AS network-level adversaries are a big threat against the anonymity of the users in TOR. They can deidentify the users via BGP hijacking attacks. Because these attacks occur between the entry and exit points in the TOR network, they are called

Table 7 Prefix length of prefixes in TOR relays which were used in the Indosat 2014 hijack [31]

Prefix length	13	14	16	18	19	20	21	23	24
Number of prefixes	3	1	3	1	1	2	5	2	5

as RAPTOR attacks. That is why Sun et al. [31] proposed a kind of proactive defence against BGP attacks. They have built a live monitoring system. This monitoring system could monitor TOR relays in real time which is a big advantage to detect and defend the BGP hijack attacks. That is why Sun et al. [31] tried to find the resilience of TOR to prefix BGP hijack attacks and interception attacks. The recent research has considered passive AS-level attackers to adopt new path selection algorithm which considered asymmetric routing, relay capacity of TOR network and colluding ASes. Nithyanand et al. [24] have developed such a TOR client named Astoria, but this client considers only passive AS-level attackers. Therefore, the calculation of the resilience against active routing attacks is important. Sun et al. [31] have investigated the Indosat 2014 hijack attack. This attack has affected most TOR relays until now. It was a BGP prefix attack with 23 prefixes. All these 23 prefixes have attacked 44 TOR relays. Sun et al. [31] have found out that these prefixes had their false origin as Indosat (ASN 4761), but their true origin was ASes of TOR network. The following table shows the true lengths of these 23 prefixes. Here it is possible to see this kind of equally specific BGP prefix attacks which are shorter path attacks and more dangerous because the traffic would go to the false origin AS easily, if the path attacks are shorter.

Hence, it is important to know the resilience of the AS against BGP hijacks (Table 7).

How Can We Define the Resilience of an AS?

Let's assume that we have source AS as s; a false origin AS as f, which is launched by a BGP hijack attack; and a true origin AS as t. If s is not deceived by f and sends the traffic to t, then this AS is resilient against a hijack attack. Sun et al. [31] have also tried to calculate the resilience of ASes in different situations as follows:

To know the probability of resilience of a source AS, it is important to predict the AS-level paths of false origin and true origin of the destination AS-level path. Gao and Rexford [16] have proved that AS-level paths have two preferences: *Local preference*: according to Gao and Rexford [16], customer root is suggested over peer route, and this peer route is suggested over provider route. Shortest path: If the paths with the highest local preference are calculated, then the paths of these selected paths with the shortest hops will be picked [16].

Sun et al. [31] have developed an algorithm to calculate origin-source-attacker resilience for the given parameters s, f and t, and they have had the list of TOR relays from the TOR network data of 1.1.2016. They have used the AS topology which is published by CAIDA (Center for Applied Internet Data Analysis) in January 2016. The details can be read from Sun et al. [31]. The following diagrams show the hijack

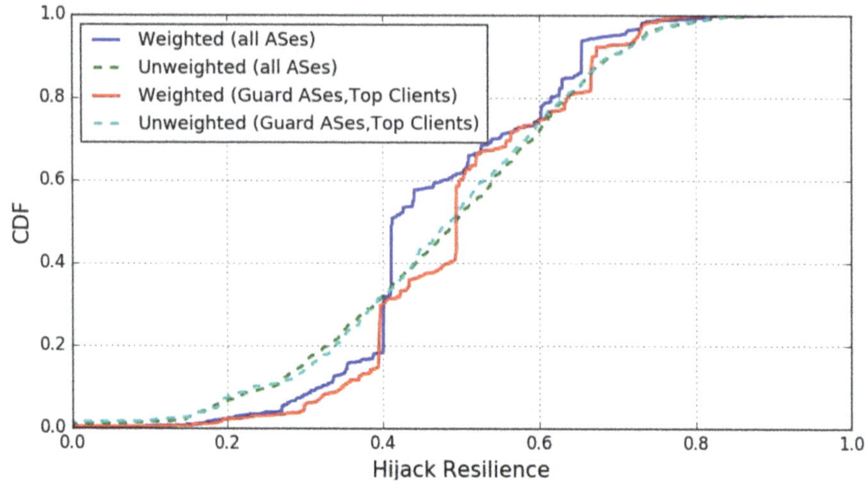

Fig. 1 Hijack resilience of TOR-related ASes (Ref. [31])

resilience for TOR-related ASes and hijack resilience and corresponding bandwidth per AS. The details can be found in the paper by Sun et al. [31] (Figs. 1 and 2).

Sun et al. [31] have the following design goals for the guard relay selection algorithm which they have developed:

1. Mitigation of equally specific prefix attacks on TOR
2. Protection of the anonymity of TOR users (clients)
3. Load balancing and performance

The guard relay selection algorithm which Sun et al. [31] have developed has two important aspects:

1. Selection of resilience metric
2. Incorporating resilience

How Could Sun et al. [31] Implement on TOR to Support Their Relay Selection Algorithm?

Sun et al. [31] must map the IP addresses of the TOR relays and TOR client to compute AS resilience. The client should perform the IP to AS mapping locally by handling the Maxmind Database which is included in TOR packages [23]. The client also uses the AS topology database from CAIDA [11]. The advantage of CAIDA database is that it updates the AS topology database monthly. That is why the overhead by downloading the latest updates does not take long. The implementation on TOR and the workflow of the algorithm are defined in five steps:

1. If Maxmind and AS topology files are not downloaded yet, they will be downloaded from Maxmind and CAIDA in the local data directory. If they are already installed, then the TOR client should check if they are up-to-date.

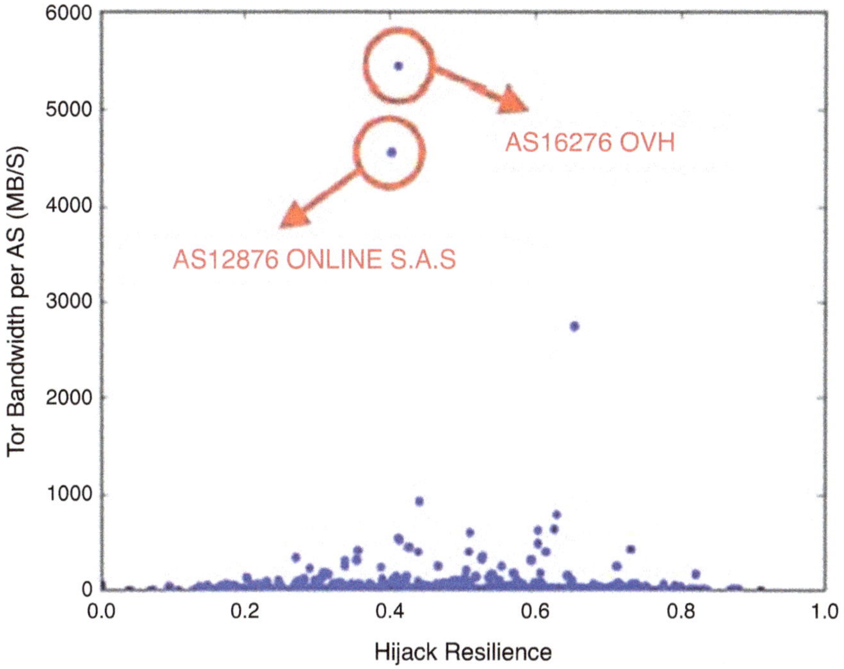

Fig. 2 Hijack resilience and corresponding bandwidth per AS. OVH and S.A.S are clear outliers (Ref. [31])

2. The TOR client maps IP with ASN and then the AS resilience will be computed.
3. The TOR client takes random samplings on all guarding relays which are the candidates and adjusts the resilience value on these samplings.
4. Then the TOR client computes the weight for each candidate relay.
5. The TOR client selects the path. The other parts of the circuit construction process will remain the same.

The process of these steps and the algorithms which are applied to these steps are clearly defined in Sun et al. [31].

2.4 *The Dark Web Activity Detection Methods*

In this subsection the recently used detection methods are introduced. These following methods are based on machine learning and deep learning methods.

2.4.1 MLP (Machine Learning Perceptron)

Kadoguchi et al. [20] explained in their paper that MLP (machine learning perceptron) could be applied to learn and train data from the Dark Web. The doc2vector method with Python is used for the NLP and feature extraction. The authors found it interesting to see the application of different machine learning areas being used to compare and see results that could detect illegal forum activities. They optimised the MLP hyperparameters by a grid search to divide hyperparameter search space into grids and test all the hyper meter combinations in intersection points of these grids. Stratified k-fold cross-validation is applied to evaluate datasets. Such a cross-validation method divides data so that the ratio of each label of evaluated data and trained data stays the same.

Preparation of data: Kadoguchi et al. [20] used Sixgill (https://www.cybersixgill.com) to collect data from the darknet. It is possible to collect hacker activity and SNS data from Sixgill. This site includes a particularly good search functionality. They had selected 350 posts which are related to malware offers and 350 posts other than malware offers. They have applied the MLP to this data and they found a classification result with 79.4% accuracy.

2.4.2 Hadoop-Based Dark Web Threat Intelligence Analysis Framework

Another interesting approach is from Yang et al. [38] with their Hadoop-based Dark Web threat intelligence analysis framework. Apache Hadoop is an open-source framework written in Java. Hadoop solves problems including the computation of big data structures. It includes the following components:

- Hadoop common: This includes common utilities and libraries which are generally used by Hadoop modules.
- Hadoop Distributed File System (HDFS): This distributed file system stores data to the machines which provide high bandwidth across the clusters.
- Hadoop YARN: This platform manages to compute the resources of the clusters and it uses them to schedule user applications.
- Hadoop MapReduce: It is a MapReduce program for large-scale data processing.

HDFS distributes the Dark Web data to multiple clusters. This can make it possible to read multiple data at the same time. HBase is a kind of column-based NoSQL database, which provides fast read and stores data. This may be helpful to work with the trained Dark Web data. MapReduce is used to distribute parallel data processing. Fudan NLP is used for the pre-processing of web page pre-processing and TF-IDF is applied for the text processing. With the help of data collection and examination of Hadoop framework, it is possible to apply text clustering and text categorisation machine learning methods to detect the different illegal activities.

2.4.3 Black Widow

There is another interesting tool like Black Widow [29], which can extract and parse raw data into HTML form, translate its language via Google Translate and extract the information. These kinds of methods help us to extract the raw data, learn from it and train this data (model) to detect the illegal activities in TOR routing or the other private networks. Our focus will be on TOR because TOR is better used to enhance privacy. Black Widow has the following life cycle:

1. Planning and requirements: Although Black Widow has generally an automated process, this step is manual. The integration of targeted darknet forums in the initial phase should be defined and managed manually. The manual steps of this phase are as follows:
   (a) Identifying Dark Web forums
   (b) Gaining access
2. Collection: This phase is an automated phase which deals with the anonymised access and collection of raw data through the TOR network.
3. Processing: This phase parses the raw HTML data, translates data in foreign languages if needed and extracts the information from this data.
4. Analysis: The main goal of Black Widow is to find the relationship between the authors and the context from different forums.
5. Dissemination: This is the last phase of Black Widow. The dissemination of the extracted information is important so that human intelligence analysts can work with the data easier.

Schäfer et al. [29] had chosen seven different forums in English, French and Russian languages. These forums have different domains for the communication. Because of the privacy, there is no mention about the names of these forums. Here is a table which they have created (Table 8).

The size of each forum has been calculated. There are two parameters for that: number of users in a forum and the size of the content. According to the results, Forum 5 has the most users with 67,535 registered users, and Forum 3 has the most content with 288,000 posts. That means Forum 3 also has the most activity. Each user has posted 22.74 messages. Forum 5 has the most passive users; each of them has 2.28 posts on average. Another point for Forum 5 is that it is a deep web forum; that is why they don't need to use additional software such as TOR to sign up.

2.4.4 Vector Space Model

Alnabulsi and Islam [3] proposed a vector space model to detect illegal forum activities. This analysis is done in two steps:

1. Collection of the Dark Web URLs: The URLs from the Dark Web are collected, which has the .onion extensions. These pages are reachable from the TOR

Table 8 Categories of Dark Web forum [29]

Forum	Forum type	Online as of December 2018	Language	Categories
Forum 1	Deep Web	Yes	English	News, porn, software, drugs
Forum 2	Deep Web	No	Russian	Marketplace, electronic money, hacking
Forum 3	Dark Web	No	French	Drugs, news, porn, technology
Forum 4	Dark Web	Yes	Russian	Marketplace, general discussions, hacking, security
Forum 5	Deep Web	Yes	English	Gaming, leaks, cracking, hacking, monetizing techniques, tutorials
Forum 6	Dark Web	No	French	News, frauds, conspiracy theories, drugs, crime
Forum 7	Deep Web	Yes	Russian	Software, security and hacking, DDoS services, marketplace

Table 9 Documents [3]

Document 1	http://zw3crggtadila2sg.onion/imageboard
Document 2	http://oxwugzccvk3dk6tj.onion/index.html
Document 3	http://answerstedhctbek.onion

Table 10 Terms related to the subjects [3]

Term	Subjects
Term 1	Piracy, hacking
Term 2	Drugs
Term 3	Scams, fake, forgery
Term 4	Politics, revolution
Term 5	Hitmen, kidnap, rape
Term 6	Philosophy, religion
Term 7	Pornography, gay, sex
Term 8	Weapons, guns

network. VSM is used to represent the number of posts related to the selected subject terms and to apply them in data mining operations.

2. Dark Web classification: After the selection of the terms which are defined in the first step, it is possible to classify the most common dialogue subject on each Dark Web forum. It is good to detect the subject weight of each forum to follow their illegal activities.

Here three forums are used to get three documents. These are (Table 9):

Alnabulsi and Islam [3] have searched many terms to classify the data domains or subjects of these forums. It is decided to select eight terms to group subjects for each forum. These were (Table 10):

The TF as of the count of posts for each document is shown in the following three tables (Tables 11, 12 and 13).

Table 11 TF of Document 1 [3]

Term	Subjects	Number of posts
Term 1	Piracy, hacking	19,875
Term 2	Drugs	4626
Term 3	Scams, fake, forgery	6606
Term 4	Politics, revolution	9511
Term 5	Hitmen, kidnap, rape	11,783
Term 6	Philosophy, religion	2642
Term 7	Pornography, gay, sex	0
Term 8	Weapons, guns	0

Table 12 TF of Document 2 [3]

Term	Subjects	Number of posts
Term 1	Piracy, hacking	1579
Term 2	Drugs	6878
Term 3	Scams, fake, forgery	0
Term 4	Politics, revolution	6771
Term 5	Hitmen, kidnap, rape	0
Term 6	Philosophy, religion	0
Term 7	Pornography, gay, sex	3880
Term 8	Weapons, guns	5932

Table 13 TF of Document 3 [3]

Term	Subjects	Number of posts
Term 1	Piracy, hacking.	12,895
Term 2	Drugs	1857
Term 3	Scams, fake, forgery	2651
Term 4	Politics, revolution	1176
Term 5	Hitmen, kidnap, rape	0
Term 6	Philosophy, religion	0
Term 7	Pornography, gay, sex	1711
Term 8	Weapons, guns	0

2.4.5 Dark Web Forum Visual Analysis Platform

Ying et al. [39] developed a visual analysis platform to handle large Dark Web forum information. It is possible to detect the relationship between the users and the forums through this platform. It is easier to find the information on a visual platform. This platform has the following modules:

1. Data acquisition module: This module uses Python + OnionScan's focused crawler framework, and it manages the Dark Web forums. The data information is stored in an SQL Server and can be shown in the platform interface easily.
2. Forum information inquiry module: This module queries from the forum message database according to the poster's name, subject, body, content, etc., and it filters out the qualified forum information from the database.

Table 14 Category-sample correlation for four samples [15]

Category	Samples 1	2	3	4	Total
Drug	651	412	517	476	2056
Market	724	833	863	854	3274
Money	833	1081	1065	1105	4084
Crime	439	129	120	179	867
Virus	1202	1302	1895	1603	6002
Adult	2377	342	311	2738	5768
Other	23	26	17	30	96
Number of analysed sites	102	125	98	100	425

3. Visualisation module: This module is responsible to visualise the filtered forum information.

2.4.6 The Methodology of Dark Web Monitoring

Ferry et al. [15] have developed a methodology to monitor Dark Web forums based on semantic analysis of sites to extract their keywords and where these keywords occur. Another algorithm which they have developed uses an API named Alyze (https://alyze.info). It is possible to open an SSH tunnel to communicate between the Alyze API and Dark Web site which is temporarily hosted on an Apache server. The algorithms can be found in their paper. To analyse the Dark Web monitoring algorithms which Ferry et al. [15] have considered, they have defined the categories as follows: drug, market, money, virus, crime, adult and other categories.

They have used TOR Network to collect 3000 Dark Web sites, and they have created 4 samples for each 100 websites, and they have tried to categorise the data and calculate the number of occurrences of each category. Here is the result of what Ferry et al. have found out (Table 14).

3 Summary of the Findings

This section provides the summary of findings about the Dark Web illegal forum activity detection methods and frameworks.

The following table shows the findings of the chosen methods and frameworks (Tables 15 and 16).

Table 15 The findings of the attacks on TOR

Paper	Technology	Methodology	Findings
Qingfeng Tan [32]	Eclipse attacks against TOR hidden services	Computation of the hidden service's descriptor IDs and grab the HSDirs	The attacker saves the private key in a file which is called as secret ID key. This file is saved in a key directory of a data directory (DataDir), and TOR instance is restarted with this directory. The TOR instance should run with the fingerprints of the ORs which are expected through the descriptor IDs. This process runs for 96 h, and the attacker needs to have 12 TOR instances to run the DoS attack. When it is done, the users cannot receive the data they need from the websites which they wanted to connect
Attarian and Hashemi [4]	Website fingerprinting attack	Stream mining algorithms based on Hoeffding trees	A website fingerprinting attack tries to identify the URLs of the visited websites by using the TOR browser. A website fingerprinting attack can be considered as monitoring of the communication to the web pages to define the traffic characteristic pattern of the communication. Adaptive Hoeffding tree algorithm has the best accuracy for the website fingerprinting attack

4 Gap Analysis of the Existing Techniques

This study explores the existing technologies, which monitor and combat the illegal forum activities inside the Dark Web. It is also important to explore the technology which attacks and defends the anonymous users of the Dark Web, which are attacked by the adversaries. For example, Attarian and Hashemi [4] have explained the streaming algorithm techniques for the website fingerprinting attacks. Routing attacks on privacy in TOR (RAPTOR) make it possible to deidentify the user inside the TOR Network, but Sun et al. [31] tried to counter these attacks to protect the anonymity of the users in the TOR network. TOR Network doesn't include only the users, who abuse the anonymity of .onion routing for their criminal activities, but also the users, who want to search for the information they needed without exposing their identities.

However, the solutions and the techniques, which are investigated in this study, have also their limitations. That's why it is important to compare and analyse the technologies to find their gaps. For example, mostly static data is used to analyse

Table 16 The findings of the monitoring Dark Web techniques

Paper	Technology	Methodology	Findings
Nicolas Ferry [15]	Monitoring Dark Web forums	Semantic analysis using Alyze Framework	Six categories are defined (drug, market, money, crime, virus, adult), and semantic analysis is applied for these categories
Alnabulsi and Islam [3]	Vector space model	ML vector space model	Terms are categorised by subjects through VSM, and the number of posts for each subject is calculated through TF. Document frequency is calculated to find TF-IDF
Schäfer [29]	Black Widow	Single-page application with angular and node JS	Black Widow has its own life cycle. It finds the relationships between users and forums through this life cycle. These relationships can be visualised
Yang et al. [38]	Hadoop-based Dark Web threat intelligence analysis framework	Apache Hadoop framework for clustering data. Researchers are inspired by DICE-E framework to identify, collect and evaluate Dark Web data with ethics (Victor Benjamin, 2012).	Web page extractions are based on CSS selectors or XPath, and these require different rules for different structures
		HDFS is used to cluster data into different clusters. Fudan NLP, a Chinese NLP, is used for web page information pre-processing. Text categorisation is done by TF-IDF	That is why they tried to use unsupervised learning algorithm because this algorithm can exist in many other ways according to the language and different structure of web pages
Kadoguchi [20]	MLP (machine learning perceptron)	doc2web with Python, an NLP Tool, and MLP are used.	There was not enough data. That is why the accuracy was 79.4% in the unseen dataset category. Nevertheless, the mean accuracy was more than 90% which is obtained by MLP model (with stratified k-fold cross-validation)
		A commercial tool Sixgill is used to collect data from Dark Web	doc2web has a high performance as NLP, and feature extraction in ML is faster with doc2web tool
Yang et al. [39]	Dark Web forum visual analysis framework	Python, OnionScan framework, SQL Server	Data acquisition is managed by Python and OnionScan tool which is written in GoLang. Forum information inquiry forum has the persistence layer and business logic of the data management, and this data is visualised in different types (statistical, network tree, table)
		No information about how they built the frontend (probably with Python)	

and apply the techniques in the Dark Web to monitor the illegal forum activities or to perform attacks on TOR to deidentify the user inside the Dark Web. There are few resources, which apply streaming data algorithms. There are also other factors which are caused by the limitations of the applied techniques or processes. Here is a summary of all these issues, which the authors have remarked:

- Ferry et al. [15] have applied their semantic analysis with an external open-source framework (Alyze Framework). That's why it is better to analyse their research through other frameworks to comply their research. Another gap is that this research group tried to do the semantic analysis locally. They have grabbed an .onion site once and saved it locally, and then they have run their semantic analysis modelling. The authors have not read any approach with streaming data.
- Alnabulsi and Islam [3] had only TOR Network in their research. The other networks such as ISP and Freenet should also be applied for the related work. The developed algorithm (vector space model) is based on TF-IDF. The most difficult problem is the text frequency. Most users cyphered their text and voice communication and avoid using general words, which can be caught easily by adversaries through text and voice recognition. For example, they can use beans instead of bullets or munition. That could be a bit misleading.
- Schäfer et al. [29] have developed Black Widow with a half-automated system. The first step of the life cycle (planning and requirements) is manual. The language translation is made by Google Translate. The authors also prefer another online translator APIs such as deepl. This research is also done with local data structure. Streaming was out of scope. If Black Widow follows an iterative way, it may be possible to automatise the requirement phase through the results of the first iteration. It can be possible to prioritise the requirements according to the relationship between the forums and users. That was the missing point that (Schäfer et al. [29] have also described.
- Yang et al. [38] have used Hadoop framework to cluster the data, and they have applied Fudan NLP for the pre-processing. They have developed an NLP framework (Fudan NLP) which performs the pre-processing of web page information. This framework uses word and character annotations and backward text segmentation. It is not mentioned clearly how accurate results this framework could deliver. That is why web page information pre-processing is limited with the limitation of this NLP framework.
- Kadoguchi et al. [20] have applied Sixgill (Cybersixgill) products as the data collection method. Cybersixgill is an Israeli company which produces tools for cybersecurity. This research group didn't give the detailed information which tools of Cybersixgill they have applied to their data. That would be helpful for the readers to understand how they could build their workflow up. Another problem is that they did not have enough data to support their analysis. More test data may have given more accurate results.
- Ying et al. [39] couldn't define well if the NoSQL feature of SQL Server is used or not. If SQL Server is used as a RDBMS, then the NoSQL feature of the SQL Server should be used to query the data information faster. The data extraction is

relied on OnionScan. It may be also possible in the future to support OnionScan with extra other NLP and text recognition tools or algorithms. Other limitation is that the query searches for the username. What happens if more users use the same username? What is about the IP addresses? The Dark Web forum visual analysis platform is briefly explained here. It is not possible to know from this survey what kind of machine learning or artificial intelligence algorithm is hidden behind this platform workflow. That is why the authors found the explanation of this survey weaker than the others in the literature.

5 Research Design and Methodological Approach

This chapter has investigated the existing research and identified the research gaps that remain with this research problem. Ferry et al. [15] used the existing research of a self-designed framework which only used semantic analysis, but it does not support full categorisation. The technology which Alnabulsi and Islam [3] have applied can be extended to search the terms in other network types. If Black Widow follows an iterative way, it may be possible to automatise the requirement phase through the results of the first iteration. It can be possible to prioritise the requirements according to the relationship between the forums and users. That was the missing point which Schäfer et al. [29] have also described.

The research of Yang et al. [38] was focused on Chinese web data. It can be extended to supply other languages, esp. English.

Kadoguchi et al. [20] applied Sixgill to support their algorithm. An extension of Sixgill might improve the data collection, or the researchers should find another way to collect more data for the categorisation of the datasets. Ying et al. [39] have used SQL Server for their research, but they didn't define how they have worked with the database. If SQL Server is used as a RDBMS, then the NoSQL feature of SQL Server should be used to query the data information faster. The data extraction is relied on OnionScan. It may be also possible in the future to support OnionScan with extra other NLP and text recognition tools or algorithms.

As a first step, the techniques and methods will be grouped which are applied for monitoring the Dark Web forums and which are used for the attacks on TOR network. Then it will be possible to define the gaps of these methods according to their usage. After the gap analysis of the previous research methods, it is possible to try to extend these algorithms or the datasets which the techniques are applied to, to see if the gaps are closed.

Another issue is about obtaining existing and new data. Existing data from the previous research activities can be tested to see if the applied method will achieve a better accuracy and better speed to monitor the activities in Dark Web or not. The second step is to obtain new data from the Dark Web to be tested. That is another milestone of this research.

It is important to build a cyber forensics maturity model, which can be considered as an adaptable and expandable framework. This can also support the open-closed principle of software architecture (open for extension and adaptation, close for change). This maturity model includes the following steps:

1. Timely collection of data (e.g. incident reports, network traffic and Dark Web forum data)
2. Observation of patterns and anomalies from cybersecurity data science (CSDS) methods
3. Use of machine learning algorithms to train a model response as an effective countermeasure

Each of these steps can be supported by a PDR^2 model in an iterative way. Eustace et al. [13] described this PDR^2 model which is Personal Cybersecurity Model of Self-Awareness and Intervention. PDR^2 includes the following iterative steps:

- Prevention
- Detection
- Response
- Recovery

Prevention is the first step which companies and organisations concerned at the first step by protecting their systems via firewalls and managing password protections. Of course, security and ethics are also other issues of prevention.

Detection is an ongoing process after prevention. It is not right to secure or prevent a system and leave it unsupervised. That's why it is important to focus on the detection phase in a frequent way. The main question here is how often a system should continuously be controlled for the detection of input and output data. These two phases are also related to the ML or AI techniques which are chosen to be applied to data traffic.

Response phase can be achieved by CERT (computer emergency response team). CERT can offer users active and up-to-date responses on the network which include system vulnerabilities. These responses can be sent as alerts. For example, if someone tries to be analysed by Wireshark and if CERT detects this traffic, it can alert this activity on time. After the response phase, it is important to recover the system if vulnerabilities in a system are detected. It is important to update the security and privacy level of the system.

These four steps can be applied to the network packet traffic and forum monitoring techniques. It is also important to prevent the user identity during a monitoring process which is ethically very important. Some other research areas were focused on a specific topic. It is possible to extend this research for other forum activities in other languages. Then we can say that if the method can be generalised for multiple languages or not.

6 Conclusion

Since the last decade, there are various research projects to detect illegal activities in both surface and Dark Web. It is important to detect the information at the right time to protect users and institutes from illegal activities, so the aim of our research into this topic is described in more detail in this project. As regular web users are facing the ever-increasing menace of attacks by threat actor from the Dark Web, further research involving machine learning and artificial intelligence algorithms for detecting patterns in a dataset and then building and training an associated machine learning model is important across many contexts, personal, business and government agencies.

With the large number of illegal forum activities happening inside the Dark Web, the main aim of this project is to make it easier and faster to extend the previous machine learning algorithms which are applied to detect and monitor forum activities or gather the web fingerprints of the illegal web sites which are visited frequently by Dark Web users. The algorithms which are included in this paper are the most updated and relevant ones to detect and extract data from Dark Web forums. Unfortunately, the hackers are also trying their best to not to be caught by the government agencies. That is why research is not going to have an end.

The most important problem is identifying the gaps which are detected in various research activities, like those listed in this paper. These gaps provide an opportunity either to build up a new algorithm or to find the best combination of effective algorithms to detect the illegal forum activities on the Dark Web. The technology evolves rapidly and that is why the types of illegal activities will vary during the evolution of the technology. It is very important to find new ways in this research to build new algorithms that support the security of the evolving technology that will help people to live in a more secure information superhighway. If the outcomes of this research can detect Dark Web threats faster, then this will help to protect all victims against future threats, through improved policies and procedures that enable quicker detection and response strategies and tactics by the IT professionals involved in frontline cybersecurity of their workplace.

Finally, the main aim of the research is to develop another machine learning algorithm which can analyse and mix up the deep learning techniques [such as term frequency/inverse document frequency (TF-IDF)]. This research will be informed by deep learning techniques, which are applied by Alnabulsi and Islam [3]; adaptive Hoeffding trees, which Attarian and Hashemi [4] have found out as the most accurate method for web site fingerprinting attacks; and other machine learning algorithms and open-source tools such as Hadoop for the detection of illegal activities on the Dark Web. The cited papers in this research give inspiration and information of latest deep learning and machine learning algorithms which make it possible to develop other accurate methods.

References

1. V. Adewopo, B. Gönen, S. Varlioglu, M. Özer, *Plunge into the Underworld: A Survey on Emergence of Darknet*. International conference on computational science and computational intelligence (CSCI), Las Vegas, NV, USA (2019)
2. O. Akyıldız, *Information Analysis and Cyber Crimes in Deep Web & Dark Web* 2018. 6th international symposium on digital forensic and security (ISDFS), 22–25 March 2018, Antalya, Turkey (2018)
3. H. Alnabulsi, R. Islam, *Identification of Illegal Forum Activities Inside the Dark Net*. International conference on machine learning and data engineering (iCMLDE), Sydney, Australia (2018)
4. R. Attarian, S. Hashemi, *Investigating the Streaming Algorithms Usage in Website Fingerprinting Attack Against TOR Privacy Enhancing Technology*. 16th international ISC (Iranian Society of Cryptology) conference on information security and cryptology (ISCISC), Mashhad, Iran (2019)
5. A. Baravalle, M.S. Lopez, S.W. Lee, *Mining the Dark Web: Drugs and Fake Ids*. IEEE 16th international conference on data mining workshops (ICDMW), Barcelona, Spain (2016)
6. L. Basyoni, N. Fetais, A. Erbad, A. Mohamed, M. Guizani, *Traffic Analysis Attacks on TOR: A Survey*. 2020 IEEE international conference on informatics, IoT, and enabling technologies (ICIoT), Doha, Qatar (2020)
7. S. Bhat, D. Lu, A. Kwon, S. Devadas, Var-CNN: a data-efficient website fingerprinting attack based on deep learning. Proc. Priv. Enhanc. Technol. **2019**, 292–310 (2018)
8. A. Bifet, G. Holmes, R. Kirkby, B. Pfahringer, MOA: massive online analysis. J. Mach. Learn. Res. **11**, 1601–1604 (2010)
9. A. Biryukov, I.P. Weinmann, *Trawling for TOR Hidden Services: Detection, Measurement, Deanonymization*. 2013 IEEE symposium on security and privacy, Berkeley, CA, USA (2013)
10. X. Cai, X.C. Zhang, B. Joshi, R. Johnson, *Touching from a Distance: Website Fingerprinting Attacks and Defenses*. CCS '12: Proceedings of the 2012 ACM conference on computer and communications security, New York, NY, USA (2016)
11. CAIDA, *A Real-time Lens into Dark Address Space of the Internet*. A CAIDA Project Summary, 2022, January 28. https://www.caida.org/funding/cri-telescope/cri-telescope_proposal/cri-telescope_proposal.pdf
12. CERN, *The birth of the Web*, 2016. https://home.cern/science/computing/birth-web
13. K. Eustace, R. Islam, P. Tsang, G.H. Fellows, Human factors, self-awareness and intervention approaches in cyber security when using mobile devices and social networks, in *Security and Privacy in Communication Networks: SecureComm 2017 International Workshops Proceedings*, Lecture Notes of the Institute for Computer Sciences, Social Informatics and Telecommunications Engineering, ed. by X. Lin, A. Ghorbani, K. Ren, S. Zhu, A. Zhang, vol. 239, (Springer, 2018), pp. 166–181. https://doi.org/10.1007/978-3-319-78816-6_13
14. S. Feghhi, D.J. Leith, A web traffic analysis attack using only timing information. IEEE Trans. Inf. Forensics Secur., 1747–1759 (2016)
15. N. Ferry, T. Hackenheimer, F. Herrmann, A. Tourette, *Methodology of Dark Web Monitoring*. 11th international conference on electronics, computers and artificial intelligence (ECAI), Pitesti, Romania (2019)
16. L. Gao, J. Rexford, Stable Internet routing without global coordination. IEEE/ACM Trans. Netw. **9**(6), 681–692 (2001)
17. K. Godawatte, M. Raza, M. Murtaz, A. Saeed, *Dark Web Along with The Dark Web Marketing and Surveillance 2019*. 20th international conference on parallel and distributed computing, applications and technologies (PDCAT), Gold Coast, QLD, Australia (2019)
18. J.T. Harviainen, A. Haasio, L. Hämäläinen, *Drug Traders on a Local Dark Web Marketplace*. Proceedings of the 23rd international conference on academic mindtrek, Tampere, Finland (2020)

19. G. Hulten, L. Spencer, P. Domingos, *Mining Time-Changing Data Streams*. KDD '01: Proceedings of the seventh ACM SIGKDD international conference on knowledge discovery and data mining, San Francisco, CA, USA (2001)
20. M. Kadoguchi, S. Hayashi, M. Hashimoto, A. Otsuka, *Exploring the Dark Web for Cyber Threat Intelligence Using Machine Learning*. 2019 IEEE international conference on intelligence and security informatics (ISI), Shenzhen, China (2019)
21. R. Koch, *Hidden in the Shadow: The Dark Web – A Growing Risk for Military Operations?* 11th international conference on cyber conflict (CyCon), Tallinn, Estonia (2019)
22. C. Manapragada, G.I. Webb, M. Salehi, *Extremely Fast Decision Tree*. KDD '18: Proceedings of the 24th ACM SIGKDD international conference on knowledge discovery & data mining, London, UK (2018)
23. Maxmind *GeoLite ASN database*
24. R. Nithyanand, O. Starov, A. Zair, P. Gill, M. Schapira, *Measuring and Mitigating AS-Level Adversaries Against. TOR* NDSS 2016, San Diego, CA, USA (2015)
25. D. O'Cearbhaill, *Trawling TOR Hidden Service – Mapping the DHT* (2013). https://donncha.is/2013/05/trawling-TOR-hidden-services/
26. G. Pantelis, P. Petrou, S. Karagiorgou, D. Alexandrou, *On Strengthening SMEs and MEs Threat Intelligence and Awareness by Identifying Data Breaches, Stolen Credentials and Illegal Activities on the Dark Web*. ARES 2021: The 16th international conference on availability, reliability and security, Vienna, Austria (2021)
27. B. Pfahringer, G. Holmes, R. Kirkby, *New Options for Hoeffding Trees*. Australasian joint conference on artificial intelligence, Goldcoast, QLD, Australia (2007)
28. S. Raaijmakers, Artificial intelligence for law enforcement: challenges and opportunities. IEEE Secur. Priv. **17**(5), 74–77 (2019)
29. M. Schäfer, M. Fuchs, M. Strohmeier, M. Engel, M. Liechti, V. Lenders, *BlackWidow: Monitoring the Dark Web for Cyber Security Information*. 11th international conference on cyber conflict (CyCon), Tallinn, Estonia (2019)
30. P. Sirinam, M. Imani, M. Juarez, M. Wright, *Deep Fingerprinting: Undermining Website Fingerprinting Defenses with Deep Learning*. CCS '18: Proceedings of the 2018 ACM SIGSAC conference on computer and communications security, New York, NY, USA (2018)
31. Y. Sun, A. Edmundson, N. Feamster, M. Chiang, P. Mittal, *Counter-RAPTOR: Safeguarding TOR Against Active Routing Attacks*. IEEE symposium on security and privacy, San Jose, CA, USA (2017)
32. Q. Tan, Y. Gao, J. Shi, X. Wang, B. Fang, *A Closer Look at Eclipse Attacks Against TOR Hidden Services*. 2017 IEEE international conference on communications (ICC), Paris, France (2017)
33. T. Wang, I. Goldberg, *Improved Website Fingerprinting on TOR*. WPES '13: Proceedings of the 12th ACM workshop on workshop on privacy in the electronic society, Berlin, Germany (2013)
34. T. Wang, X. Cai, R. Nithyanand, R. Johnson, I. Goldberg, *Effective Attacks and Provable Defenses for Website Fingerprinting*. Proceedings of the 23rd USENIX security symposium, San Diego, CA, USA (2014)
35. M. Wang, X. Wang, J. Shi, Q. Tan, Y. Gao, M. Chen, X. Jiang, *Who Are in the Darknet? Measurement and Analysis of Darknet Person Attributes*. IEEE third international conference on data science in cyberspace (DSC), Guangzhou, China (2018)
36. K. Williams, Untangling the dark web: taking on the human sex trafficking industry. IEEE Women Eng. Mag. **7**(2), 23–26 (2013)
37. M. Yang, X. Gu, Z. Ling, C. Yin, J. Luo, An active de-anonymizing attack against TOR web traffic. Tsinghua Sci. Technol. **22**(6), 702–713 (2017)
38. Y. Yang, H. Yu, L. Yang, M. Yang, L. Chen, G. Zhu, L. Wen, *Hadoop-Based Dark Web Threat Intelligence Analysis Framework*. IEEE 3rd advanced information management, communicates, electronic and automation control conference (IMCEC), Chongqing, China (2019a)
39. Y. Yang, L. Yang, M. Yang, H. Yu, G. Zhu, Z. Chen, L. Chen, *Dark Web Forum Correlation Analysis Research*. 8th joint international information technology and artificial intelligence conference (ITAIC), Chongqing, China (2019b)

A Secured 5G Network Slices Auction Broker

João Marques Silva and Nuno Souto

1 Introduction

Network softwarization and programmability is the new jargon used for the junction of software-defined networking (SDN), network function virtualization (NFV), and cloud-edge-fog computing – the technologies that are driving an unprecedented techno-economic shift in the telecom and ICT industries. Molding the network resources to the customer's needs is the best way to reduce operational costs, provide better flexibility, and bring new service paradigms. These measures will enable the deployment of 5G infrastructures and services, spanning from high data rate fixed-mobile communications to the Internet of Things, which is expected to accelerate the digital transformation that all the industry is witnessing. As a result, new service and business models will emerge, leading to a significant socioeconomic impact.

This business model revolution will have vertical segment operators pay for a dedicated end-to-end slice of the mobile network, made from several sub-slices from different network operators, to deliver added-value services to their customers. The concept of network slicing [18] emerged as an innovative network sharing solution designed to increase the revenues of infrastructure providers in 5G and beyond 5G. The dynamically attributed (via virtualization) slices of physical infrastructure can be leased to vertical operators/industries (automotive, e-health, etc.) or over-the-top (OTT) service providers leveraging on this new business model. These "verticals" will need to cope with the seamless provision of (shared) network resources (aka slice stitching) to meet the overall service-level agreement (SLA) contracted by their

J. M. Silva (✉)
Instituto Universitário de Lisboa (ISCTE-IUL), ISTAR, Lisbon, Portugal
e-mail: joao.silva@iscte-iul.pt

N. Souto
Instituto Universitário de Lisboa (ISCTE-IUL), IT, Lisbon, Portugal
e-mail: nuno.souto@iscte-iul.pt

K. Daimi et al. (eds.), *Emerging Trends in Cybersecurity Applications*,
https://doi.org/10.1007/978-3-031-09640-2_6

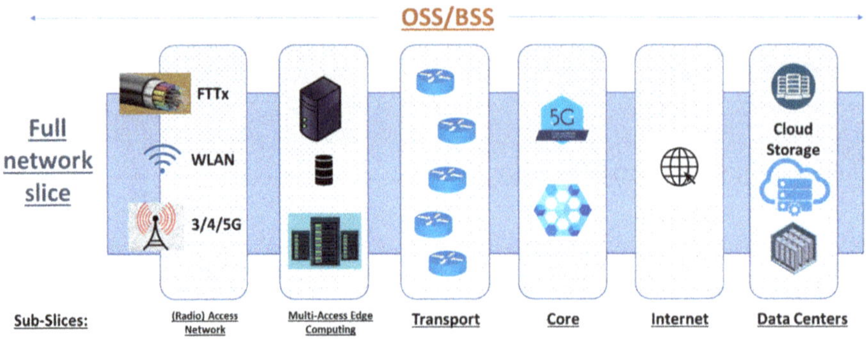

Fig. 1 Composition of a network slice

customers. This SLA must be considered when contracting all the necessary sub-slices to provide the full end-to-end network slice [10]. The authors in Habibi et al. [12] have introduced several metrics regarding the slice-based network SLAs including throughput, penalty, cost, revenue, profit, and QoS-related metrics. The enforcement of these SLAs is still an open issue, as no automated process has been proposed to address the problem

This complex network orchestration needs to be done in a timely manner, ensuring that all network slice requests are mapped into the appropriate physical network resources. The OSS/BSS (operations support system/business support system) is thus guaranteed through a custom-made network slice, as depicted in Fig. 1.

Network slices are composed/stitched with several sub-slices, namely, cloud data centers, the core domain (where the EPC (evolved packet core) operates), backhaul links to provide transport from the core network to the multi-access edge data centers (supporting edge computing services [1]), and the (radio) access network, (R)AN (with eventual fronthaul links).

When an end-to-end network slice is required by a vertical, two main operations must be considered:

(i) Procurement/brokering phase – find the necessary resources (sub-slices) to meet the required SLAs and price – in this part, we recommend an auction-type negotiation between the several sub-slice providers and to have all compete to supply their best offer at the best price.
(ii) Network and computational resource management, to stitch the entire slice together, ensuring that the end-to-end service complies with the requisites. This implies that several issues are addressed, such as cross-domain latency and throughput performance guarantees.

These operations must be handled by an NSB (Network Slice Broker), envisioned as an automated procurement and management solution that works in conjunction with multiple providers from various sub-slices, whose function is to collect network statistics and select the best sub-slices (via auction or some other means, which may involve machine learning techniques) in real time. The work in Sciancalepore et al. [24] studied a preliminary 5G Network Slice Broker supporting network slice instantiation, though only on the radio access network (RAN) level; in order to compose a full network slice, logical isolation between sub-slices tailored to specific service templates must be devised [10]. The NSB could also be responsible for ensuring that the requested SLAs are honored.

Some works [27] showed the feasibility of instating different network slices, though initially only at the (long-term evolution) LTE's RAN level – this was accomplished through the local maintenance terminal (LMT) of the evolved NodeBs (eNBs), which is programmable and allows to configure the allocation of the physical network resources (in terms of physical resource blocks – PRBs) of a specific mobile network operator. Orchestrating the whole slice will be somewhat similar, though with a higher degree of complexity due to the different (configurable) interfaces of each sub-slice (and operator, though it is expected that normalization will solve this).

2 Brokerage Model

The brokerage model for the 5G slicing method is complex and needs to be broken down into its main components (Fig. 2). So being, initially, each vertical (e.g., a public emergency network, a high-speed network, or an ultra-low delay network) will ask the 5G Network Slice Broker (NSB) for a slice to accommodate their needs (where a certain number of parameters are described, such as bandwidth, latency, jitter, duration of the service, etc.) The resources that are traded in these resource markets range from spatial streams [3] to spectrum/antennas [17] and resource blocks [7]. The requests will all have to abide by certain (not yet defined) formats, where everything that the service needs is described explicitly.

The NSB will accommodate all requests and compile a series of sub-slice requests to comply with the overall requests from each vertical. The sub-slices' template (format and parameters) will vary according to the sub-slice type, in a format in which providers must agree upon and adhere.

Each request from each vertical will be composed of a series (1–6) of sub-slices – this is the first opportunity to employ some economy of scale since it may be possible for a certain sub-slice to accommodate (simultaneously or not) the requests for two or more verticals. If a certain sub-slice can accommodate the requisites of two (or more) verticals simultaneously, then the NSB may try to exploit overbooking to maximize the use of resources.

The sub-slices will have to be negotiated separately with each sub-slice provider, and this can be done in various ways; either the NSB monitors the whole network and selects the best (available) sub-slices for the verticals' requirements according to traffic forecasting (achievable, for instance, via machine learning), placing a very big strain on the NSB, or the NSB "asks" the sub-slice providers for their best offers to meet the demands.

Game theory has been used for similar markets in communications research where multiple players are involved on both sides of the market [6]. Among the

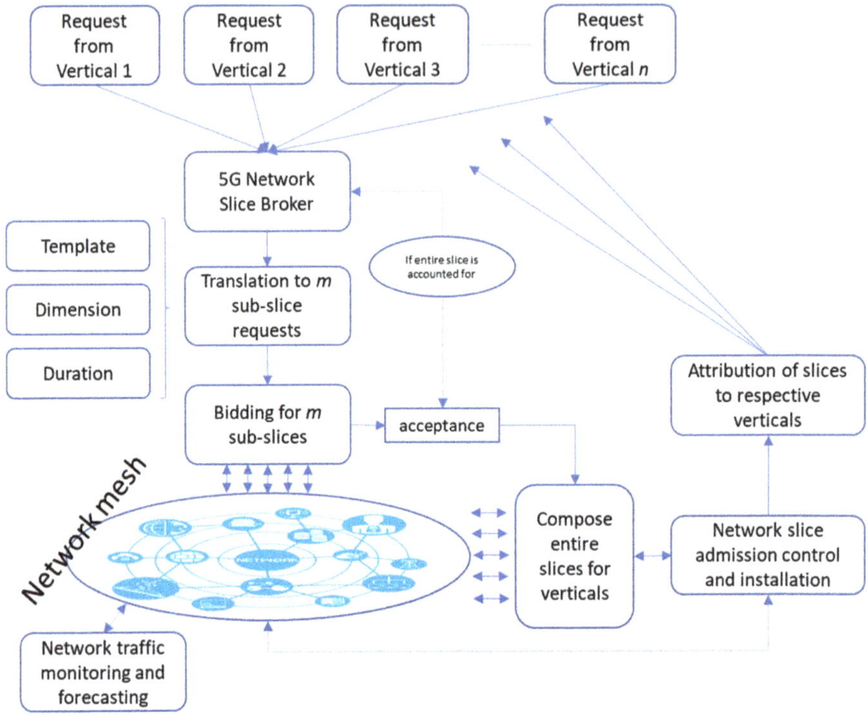

Fig. 2 Brokerage model of 5G slices

various game theory-based solutions, auctions are very popular as market resolution tools [16]. This bidding model, where each sub-slice is selected according to an auction designed specifically for the required sub-slice for a specific vertical (or a series of verticals, if the NSB opts to compile a sub-slice to accommodate more than one vertical in order to save resources), will demand that the NSB host many sub-slice auctions for each particular slice.

Acceptance for the sub-slices will only be issued if the entire slice they belong to is orchestrated else the NSB will inform the sub-slice provider that its sub-slice wasn't used and is once again available.

Once the entire slices are orchestrated, information is sent to the sub-slice providers confirming that their sub-slice will be used, and thus payment will be issued and authorization for the sub-slice's use will be issued. The 5G broker will next be responsible for issuing the necessary commands for the network slice admission control and installation, and once everything is prepared, the verticals will be informed of the slice that was attributed to them (and confirm/issue payment to the NSB). Once everything is set up, the verticals can start using their network slices.

2.1 The Business Model of the Network Slice Broker

The business model for this Network Slice Broker (NSB) scheme is peculiar, in the sense that the sub-slice providers (suppliers) are also customers of the NSB, alongside the verticals (the full business model framework is found in Osterwalder [20], but in this chapter, we will focus on the main business aspects). The NSB brokerage is merely the network and access provider part of larger business models involving the verticals. However, it is a crucial part since it makes the connections, according to specific service-level agreements (SLAs), to enable the communication and access to information required by the verticals' businesses. The relationship type, in this case, is business-to-business (B2B).

The NSB interconnects several verticals to several sub-slice providers and thus plays intermediates an $n:m$ relation, where n buyers meet m sellers (Fig. 3). Note from the figure that each sub-slice may have many providers, and thus the NSB will have to decide (via auction or some other means) which provider to select the current sub-slice from (according to price, availability, and SLAs). In the case of network auctions, a separate network auction may be hosted for each sub-slice, with the NSB hosting several parallel auctions.

Transactions are executed in parallel with other transactions, competing for resources, while others may be complementary. Another example of n:m relations are stock markets.

The market model is an electronic marketplace/intermediary solution [4, 8], which relieves sellers and buyers from administering the catalog's content and facilitates integration and market transparency through product data harmonization. Marketplaces are run by third parties to provide an online means for buyers and sellers to establish business transactions. The whole system can be seen as an inter-

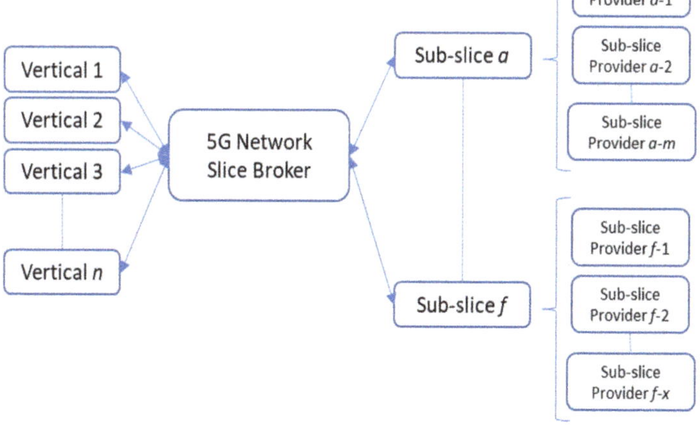

Fig. 3 n:m relationship of the brokered market

collaboration model, where two or more parties belonging to different organizations collaborate to achieve a specific goal, such as creating a product or service – in this case, the service is the use of the network slice. If there is a systematic use of the same/similar resources between the same players (in case the NSB doesn't have a big diversity of sub-slice providers and/or verticals), then we could almost name it as a virtual enterprise (which is the case when different companies join forces and share resources and policies).

2.2 Broker's Business Plan Executive Summary

The network broker's business plan is, essentially, to bring together vertical operators and sub-slice providers (SSP), providing all needed services so that the vertical's clients are satisfied by the network resources of the sub-slice operators. The broker will not only compile and translate the requirements of the verticals into sub-slice particular needs, but it will also negotiate the best price with the SSPs and intermediate and finalize the deal with the verticals. It will also aid in the connection requirements between SSPs and verticals.

The sector the broker operates in is computer networks and telecommunications essentially since this is the basis for communications between all other sectors; from the broker's viewpoint, it only connects verticals to SSPs. The target audience will thus be verticals from all sectors and the communication network SSPs. The outlook of the marketplace is bright since 5G networks will soon be standard in developed countries and the requirements for data connections and high bandwidth are ever-increasing.

Brokers will face competition from other brokers, and thus value must be offered to the broker's main users (SSPs and verticals). In the first generation of brokers, webs of trust between SSPs, verticals, and the broker could be formed, and thus the whole process would work as an association between trusting partners. In a future second generation of brokers, the whole process could be more liberal, and verticals could opt to choose brokers based on their SSP connections and/or financial/technical conditions, among other aspects.

The broker's initial investment to enter the business primarily focuses on the interconnection with the SSPs and the interfaces made available to the verticals. Such integration amounts could range from a couple to tens of thousands euros to over 100.000 €; it depends mostly on the connections that the broker must establish with the SSPs – the brokers with the most connections will be first to the market and profit greatly from it – profits will be provided by service commission fees paid by the verticals, included in the SSPs' network slice package. It is strongly recommended that network brokers be independent of SSPs, though it is expected that many SSPs will try to push their brokers to market – this is something that the competent competition authorities should be on the lookout for. The whole system should be automated, and its rules should be clear to all involved parties.

2.3 Use of Software-Defined Networking

Several research papers deal with the automated management of network resources [2, 15] to comply with the new network demands. Some authors proposed auction procedures [2, 11, 14, 15] to attribute the network resources with machine learning techniques to deal with the high complexity of the system.

Within these recent works in the literature, some propose auction mechanisms [2, 11, 14, 15] to manage the network resources, mostly in mobile access cases [5, 22, 23]. To deal with the high complexity of managing the network resources in a more efficient way, machine learning techniques have also been used [25].

Although a few studies are based on auctions for selecting the network resources, the use of SDN to control the auctions by acting as a broker has not been studied yet. Our contribution builds on previous work [5], where it was recognized that research for new business models based on sharing resources of 5G slices was required, alongside the need to investigate fairness in the attribution of network resources requested by mobile virtual network operators (MVNOs) [26].

This section points out how the SDN-based broker could manage the 5G slice attribution to verticals. To accomplish this, three layers are used, namely, data transfer, control, and management, as shown in Fig. 4. These layers have the following functions:

(a) Data layer: information transfer between the nodes
(b) Control layer: routing through the data layer
(c) Management layer: admission control of new clients (see Fig. 4, steps 1 and 4) and use of all available resources in the network infrastructure

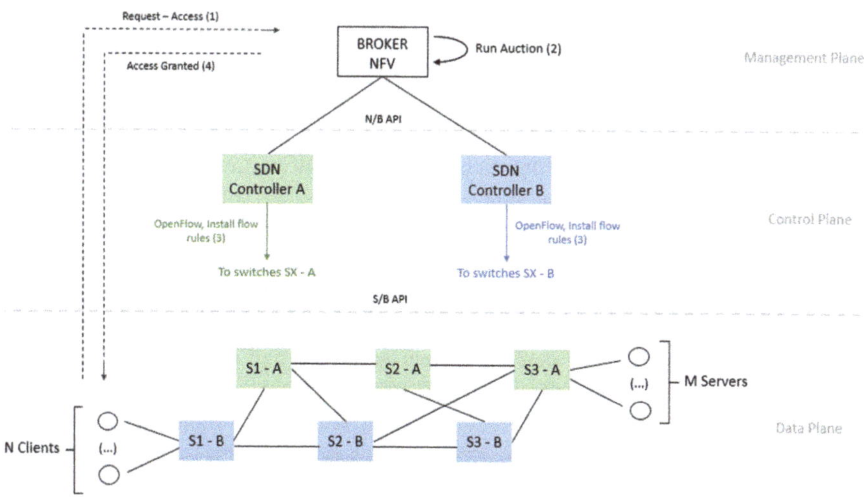

Fig. 4 Proposal of an SDN-based broker

After a new client is accepted, the NFV in the management layer would hold an auction using the control layer's SDN controllers (see Fig. 4, step 2). The outcome of the auction would be a combination of selected switches from the different sub-slice providers, creating an overall network path (see Fig. 4, step 3) that could be either a sub-slice or slice, depending on circumstances.

3 Blockchain for Registering Slice Deals

Blockchain is a concept for a database of digital contracts. The concept was introduced alongside the cryptocurrency Bitcoin to promote secure transactions between (anonymous if desired) participants using a decentralized system [9], where all operations were validated by different entities (known as miners, which are participants of the blockchain system that earn certain types of resources as a reward; in the bitcoin system, they naturally earn bitcoins). Blockchain got its name basically because it's a distributed chain of blocks, where each block contains information about a certain number of made deals. The miners compile blocks from aggregating several digital transactions, which are added sequentially to the ever-growing blockchain after peers' validation. Once a block is added to the chain, it is impossible to change its content or remove it (immutability) since it becomes part of the chain's global hash value.

In Hewa et al. [13], a blockchain-based solution was presented to prevent DoS (denial-of-service) attacks on the slice broker, which significantly reduced latency under a DoS attack scenario. The paper highlights the need for blockchain in these scenarios.

Blockchain can operate in one of two different modes: permission-less or permissioned [21]. The permission-less mode, also known as a public blockchain network, allows any user to create and add transactions to the chain, whereas the permissioned blockchain is a private centralized blockchain network that requires authentication and authorization to use.

Since in enterprise ecosystems such as telecoms, all parties involved are known to each other (participants and miners/managers are the sub-slice operators and verticals), there is no need for anonymity, and as such, permissioned blockchains are considered a better fit. This is because permissioned blockchains are designed for enterprise ecosystems and use simpler and less resource-consuming consensus protocols (such as Raft [19]) than public blockchains.

Each sub-slice attribution is a small smart-contract, and the whole orchestrated slices are a series of those small contracts. Every small contract should have a unique identifier and data fields that represent what resources were transitioned. Using a blockchain to register and validate the contracts once the slices are orchestrated should be done in such a way that the whole slice (group of small contracts) is represented – so being, and in order to maintain the operation method of the blockchain, where miners compose blocks out of contracts that are still "free" (not yet placed in a block), it is wise to keep all of the sub-slice contracts that belong to

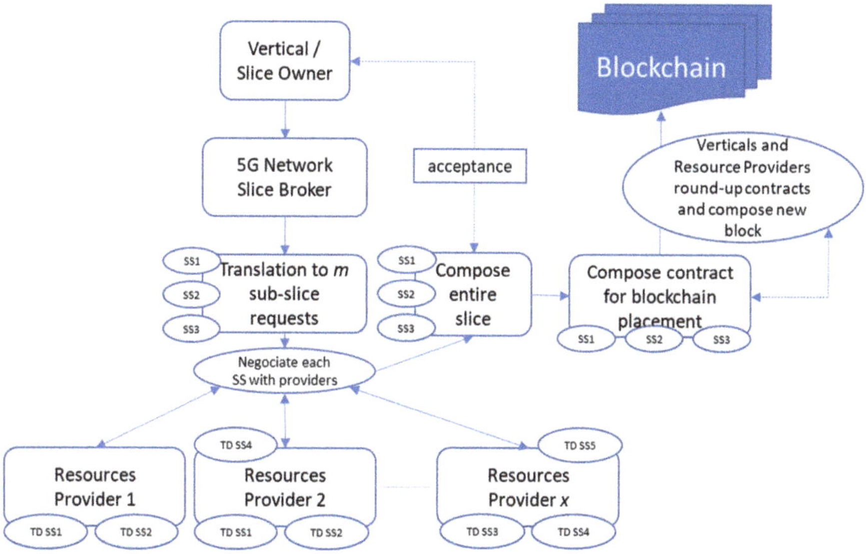

Fig. 5 Registration of network slice deal

a particular slice together. So being, the NSB should issue smart slice contracts for block placement, joining all the small sub-slice contracts into one slice contract. In the case that the sub-slice contract is common to two or more slices, this should be mentioned in the sub-slice contract itself, and all the slice contracts containing the common sub-slice should incorporate it.

Figure 5 portrays the business process of conceptually registering the network slice deal in the blockchain.

As stated previously, the vertical's request (containing high-level information) is translated into smaller sub-slice requests by the NSB. Each sub-slice represents a different domain, and each domain will have its specific resources and instantiations. We can say that each sub-slice belongs to a different technological domain (TD), which can be a computing resource domain (such as CPU, I/O), a storage domain, a radio domain (eNB, central unit (CU), distributed unit (DU), remote radio head (RRH)/remote radio unit (RRU)), and transport domain (e.g., VLAN, VPN). A slice request may be composed of one sub-slice or many sub-slices, and thus the slice contract is a composition of the small sub-slice contracts that constitute it. It is the whole slice contract that is composed and advertised by the NSB, so that the "miners" of the system (verticals and resource providers that are part of the system's blockchain management) group these contracts into blocks and register/validate them.

4 Dispute for Resources

Under normal circumstances, slices are orchestrated by the required sub-slices and leased to verticals; but sometimes it is possible that there aren't enough sub-slices to accommodate all requests. In this situation, the NSB must decide which slices should be orchestrated and which should be dropped. This decision must be made according to pre-defined rules so that no vertical operator is favored to the detriment of another.

When a decision must be made due to having just one sub-slice that is needed for two (or more) slices, several criteria could be used to select the slice deal to be served first:

1. Slice whose order has the lowest timestamp.
2. The slice with the highest number of sub-slices gets "served" first.
3. Slice whose total order is highest gets served first.
4. Verticals with the highest number of requests rejected in a specific time frame get served first (requires NSB to keep track of verticals whose requests weren't satisfied in the previous *n* transactions).
5. Verticals could dispute the sub-slices in a posterior auction held by the NSB.
6. Slice requests could have priorities associated, with the highest priorities served first.

Since the NSB should be neutral in accepting requests, it is up to the verticals that use the NSB to formulate and/or accept the pre-determined criteria for sub-slice disputes.

4.1 Simulations

Simulations were made to illustrate possible outcomes due to resource disputes in the context of request overload/scarcity of (sub-slice) resources. Six strategies were devised considering the previously stated criteria and are described in Table 1. The first strategy simply selects the slice deal whose request was made first (lowest timestamp), and strategy 6 takes the verticals to a posterior auction to select the slice deal to fulfill. Strategies 2–5 all have two conditions, in case there still is a tie after the first condition. Strategy 2 opts to favor the vertical whose slice request has the highest number of required sub-slices, which is similar to strategy 3. Strategy 3 considers the total financial amount of the deal if we simplify things and state that every sub-slice has a similar base value (something which isn't necessarily true since different sub-slices can have significant variations in value and complexity). Strategy 4 has the NSB which registers the number of slice deal rejections of each vertical in order to prioritize them in further deals, and strategy 5 assumes that verticals can be prioritized, either by type of service or by means of buying prime management benefits from the NSB.

Table 1 Strategies for selecting which slice deal to select in case of resource dispute

Strategy 1	Lowest timestamp
Strategy 2	The highest number of sub-slices
	Lowest timestamp
Strategy 3	Highest financial deal
	Lowest timestamp
Strategy 4	The higher number of previous rejections
	Lowest timestamp
Strategy 5	Higher priority
	Lowest timestamp
Strategy 6	Posterior auction to select the winner

Table 2 Successful slice deals (%) for each strategy (S1–S5), per vertical (V1–V5). Includes % of SSW (sub-slices withdrawn) from failed deals

	V1	V2	V3	V4	V5	SSW
S1	37%	34%	33%	30%	27%	42%
S2	29%	34%	35%	38%	41%	31%
S3	29%	30%	34%	39%	40%	32%
S4	32%	31%	33%	32%	31%	40%
S5	96%	76%	45%	17%	2%	54%

A simulator was devised in MATLAB©. It was considered that the NSB would wait the necessary amount of time to collect slice requests from all verticals and that every vertical would request the same service once again before giving up in case it wasn't served at first.

Five verticals were considered, issuing requests for the NSB. The NSB was in close contact with 20 sub-slice providers, each with a varying amount of available sub-slices and sub-slice types. A total of five sub-slice types were considered.

The five verticals considered each issued a request with a constant number of sub-slices of varying types; vertical 1 issued a request of only one sub-slice, vertical 2 of two sub-slices, . . . , and vertical 5 issued a request requiring all five different types of sub-slices. In strategy 5, we assumed that vertical 1 had the highest priority and vertical 5 had the lowest priority, with the intermediate verticals having decreasing priority from vertical 2 to 4. Strategy 6 has the NSB hosting a posterior auction between the verticals disputing the sub-slices and thus is a clear case of favoring whoever pays more for the service.

Since an overloaded situation was devised, the sub-slice providers would never have enough available sub-slices to cover all the requests, having between 6 and 12 available sub-slices in total (including different sub-slice types); thus, only a portion of the requests was covered for each run. Table 2 registered the successful slice deals handled for each strategy and vertical in percentage. The percentage of sub-slice deals that had to be withdrawn was also calculated in total for each strategy. Strategy 6 wasn't simulated since it depends solely on a posterior financial dispute between verticals for the scarce sub-slices available.

The results are depicted graphically in Fig. 6. From the results, we can infer the following:

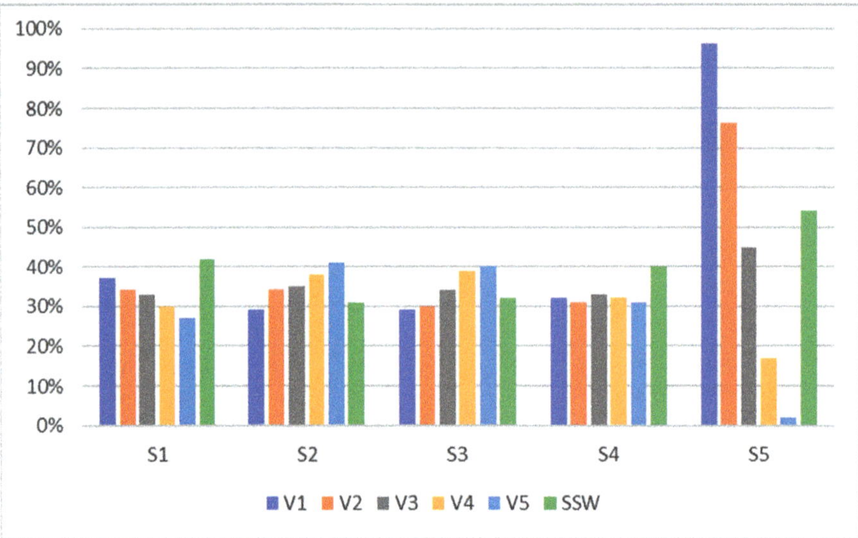

Fig. 6 Graphical results of successful slice deals per strategy and vertical, including SSW per strategy

(a) The random setting (equivalent to the lowest timestamp of the request) of strategy 1 favors the verticals with fewer sub-slices. When a vertical that has many sub-slices isn't chosen at first, the chance that there will be enough sub-slices to fulfill its request is lower than for other smaller slices and thus will have a lower success rate.
(b) Strategies 3 and 4 are similar since the highest financial deal and the deal with the highest number of sub-slices are strongly correlated. These strategies yield the lowest values of sub-slices withdrawn (SSW) since the major slice deals will make use of most of the available sub-slices, with the smaller deals taking advantage of some "leftover" sub-slices.
(c) Strategy 4 is the one that delivers equilibrium between acceptance of the different types of slice deals; as expected, the SSW rises compared to strategies 2 and 3.
(d) Strategy 5 is the spike off the chart; since vertical 1 has the highest priority with only one sub-slice, it always gets served (as long as the sub-slice providers have the requested sub-slice available; explaining why its value isn't 100% as expected). Vertical 5, with the lowest priority and highest number of sub-slices, has a miserable record of just 2% success. Note the SSW, which surpasses 50%, since the deals encompassing several sub-slices aren't being served.

Although strategy 6 wasn't simulated, we can infer that its results would be similar to strategies 2 and 3 since the highest financial deals would, in theory, be those that would stand a higher chance of winning the posterior auction disputing the much-needed sub-slices.

5 Conclusions

This chapter deals with the new business model of orchestrating a network slice using different sub-slices from different providers. This method of dividing the network into different technological domains and compiling network slices for use by vertical operators (that are going to sell the package to end users) augments the efficiency of the network, turning it much more versatile. This chapter deals with the business model of a Network Slice Broker that would manage the different pieces of each technological domain, explains how SDN could be used as a network broker and how blockchain could be used to register and validate all deals, and strategies on how to deal with resource scarcity, in case of overloading.

Acknowledgments The authors acknowledge the support given by Instituto de Telecomunicações, Lisbon, Portugal.

Funding The work of the authors is funded by FCT/MCTES through national funds and when applicable funded by EU funds under projects: UIDB/50008/2020, UIDB/04466/2020 and UIDP/04466/2020.

References

1. 5G Service-Guaranteed Network Slicing White Chapter, White chapter, China Mobile Communications Corporation, Huawei Technologies Co., Ltd., Deutsche Telekom AGVolkswagen, February 2017
2. Y. Abdulsalam, M.S. Hossain, COVID-19 networking demand: an auction-based mechanism for automated selection of edge computing services. IEEE Trans. Netw. Sci. Eng. (2020). https://doi.org/10.1109/TNSE.2020.3026637
3. H. Ahmadi et al., *Virtualization of Spatial Streams for Enhanced Spectrum Sharing* (GLOBECOM, 2016)
4. N. Archer, J. Gebauer, *Managing in the Context of the New Electronic Marketplaces*. 1st world congress on the management of electronic commerce, January 19–21, Hamilton, ON, Canada (2000)
5. A.A. Barakabitze, A. Ahmad, R. Mijumbi, A. Hines, 5G network slicing using SDN and NFV: a survey of taxonomy, architectures and future challenges. Comput. Netw. **167**(106984), 1–40 (2020)
6. D.E. Charilas, A.D. Panagopoulos, A survey on game theory applications in wireless networks. Comput. Netw. **54** (2010)
7. X. Chen et al., An auction-based spectrum leasing mechanism for mobile macro-femtocell networks of IoT. Sensors **17** (2017)
8. K. Eyholzer, *Electronic Purchasing. Arbeitsbericht Nr. 116* (Institut für Wirtschaftsinformatik, University of Bern, 1999)
9. Q. Feng, D. He, S. Zeadally, M.K. Khan, N. Kumar, A survey on privacy protection in blockchain system. J. Netw. Comput. Appl. **126**, 45–58 (2019)
10. GSMA, Smart 5G Networks: Enabled by Network Slicing and Tailored to Customers' Needs, https://www.gsma.com/futurenetworks/wp-content/uploads/2017/09/5G-Network-Slicing-Report.pdf, September 2017

11. U. Habiba, E. Hossain, Auction mechanisms for virtualization in 5G cellular networks: basics, trends, and open challenges. IEEE Commun. Surv. Tutor. **20**(3), 2264–2293 (2018). https://doi.org/10.1109/COMST.2018.2811395
12. M.A. Habibi et al., The structure of service level agreement of slice-based 5G network. CoRR **abs/1806.10426** (2018)
13. T. Hewa, A. Kalla, P. Porambage, M. Liyanage, M. Ylianttila, How DoS attacks can be mounted on Network Slice Broker and can they be mitigated using blockchain? in *2021 IEEE 32nd Annual International Symposium on Personal, Indoor and Mobile Radio Communications (PIMRC)*, (2021), pp. 1525–1531. https://doi.org/10.1109/PIMRC50174.2021.9569375
14. D.H. Kim et al., Pricing mechanism for virtualized heterogeneous resources in wireless network virtualization, in *2020 International Conference on Information Networking (ICOIN), Barcelona, Spain*, (2020), pp. 366–371. https://doi.org/10.1109/ICOIN48656.2020.9016477
15. T.H.T. Le et al., Auction mechanism for dynamic bandwidth allocation in multi-tenant edge computing. IEEE Trans. Veh. Technol. **69**(12), 15162–15176 (2020). https://doi.org/10.1109/TVT.2020.3036470
16. N. C. Luong et al., "Applications of economic and pricing models for resource management in 5G wireless networks: a survey.," IEEE Commun. Surv. Tutor., 2018.
17. J. McMenamy, A. Farhang, N. Marchetti, I. Macaluso, *Enhanced Auction-assisted Lsa*. ISWCS conference (2016)
18. NGMN Alliance, *Description of Network Slicing Concept* (Public Deliverable, 2016)
19. D. Ongaro, J. Ousterhout, *In Search of an Understandable Consensus Algorithm*. USENIX conference (2014)
20. A. Osterwalder, Y. Pigneur, *Business Model Generation* (Wiley, Hoboken, 2010)
21. T. Salman, M. Zolanvari, A. Erbad, R. Jain, M. Samaka, Security services using blockchains: a state of the art survey. IEEE Commun. Surv. Tutor. **21**(1), 858–880 (2019)
22. K. Samdanis, X. Costa-Perez, V. Sciancalepore, From network sharing to multi-tenancy: the 5G network slice broker. IEEE Commun. Mag. **54**(7), 32–39 (2016). https://doi.org/10.1109/MCOM.2016.7514161
23. Y. Sandhya, K. Sinha, Haribabu, A survey: hybrid SDN. J. Netw. Comput. Appl. **100**(2017), 35–55 (2017)
24. V. Sciancalepore et al., Mobile traffic forecasting for maximizing 5G network slicing resource utilization, in *IEEE INFOCOM '17*, (2017)
25. V. Sciancalepore, X. Costa-Perez, A. Banchs, RL-NSB: reinforcement learning-based 5G network slice broker. IEEE/ACM Trans. Netw. **27**(4), 1543–1557 (2019). https://doi.org/10.1109/TNET.2019.2924471
26. Y.K. Tun, N.H. Tran, D.T. Ngo, S.R. Pandey, Z. Han, C.S. Hong, Wireless network slicing: generalized Kelly mechanism-based resource allocation. IEEE J. Sel. Areas Commun. **37**(8), 1794–1807 (2019). https://doi.org/10.1109/JSAC.2019.2927100
27. L. Zanzi, V.I. Sciancalepore, A. Saavedra, X. Pérez, *OVNES: Demonstrating 5G Network Slicing Overbooking on Real Deployments*, IEEE conference on computer communications, Infocom'18, Poster and Demo, 2018

Applying Zero Trust Architecture and Probability-Based Authentication to Preserve Security and Privacy of Data in the Cloud

Yvette Colomb, Peter White, Rafiqul Islam, and Abeer Alsadoon

1 Introduction

Many companies are moving from internal on-premises data centres to Cloud to save on capital and operational expenditure, often retaining some on-premises applications and data. Security is implemented with the traditional perimeter-style defence with some of the recommended Cloud Service Provider's (CSP) built-in security; however, data breaches are still occurring [1, 3, 4]. Traditional security implementations for on-premises organisations must be examined and compared to the ZT CC security alternative. An unqualified combination of the two different styles can leave organisations vulnerable to threats, and it can be difficult to sift through such a security style to ensure that the organisation has the best available CC security posture to protect the organisation.

A ZT architecture safeguards against bad actors moving laterally within a previously defenceless intranet, gaining unauthorised access to applications and data, or exfiltrating data from behind firewalls that, historically, did not monitor outgoing traffic. ZT aims to encrypt data efficiently wherever possible and uses secure protocols to protect data in transit and storage.

PBA or risk-based authentication is relatively new, but the research demonstrates that this is more secure than simple two-factor (2FA) or multi-factor authentication (MFA) [2, 5, 6]. Using PBA, increasing identifiers when authenticating users, there is an increased efficacy in preventing bad actors from accessing applications [2]. PBA helps prevent unauthorised access from the login stage, and ZT secures the data once a user is authorised and authenticated.

Y. Colomb (✉) · P. White · R. Islam · A. Alsadoon
Charles Sturt University, Bathurst, Australia
e-mail: ycolomb@csu.edu.au; pewhite@csu.edu.au; mislam@csu.edu.au; abalsadoon@csu.edu.au

K. Daimi et al. (eds.), *Emerging Trends in Cybersecurity Applications*,
https://doi.org/10.1007/978-3-031-09640-2_7

There is also confusion between data security and data privacy [7]. CC affects data privacy in unexpected ways due to the varied locations of Cloud services and the variation of laws affecting the data. A thorough understanding of Cloud networking and data privacy laws can enable organisations to comply with data privacy laws. The implementation of ZT will provide the best defence against data breaches, ensuring the most effective means of maintaining data privacy.

The security needs are determined by examining the type of data, business locations, and location of the Cloud services. Stakeholders need to be educated and informed about the consequences of data compromises in terms of economic loss, reputation, and legal issues to perform realistic risk analysis and prioritise security needs.

This chapter will discuss the changes in network structure due to CC, how previously used security methods need to be reapplied to CC, some of the pitfalls, and most importantly, how to structure a ZT architecture within any CC network.

2 Cloud Computing: Rethinking Security

2.1 Traditional Network Structure

Traditionally security has relied on encryption, firewalls, antivirus scanning, access control, segmentation, intrusion detection systems, and backups; the issue with the defence style is not the tools but how they are applied in a CC architecture.

One of the biggest issues with traditional security methods is implicit trust within the organisation. The assumption is that the organisation is protected by preventing the passage of malware into an organisation's network. This traditional method of firewalls, using a perimeter, is frequently likened to a moat and wall defence around a castle containing a village with only one entrance through a drawbridge. The traditional network was focused on preventing access through this drawbridge or over the castle walls and based on the assumption that the internal intranet was a safe zone, as shown in Fig. 1. Traditional perimeter defence focuses on inbound traffic, usually with minimal consideration for inspecting outgoing traffic. It is assumed that everyone within the perimeter is trusted; outgoing traffic inspections are considered unnecessary. A bad actor only needs to obtain entrance to the network to exfiltrate data, often without notice, sometimes covering up footprints through log deletion [3].

The other oversight with perimeter defence is the threat from within an organisation. This does not include only bad actors but the threat via email, USB ports, etc. This chapter offers guidance on implementing a strong security posture in changing networks and attempts to change mindsets about approaching security.

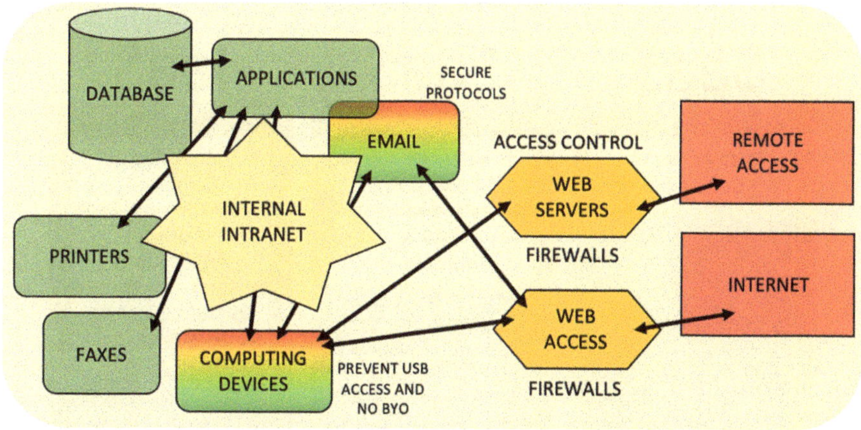

Fig. 1 Traditional network structure

2.2 Cloud Network Structure

Securing Cloud applications requires understanding the network structure and how it differs from a traditional network structure. There is no longer a definable or, relatively, secure intranet. Users access both on-premises and Cloud applications from organisational premises and remotely, as shown in Fig. 8.

2.2.1 Shared Responsibilities

Data security was entirely the data controller's responsibility until the EU General Data Protection Regulation (GDPR) introduced shared responsibilities between the CSP and Cloud client. While the Cloud Service Provider (CSP) is responsible for providing services under these models, the responsibility for the data remains with the organisation contracting to the CSP.

Cloud Computing offers three types of service models: Software as a Service (SaaS), Platform as a Service (PaaS), and Infrastructure as a Service (IaaS). There are security issues in common with all these models, but also security issues specific to each model. Between these models there is a division of responsibility between the organisation and CSP, as shown in Fig. 2. This discussion refers to the responsibility of hardware infrastructure, middleware, and software and the actual data as it is processed through Cloud services, but not the location. All these models use remote login for users and eliminate the traditional intranet; this is discussed in detail in Cloud Network Security.

For all types of Cloud service models, there is always a risk with the end user of any application; this falls beyond the scope of the CSP and is where the security responsibility ends for SaaS. All other layers of security from the end-user

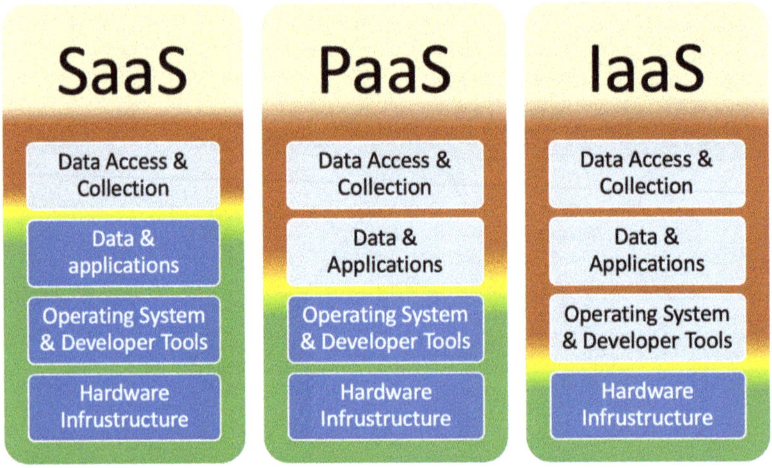

Fig. 2 Cloud model shared responsibilities

applications through to middleware, the operating system, and down to the hardware are the responsibility of the CSP. SaaS offers the customer the entire package of software needs for their business. It may include email, storage, a public-facing interface, an internal interface, and other services. Some examples are Microsoft Office, Amazon WorkSpaces, and Salesforce. Desktop as a Service (DaaS), such as Amazon WorkSpaces, is becoming increasingly popular and replaces the need for traditional software downloads, and these services take care of all patching and software updates. The issue with such services is the need for Internet access to work and the security needs surrounding remote access.

PaaS includes the operating system (OS) and relevant development frameworks so that applications can be built and run in this virtual Cloud environment. CSPs are responsible for the operating system and base developer tools (frameworks, SDKs, etc.), communication services, databases, and the hardware. The client is responsible for the end-user applications developed within the system. The CSP is responsible for keeping the OS patched and any updates to all other provisions.

IaaS offers the bare virtual metal upon which the customer can run all their software including the operating system and applications. It saves the customer from costly hardware purchase and maintenance, but it also places the responsibility of security in the client's lap. The client is responsible for all software updates, including patches and any possible customisation of software applications running on the operating system. The only real responsibility the CSP has when providing IaaS is for the underlying physical or virtual hardware infrastructure.

When using PaaS and IaaS, organisations are responsible for implementing security within their software applications and can implement access control to their software, whereas SaaS is reliant on the CSP's access controls, which have some limitations in terms of customisation. When migrating to CC, it is important that software access controls and security protocols are updated to suit a Cloud environment; this is discussed further in Cloud Network Security.

Many other Cloud services are developing to replace traditional software and hardware needs. Some examples are Functions as a Service (FaaS) and Infrastructure as code (IaC), but this is beyond the scope of this chapter. It is noteworthy that all these platforms have one thing in common – they require remote access.

2.2.2 Cloud Deployment

Cloud deployments generally come in the form of private, community, and public Clouds. Private Clouds were traditionally maintained by the organisation using the private Cloud. There are now colocation services that provide space, power, and cooling for the hardware for private Clouds, the benefit of which is the shared cost of a server location and upkeep [8]. Some such services are offering the server racks within a colocation setting. The customer may provide their own racks, or the colocation may offer racks, depending on the arrangement. Community Clouds are effectively a collective of private Clouds used for organisations with a common purpose or scope, for instance, government organisations or medical organisations. This provides the privacy required for data-sensitive agencies while sharing the overheads of hardware and network maintenance. Public Clouds are what many consumers are familiar with and offer services for organisations and the public; the three major global public Cloud providers are Amazon Web Services (AWS), Microsoft Azure (Azure), and the Google Cloud Platform (Google). Some organisations use a combination of private and public Clouds, known as hybrid deployment, which is becoming increasingly popular – a 2019 survey showed that over 50% of Cloud consumers use a hybrid deployment and multiple CSPs [9]. This results in a network structure with many access points that are vulnerable to attack.

2.2.3 Cloud Geography

Large CSPs divide the services into a top-level geographical division called regions. These regions are then divided into smaller sections known as availability zones (AZ). Organisations can select where data is stored and regions where scaling will occur and other features such as backups. This functionality can give organisations a false sense of control over data sovereignty.

Other services are becoming increasingly popular and one of these is edge computing. Edge locations are close to the end user and are not guaranteed to fall within an organisation's selected AZs or region; see Fig. 3. It is within these edge locations that CSPs offer a wavelength zone (WZ) to run microservices and lambda functions for rapid communications required at the edge as one example [10]. This completely changes the notion of having data collected or even stored in one location, as data is collected, modified, deleted, and temporarily stored at edge locations. This

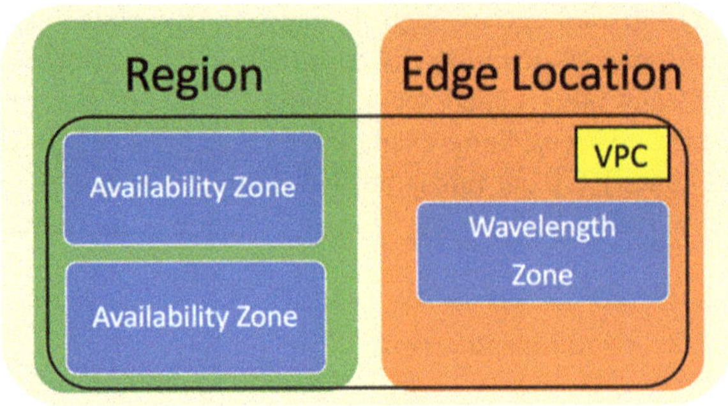

Fig. 3 Cloud regions and edge locations

can be further complicated with processing data and logging processes possibly occurring in yet another location. It cannot be assumed that all data processing occurs within a selected region, which affects data sovereignty and privacy laws. Data privacy laws between countries can vary, and some data privacy laws, such as the GDPR, do not permit the transfer of PII data into countries that do not apply similar rigour to data privacy; this is discussed further in Geopolitical Issues.

2.3 Edge and Fog Computing

With the increasing growth of the Internet of Things (IoT), there is an increasing demand to decentralise CC to reduce latency and improve performance with services such as edge and fog computing. The main differences between edge and fog computing are where the data processing occurs. Data processing is dispersed geographically near IoT devices, as shown in Fig. 4. Edge computing processes data on the device or sensor where the data is generated. Fog computing processes data on the LAN close to where the data is generated. The fog consists of nodes that cache data locally, performing some data processing activities before transmitting with the centralised Cloud applications [11, 12].

Decentralisation further complicates CC with challenges for developers to write secure microservices and lambda functions to hook into Cloud databases and applications. These smaller functional programs allow for greater growth and less downtime than monolithic applications, but the diversity of programming languages and platforms and the calls for privileged data create a security weakness in the CC dynamics. The edge services occur outside of the centralised Cloud environment and are not secured by the controls maintained within the main virtual frame of the Cloud system. The IoT has left many end users exposed to poor practices such as communication over Telnet, which is not encrypted. This leaves passwords exposed in plain text to sniffers on the network [13].

Fig. 4 Cloud, fog, and edge computing

2.4 Cloud Threats

Lockdown laws due to the COVID-19 pandemic have caused an increase in people working from home; with an increase in social networking to connect isolated people, there is a corresponding increase in Cloud cybercrime [14–16]. This remote workforce has only further exacerbated already challenged CC security postures; users are logging in from remote networks, and this further complicates identifying between genuine logins and where a bad actor is attempting a login. A strong CC security posture helps defend against this.

While the largest global CSPs have forged paths of ZT security architecture and there are many security holes with organisations using CC, there has been a degree of confusion surrounding ZT and the notion of a perimeterless network, which has left organisations open to many attacks. Surely if there are advancements in security, there will be advancements in attack methods, but many vulnerabilities can be relatively easily mitigated.

Some of the most common threats to CC are:

- Insider threats are a substantial threat to CC [17, 18].
- Cloud misconfigurations lead to vulnerability in APIs [19].
- Security configuration such as not changing default passwords is a common and easily exploited vulnerability [20].
- Malware such as XSS causes data breaches and data losses [19].
- DOS and DDOS attacks [19, 21].
- Identity theft [20].
- Phishing attacks [20].
- End device vulnerabilities [13].

While this list is not comprehensive, implementing security protocols designed for Cloud architecture is the best defence organisations have against threats. For any threat, a ZT architecture can prevent that threat from moving laterally through a network.

3 Data Privacy in the Cloud

3.1 Data Security vs. Data Privacy

The CIA Triad is useful to determine if there has been a data security breach. All successful cyberattacks compromise one or more of the CIA pillars of confidentiality, integrity, and availability. Data privacy concerns the individual and differs from organisational privacy, such as intellectual property or trade secrets.

Data security and data privacy have areas of overlap, but many people struggle with understanding the difference between these. Some confusion relates to the difference between a breach and an intrusion [7]. A breach is when PII is accessed by an unauthorised individual, which includes leaking such data into the public domain – in other words, a breach of confidentiality. An intrusion is when an unauthorised actor gains access to a network.

3.2 Data Privacy Complexity

Preserving data privacy is more complex in some ways than data security, as the laws surrounding data privacy vary between countries and, sometimes, between states. What may be considered a data privacy breach in one country may not be an issue in another; see Fig. 5. Having said this, one common breach to all data privacy under all privacy laws is whenever an unauthorised actor accesses Personal Identifiable Information (PII).

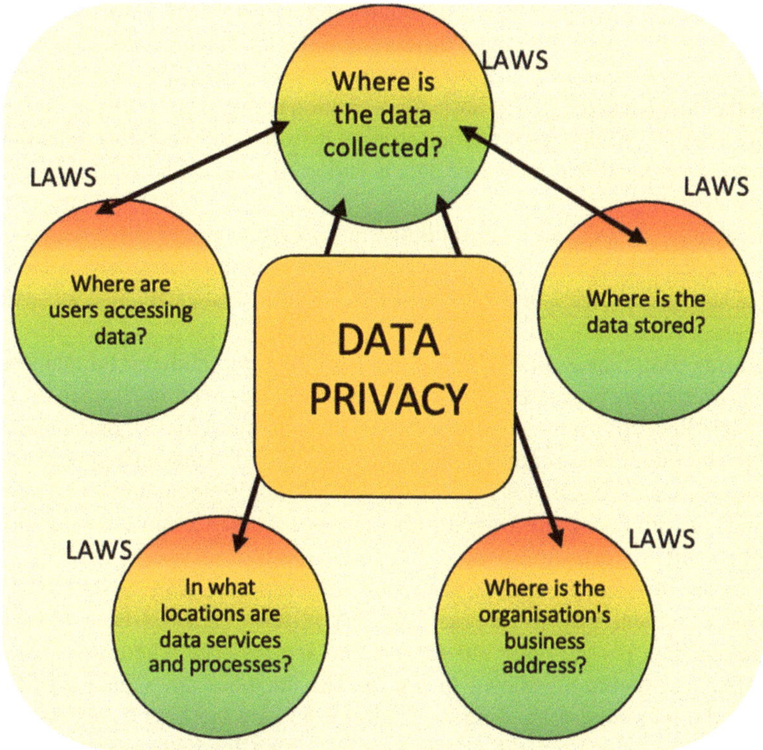

Fig. 5 Data privacy laws and governance

The complexity arises in how PII is stored, where it is stored, how long it is retained, and any potential scope creep from the original intention of the data collection. Under the GDPR, the individual has a right to have their PII deleted, and the data cannot be shared between organisations without consent. Understanding and obtaining the individual's consent on how their data is handled over time can become a grey area.

3.3 Geopolitical Issues

Many data processes can be conducted outside of a Cloud zone. For example, data can be cached on the edge or the fog, and logs may be dumped and stored in a different region. It is important to know what is being recorded in logs. When data is written to the Cloud, a result may be recorded; where is this result recorded, and what data is contained in the result? This can be a complicated network when observing data privacy laws and the permitted geographical locations for that data.

Data privacy laws vary between countries. The European Union is regulated by the General Data Protection Regulation (GDPR). The GDPR is sometimes considered a gold standard [22] for data privacy and protection; it is stricter than many other local data protection laws. The variation between laws can be challenging for organisations as data in CC is often accessible from multiple global locations and considering the local laws of where data is collected, stored, and then accessed can become a juggling act.

There are also many laws across different geographical zones, often associated with countries, but zones may also vary within a country. Data can be stored or traverse states, provinces, or regions within a country, and each can apply varying data privacy laws.

Ensuring compliance with the laws of each location is a challenge for many organisations, even larger organisations that have the resources to employ specialist teams for this job. Recently Amazon was fined \$886 m (the largest fine for data breaches to date) for alleged failure to comply with GDPR laws, which Amazon is appealing [23]. GDPR penalties are calculated at 4% of an organisation's global turnover. Facebook was fined \$5 billion by the United States Federal Trade Commission for privacy violation in 2019 [7].

As CC is increasingly global, it is vital to observe all data privacy laws, and the most expedient way to safeguard against privacy breaches is to ensure that the strictest data privacy laws are observed in any instance where data can leave a geographical region, as shown in Fig. 6. This includes any Cloud service that supports the IoT, edge computing, or where people are working or accessing data remotely from regions outside of the organisation's Cloud storage system. When preserving data privacy and considering the applicable data privacy laws, the first principle is always to observe the regulations of the locations where the data is collected and the business is established and then investigate other applicable laws. There are rigorous regulations about encrypting and storing data, the transit, and even the disposal of data, but it is impossible to cover everything in this chapter. Therefore, larger businesses are generally required to have a "data officer", and smaller businesses need to seek expert advice when establishing Cloud services.

The GDPR defines three actors involved in data regulation: the controller, the processor, and the data subject. The controller is any organisation that directs the collection of personal data; the processor is any organisation that collects that data under the direction of the controller (and may be the same entity); and the data subject is an individual residing in the EU. Under the GDPR, any business that has an establishment in the European Union (EU) or collects data from individuals in the EU must abide by Article 3 of the GDPR [24]. Check that the laws of other locations do not threaten the privacy of the data or the data sovereignty of the business. The laws of a destination country must comply with the GDPR to permit the legal transfer of PII to another country.

Data Disposal

In terms of CC, one important and possibly overlooked aspect of data collection is data disposal. Under Article 17 of the GDPR, individuals have the right to be

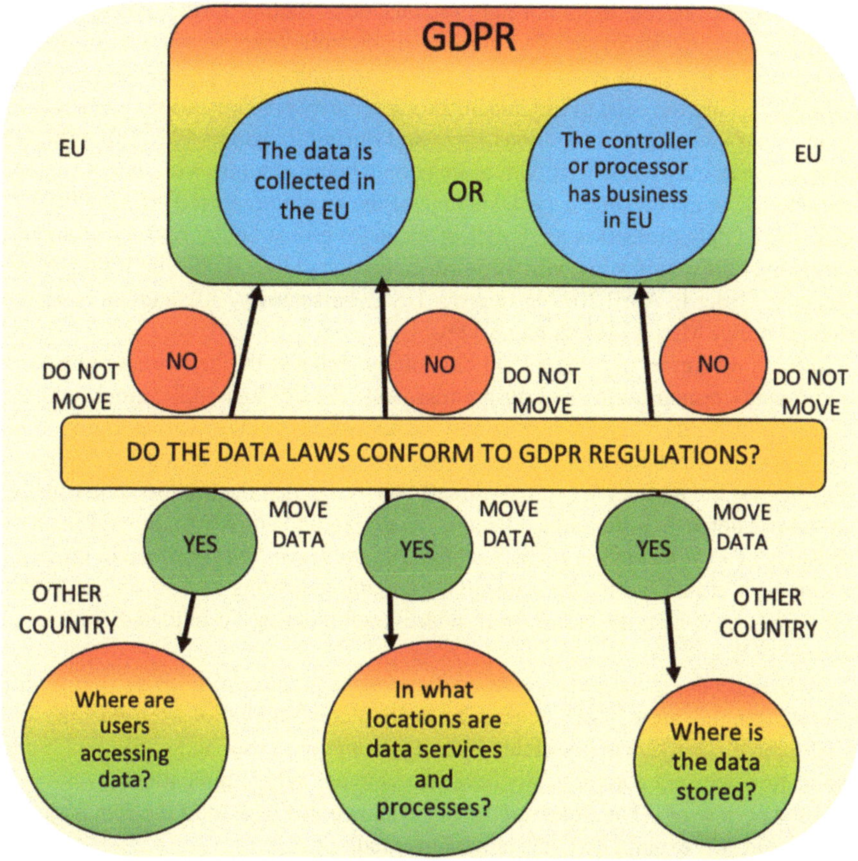

Fig. 6 The GDPR and geographical data movement

forgotten [24]. An organisation needs to have predetermined time limits on data storage and disposal, depending on the data category and the associated laws relevant to the business, both under data protection schemes and within any relevant business or organisation domain. How this data is disposed of in CC may not be transparent, and the data may not actually be deleted from all databases. This needs to be rigorously explored with the CSP and within any service-level agreement (SLA).

4 Zero Trust Architecture in the Cloud

In a world where CC is now a part of everyday life and embedded within most typical business practices, it has been a steep learning curve and challenge for organisations to secure Cloud applications. Suggesting ZT implementation blindly, as a cure-all for CC security needs, does not help organisations flailing in the advent of such change in network structure. The purpose of the following sections is to provide an application of Cloud security that is useful and efficient and, more importantly, viable. To ignore the need to implement strong security postures is folly; cybercrime is not going away [25, 26] and has increased during lockdown [27, 28], although these statistics may be skewed with more people being at home with more time and less activity, as cybercrime is typically one of the most unreported crimes [29]. CC security and data protection are issues of focus for global bodies and governments, and changes are manifesting with the implementation of the GDPR, the White House Executive Order in April 2021 to improve US Cybersecurity [30], and the Essential Eight Maturity Model [31], to name a few.

Moving resources into the Cloud has totally changed the network architecture. Traditionally organisations had on-premises data centres and applications, and the organisation centred around a relatively secure intranet. With Cloud Computing, most organisational resources are accessed over the Internet, as shown in Fig. 7. This exposes the intranet to external threats and means that the traditional perimeter, castle-style defence, is no longer effective. Attempting to employ traditional security measures in a Cloud environment is ineffective and leaves data exposed to threats.

Google is arguably a leader in designing and implementing ZT [32] and no longer determines user identity based on IP address, eliminating an outdated trust in an intranet [33].

No user access to systems and/or applications is granted without an appropriate level of authentication. Note, access is usually a combination of user/machine and environment. This also applies laterally to applications. A user authenticated for one application does not automatically have access to any other applications.

An overview of ZT architecture in CC is shown in Fig. 8, and the remainder of this section will explain how this is implemented.

4.1 Secure Protocols

4.1.1 Do Not Trust the Intranet

Do not trust internal connections without authentication. A ZT CC architecture treats intranet and Internet access the same, doing away with the complication of remote access and the bulk of applications being held in the Cloud and not on-premises. This means security specialists do not need to continually juggle network protocols as the network changes where resources are held, although it is vital to understand what resources are being held to manage access control. There is no longer an internally accessed intranet for most applications within a combined on-premises and CC environment.

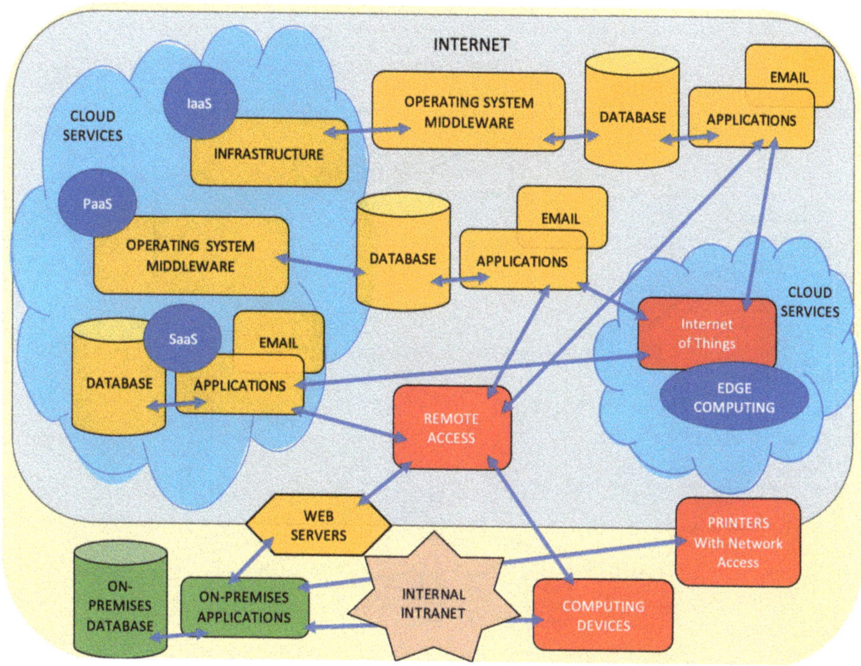

Fig. 7 Cloud network structure

4.1.2 Securing Networks

All network sessions must have authentication and authorisation, always with encrypted protocols.

Use secure protocols:

- Single sign-on protocol using technologies such as SAML.
- Hypertext Transfer Protocol Secure (HTTPS).
- FTP over TLS.
- Html5.
- Encrypt insecure protocols such as Telnet with secure protocols such as SSH.

4.2 Legacy Applications

It is unrealistic to rewrite all programs or eliminate legacy code or equipment. Applications with outdated security procedures and vulnerable code need to be in an isolated network. An identity-aware firewall or data diode can protect this isolated network [34]. Data diodes can also be used effectively to protect unsecured equipment [35]. Outdated, insecure protocols and equipment can be encapsulated to meet CC security standards [36].

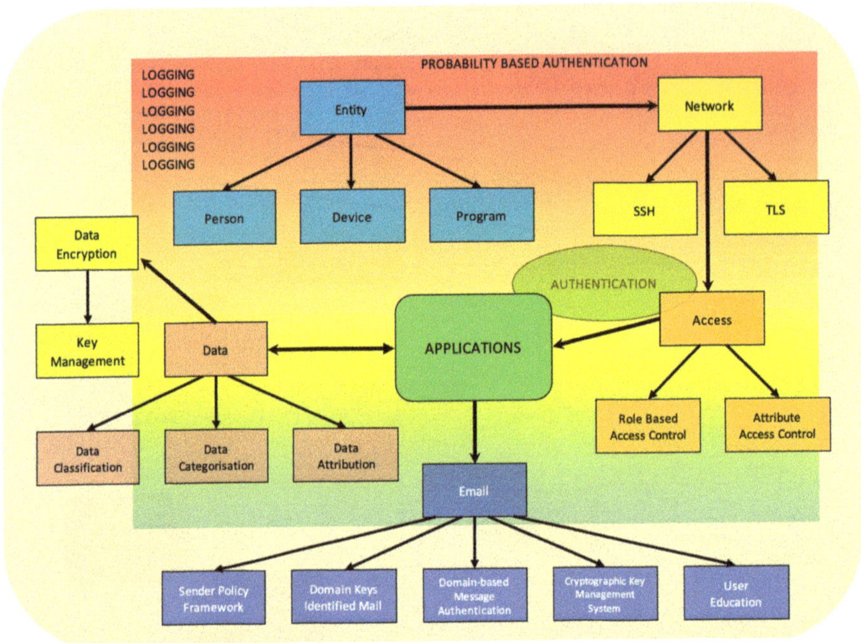

Fig. 8 Cloud zero trust architecture

4.3 Data Classification and Categorisation

This subsection includes protecting data both programmatically and within the organisational culture. In all considerations, data and the applications that use that data need to be managed methodically and logically.

4.3.1 Data Classification

One way of preventing inappropriate data sharing is to classify data across and between organisations, and this is common practice within any large organisation. Data classification is aimed at the user to acknowledge and comply with the classification used on correspondence, by letter or email and within organisation documents. In some instances, there can be penalties and possible imprisonment applied to breaching data classifications, but this applies to legal agreements the user is bound to when accessing the organisation's applications.

Data classifications are comparable across varying domain types. Between three and five classifications are typically used, for example, Top Secret, Highly Confidential, Proprietary, Internal Use Only, and Public. Many organisations will use Confidential, Internal, and Public; other terms such as classified or restricted may

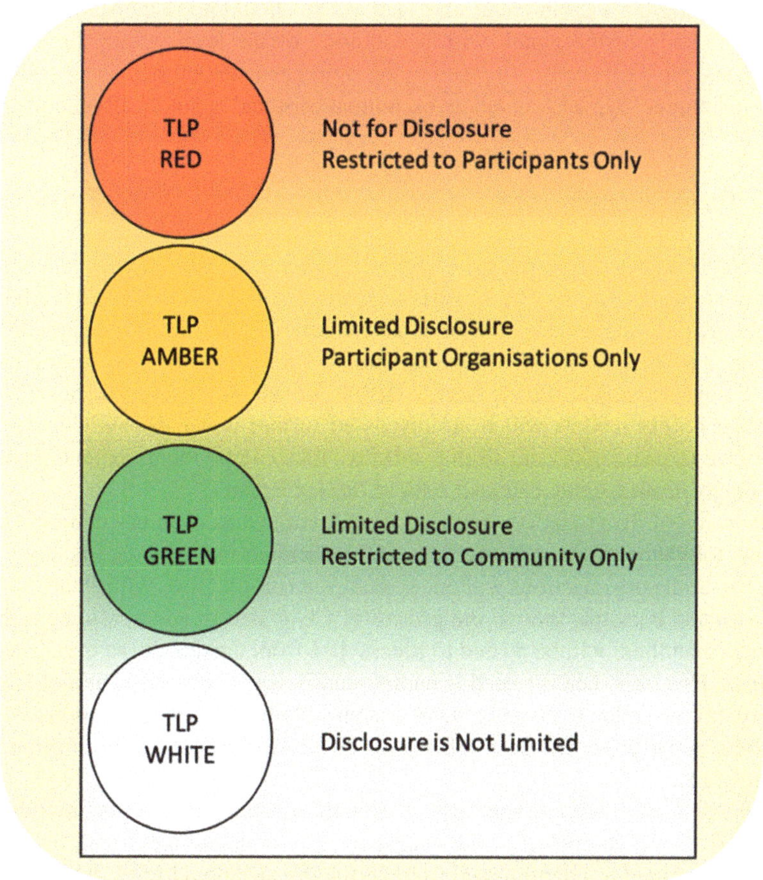

Fig. 9 Traffic Light Protocol

also be used. The Australian government uses Top Secret, Secret, Protected, Official: Sensitive, and Sensitive [37]. The defining factor is an understanding within the organisation of what the various classifications entail.

The Traffic Light Protocol (TLP) is an example of this type of classification and is used by the European Inner Six, other global bodies, many computer security incident response teams (CSIRT), and Information Sharing and Analysis Centres (ISACs) [38].

As the name implies, a traffic light schema indicates data classification, as seen in Fig. 9. Red is the most restricted, between participants only and not to be shared. The amber classification allows sharing between participating organisations. Green restricts data sharing between the organisational community, and white has no restriction, so is essentially public.

When using the TLP in email, the TLP access should be in the subject line and before the body of the email, with a definition of the level within the email. In documents, the TLP access is placed in the right hand side of the header and footer of the document [39]. TLP is reliant on mutual trust and is not legally binding [38].

4.3.2 Data Categorisation

Data categorisation is another way of protecting data, as it separates the data between PII and less sensitive data. This assists in setting up access controls. It is advisable for any organisation holding PII or sensitive data about trade secrets, intellectual property, and the like, to use data categories. Within a software system, this can be used to implement Role-Based Access (RBA) or Task-Role-Based Access (TRBA) and other access models which are discussed further under Access Control. It is imperative to categorise data so that sensitive data can be identified and protected accordingly. Data tagging can also achieve the same result.

The value of data to an organisation, based on the data category combined with the risk tolerance of the organisation, will determine how the data is stored and accessed. Some organisations will opt to retain data on-premises with clearly defined and restricted backups, though the growth of CC makes this increasingly difficult, as more often there will be a need to access data from outside of the organisational premises. How this data is stored is paramount. BeyondCorp, a division of Google, uses a data encryption key management system that illustrates well how organisations will lock down protected data. This is discussed further under Encryption, Key Management.

Using data attributes at a base level, within the database which then flows through programmatically to coincide with access roles or policies, keeps data from being mistakenly spilled to unauthorised users at the organisational level. It is important to build best practices for security and data hygiene from the ground up. One way to update existing systems is to add an extra database table for categorisation level and insert an extra attribute to the various data tables. All calls to the database would then need to be updated to include the attribute when granting access to the user.

4.4 Data Encryption

All data is held to be secure by default on a ZT network.

4.4.1 Data at Rest

Encrypting data at rest can also be referred to as server-side encryption for CC, the intended storage place being in the Cloud and not on-premises. Data written to the Cloud, at rest, is encrypted using AES-256. AES-256 is a secure symmetric

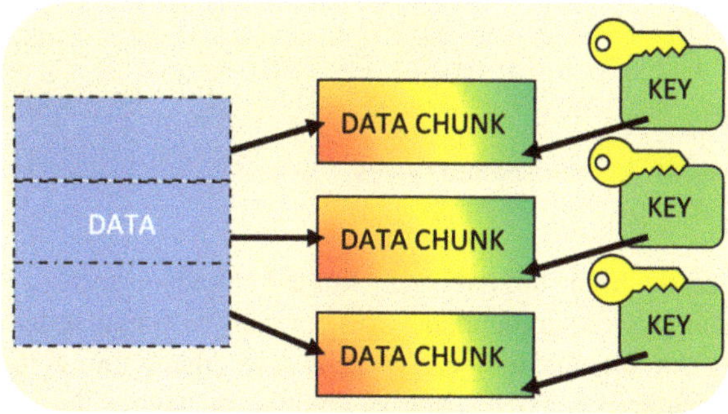

Fig. 10 Encrypting data in chunks

encryption standard and unbreakable by brute force on modern computers [40]. Data is divided into blocks or chunks, and each block is encrypted separately, as shown in Fig. 10. The biggest threat to AES-256 is key management.

4.4.2 Data in Transit

Data in transit is protected using secured connections, such as HTTPS and other protocols discussed in Secure Protocols. There is always a risk when transmitting data in plain text, and secure protocols do assist with this; however, wherever possible, sensitive data should be encrypted before transmission, which is referred to as client-side encryption for data stored in the Cloud. When data is encrypted client-side, encrypted data is transmitted and then encrypted again server-side. Providing the encryption keys are secured, this is effective protection for sensitive data.

AWS recommends using AES-256 for client-side encryption [41]. Google recommends using Tink for client-side encryption and key management [42]. Tink recommends using AES-128 encryption standard for client-side encryption [43]. This is a trade-off between security [44] and speed, and it would be faster for encrypting data streams. There have been many designs to improve the security of AES-128 [45–47].

4.4.3 Key Management

Organisations can use CSP keys, private keys kept on Cloud, or keys kept on-premises or another Cloud. Whatever method, it is vital that these keys are protected from unauthorised access.

The large global CSPs encrypt all the data that is stored by their respective organisations. A big threat to data encryption is key management. It is vital to keep keys in a controlled environment and that key management remains under a protected environment with secure redundancy. An example of protective key management is Google's BeyondCorp Key Management System; see Fig. 11. The KMS is backed up with distributed memory, which is in turn backed up on hardware. These backups

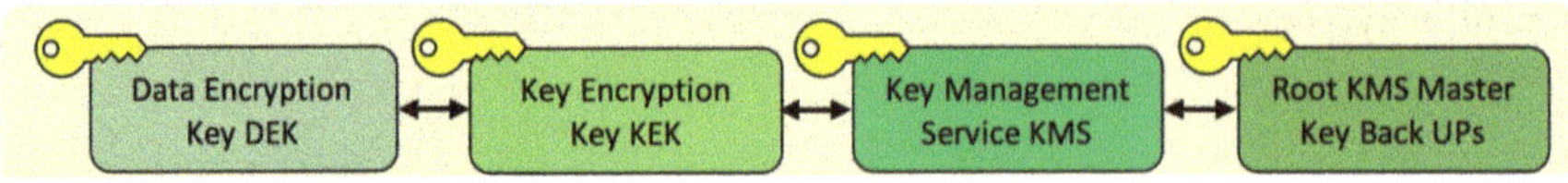

Fig. 11 Key management system

and the hardware have very limited employee access. This key management design is highly recommended for any organisation managing data encryption [48].

An organisation can benefit from maintaining its own KMS; using a KMS provided by a CSP, such as Google, means revoking the right of control over the data keys. The implication for this is that a court order can force the CSP to provide the keys, and in the case of recent laws, many governments can make these data requests without any onus on reporting these requests to the organisation.

4.4.4 Data Masking

Data masking is an effective way to help protect data. Several techniques can be used: shuffling, nulling values, scrambling, and substitution. Sensitive data is generally considered for masking if it includes PII, bank account, credit card details, or medical information. A common form of masking is when a user is presented with payment details, and the details of a stored credit card will be masked with asterisks, and only the last three digits will be exposed.

Article 4 of [24] GDPR has laws about pseudonymisation:

> 'pseudonymisation' means the processing of personal data in such a manner that the personal data can no longer be attributed to a specific data subject without the use of additional information, provided that such additional information is kept separately and is subject to technical and organisational measures to ensure that the personal data are not attributed to an identified or identifiable natural person.

4.5 Access Control

Granting access privileges via many traditional paradigms is fraught with issues in any organisation, and the vulnerabilities with data in the Cloud are only exacerbated by failure to maintain the principle of least privilege. Once access is granted, it is more difficult to remove privileges than to grant only the minimal required access in the first instance.

To begin assessing security needs, the following questions need to be addressed:

- What data is going to be stored or captured in the Cloud?
- To which categories does this data belong?
- What legal or regulatory issues come into play with this data?
- Who needs to access this data and which parts do they need to access?

A security posture is then designed based on the answers to these questions. While there are many tools required to best secure a Cloud environment, the good news is this same skill set can be used repeatedly across varying CC systems.

Some access control models are more effective in maintaining appropriate access levels in-line with the principle of least privilege and are suitable in a CC environment where scalability, versatility, and inheritance are also required [18, 49].

4.5.1 Role-Based Access Control

RBAC or TRBAC is effective in CC; it is applied based on the position within an organisation, and access is not associated with a user ID [50]. This helps to prevent employees leaving an organisation from retaining login access, and it also prevents individuals from accruing access by changing positions within an organisation. If applied correctly it conforms to the principle of least privilege or a need-to-know policy. Wherever possible this should be applied. Arguably one hurdle here is applying this type of policy retrospectively, as employees may feel stripped of authority and devalued when access privileges are removed in an organisation-wide upgrade to security.

RBAC enables the inheritance of roles, meaning this type of access control can extend to a Hierarchical Access Control (HAC), simplifying access allocation within an organisation [50]. It does fail to distinguish between data sensitivity [49].

4.5.2 Attribute-Based Access Control

An Attribute-Based Access Control (ABAC) model is also known as a Policy-Based Access Control (PBAC) or Claims-Based Access Control (CBAC) by allocating users with attributes which grant the user access to specific groups of data type. An example would be the difference between what a nurse, doctor, and clerk can see when accessing a patient record [51]. An advantage of ABAC control is it allows access to users not known to a system [49]; this works well for contractor scenarios. A contractor may be granted access as a “doctor” or a group of attributes and so access the data available for this attribute.

4.5.3 Mandatory Access Control

Mandatory Access Control (MAC) works by using two properties, no-read-up and no-write-down, on objects. This model has vulnerabilities and drawbacks. It is not versatile or dynamically scalable in offering inheritance or conforming to the principle of least privilege. There is also the ability for unclassified data to modify classified data [49].

4.5.4 Discretionary Access Control

It is ill advised to use Discretionary Access Control, as access is controlled by the object owner and this cannot be controlled by the organisation centrally, as access is passed between users, meaning sensitive data can be easily leaked [49].

4.6 Email Security

4.6.1 Secure Email Protocols

Email software needs to be protected using Sender Policy Framework (SPF), Domain Keys Identified Mail (DKIM), and Domain-based Message Authentication, Reporting and Conformance (DMARC) [52–54]. Security should also include IMAP and POP3 connecting over TLS and Cryptographic Key Management System (CKMS) [52].

4.6.2 User Education

The end user is often regarded as the weakest link in IT, with application interactions such as email. For the application to be usable, the end user needs to be educated about common email threats [55]. Regular email security sessions need to be employed and can be done in an online setting. Employees can be tested with fake phishing emails; failure to report, or worse, clicking on links within the email can automatically trigger an education session [56–58]. Despite taking every precaution, threats remain [59], and it pays to remain vigilant in cybersecurity employee programs.

4.7 Protecting Edge Devices

IoT networks are vulnerable to password attacks, but this can be remedied. Microsoft analysed raw data and attacks on its sensor network and collected data from 280,000 attacks. The trends that came from examining this MS raw data and attacks included changing default passwords, using shorter passwords of less than 10 characters, not including numbers or special characters in passwords, and using familiar words such as "admin" in admin login passwords [60].

For best practices over IoT networks:

- Force immediate change of default passwords on devices.
- Enforce strong password policies.
- Use SSH to communicate over Telnet connections (Telnet is not encrypted).

- Ensure all software is updated.
- Monitor IoT activity to log attacks and unauthorised use.
- Develop an organisation policy for vulnerability disclosure [13, 60].

5 Probability-Based Authentication

ZT entails moving away from binary authentication to a risk-based probability authentication. Authentication is no longer reliant upon the user but on the weight of all factors relating to a login attempt, as shown in Fig. 12. The decision to permit a user access is based on the probability that the attempted login is not the user. If the probability that the login attempt is not the user is low, then the user is granted access. Access is based on more features than just a user; rules can be created, for example, "if device geolocated here . . . ," or an IoT device will use device certs as credentials, rather than user credentials. If there are flags surrounding the login attempt, then conditional access may be granted or no access. Authentication follows a flow of "If Then". If this condition is met, then move to the next condition and so on [61].

This section will detail the methods required to implement a PBA, and the overall workflow can be seen in Fig. 14, which shows a decision tree for PBA and how authentication failures affect the login attempt in managing authentication failure.

5.1 Applying Authentication Everywhere

With the move away from traditional monolithic programming, CC increasingly relies on smaller software programs, such as microservices and lambda functions, to perform dedicated tasks or a range of related functions. This enables rapid deployment and makes it easy to add functionality to a Cloud system, but it comes with complex security issues. Whenever there is a connection between programs, there is the potential for a security exploit. Interconnecting APIs need to follow commands to interconnect and access a database and the user interface. Within the traditional on-premises servers of an organisation, these program calls have not required the same rigorous protection as it was occurring within a relatively secured intranet and protection from unauthorised access was limited to users within the intranet. Often, there was no consideration of the possibility that external bad actors would insert themselves into this communication protocol to gain unauthorised access to data. This is based on the presumption that employees, generally, were not experts in software development, so an internal threat was negligible. With CC a bad actor can gain access to a software application from anywhere in the world, so all API calls must be protected with separate authentication.

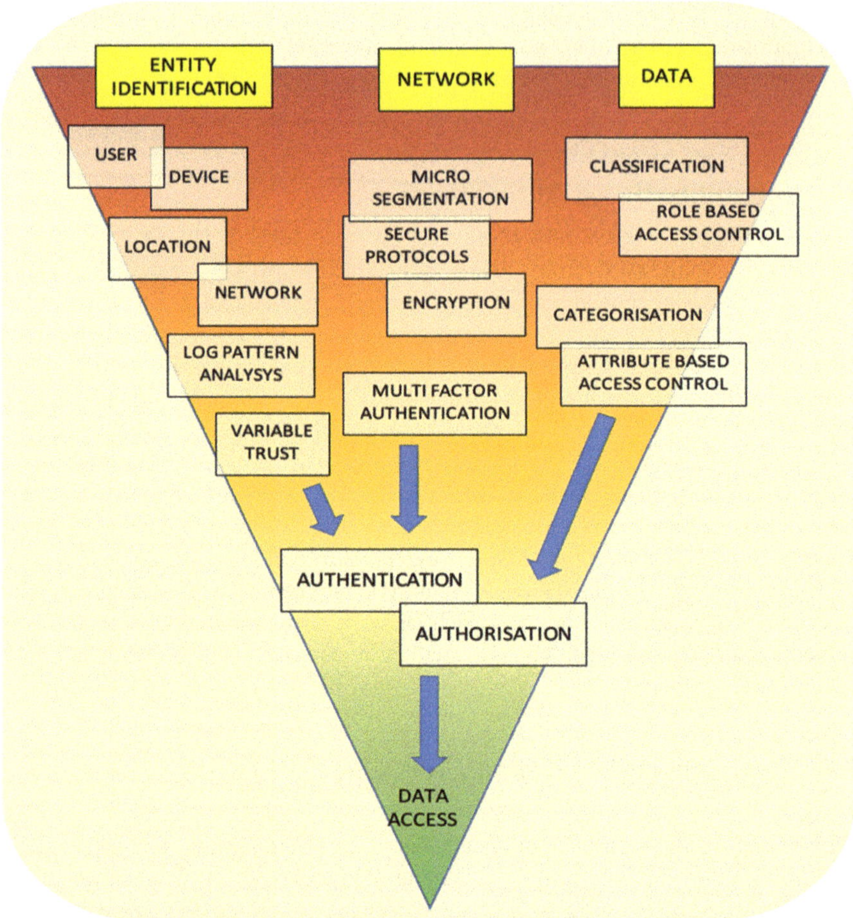

Fig. 12 Factors used to determine authentication

5.2 Logging

Logging all network activity is the basis of a PBA. It is by obtaining baselines that aberrations are noted using tools such as normative transaction profiling [62]. Log all activity: network location, time, device, and activity. Comprehensive logging of activities, logins and access details relating to user logins, network geolocation, device details, IP address, and browsing activity within the applications creates a typical baseline for entities. This aids machine learning and network monitoring to predict or flag any activity that deviates from this baseline. This is imperative in designing an identity-based security paradigm, as the previous binary determination of authentication is disposed of, and authentication based on probability and risk is used to secure ZT networks.

Machine learning can be used to detect aberrant patterns and potentially suspicious behaviour, for instance, updating programs with known activities of bad actors such as botnet IP addresses and compromised credentials and ensuring there is adequate programming to alert for suspicious activity so human evaluation can be performed promptly and prioritised to accommodate risk analysis and user experience. A relatable example of alerting an organisation to unusual or suspicious activity is using a normative transactional profile which banks use to flag unusual credit card spending activity.

5.3 User Identity or Entity

An entity is not limited to the end user but may also be a service or process running in an application seeking a database connection to complete the task. A user identity or entity can be a person, device, organisation, legal, family, group, business, a program, or an agent, either a human, delegation, AI, or learning program.

User identity can be checked beyond a login ID and password alone. Logs can be used to detect unusual patterns with the user ID by keeping a history of user transactions.

- Are there failed login attempts?
- Are the credentials for sale on the dark web?
- Is the user part of the organisation or a contractor?
- What is the user's level of access?
- Is the user a program execution? Is the code signed?

Isolate the organisational applications from the login identity provider. Ensure that the login is recognisable to help prevent phishing scams, so a phoney website or login screen is more likely to be identifiable. Always assign least privilege. Hide all applications from all the users; only allow authenticated and authorised users to see the application they are permitted to see.

Use logs to observe changes to activity, such as downloading files, and note if a user has recently acquired a new level of access and compare the activity of uses with similar roles [17, 63].

5.4 Multi-Factor Authentication

Two-factor authentication (2FA) has become a common way of securing access. 2FA is not a silver bullet, but it is still helpful in providing safeguards from many threats. 2FA can protect against replay attacks, brute force attacks, and many social engineering attacks [64], as the password or session information alone is insufficient. There are applications available to assist in 2FA, bypassing issues such as no mobile network. This can be useful in organisations that block external network connections

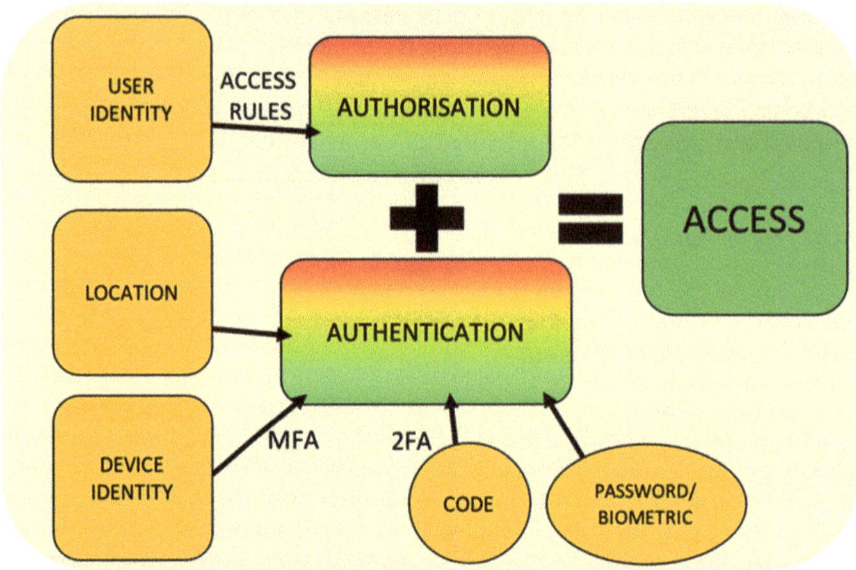

Fig. 13 Identifiers required for login

and only allow connections through a dedicated network. An example of these would be server centres, some medical facilities, and some government bodies related to the armed forces. Although 2FA has become widely accepted as safeguarding authentication, it has weaknesses, and some can be hacked without major difficulty [65, 66].

Multi-factor authentication (MFA) codes are more securely obtained via an authenticator app or authentication device or phone call, rather than SMS [66]. While authentication includes the digital identity via MFA, this may require a biometric match, a password, or an authenticator application passcode, as shown in Fig. 13. It can also include the network location and device [67]. Phishing is an issue for end users, and many times only discerning users or IT experts can spot a phishing attempt. While educating the wider public is an onerous task, the importance of end-user education within organisations cannot be emphasised enough.

5.5 *Network Access*

Remote and intranet access are treated equally due to the exposed nature of a CC network structure. A network per se is not authenticated but rather examined for authenticity to apportion the likelihood that the attempted login is from the purported user and not a bad actor. This examination uses several tools, known activity, known botnets, and Tor browser. Having said this, whitelisting IP addresses is still an effective tool, particularly when developers are accessing source code and pushing changes.

When logging user activity, there will usually be a trend – it might involve a regular home network or a propensity to login from Internet cafes, mobile data, or airports. Whatever the trend, the user will have one. An attempted login under a user ID using a network from outside the usual activity should raise a flag. This does not automatically mean that a bad actor is attempting to use the ID. However, using a combination of login identity factors will add to this probability and may trigger extra login challenges or lead to the login session being limited or prevented.

Another important aspect of logging activity is to note geospatial location. If a user has been consistently logging in from known areas, a login from an entirely different global location may warrant a flag, more so if a user has logged in from one country and then an hour later attempts to log in from the opposite side of the globe, a physical impossibility.

If login attempts are made from a known botnet, or traffic patterns are known to match bad actors, this would be a major flag in determining a denial of access. Part of the terms of usage may include not using Tor browsers or any type of browser that masks the IP address, as they are more likely to be associated with nefarious activity in preventing tracking and, if done thoroughly, preventing the identification of the login device. This would greatly reduce the identifying attributes that can be used to calculate access permissions.

When examining network access, look for:

- Known login addresses
- Tor browser
- Known botnet addresses
- IP masking
- Known addresses of attacks
- Impossible geolocation changes
- Changes in routine network activity

5.6 *Device Health*

- Organisational Devices

Device monitoring depends upon what application is being accessed and who is accessing the application. Employees accessing sensitive data will need a more rigorous screening than the end user of an IoT device. This does not imply that the data at the end of an IoT device is not sensitive or important, but users with access to large amounts of sensitive data, including PII, trade secrets, and the like, need to be rigorously controlled.

Devices may be enrolled and logged within a tight security system for users accessing sensitive data, creating a device identity trust. An organisation that requires enrolled devices can monitor devices by MAC address, using security checks to help detect spoofing attacks [68].

- Contractors

When contractors may be accessing a Cloud network without a managed machine, there can be strategies to further protect against bad actors. Using timeouts requiring passcodes, requesting device enrolment, and asking the contractor to agree to a Terms of Use when registering while maintaining a log of all activity also help secure the network.

- Consumer Networks

Larger networks, such as IoT networks, can be monitored to determine device type [69]. Another simple method is to use protocols to ask the device for identification [70].

Organisations can develop security policies that encompass device hygiene and compliance, further securing CC networks. Scans of OS kernels, firmware, and drivers can examine the health and compliance of devices. Organisational devices and BYOD can both undergo similar checks. The rigour of these policies depends on the sensitivity of the data or equipment that is being accessed.

The following device state can be checked:

- Firewalls

Computers can be checked that firewalls are active, and the user may be issued a notification stating that the firewall needs to be enabled before access will be provided. MS does this with its employee network.

- Boot State

Likewise, the boot state can be scanned, and scans for signed drivers and firmware can be implemented to produce the same effect, perhaps with a notification to contact the organisation.

- Compromised Operating Systems

Apple phones can be checked for jailbreaking and Androids if they are rooted. The organisational policy may choose to deny access for such devices or provide limited access, depending on the sensitivity of the data being accessed.

- Compromised devices

If a device is compromised with malware or ransomware, then access can be denied.

5.7 Managing Authentication Failure

When the probability that a user is likely to be the right user, but there are compromises in the authentication flow process, an organisation can make informed risk-based decisions on how to process these logins. Sometimes a risk analysis may weigh on balance that harm is unlikely from allowing access, but as the risk is raised, the access is allowed with conditions to protect the organisation's data. Much of this may be based on the sensitivity of what is being accessed. Figure 14 demonstrates the workflow and decision-making that determines the success of a login attempt.

If a user is granted conditional access, it may be helpful to inform the user before proceeding into the application that all their actions will be audited and to audit all the user's activity.

Disabling or limiting features from the application that can share data can also provide more protection. Disable share options to unmanaged Clouds, and disable or limit downloads to unauthorised devices. Disable copy and paste into unmanaged applications. There is also the potential to disable screenshots, in extreme cases, but this requires controlling features within the device's operating system and would require system permissions from the device. This may not always be possible, and perhaps if the data is sensitive, it would be prudent to deny access in this instance.

Forcing password resets, device enrolment, or contact with the organisation are ways to increase security. Using meaningful notifications at the application's login screen assists the user experience and may help to alleviate frustration.

When an account seems to be compromised, the account can be suspended until the user contacts the organisation, at which point a strict protocol needs to be

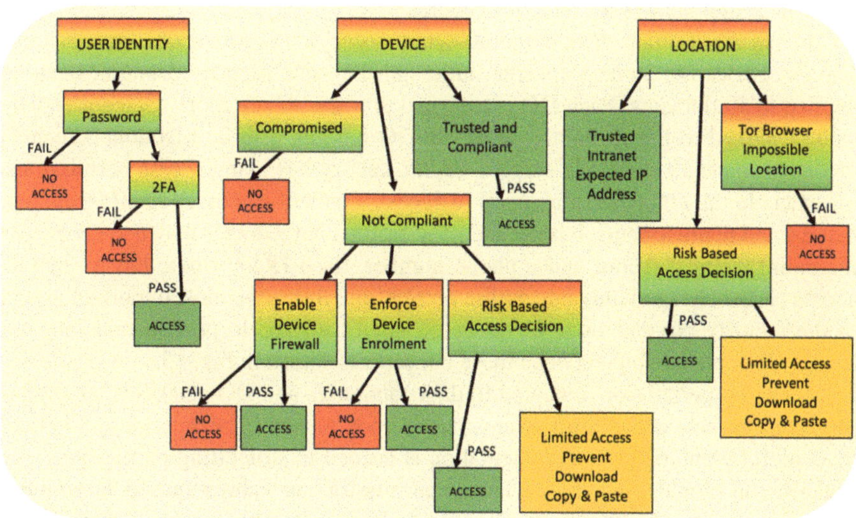

Fig. 14 Probability-based authentication decision tree

enforced to allow further account access, including basics, such as changing the password.

6 Zero Trust Data

Further to implementing a zero trust architecture to the CC and probability-based authentication, these principles can be applied to the data itself, creating a method known as zero trust data (ZTD): using a combination of Data Classification, Data Categorisation, both Attribute-Based Access Control and Role-Based Access Control, and organisational directives specific to device and network requirements. This section will focus on the data implementation. Implementation according to the PBA and ZT requirements within an organisation means that security rules must be set up based on the sensitivity and overall permission ranking of the data.

6.1 Zero Trust Data Implementation

ZTD relies upon a meticulous examination of database objects (DO) or entities. For the sake of this section, we will refer to these as data objects, not to be confused with object-oriented programming (OOP), although the same principles may apply. A DO could be, for example: “person”, “contact details”, “medical record”, or “financial record”. DOs have many attributes that should correlate to the actual attributes of the database entity, for example, a medical record may have “tests”, “test results”, “reports”, “appointments”, and “procedures”.

The organisation needs to structure data categories and data access permissions. The categories are domain specific, although the same security principles can be applied to all domains. Each DO is assigned to a specific category. Categories may well be linked to the domain type of the DO but can also be functional, such as medical, financial, and trade secrets. Each category has many access permissions, for example: an employee may need to access a patient’s contact details, payment details, and appointments, but not their test results, but a doctor may need to access the entire medical record and a phone number. Each DO attribute has a separate access permission and only one access permission; this is the most important aspect of the security process, as it serves as the final backstop to prevent unauthorised access. The relationship between these concepts is shown in Fig. 15.

By having permission allocated to data categories and DO attributes, an organisation can create permission groups based upon matching permissions. It is here that careful planning and implementation is needed in structuring permissions. An organisation should aim to have as many permissions as viably required. Examining the variety and depth of the stored data and the varying needs for access, permissions can be set up following organisational security rules, for example: access permissions can be categories in an alpha numerical sense, which is more difficult to make

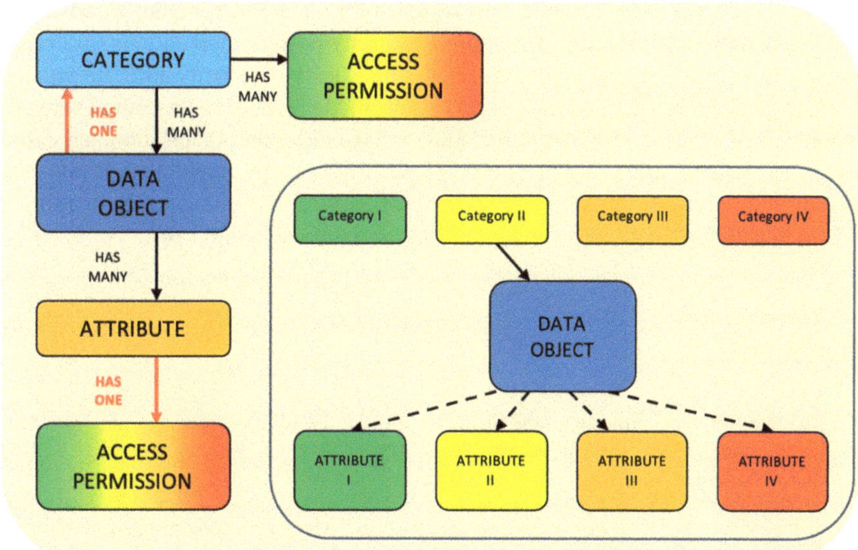

Fig. 15 Access permissions, categories, and data object relationships

sense of, but computationally simple, or catenated with a category and classification. Importantly, all permissions must be linked with a data sensitivity level, which can be implied within the security directives, but with data privacy, such a delicate area, it would be preferable to actively catenate a data sensitivity level with each permission.

In practice this would mean a user wanting to access their own records would be able to access the data objects within their assigned category, such as "Patient". This would cover data objects connected with the user's patient ID and with access permissions to all the attributes that are consistent with the "Patient" self-access. Or a doctor may have access to data objects with categorisation "Medical" and sub-categories "Professional" and then "Doctor". The doctor would have access to all data objects linked with the doctor's user ID and the attributes assigned with permissions allocated to "Doctor".

Rules can be created to enforce conditions upon authentication dependent upon the resultant permission of each data attribute. Although meticulous to set up, this system can enhance a strong security posture.

7 Conclusions

Organisations need to rethink the traditional security measures with a view to the changed network structure of CC and bring an open mind to what can be perceived as buzzwords, such as ZT, perimeterless security, and PBA. Using the same security tools that have been available in a more traditional on-premises network and embracing advances in IT security with as much vigour as advances in IT generally, organisations can improve security posture markedly by observing standards set out by global bodies and authorities on cybersecurity and data privacy laws. Organisations can raise awareness that the hidden cost of an unsecured network leaves Cloud applications vulnerable and can take down an organisation and cause havoc to the end users. Responsible CC security can be implemented into any organisation regardless of age and security of applications and equipment using a ZT architecture designed for Cloud applications combined with PBA.

References

1. P.J. Sun, Privacy protection and data security in cloud computing: A survey, challenges, and solutions. IEEE Access **7**, 147420–147452 (2019). https://doi.org/10.1109/ACCESS.2019.2946185
2. S. Wiefling, L. Lo Iacono, M. Dürmuth, in *ICT Systems Security and Privacy Protection*. Is This Really You? An Empirical Study on Risk-Based Authentication Applied in the Wild (Cham, 2019), Springer International Publishing, pp. 134–148. https://doi.org/10.1007/978-3-030-22312-0_10
3. E. Gilman, D. Barth, in *Zero Trust Networks* (O'Reilly, 2017). Available: https://learning.oreilly.com/library/view/zero-trust-networks/9781491962183/. Accessed 18 Dec 2021. [Online]
4. P. Suryateja, Threats and vulnerabilities of cloud computing: A review. Int. J. Comput. Sci. Eng. **6** (2018). https://doi.org/10.26438/ijcse/v6i3.298303
5. M. Sepczuk, Z. Kotulski, A new risk-based authentication management model oriented on user's experience. Comput. Secur. **73**, 17–33 (2018). https://doi.org/10.1016/j.cose.2017.10.002
6. S. Wiefling, M. Dürmuth, L.L. Iacono, What's in score for website users: A data-driven long-term study on risk-based authentication characteristics. arXiv:2101.10681 [cs] **12675**, 361–381 (2021). https://doi.org/10.1007/978-3-662-64331-0_19
7. C. Cunningham, D. Holmes, J. Pollard, in *The Eight Business and Security Benefits of Zero Trust* (2019), p. 18
8. M.A. Islam, H. Mahmud, S. Ren, X. Wang, in *2015 IEEE 21st International Symposium on High Performance Computer Architecture (HPCA)*. Paying to Save: Reducing Cost of Colocation Data Center Via Rewards (2015), pp. 235–245. https://doi.org/10.1109/HPCA.2015.7056036
9. H. Baron, S. Heide, S. Mahmud, J. Yeoh, in *Cloud Security Complexity*. Cloud Security Alliance (2019). Available: https://cloudsecurityalliance.org/artifacts/cloud-security-complexity/. Accessed 01 Dec 2021. [Online]
10. Amazon Web Services, Regions and Zones – Amazon Elastic Compute Cloud. Available: https://docs.aws.amazon.com/AWSEC2/latest/UserGuide/using-regions-availability-zones.html. Accessed 25 Nov 2021. [Online]
11. S. Yi, Z. Hao, Z. Qin, Q. Li, in *2015 Third IEEE Workshop on Hot Topics in Web Systems and Technologies*. Fog Computing: Platform and Applications (2015), pp. 73–78. https://doi.org/10.1109/HotWeb.2015.22
12. L. Shooshtarian, D. Lan, A. Taherkordi, in *Pervasive Systems, Algorithms and Networks*. A Clustering-Based Approach to Efficient Resource Allocation in Fog Computing (Cham, 2019), Springer International Publishing, pp. 207–224. https://doi.org/10.1007/978-3-030-30143-9_17
13. Global Cyber Alliance, IoT Policy and Attack Report (2021). Available: https://www.globalcyberalliance.org/reports_publications/iot-policy-and-attack-report/. Accessed 19 Dec 2021. [Online]

14. R. Sobers, 98 Must-Know Data Breach Statistics for 2021 | Varonis (2020)
15. Australian Cyber Security Centre, ACSC Annual Cyber Threat Report 2020–21. Available: https://www.cyber.gov.au/acsc/view-all-content/reports-and-statistics/acsc-annual-cyber-threat-report-2020-21. Accessed 25 Nov 2021. [Online]
16. IBM Corporation, Cost of a Data Breach Report 2021 (2021). Available: https://www.ibm.com/au-en/security/data-breach. Accessed 16 Feb 2022. [Online]
17. P.A. Legg, in *2015 IEEE Symposium on Visualization for Cyber Security*. Visualizing the Insider Threat: Challenges and Tools for Identifying Malicious User Activity (2015), pp. 1–7. https://doi.org/10.1109/VIZSEC.2015.7312772
18. P. Sun, Security and privacy protection in cloud computing: Discussions and challenges. J. Netw. Comput. Appl. **160**, 102642 (2020). https://doi.org/10.1016/j.jnca.2020.102642
19. A. Aljumah, T.A. Ahanger, Cyber security threats, challenges and defence mechanisms in cloud computing. IET Commun. **14**(7), 1185–1191 (2020). https://doi.org/10.1049/iet-com.2019.0040
20. A. Singh, K. Chatterjee, Cloud security issues and challenges: A survey. J. Netw. Comput. Appl. **79**, 88–115 (2017). https://doi.org/10.1016/j.jnca.2016.11.027
21. I. Gul, M. Hussain, Distributed cloud intrusion detection model. Int. J. Adv. Sci. Technol. **34**, 71–82 (2011)
22. A. Mantelero, The future of data protection: Gold standard vs. global standard. Comput. Law Secur. Rev. **40**, 105500 (2021). https://doi.org/10.1016/j.clsr.2020.105500
23. C. Nast, in *Wired UK*. Why Amazon's £636m GDPR Fine Really Matters. Available: https://www.wired.co.uk/article/amazon-gdpr-fine. Accessed 26 Nov 2021. [Online]
24. European Parliament, in *Regulation (EU) 2016/679*. Council of the European Union (2016). [Online]. Available: http://data.europa.eu/eli/reg/2016/679/oj
25. A. Bendovschi, Cyber-attacks – Trends, patterns and security countermeasures. Procedia Econ. Financ. **28**, 24–31 (2015). https://doi.org/10.1016/S2212-5671(15)01077-1
26. A. Shalaginov, J.W. Johnsen, K. Franke, in *2017 IEEE International Conference on Big Data (Big Data)*. Cyber Crime Investigations in the Era of Big Data (2017), pp. 3672–3676. https://doi.org/10.1109/BigData.2017.8258362
27. D. Buil-Gil, F. Miró-Llinares, A. Moneva, S. Kemp, N. Díaz-Castaño, Cybercrime and shifts in opportunities during COVID-19: A preliminary analysis in the UK. Eur. Soc. **23**(sup1), S47–S59 (2021). https://doi.org/10.1080/14616696.2020.1804973
28. S. Monteith, M. Bauer, M. Alda, J. Geddes, P.C. Whybrow, T. Glenn, Increasing cybercrime since the pandemic: Concerns for psychiatry. Curr. Psychiatry Rep. **23**(4), 18 (2021). https://doi.org/10.1007/s11920-021-01228-w
29. S.G.A. van de Weijer, R. Leukfeldt, W. Bernasco, Determinants of reporting cybercrime: A comparison between identity theft, consumer fraud, and hacking. Eur. J. Criminol. **16**(4), 486–508 (2019). https://doi.org/10.1177/1477370818773610
30. The White House, Executive Order 14028, Improving the Nation's Cybersecurity (2021). https://www.whitehouse.gov/briefing-room/presidential-actions/2021/05/12/executive-order-on-improving-the-nations-cybersecurity/. Accessed 19 Dec 2021
31. The Australian Cyber Security Centre, Essential Eight Maturity Model | Cyber.gov.au (2021). Available: https://www.cyber.gov.au/acsc/view-all-content/publications/essential-eight-maturity-model. Accessed 29 Oct 2021. [Online]
32. BeyondCorp, Run Zero Trust Security Like Google. http://www.beyondcorp.com/. Accessed 15 Dec 2021
33. R. Ward, B. Beyer, BeyondCorp: A new approach to enterprise security. Google Res. **39**(6), 6–11 (2014)
34. H. Okhravi, F.T. Sheldon, in *Proceedings of the Sixth Annual Workshop on Cyber Security and Information Intelligence Research*. Data Diodes in Support of Trustworthy Cyber Infrastructure (New York, 2010), pp. 1–4. https://doi.org/10.1145/1852666.1852692
35. B.-S. Jeon, J.-C. Na, in *2016 18th International Conference on Advanced Communication Technology (ICACT)*. A Study of Cyber Security Policy in Industrial Control System Using Data Diodes (2016), pp. 314–317. https://doi.org/10.1109/ICACT.2016.7423374

36. Y. Zhang, G. Zhang, Y. Liu, D. Hu, Research on services encapsulation and virtualization access model of machine for cloud manufacturing. J. Intell. Manuf. **28**(5), 1109–1123 (2017). https://doi.org/10.1007/s10845-015-1064-2
37. Attorney-General's Department, Policy 8: Sensitive and Classified Information. Australian Government 2021. Available: https://www.protectivesecurity.gov.au/system/files/2021-11/pspf-policy-8-sensitive-and-classified-information.pdf. Accessed 16 Feb 2022. [Online]
38. European Union Agency for Cybersecurity, Considerations on the Traffic Light Protocol. https://www.enisa.europa.eu/topics/csirts-in-europe/glossary/considerations-on-the-traffic-light-protocol. Accessed 15 Dec 2021
39. Cybersecurity & Infrastructure Security Agency, Traffic Light Protocol (TLP) Definitions and Usage. Available: https://www.cisa.gov/tlp. Accessed 15 Dec 2021. [Online]
40. S. Rao, D. Mahto, D. Yadav, D. Khan, The AES-256 cryptosystem resists quantum attacks. Int. J. Adv. Res. Comput. Sci. **8**, 404–408 (2017)
41. Amazon Web Services, Protecting Data Using Client-Side Encryption (2022). Available: https://docs.aws.amazon.com/AmazonS3/latest/userguide/UsingClientSideEncryption.html. Accessed 21 Dec 2021. [Online]
42. Google Cloud, Client-Side Encryption Keys. Available: https://cloud.google.com/storage/docs/encryption/client-side-keys. Accessed 21 Dec 2021. [Online]
43. Google Developers, I Want to Encrypt Data (2021). Available: https://developers.google.com/tink/encrypt-data. Accessed 21 Dec 2021. [Online]
44. P. Arpaia, F. Bonavolontà, A. Cioffi, in *2020 IEEE International Workshop on Metrology for Industry 4.0 IoT*. Security Vulnerability in Internet of Things Sensor Networks Protected by Advanced Encryption Standard (2020), pp. 452–457. https://doi.org/10.1109/MetroInd4.0IoT48571.2020.9138236
45. M. Forhad, S. Riaz, M. Hossain, M. Das, An improvement of advanced encryption standard. **18**, 159–166 (2018)
46. R. Saha, G. Geetha, G. Kumar, T. Kim, RK-AES: An improved version of AES using a new key generation process with random keys. Secur. Commun. Netw. **2018**, e9802475 (2018). https://doi.org/10.1155/2018/9802475
47. I.A. Awan, M. Shiraz, M.U. Hashmi, Q. Shaheen, R. Akhtar, A. Ditta, Secure framework enhancing AES algorithm in cloud computing. Secur. Commun. Netw. **2020**, e8863345 (2020). https://doi.org/10.1155/2020/8863345
48. Google Cloud, Encryption at Rest in Google Cloud (2020). Available: https://cloud.google.com/security/encryption/default-encryption. Accessed 14 Dec 2021. [Online]
49. A. Younis, K. Kifayat, M. Merabti, An access control model for cloud computing. J. Inf. Secur. Appl. **19**(1), 45–60 (2014). https://doi.org/10.1016/j.jisa.2014.04.003
50. D.F. Ferraiolo, R. Sandhu, S. Gavrila, D.R. Kuhn, R. Chandramouli, Proposed NIST standard for role-based access control. ACM Trans. Inf. Syst. Secur. **4**(3), 224–274 (2001). https://doi.org/10.1145/501978.501980
51. V.C. Hu, D.R. Kuhn, D.F. Ferraiolo, J. Voas, Attribute-based access control. Computer **48**(2), 85–88 (2015). https://doi.org/10.1109/MC.2015.33
52. R. Chandramouli, S.L. Garfinkel, J.S. Nightingale, S.W. Rose, Trustworthy Email (2016). Available: http://www.nist.gov/publications/trustworthy-email. Accessed 25 Nov 2021. [Online]
53. S.J. Nightingale, Email Authentication Mechanisms: DMARC, SPF and DKIM. National Institute of Standards and Technology, Gaithersburg, MD, NIST TN 1945 (2017). https://doi.org/10.6028/NIST.TN.1945
54. G. Kambourakis, G.D. Gil, I. Sanchez, What email servers can tell to Johnny: An empirical study of provider-to-provider email security. IEEE Access **8**, 130066–130081 (2020). https://doi.org/10.1109/ACCESS.2020.3009122
55. S. Bax, T. McGill, V. Hobbs, Maladaptive behaviour in response to email phishing threats: The roles of rewards and response costs. Comput. Secur. **106**, 102278 (2021). https://doi.org/10.1016/j.cose.2021.102278
56. Z. Durumeric et al., in *Proceedings of the 2015 Internet Measurement Conference*. Neither Snow nor Rain nor MITM . . . : An Empirical Analysis of Email Delivery Security (New York, 2015), pp. 27–39. https://doi.org/10.1145/2815675.2815695

57. I.D. Foster, J. Larson, M. Masich, A.C. Snoeren, S. Savage, K. Levchenko, in *Proceedings of the 22nd ACM SIGSAC Conference on Computer and Communications Security*. Security by Any Other Name: On the Effectiveness of Provider Based Email Security (New York, 2015), pp. 450–464. https://doi.org/10.1145/2810103.2813607
58. M. Haider, H. Mohammed, A survey of email service; attacks, security methods and protocols. Int. J. Comput. Appl. **162**, 31–40 (2017). https://doi.org/10.5120/ijca2017913417
59. J. Chen, V. Paxson, J. Jiang, in *Composition Kills: A Case Study of Email Sender Authentication*, p. 18
60. M. Braverman-Blumenstyk, Learn how Microsoft strengthens IoT and OT security with Zero Trust. *Microsoft Security Blog* (2021). https://www.microsoft.com/security/blog/2021/11/08/learn-how-microsoft-strengthens-iot-and-ot-security-with-zero-trust/. Accessed 19 Dec 2021
61. Microsoft, Implementing a Zero Trust security model at Microsoft. *Microsoft | Inside Track*, 2022. https://www.microsoft.com/en-us/insidetrack/implementing-a-zero-trust-security-model-at-microsoft. Accessed 19 Dec 2021
62. S. Wachter, Normative challenges of identification in the Internet of Things: Privacy, profiling, discrimination, and the GDPR. Comput. Law Secur. Rev. **34**(3), 436–449 (2018). https://doi.org/10.1016/j.clsr.2018.02.002
63. W. Wang, J. Han, M. Song, X. Wang, in *2011 6th International Conference on Pervasive Computing and Applications*. The Design of a Trust and Role Based Access Control Model in Cloud Computing (2011), pp. 330–334. https://doi.org/10.1109/ICPCA.2011.6106526
64. A. Ometov, S. Bezzateev, N. Mäkitalo, S. Andreev, T. Mikkonen, Y. Koucheryavy, Multi-factor authentication: A survey. *Cryptography*. **2**(1), Art. no. 1 (2018). https://doi.org/10.3390/cryptography2010001
65. R.A. Grimes, *Hacking Multifactor Authentication* (Wiley, Newark, 2020)
66. C. Jacomme, S. Kremer, An extensive formal analysis of multi-factor authentication protocols. ACM Trans. Privacy Secur. **24**(2), 1–34 (2021). https://doi.org/10.1145/3440712
67. E. Grosse, M. Upadhyay, Authentication at scale. IEEE Secur. Privacy **11**(1), 15–22 (2013). https://doi.org/10.1109/MSP.2012.162
68. M. Anathi, K. Vijayakumar, An intelligent approach for dynamic network traffic restriction using MAC address verification. Comput. Commun. **154**, 559–564 (2020). https://doi.org/10.1016/j.comcom.2020.02.021
69. M.R. Shahid, G. Blanc, Z. Zhang, H. Debar, in *2018 IEEE International Conference on Big Data (Big Data)*. IoT Devices Recognition Through Network Traffic Analysis (2018), pp. 5187–5192. https://doi.org/10.1109/BigData.2018.8622243
70. M. Laštovička, P. Čeleda, in *Security of Networks and Services in an All-Connected World*. Situational Awareness: Detecting Critical Dependencies and Devices in a Network (Cham, 2017), pp. 173–178. https://doi.org/10.1007/978-3-319-60774-0_17

DataCookie: Sorting Cookies Using Data Mining for Prevention of Cross-Site Scripting (XSS)

Germán E. Rodríguez, Jenny G. Torres, and Eduardo Benavides-Astudillo

1 Introduction

In the last decade, innovation and the use of information technologies in organizations have been one of the most used strategies to improve customer service in developed countries [1]. However, the use of technologies also implies a series of vulnerabilities, such as cyber-attacks [2], which often go unchecked due to weak network infrastructure, affecting the productivity and credibility of any organization.

According to OWASP 2017 Top Ten Application Security Risks, the Cross-Site Scripting (XSS) vulnerability ranked seventh. By 2021, this vulnerability rose to third place with the name A03:2021-Injection. One of the objectives of this attack is the theft of cookies stored on the computer of any victim to extract sensitive data.

In this context, the security solutions proposed by the scientific community deploy new mechanisms that selectively allow the blocking of cookies in the browser. Nevertheless, the different mechanisms focus on blocking cookies at the domain level are insufficient to protect the user [3]. On the other hand, in the literature we have found several proposals that allow the analysis of cookies [4–16]. However, the studies mentioned before can be evidenced by the absence of works that propose a solution to classify cookies by analyzing their attributes.

G. E. Rodríguez (✉) · E. Benavides-Astudillo
Facultad de Ingeniería en Sistemas, Escuela Politécnica Nacional, Quito, Ecuador

Departamento de Ciencias dela Computación, Universidad de las Fuerzas Armadas-ESPE, Santo Domingo, Ecuador
e-mail: german.rodriguez@epn.edu.ec; gerodriguez10@espe.edu.ec; diego.benavides@epn.edu.ec; debenavides@espe.edu.ec

J. G. Torres
Facultad de Ingeniería en Sistemas, Escuela Politécnica Nacional, Quito, Ecuador
e-mail: jenny.torres@epn.edu.ec

K. Daimi et al. (eds.), *Emerging Trends in Cybersecurity Applications*,
https://doi.org/10.1007/978-3-031-09640-2_8

This paper presents DataCookie, a model formed by 6 phases based on the CRISP-DM methodology to analyze cookies generated in the user's computers using data mining. The information collected is structured in a data set to execute decision tree classification algorithms using the Weka tool. In the case of decision trees algorithms, we have compared the performance of all available algorithms to select the most appropriate. Finally, we obtained the rules to classify new cookies according to the influence of the selected parameters.

The rest of the article proceeds as follows. Section 1 presents a brief description of the related work; Sect. 2 details the model proposed, DataCookie. Section 3 analyzes the results obtained, and finally, Sect. 4 concludes the paper and presents future work.

This proposal focuses on analyzing cookies using data mining techniques. In the literature, we found studies focused on analyzing cookies from different perspectives. A proposal called Cookie Miner [4] is a novel online system that aims to automatically reconstruct the web downloading chains and find the entry points. Another framework [11] uses a technique called multi-platform memory mapping to build data exploits against Internet Explorer and Chrome. Even if both proposals analyze cookies, none of them uses any data mining technique.

In [5], the authors evaluate the threats through a significant analysis of anonymous cookies provided by Cookiepedia.co.uk. The main objective of this proposal is to improve the general understanding of the behavior of its configuration and its privacy implications. Extracting information, such as visited links, user credentials, and session cookies [6]. However, despite extensive Internet traffic data, existing network data mining tools do not support encrypted network traffic or new protocols.

Although no solutions have been found that use data mining to analyze cookies, we have found solutions that make use of data mining of networks. Is the case of ClickMiner [7] and NetworkMiner [8]. These do not admit all the protocols and can provide inaccurate results, mainly because of the analysis of large amounts of data extracted from the network. Under the same context, there is a novel unsupervised methodology [13], which takes advantage of application-level traffic logs to automatically detect services running a tracing activity, thus allowing the generation of blocklists.

Another approach to solving the analysis of large amounts of data extracted from the network is proposed by [15] which shows the use of HTTP cookies using network traces through a data set consisting of more than 5.6 billion HTTP connections. They developed TrackAdvisor as a method that uses automatic learning to identify HTTP requests that transmit confidential information to third parties with very high accuracy. In the same context, there is a modular framework [10], which is designed to assist forensic investigators in their procedures. Its purpose is to extract data from a user's social network profiles, taking advantage of stored credentials and session cookies.

Sivakorn et al. [12] propose to evaluate different websites and explore the functionality and information that is exposed to attackers who have stolen users' HTTP cookies. Another research was found based on hiding the identity of the user using the Tor browser [16]. This study revealed a number of serious shortcomings, the attackers were able to obtain information such as: home address, work address, also they were able to visit Google or Bing websites that expose the user's entire search history and send emails from the victim's account. Finally, cookie management protocols were also found. In this sense, TTPCookie [14] evaluates a set of data obtained from automated visits to 5000 websites, which allowed them to obtain a behavioral orientation with low privacy risks.

After analyzing these works we have not found any that make cookies mining using the history of users browsing in a public institution, so our work contributes in the analysis of XSS vulnerabilities, with this will generate rules through cookie mining to make decisions.

2 DataCookie Model

To propose the DataCookie model, the methodology CRISP-DM was chosen as a framework for executing the data mining process. Our goal with this proposal was to build a model for sorting cookies based on different attributes selected from users' navigation databases. DataCookie consists of 6 phases, shown in Fig. 1:

a. **System Setup**: objectives and requirements are defined. The context is described to set goals.
b. **Processing and cleaning**: the initial data collection and exploration is performed.
c. **Data Analysis**: the data is prepared to obtain the data set that will be analyzed.
d. **Modeling**: this phase is concerned with applying the techniques of data mining to the structured data set.
e. **Performance Evaluation**: in this phase, the results are evaluated. A review of the process is made and subsequent steps or actions to be taken are established.
f. **Deployment**: this last phase takes advantage of the utility of the models used to make decisions.

2.1 System Setup

This analysis was executed in a public institution. The navigation history of the 400 users of a typical workday was obtained to analyze all the domains and subdomains generated; the WSA (Cisco Ironport) Web Security Control data was used. The privacy of the users was not affected because, as network administrators, we can review the browsing history of the users in order to create new rules or check the non-compliance of the already configured ones.

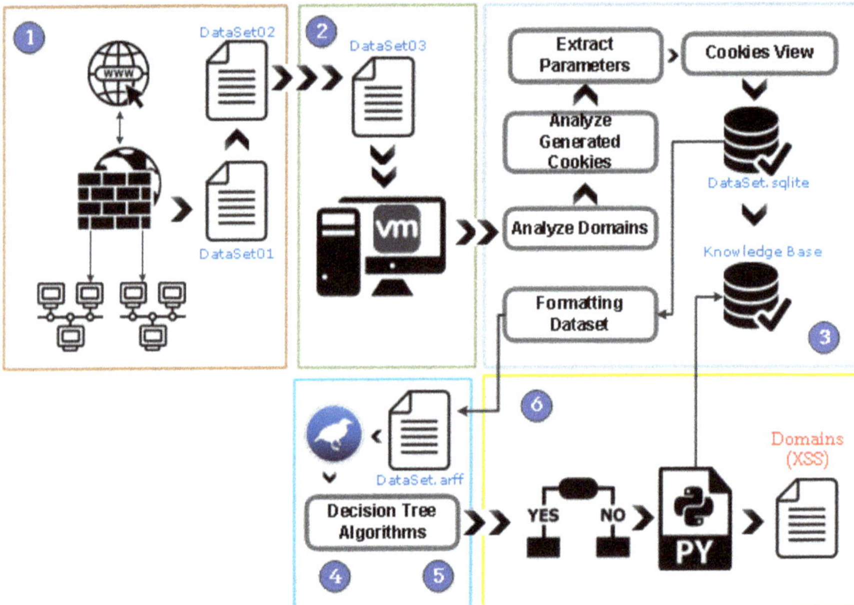

Fig. 1 DataCookie: proposed model for obtaining rules from cookie mining

URL	URL CATEGORY	Disposition	Threat Type
http://178.255.156.106/dou	Infrastructure and C	Allow	-
"https://tt.onthe.io:443/?k	Business and Indust	Allow	othermalware
http://ctldl.windowsupdat	Software Updates	Block - URL Cat	-
https://a.wunderlist.com:4	Computers and Inte	Allow	-
"https://graph.facebook.cc	Social Networking	Block - URL Cat	-
http://comcluster.cxense.(	Computers and Inte	Allow	-
http://ds.download.windo	Software Updates	Allow	-
https://4-edge-chat.facebc	Social Networking	Allow	-
https://microsoft.stream1.	Infrastructure and C	Allow	-

Fig. 2 Resume of records obtained from the web security appliance (DataSet01)

2.2 *Processing and Cleaning*

This first record contains the browsing history of all users. As can be seen in Fig. 2, it contains the URL, URL Category, Disposition, and Threat Type fields. According to the preset rules of the appliance, there are 77 categories of navigation. A total of 231,094 records were found.

This first **DataSet01** has repeated domains, for it was programmed a script in Python to clean the file and obtain all the domains without repetition. A total of

URL	URL CATEGORY	Disposition	Threat Typ
https://settings.luckyorange.net:4	Infrastructure and C	Allow	adware
"https://tt.onthe.io:443/?k[]=3937	Business and Indust	Allow	othermalware
https://static.ads-twitter.com:443/	Advertisements	Allow	adware
https://settings.luckyorange.net:4	Infrastructure and C	Allow	adware
http://cdn.luckyorange.com/w.js	Computers and Inte	Allow	othermalware
"https://tt.onthe.io:443/?k[]=3937	Business and Indust	Allow	othermalware
https://settings.luckyorange.net:4	Infrastructure and C	Allow	adware
"https://tt.onthe.io:443/?k[]=3937	Business and Indust	Allow	othermalware
https://static.ads-twitter.com:443/	Advertisements	Allow	adware
https://settings.luckyorange.net:4	Infrastructure and C	Allow	adware

Fig. 3 Resume of records obtained to structure the DataSet03

134951 records were obtained for the second dataset **DataSet02**. For each category, the disposition rules are: *allow*, *block*, and *monitor*.

For the following dataset **DataSet03** in the Fig. 3, the records that had the Disposition field with the Allow rule and that present in its Threat Type field some threat (Adware, other Malware, phishing, Spam, Trojan) were filtered. Recall that the same appliance. With 6432 records, the third data set was formed to continue with the analysis of the domains found. The Python script was again applied to eliminate duplicate records, and a final data set with 5113 records was obtained. In this process, it was found that 8 of the 77 categories were allowing access to pages with some threat. The categories found were: Advertisements, Astrology, Block, File Transfer, Games, Hacking, Illegal Activities, Uncategorized.

The next step was to find out if any of these domains had XSS vulnerability. The cookies generated by these domains in a virtual machine configured with a Windows 7 operating system with 1GB of RAM and one processor were analyzed. The NAT mode in the network configuration was set to protect the virtual environment. This machine was used to manually access all the domains of the DataSet03 through the Mozilla Firefox browser. As a result, 1666 cookies were obtained. It is worth mentioning that *not all registered domains generated cookies* because these correspond to those that require a user and password to log in, for example, social networks and email (session cookies).

2.3 *Data Analysis*

The RFC 6265 (HTTP State Management Mechanism) provides the normative reference to interpret each parameter of the cookies:

Name: the name of cookie

Value: the value of the cookie. This value is stored on the client computer. It does not store sensitive information.

Expire: the expiration time of the cookie. This is a UNIX time stamp. It is a number represented in seconds.
Path: the path on the server where the cookie will be available. If set to the root (/), the cookie will be available throughout the domain.
Domain: the domain for which the cookie is available. By setting this to a sub domain (www.example.com), the cookie will be available for that sub domain and all other sub-domains within this (w2.www.example.com).
Secure: shows that the cookie should only be transmitted over a secure connection from the client. When set to TRUE, the cookie will only be set if a secure connection exists.
HttpOnly: when set to TRUE, the cookie will only be accessible through the HTTP protocol. This means that the cookie will not be accessible by scripting languages, such as JavaScript. This configuration can effectively help to reduce identity theft through XSS attacks (although it does not support all browsers).

To complete this phase, CookieView [17] and SQLite Studio tools were used. The first one analyzes the parameters of registered cookies acting as a cookie viewer. With the second tool, a central database (**DataSet.sqlite**) was structured to perform the mining work. Initially, these values were exported to a minable data set. The main fields selected were: *domain, name, value, isSecure*, and *isHttpOnly*. Because the original fields corresponding to *ExpirationDate, LastAccess*, and *CreationDate* were in Coordinated Universal Time (UTC) format, we did an operation to transform to format *dd/mm/yr–hh:mm:ss*. However, only the *ExpirationDate* field was used, and an operation was added to have a value of TRUE if the expiration date was greater than two years; otherwise, FALSE.

2.4 Modeling

In order to analyze this data in Weka, the records of the data set were exported to .CSV format and later edited in a notepad to format them as follows:

@attribute Category: It corresponds to the categories of Table 1 whose access policy was defined in *"allow."* This attribute was added as a new column in the database to establish the classes according to the cookies generated by the domains corresponding to each category.
@attribute Path (root, noroot): If the cookie is saved in the root directory or a subdirectory within the root.
@attribute isSecure (TRUE, FALSE): if the cookie is sent via HTTP (FALSE) or HTTPS (TRUE). The original values of this parameter were 0 and 1.
@attribute isHttpOnly (TRUE, FALSE): the value is TRUE if it does not allow XSS attacks, otherwise is FALSE. The original values were 0 and 1.
@attribute Mayor2A (TRUE, FALSE): if the original expiration date is greater than 2 years it will be TRUE, otherwise FALSE.

Table 1 Final data set attribute description

Attribute	Description	Possible values
Category	Set the classes according to the cookies generated by the corresponding domains	Advertisements, Astrology, Block, File Transfer, Service, Games, Hacking, Illegal Activities, Uncategorized URL, etc.
Path	If the cookie is saved in the root directory	root, noroot
isSecure	If the cookie is sent over HTTP or HTTPS	TRUE, FALSE
isHttpOnly	If it does not allow XSS attacks	TRUE, FALSE
High2A	If expiration date is greater than 2 years	TRUE, FALSE
Suspect	If there is redirection between domains, or have JavaScript files	YES, NO

@attribute Sospechoso (YES, NO): If there is any redirection between domains or files with JavaScript extension **.js*, the cookie will be qualified as suspicious with the value YES. This investigation [18] was the basis to classify as Suspicious the records of our **DataSet.arff**, which fulfilled the conditions proposed in the conceptual model of this first study.

Finally, the file already formatted with extension ***.arff** is saved, which is the format that the Weka software understands. A preview of the final data set is shown in Fig. 4.
Table 1 presents a summary of the description of the attributes and the values that have been established.

In the next phase, the following algorithms in Weka software were selected to run data mining:

- RepTree
- J48
- RandomForest
- DesicionStump
- HoeffdingTree
- LMT
- RandomTree

The comparison was made with all the models of the Tree section in this software. The objective was to analyze the performance of each one and make a comparison to choose the best algorithm. The rules were extracted from the resulting decision tree to make decisions to classify new cookies.

The objective of the selected algorithm was to predict if a new cookie is suspicious or not, and we wanted to know how good the prediction would be with future data; for this, we used the Cross-Validation technique with 10 Iterations. This technique

```
%COOKIES CLASIFIER FOR XSS PREVENTION
@relation 'Clasificador de Cookies'

%PARAMETROS DE LAS COOKIES DE ACUERDO A RFC
@attribute Categoria {Advertisements,Astrology,
@attribute Path {root,noroot}
@attribute isSecure {TRUE,FALSE}
@attribute isHttpOnly {TRUE,FALSE}
@attribute Mayor2A {TRUE,FALSE}
@attribute Sospechoso {YES,NO}

%DATOS DE LAS COOKIES RECOGIDAS
@data
'Illegal Activities',root,FALSE,TRUE,FALSE,YES
'Illegal Activities',root,FALSE,FALSE,FALSE,YES
'Illegal Activities',root,FALSE,FALSE,FALSE,YES
'Illegal Activities',root,FALSE,FALSE,FALSE,YES
'Illegal Activities',root,FALSE,FALSE,FALSE,YES
```

Fig. 4 Preview of the final DataSet.arff format for to analyze in Weka software

aims to overcome the over-fitting problem and make the predictions more general. Thus the data set is divided into ten equal parts (folds); each fold is used once for testing, nine times for training.

2.5 Performance Evaluation

In this phase, we evaluated and compared the performance of the decision trees used in Weka. As seen in Tables 2 and 3, we have selected the following parameters for analysis:

- Correctly Classified Instances (Correctly)
- Incorrectly Classified Instances (Incorrectly)
- True Positive Rate (TP Rate)
- False Positive Rate (FP Rate)
- Precision
- ROC Area (AUC)

The number of correctly and incorrectly instances is the basis for determining the model's accuracy. Instances that are correctly classified equal the sum of True

Table 2 Results output for WEKA Decision Tree algorithms (YES class)

Model	Correctly	Incorrectly	TP rate	FP rate	Precision	ROC area
Reptree	1578	88	0.714	0.005	0.967	0.832
J48	1583	83	0.714	0.001	0.990	0.834
Random Forest	1583	83	0.714	0.001	0.990	0.927
DesicionStump	1577	89	0.714	0.006	0.952	0.829
HoeffdingTree	1581	85	0.714	0.003	0.981	0.832
LMT	1583	83	0.714	0.001	0.990	0.921
RandomTree	1581	85	0.714	0.003	0.981	0.929

Table 3 Results output for WEKA Decision Tree algorithms (NO class)

Model	Correctly	Incorrectly	TP Rate	FP rate	Precision	ROC area
Reptree	1578	88	0.995	0.286	0.944	0.832
J48	1583	83	0.999	0.286	0.945	0.834
Random Forest	1583	83	0.999	0.286	0.945	0.927
DesicionStump	1577	89	0.994	0.286	0.944	0.829
HoeffdingTree	1581	85	0.997	0.286	0.945	0.832
LMT	1583	83	0.999	0.286	0.945	0.921
RandomTree	1581	85	0.997	0.286	0.945	0.929

Positive (TP) and True Negative (TN). In the same way, those classified incorrectly are the sum of the instances that are False Positive (FP) and False Negative (FN). The total number of instances correctly divided by the number of instances provides the precision. In other words, precision is the proportion of positive cases that predicted that were correct.

Each algorithm belonging to the Trees group of algorithms was executed. In each execution, the data of the following variables were collected: Correctly Classified Instances, Incorrectly Classified Instances, Mean absolute error, root mean squared error, relative absolute error, Root relative squared error, Precision, and Time to build a model.

In Tables 2 and 3 we present the results of the values obtained for each decision tree model after running the tests with the DataSet.arff.

The ROC curves are two-dimensional graphs in which the true positive rate (TPR) is plotted on the Y-axis, and the false positive rate (FPR) is plotted on the X-axis (Figs. 5, 6, and 7). The excellent performance of our sorter is reflected by a ROC curve which lies in the upper left triangle of the square (Fig. 6a and b). AUC provides a value description for the performance of the ROC curve. Our AUC value is a portion of the area of the unit square, so its value will always be between 0, and 1 and usually more significant than 0.5 [19], according to [20] the area under a ROC curve (AUC) is a criterion used in many applications to measure the quality of a classification algorithm. A very poor sorter has an AUC of around 0.5 (no discrimination, it is chosen at random) and 1 (perfect discrimination).

	RepTree	
Correctly Classified Instances	1578	94,7179%
Incorrectly Classified Instances	88	5,2821%
Mean absolute error	0,0978	
Root mean squared error	0.2233	
Relative absolute error		34,6592%
Root relative squared error		59.4562 %
Precision		94,8%
Time to build model	0.01s	

Fig. 5 Results output For WEKA RepTree algorithm

	J48	
Correctly Classified Instances	1583	95.018%
Incorrectly Classified Instances	83	4.982 %
Mean absolute error	0.0942	
Root mean squared error	0.2172	
Relative absolute error		33,3844 %
Root relative squared error		57.834 %
Precision		95.2%
Time to build model	0.04s	

Fig. 6 Results output For WEKA J48 algorithm

	Random Forest	
Correctly Classified Instances	1583	95.018%
Incorrectly Classified Instances	83	4.982%
Mean absolute error	0.0898	
Root mean squared error	0.2127	
Relative absolute error		31.8139%
Root relative squared error		56.6393%
Precision		95,2%
Time to build model	0.15s	

Fig. 7 Results output For WEKA RandomForest algorithm

To select the appropriate model, we analyze the values of Tables 2 and 3. The Decision Stump, HoeffdingTree, LMT, and RandomForest models did not offer a visible result of their final tree structure in the Weka software, so these models were discarded.

We are left with the J48, RepTree, and RandomTree models; according to Tables 2 and 3, the RandomTree model has an AUC of 0.929, which is higher than the J48 and RepTree models; however, its accuracy is lower than the J48 model. Also, the J48 model is more accurate than the RandomTree and RepTree models.

We also analyze the correctly classified instances; model J48 has a higher value (1583) than RepTree (1578) and RandomTree (1581). Likewise, the J48 model incorrectly classifies fewer instances (83) than the RepTree (88) and RandomTree

	DesicionStump	
Correctly Classified Instances	1577	94.6579%
Incorrectly Classified Instances	89	5.3421%
Mean absolute error	0.1011	
Root mean squared error	0.225	
Relative absolute error		35,8184%
Root relative squared error		59,9205%
Precision		94,7%
Time to build model	0.01s	

Fig. 8 Results output For WEKA DesicionStump algorithm

	HoeffdingTree	
Correctly Classified Instances	1581	94.89%
Incorrectly Classified Instances	85	5.102%
Mean absolute error	0.0967	
Root mean squared error	0.2195	
Relative absolute error		34,24%
Root relative squared error		58,45%
Precision		95,1%
Time to build model	0.02s	

Fig. 9 Results output For WEKA Hoeffding algorithm

(85) models. The selected model will then be J48 to obtain the programming rules of our script. Figure 8 shows the decision tree structure obtained with the J48 model.

2.6 *Deployment*

Our goal was to create a data set from the cookies generated by analyzing users' browsing history. This data set helped to train our model responsible for classifying new cookies from a new data set of navigation obtained from the web control appliance. The rules obtained from the decision tree algorithms were applied in phase 3 of our model (Data Analysis) because the navigation records and the cookies information that had been generated were stored there (DataSet.SQLite). In addition, in phase 3, we structure our data set with the attributes of selected cookies and those we add (Category).

The rules obtained from the decision tree, using the J48 algorithm, allowed us to develop a script using Python to obtain the domains to which the cookies classified as suspicious belong. In Fig. 9, a preview of the code developed to structure these rules is shown, while Fig. 10 shows a view of the output obtained when executing this script.

This script was executed with a new structured data set obtained following the sequence of phase 1 but this time obtaining the navigation record of a different day.

Fig. 10 Results output For WEKA LMT algorithm

	LMT	
Correctly Classified Instances	1583	95.018%
Incorrectly Classified Instances	83	4.98%
Mean absolute error	0.0968	
Root mean squared error	0.2131	
Relative absolute error		34,28%
Root relative squared error		56,75%
Precision		95,2%
Time to build model	0.68s	

Subsequently, the cookies were registered. The final data set was joined by 1185 new cookies registered and managed to rate 26 domains as suspicious through the analysis of their cookies.

3 Analysis of Results

3.1 Security Analysis

In a previous research [18], we propose an analytic model to classify cookies in order to prevent XSS. Our objective was to analyze the information of the cookies generated by a website committed to a controlled attack, classifying them according to their attributes. In that research, we showed how XSS attacks could be executed through compromised JavaScript files (*.js).

In this new research, the data collection performed in phase 2 showed that the pages visited generated suspicious cookies and JavaScript files. As shown in Fig. 11, we found a script that can be used to execute an XSS attack. The disadvantage, in this case, is that it is sent in plain text because of HTTP protocol, which means that with a network sniffer like Wireshark, it can be easily captured.

Figure 12 shows one of the domains found in the *Advertisements* category that contains a phishing attack and whose cookie can be seen in Fig. 13. It was observed that this page was redirected by an original domain called codeonclick.com.

The results obtained in phase 3 let us to structure the data set with the analyzed The results obtained in phase 3 let us structure the data set with the analyzed attributes of the cookies. SQLite offered an easy administration of the cookie database that the Mozilla Firefox browser generated. In phases 4 and 5, RepTree, J48, RandomForest, DesicionStump, HoeffdingTree, LMT, and RandomTree algorithms were evaluated. It was observed that the J48 algorithm is more accurate than the RepTree and RandomTree models; it also offers a more significant number of correctly classified instances and a smaller number of incorrectly classified instances.

With the script developed in phase 6, it was found that for a new data set of 1185 records, only 26 were classified as suspicious cookies, which means that 2.19% of the websites visited by users in a day of navigation present an XSS vulnerability.

Fig. 11 Results output For WEKA RandomTree algorithm

	RandomTree	
Correctly Classified Instances	1581	94.89%
Incorrectly Classified Instances	85	5.10%
Mean absolute error	0.08	
Root mean squared error	0.2123	
Relative absolute error		31,32%
Root relative squared error		56,52%
Precision		95,1%
Time to build model	0.1s	

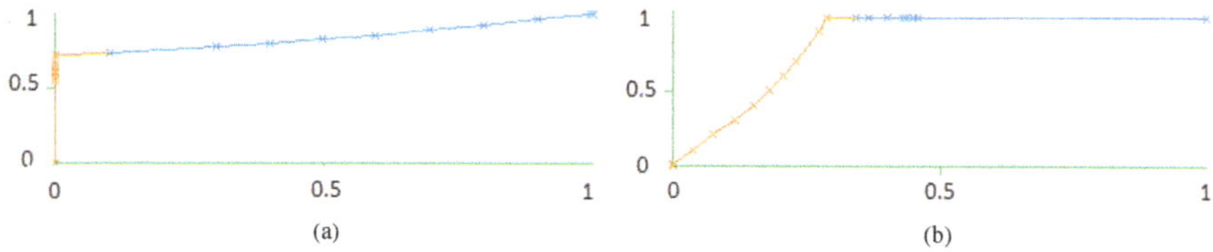

Fig. 12 ROC area for J48 model—YES and NO class. (**a**) J48 YES class. (**b**) J48 NO class

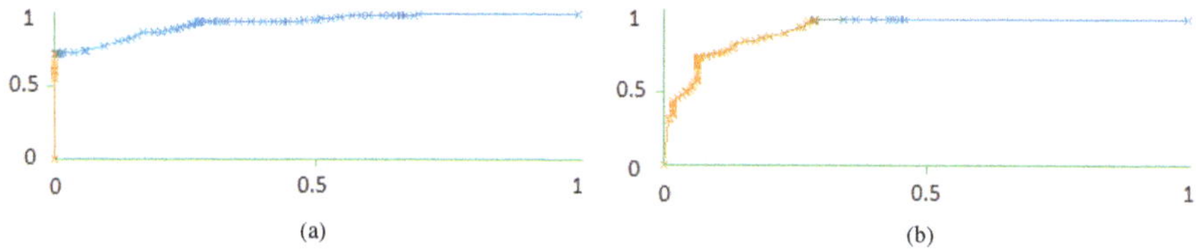

Fig. 13 ROC Area for RandomTree model—YES and NO Class. (**a**) RandomTree YES class. (**b**) RandomTree NO class

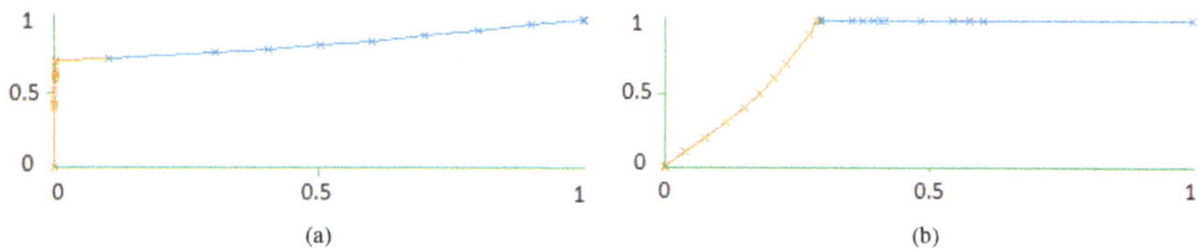

Fig. 14 ROC area for RepTree model—YES and NO class. (**a**) RepTree YES class. (**b**) RepTree NO class

This script helped us improve user browsing categories and block the suspicious domains to which the analyzed cookies belonged. The results obtained in each phase of the proposed model show that the selected classification algorithm is a good rule generator for classifying cookies.

The results show statistics of users of a public institution whose infrastructure is protected by security equipment for a web control. With these results, new rules were generated in the security appliance that allows filtering the domains of the web pages that the script qualified as suspicious. An attacker could take advantage of a Wi-Fi connection and enter the corporate network to be able to steal these cookies

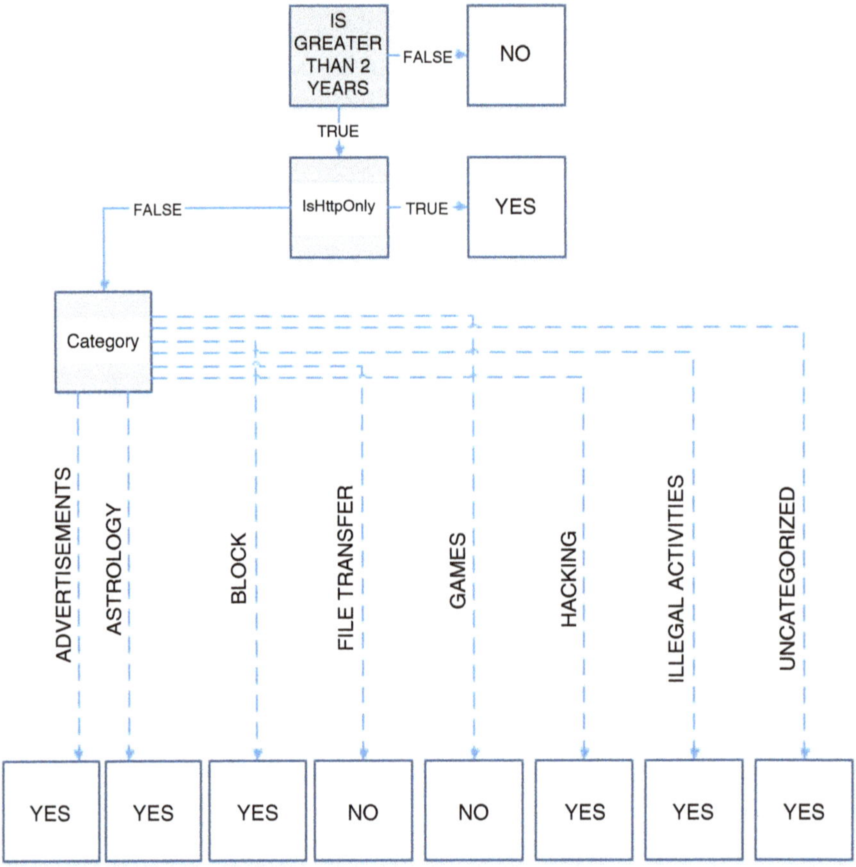

Fig. 15 Structure of the decision tree obtained with the J48 algorithm

from the same LAN of the victims. With our algorithm developed in Python, the appliance could identify cookies from suspicious domains and delete them, so they are not stolen.

3.2 Efficiency Analysis

Regarding the minimum amount of resources used to fulfill the goal of this research, we can say that our script is efficient. The result discovers domains vulnerable to XSS attacks simply by analyzing their cookies. We use tools that can later be replaced by code developed in any programming language. The basis of our script was the rules extracted by running the different algorithms in Weka, which did not represent acquisition costs of licenses or additional resources.

Fig. 16 View of Python algorithm to obtain the domain of cookies classified as suspicious according to the rules obtained with Weka

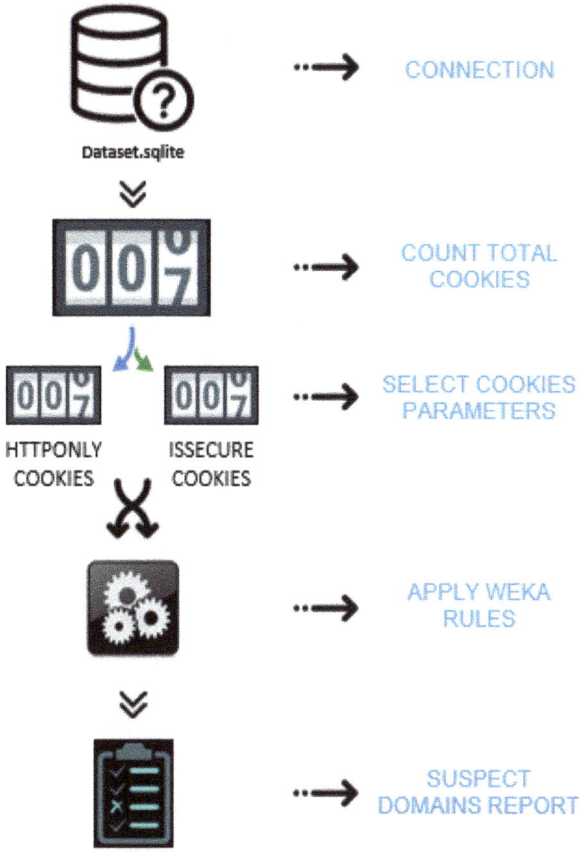

If we compare our algorithm with free tools like Galleta, which was developed to examine the content of cookies, we can see that this tool is oriented only to Internet Explorer. Instead, our algorithm can connect to the databases of cookies of any browser. On the other hand, this software does not make data mining with cookies, and use documentation is limited.

We have analyzed information on tools to erase cookies, for example, on the official sites of the most used browsers (Chrome, Firefox, Edge, Opera) and showed tutorials on cleaning the cookies that are created on our computers, that is, this cleaning does not make extraction of data from cookies or an analysis of its attributes. As we mentioned in our analysis of related works, some proposals analyze the attributes of cookies, but we have not found information related to the mining of cookies to prevent XSS attacks.

```
Shell

>>> %Run sqlite-pythonControl.py

  Cookie database opened successfully
  Number of cookies found: 1185
  Number of cookies isHttponly type found: 93
  Number of cookies isSecure type found: 27

  Applying rules obtained from WEKA...

  Suspect Domain found: codeonclick.com
  Suspect Domain found: codeonclick.com
  Suspect Domain found: switchads.com
  Suspect Domain found: yandex.ru
  Suspect Domain found: yandex.ru
  Suspect Domain found: powerlinks.com
  Suspect Domain found: powerlinks.com
  Suspect Domain found: pubmatic.com
  Suspect Domain found: deployads.com
  Suspect Domain found: afsanalytics.com
  Suspect Domain found: admedo.com
  Suspect Domain found: adriver.ru
  Suspect Domain found: adroll.com
```

Fig. 17 Suspect domains report, output after executing the script developed in Python

URL	URL CATEGORY	Disposition	Bandwidth
http://pagead2.googlesyndication.com/pagead/sho	Advertisements	Allow	696

Fig. 18 Domain which contains JavaScript files extension susceptible to XSS attacks

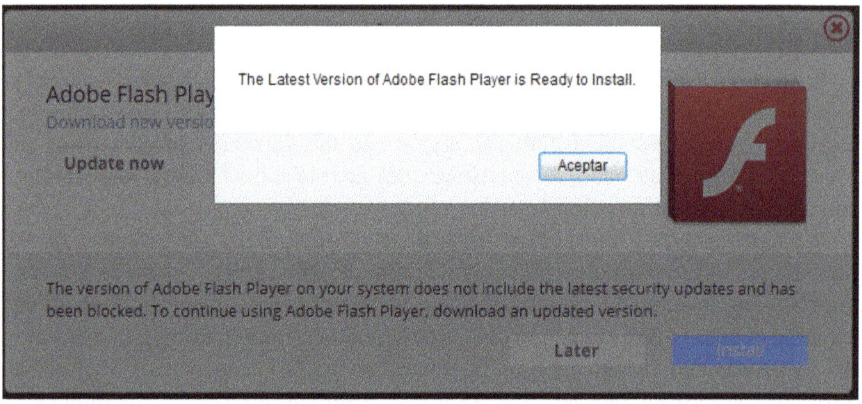

Fig. 19 Domain of the advertisements category that presents a phishing attack with an update message of the program Adobe Flash Player

Domain/Host	Path	Name	Value	Expiration Date	Secure
installupgradenow.bigtrafficsystemsforupgrades.date	/	channel	ronn_pc_ach_a1	9/16/2017 7:05:14 PM	No

Fig. 20 Cookie generated for the domain codeonclick.com and redirected to the domain http://installupgradenow.bigtrafficsystemsforupgrades.date

4 Conclusions and Future Work

This paper presents a model based on the CRISP-DM methodology for data mining used for analyzing cookies. The data were obtained using navigation reports from a Web Security Appliance (IT equipment) in a public institution. All the cookies generated by these domains were registered through a virtual environment. The following parameters were evaluated: its date of creation, date of expiration, isHttpOnly, isSecure and *isRoot*; and a training data set was constructed with a new attribute called "Category."

The DataCookie model was structured in 6 phases where different tasks were performed to meet each objective. An essential phase allowed the design of the data set with information on the cookies generated based on the users' browsing history. With this data set generated, the Weka tool was used to apply algorithms based on decision trees to obtain rules that allow classifying new cookies as suspicious or not. With the rules obtained from the J48 tree decision algorithm, a script was developed in Python to classify a new cookie registry and register the domain of those classified as suspicious.

In future work, we will think about the automation of the proposed framework using Python. In this way, the rules would be applied through an agent installed directly on each user's computer. Our vision is to develop a central system that can manage all the database files where the cookies of all users are stored in real-time. This system would have a knowledge base of classified cookies that would provide feedback each day with new information.

Acknowledgments For the development of this work, the data were obtained from the Instituto Nacional de Estadísticas y Censos (INEC) from Ecuador, which served as the basis for the analysis and search of the parameters obtained, thus motivating innovation in the public sector and the proposal of new blocking rules for the computer security area.

References

1. N.S. Alkhatri, N. Zaki, E. Mohammed, M. Shallal, The use of data mining techniques to predict the ranking of e-government services, in *2016 12th International Conference on Innovations in Information Technology (IIT)*, pp. 1–6 (2016)
2. Upf.edu-UPF, The hidden face of the internet (2017). https://www.upf.edu/hipertextnet/numero-1/internet.html, September 2017

3. F. Roesner, T. Kohno, D. Wetherall, Detecting and defending against third-party tracking on the web, in *Proceedings of the 9th USENIX Conference on Networked Systems Design and Implementation*, NSDI'12, pp. 12–12 (2012)
4. Z. Chen, P. Zhang, C. Zheng, Q. Liu, Cookieminer: Towards real-time reconstruction of web-downloading chains from network traces, in *2016 IEEE International Conference on Communications (ICC)*, pp. 1–6 (2016)
5. A. Cahn, S. Alfeld, P. Barford, S. Muthukrishnan, What's in the community cookie jar? in in *2016 IEEE/ACM International Conference on Advances in Social Networks Analysis and Mining (ASONAM)*, pp. 567–570 (2016)
6. A. Amro, S. Almuhammadi, S. Zhioua, Netinfominer: High-level information extraction from network traffic, in *2017 IEEE International Conference on Big Data and Smart Computing (BigComp)*, pp. 143–150 (2017)
7. C. Neasbitt, R. Perdisci, K. Li, T. Nelms, Clickminer: Towards forensic reconstruction of user-browser interactions from network traces, in *Proceedings of the 2014 ACM SIGSAC Conference on Computer and Communications Security*
8. E. Hjelmvik, Passive network security analysis with NetworkMiner. Secure (18), 1–100 (2008)
9. R. Diaz-Morales, Cross-device tracking: Matching devices and cookies, in *2015 IEEE International Conference on Data Mining Workshop (ICDMW)*, pp. 1699–1704 (2015)
10. I. Polakis, P. Ilia, Z. Tzermias, S. Ioannidis, P. Fragopoulou, Social forensics: Searching for needles in digital haystacks, in *2015 4th International Workshop on Building Analysis Datasets and Gathering Experience Returns for Security (BADGERS)*, pp. 54–66 (2015)
11. R. Rogowski, M. Morton, F. Li, F. Monrose, K.Z. Snow, M. Polychronakis, Revisiting browser security in the modern era: New data-only attacks and defenses, in *2017 IEEE European Symposium on Security and Privacy (EuroS P)*, pp. 366–381 (2017)
12. S. Sivakorn, I. Polakis, A.D. Keromytis, The cracked cookie jar: Http cookie hijacking and the exposure of private information, in *2016 IEEE Symposium on Security and Privacy (SP)*, pp. 724–742 (2016)
13. H. Metwalley, S. Traverso, M. Mellia, Unsupervised detection of web trackers, in *2015 IEEE Global Communications Conference (GLOBECOM)*, pp. 1–6 (2015)
14. A. Javed, C. Merz, J. Schwenk, TTPCookie: Flexible third-party cookie management for increasing online privacy, in *2014 IEEE 13th International Conference on Trust, Security and Privacy in Computing and Communications*, pp. 37–44 (2014)
15. R. Gonzalez, L. Jiang, M. Ahmed, M. Marciel, R. Cuevas, H. Metwalley, S. Niccolini, The cookie recipe: Untangling the use of cookies in the wild, in *2017 Network Traffic Measurement and Analysis Conference (TMA)*, pp. 1–9 (2017)
16. T.G. Abbott, K.J. Lai, M.R. Lieberman, E.C. Price, Browser-based attacks on tor, in *International Workshop on Privacy Enhancing Technologies* (Springer, 2007), pp. 184–199
17. Nirsoft, Mzcookiesview v1.56. https://www.nirsoft.net/utils/mzcv.html, september 2017
18. G.E. Rodríguez, D.E. Benavides, J. Torres, P. Flores, W. Fuertes, *Cookie Scout: An Analytic Model for Prevention of Cross-Site Scripting (XSS) Using a Cookie Classifier* (Springer International Publishing, Cham, 2018), pp. 497–507
19. T. Fawcett, An introduction to roc analysis. Pattern Recogn Lett., 861–74 (2006)
20. C. Cortes, M. Mohri, AUC optimization vs. error rate minimization. Adv. Neural Inf. Process. Syst., 313–320 (2004)

Part III
Vehicle Applications Security

Analysing Cyber Attacks and Risks in V2X-Assisted Autonomous Highway Merging

Chao Chen, Ugur Ilker Atmaca, Konstantinos Koufos, Mehrdad Dianati, and Carsten Maple

1 Introduction

Progresses in highly reliable Vehicle-to-Everything (V2X) communication systems are expected to usher in the era of Connected Autonomous Vehicles (CAVs) [1, 2], which create opportunities for safer and more efficient implementations of autonomous driving functions. However, wireless connectivity can also open new attack surfaces in V2X-assisted autonomous vehicles, which must be understood and addressed before the commercialisation of CAV functions. While generic countermeasures (e.g., encryption and authentication) are considered in various forms of V2X systems, lessons from the past show that each CAV function should be separately analysed, and countermeasures should be customised for it to ensure its security and efficiency [3].

To this end, this paper focuses on security threats and risk assessment in a V2X-assisted autonomous highway merging function. The goal of such a function is to help an autonomous vehicle to safely merge into the highway from a slip road (i.e., freeway on-ramp merging) without unnecessary slowdowns or delays. In a typical implementation of such functionality, the CAVs are considered to be equipped with their own sensors. However, in many complex and safety-critical scenarios, on-board sensors can be impaired by obstructions, such as vegetation, or other road features, as illustrated by Fig. 1, that block the view of the on-board sensors. In such scenarios, the infrastructure sensors can make a crucial difference by detecting the oncoming vehicles along the highway and broadcasting their locations, velocities, etc., through a V2X system to the surrounding CAVs approaching the merging point. Such broadcast messages can then be fused by the CAV's on-board perception system

C. Chen · U. Ilker Atmaca · K. Koufos · M. Dianati (✉) · Carsten Maple
Warwick Manufacturing Group (WMG), University of Warwick, Coventry, UK
e-mail: c.chen.27@warwick.ac.uk; ugur-ilker.atmaca@warwick.ac.uk; konstantinos.koufos@warwick.ac.uk; m.dianati@warwick.ac.uk; cm@warwick.ac.uk

K. Daimi et al. (eds.), *Emerging Trends in Cybersecurity Applications*,
https://doi.org/10.1007/978-3-031-09640-2_9

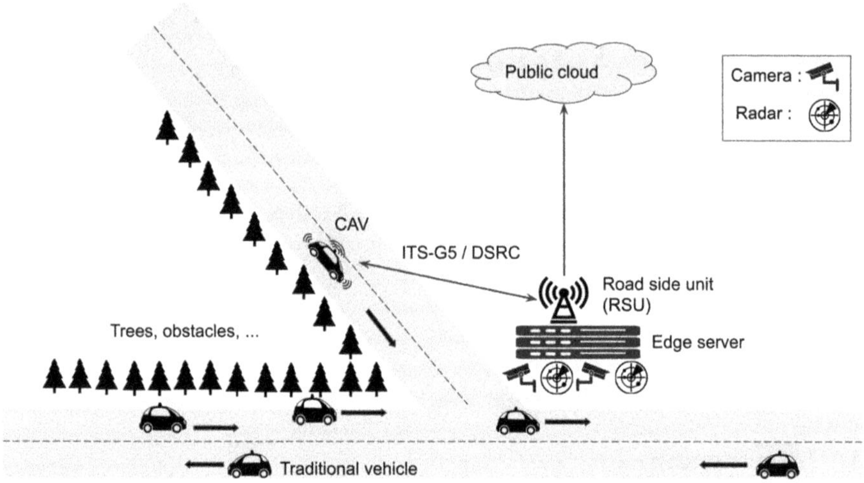

Fig. 1 An exemplary use case for V2X-assisted highway merging, where roadside plantation obscures the vision of on-board vehicular sensors

to create a highly accurate and reliable perception of the road segment, enabling a safer and efficient merging manoeuvre. However, from a security perspective, such a system has exposed entry points for malicious actors.

The existing literature on cyber security for CAVs has mostly considered the threats at the CAV side [4–7]. At the V2X-system side, the threat analysis has been conducted merely at the communication layer, see, for instance, [8], without considering potential attacks on the infrastructure sensors, as we will do in this paper. Also, no precise merging scenario has been analysed thus far. Therefore the existing cyber security analysis for CAVs is rather incomplete, and it does not provide any practical insights into ways of evaluating and ranking the risk of relative threats and their mitigation schemes. To the best of our knowledge, this paper is the first that applies an abuse case that combines multiple attacks to completely "blind" the road infrastructure within the V2X system. Our contributions are: (i) Proposition of a generic reference architecture (RA) for implementation of V2X-assisted autonomous highway merging, and (ii) analysis of the potential cyber security threats to the RA and assessment of their risks. Particularly, we analyse potential abuse cases against the road infrastructure consisting of radar and cameras. We discuss how a malicious actor can jam the radar and/or tamper with the image object detector in the camera. Finally, we suggest related mitigation schemes.

2 Related Work

The V2X functionality is expected to enhance road traffic efficiency and safety, and because of that, it has already attracted a considerable interest in the academic [2, 9] and industrial research communities [10]. The study in [2] has discussed the use of messages transmitted from the road side unit (RSU) over V2X to the CAV for helping to execute a lane merge. The messages contain the speeds and locations of surrounding vehicles to the CAV. Furthermore, the study in [9] has suggested using a camera and image object detector to build the semantic map of the environment which is sent over V2X to the merging CAV. Finally, the study in [10] has proposed to install both cameras and radars at the RSU. While the above studies conclude that merges can be executed efficiently and safely with the V2X system, unfortunately, they do not address the issues due to cyber security, which is the main topic of our work.

The literature on threat analysis and risk assessment for CAVs is vast, see [5, 7, 11–13] and the references therein. The study in [5] has defined potential threats, e.g., various types of spoofing, tampering, etc., identified what kind of expertise, knowledge, equipment and, window-of-opportunity is needed to realise the threat, and they have finally quantified and ranked the priority for each attack. The studies in [11, 13] have also used the controllability criterion while ranking the threats, e.g., a threat that can be easily controlled and avoided by the driver and/or by the CAV should be associated with low risk. The above studies have considered cyber security analysis only on the CAV's side. In addition, they have not addressed a specific autonomous driving function, e.g., V2X-aided highway merging, as we will do here.

Due to the high economic cost along with the technology readiness level for the lidar, we consider in our RA only cameras and radars as the primary sensors at the RSU. Radar jamming is widely used for military purposes, but it has recently started to get attention to exploit it for jamming on CAVs. For instance, the study in [4] has considered the transmission of artificial noise for jamming. The study in [14] has applied a radar signal analyser, frequency multiplier, and signal generator to successfully jam Tesla's radar, rendering it unable to detect any surrounding objects.

Exploiting the vulnerabilities of a machine learning model, particularly for the image object detector applied in the camera, has existed for long time at a theoretical level, but it has only recently become a reality [15, 16]. The studies in [15, 17] have printed an adversarial patch, and the study in [16] has attached a poster and a patch to the target object, making it "invisible" to the detector. In our abuse case, we will combine these methods of tampering with the image object detector and the jamming of the radar, to achieve the goal of the attack, which is a totally "blind" RSU.

3 Reference Architecture for Autonomous Highway Merging

Given the scenario depicted in Fig. 1, we have investigated related academic and industrial projects [2, 9], to come up with a proposal about the reference architecture (RA) for autonomous V2X-assisted highway merging. A fairly generic model is illustrated in Fig. 2.

The RSU is installed near the merging point, and the radars and cameras observe the traffic along the highway. Firstly, they stream raw data, i.e., images and radar signals, to the edge server via cable. Secondly, the edge server processes the received data using models such as the object detector and the radar signal analyser. The edge server fuses the outputs of the processing models and produces a list of objects including, e.g., their type, size, and speed. This information can then be used to generate the collective perception messages (CPMs) [18]. Finally, the RSU broadcasts the CPMs via ITS-G5 or dedicated short-range communications (DSRC) to the CAV. We would like to note that vehicle-to-vehicle cooperative driving messages for platooning control and manoeuvre coordination are not relevant to our RA.

The CAV at the slip road becomes aware of its surroundings and the situation along the highway through its on-board sensors, such as camera, radar, lidar, GNSS, etc. In addition, it establishes wireless connectivity with the RSU through its on-board unit (OBU) and receives the CPMs over V2X. The on- and off-board information is fused to construct the overall perception, which is then fed into the machine learning autonomous driving application to infer the most suitable trajectory and speed to perform the merge (vehicle control). While approaching the highway, the CAV keeps on receiving the CPMs and continuously updates its

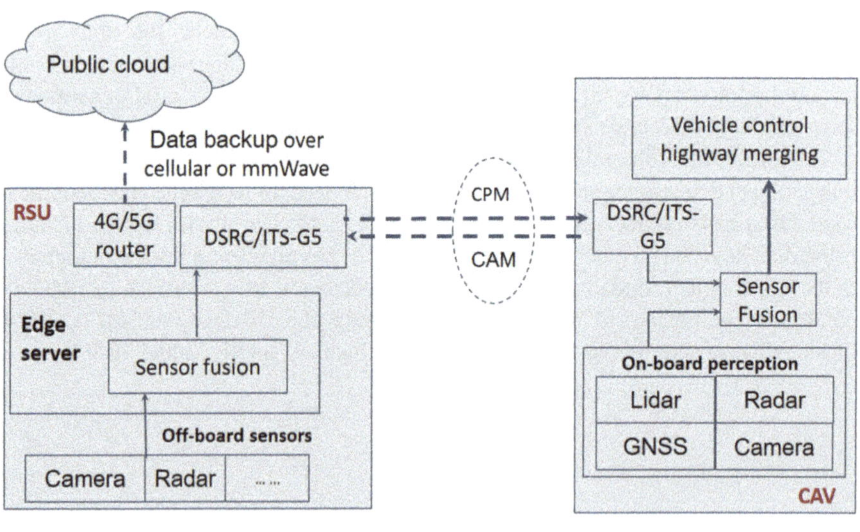

Fig. 2 Generic RA for V2X-assisted highway merging

knowledge about the highway traffic and its surroundings, and uses this to adjust its merging manoeuvre. Additionally, it keeps on broadcasting cooperative awareness messages (CAMs), and the RSU, in its turn, logs all the received data to the public cloud as a backup. For that, the RSU must also have cellular and/or mm-wave wireless connectivity to the cloud.

Unlike CPM, the CAM contains only local information about the location and the speed of the CAV. The installation of radars and cameras near the intersection essentially bypasses the need for sending CAMs to the RSU, hence, the proposed RA is also relevant for early penetration of CAVs.

4 Abuse Cases

In this section, we will describe potential abuse cases attacking the RA. This is to highlight the importance of understanding and mitigating potential threats for such systems before their commercial rollout. In order to do that, firstly, we will identify all attack surfaces. Subsequently, we will concentrate on attack surfaces at the RSU, because this subject has not been treated so far in the literature. In particular, we will consider attacking the processing models for the image object detector at the edge server, as well as jamming the radar sensor on the RSU. These abuse cases jeopardise road safety by resulting in the broadcast of imprecise CPMs, and subsequently lead to the execution of unsuitable, perhaps risky, manoeuvres at the CAV approaching the merging point. We illustrate these abuse cases in the form of an attack tree in Fig. 3, and explain them in detail next.

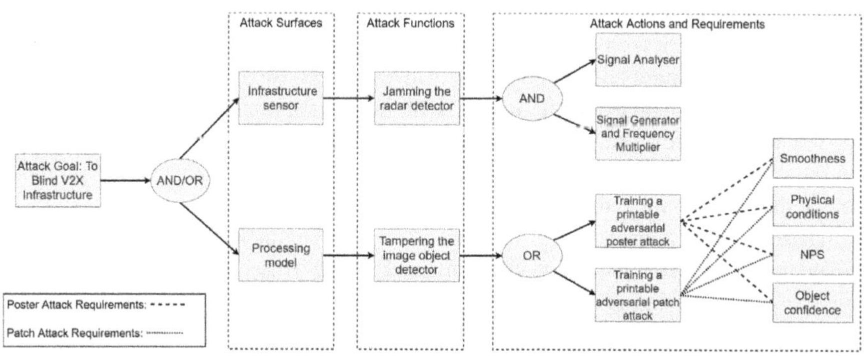

Fig. 3 The attack tree for the abuse cases

4.1 Attack Surface Analysis

The attack surface of a system is the set of vulnerable components (entry points), which a threat actor could exploit to attack. The RA in Fig. 2 has provided an abstract model of the high-level system components and their interactions, which can be used to identify the following attack surfaces in our system setup.

- **Vehicular sensors**: All on-board sensors (e.g., cameras, radar, lidar and GNSS, etc.) equipped on the CAV.
- **V2X communication**: The OBU at the CAV and the wireless communication router at the RSU.
- **Infrastructure sensors**: The off-board sensors at the RSU, such as cameras and radars.
- **Processing models**: The models within the edge server for processing the raw sensor data, e.g., the received radar signal and the raw image data.
- **Infrastructure edge server**: The computing unit at the edge server for hosting the sensor fusion models, e.g., to fuse the outputs from the radar signal analyser and the image object detector.

We have excluded the public cloud from the attack surfaces, because its communication to the RSU is unidirectional, see Fig. 2. Security issues at the public cloud do not affect the RSU and the CAV in our RA.

4.2 Tampering with the Input to the Image Object Detector

There are many ways to attack the sensor camera at the vehicle and/or the RSU, e.g., sensor blinding with light emitting diode [14]. However, tampering the objector detector in the camera has not been so far investigated. Next, we will firstly explore how does the image object detector work with images captured by the cameras. Then, we will describe how the threat actor can cast the attack as an optimisation problem on the pixels of the adversarial image. The goal is to construct an adversarial input image that may confuse the object detection system on the RSU.

There are two approaches to the execution of this attack. The first approach is called the “poster attack”, which generates a subtle adversarial poster confined to the surface area of the target object. The poster usually has the size of the target object, and it is inconspicuous, e.g., it looks like a graffiti, to the human eye. The second approach is called “patch attack”, which generates a small-sized patch attached on the target object. For example, adversaries may stick a misleading patch on the surface of an oncoming vehicle to prevent its detection on the camera; or they can utilise drones to reflect the patches and false traffic signs on vehicles or the road to induce accidents or stop the traffic flow. To generate the adversarial image, the threat actor needs to know the full details of the machine learning model of the object detector.

The risk of such threats can be mitigated by ensuring confidentiality of the utilised object detector model.

(1) Background on Image Object Detectors Image object detection is a classical task in computer vision. It does not only classify the objects in an image, but it also generates the coordinates of the bounding boxes for each detected object. In this work, we will focus on the state-of-the-art Yolov2 detector, see [19] which is widely employed nowadays because it is faster than other detectors without compromising on accuracy.

The Yolov2 firstly divides the input image in a 19×19 grid. For each grid cell, it generates five anchor boxes of different sizes, which can capture various objects within the cell, see Fig. 4. Each anchor box produces a prediction box, which contains: (i) the coordinates of the bounding box, (ii) a confidence score which measures the certainty that the given anchor box contains an object, and (iii) the probability values that the detected object belongs to each of the 80 object classes. The object detector will finally discard all prediction boxes whose confidence score is lower than a predefined threshold, e.g., the default Yolov2 uses 70% confidence level to produce the final prediction result.

(2) Physical Adversarial Image In order to generate an adversarial image, the threat actor must be able to evaluate the following metrics: The smoothness of pixels, the object confidence score, and the non-printability score (NPS) [20], see the bottom branch of the attack tree in Fig. 3. The threat actor aims to minimise a function of these metrics, usually their linear weighted sum, along with some physical condition transformations. Here, different transformations reflect different physical conditions on the input image, such as varying distance, angle, rotation, and brightness [21]. By

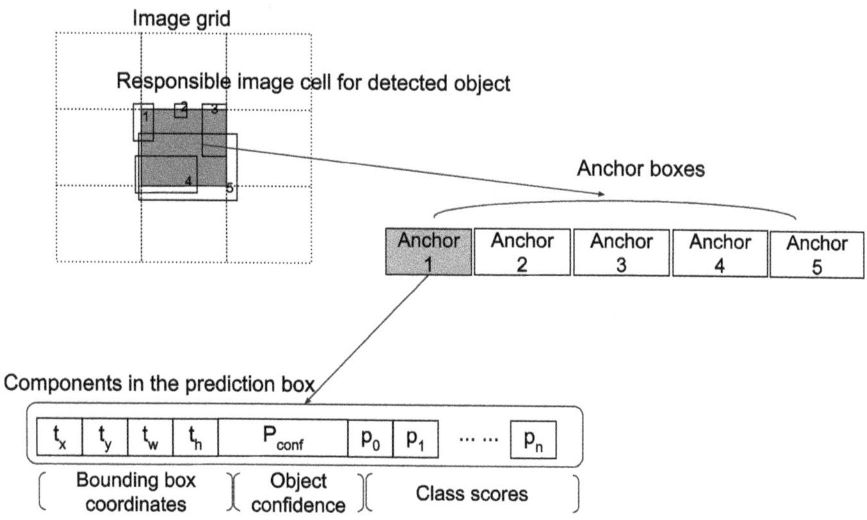

Fig. 4 Yolov v2 image object detector

minimising the function of these metrics, hereafter referred to as the loss function, the generated adversarial image is likely to confuse the object detector.

One property of natural images is that neighbouring pixels share similar colours, which is known as smoothness. To make the adversarial image as natural as possible, the colour difference between neighbouring pixels should be small, preventing the generation of noisy images. The object confidence score is the second metric that the threat actor wishes to reduce because a low score can mislead the object detector to ignore the target object with the adversarial image attached to it. If the threat actor wants to hide a vehicle, the adversarial image must lower the score of vehicle detection. Finally, the NPS represents the ability of the printer to reproduce the colours of the adversarial image. Naturally, the threat actor aims at a low NPS, which means that the printed image is close, in terms of colours, to the adversarial image.

To sum up, the workflow of this physical adversarial attack to the image object detector consists of four steps: In the first step, the threat actor has to define the loss metrics and select a loss function. Secondly, he initialises the optimisation process with a random adversarial image and applies it to the top of the image of the target object. If this is a "patch attack", he fixes the position of the patch at the centre of the target object, and otherwise, he covers the full object. He then applies the required physical condition transformations. Thirdly, the superposition of the adversarial image on that of the target object is fed into the image object detector to predict the bounding boxes and evaluate the class probabilities and the object confidence score. Based on the selected loss function, the convolutional neural network back-propagates and updates the pixels of the initial adversarial image using stochastic gradient descent (SGD). Eventually, the loss function converges after some iterations. In the fourth step, the threat actor prints the last adversarial image generated from the neural network and attaches it onto the target object.

4.3 Jamming Infrastructure Radar

Jamming a radar requires to generate electromagnetic waves at the same (or nearly adjacent) frequency spectrum band to that of the target radar. These waves will act as additional noise, possibly lowering its signal-to-noise-ratio, and compromising its detection performance. To jam the radar at the RSU, we will apply the exemplary methodology from [14], where the mm-wave radar from Tesla was jammed. The jamming device must include the following components: An oscilloscope, a signal analyser, a signal generator, and a frequency multiplier. The signal analyser and the oscilloscope are needed to identify the operational frequency of the radar. The signal generator working together with a frequency multiplier is a low-cost alternative to signal generators with a higher maximum frequency range.

5 Threat Modelling Methodology

In this section, we will first identify the cyber security requirements relevant to the V2X-assisted highway merging scenario. After that, we will map these requirements to threats and discuss how to evaluate their impact and risk ranking. In order to associate the cyber security requirements with threat classes, we will use the STRIDE model, because it is a systematic and lightweight approach. STRIDE stands for spoofing (S), tampering (T), repudiation (R), information disclosure (I), denial-of-service (D), and elevation-of-privilege (E). For risk analysis, we will use the TARA+ methodology, as it quantifies the risk ranking based on the attack potential and its related impact along with controllability [11].

5.1 *Cyber Security Requirements*

Various research projects have conducted a general analysis of cyber security requirements for CAVs with V2X connectivity and divided the cyber security requirements into the following categories: confidentiality, integrity, availability, authenticity, privacy, trace-ability, authorisation, non-repudiation, and robustness against external threats. We will consider the above requirements and identify those that can be violated by a threat actor in our system's RA, see the rightmost column of Table 1. Jamming the radar will affect the availability of the infrastructure sensors, and tampering with the input images to the object detector will violate the integrity of the processing models. On the other hand, the identification of threats has a higher priority than the tracing of individual threat actors in our system setup, making non-repudiation irrelevant.

In Table 1, we have mapped the cyber security requirements for V2X-assisted CAVs to threat classes according to STRIDE. In the second column of Table 2, we have investigated the vulnerability of the five attack surfaces presented in Sect. 4.1 concerning these threat classes. To give an example, while spoofing, tampering, and denial-of-service apply to almost all attack surfaces in our RA, the elevation-of-privilege is relevant only to the infrastructure edge server. The threat actor can

Table 1 STRIDE and cyber security requirements for V2X-assisted highway merging

Threat classes	Cyber security requirements	Highway merging functionality
Spoofing (S)	Authenticity	✓
Tampering (T)	Integrity	✓
Repudiation (R)	Trace-ability, Non-repudiation	×
Information Disclosure (I)	Confidentiality Privacy	✓ ×
Denial-of-Service (D)	Availability	✓
Elevation-of-Privilege (E)	Authorisation	✓

Table 2 Attack surface analysis for V2X-assisted highway merging

Attack surface	STRIDE	Expertise (E_x)	Knowledge (K_t)	Equipment (E_m)	Window-of-opportunity (W_o)	Controllability factor (C)
Vehicular (on-board) sensors	S, T, D	Proficient, expert and layman	Public and sensitive	Standard, specialised, and bespoke	Medium and large	The attacks can be controlled by anomaly detection algorithms in the sensors, or by an intrusion detection system (IDS)
V2X communication	S, T, D, I	Expert and layman	Restricted	Specialised and bespoke	Medium and large	The attack is controllable by a plausibility check, e.g., certification-based V2X security framework, which helps to build trustworthiness between the vehicle and the infrastructure
Infrastructure (off-board) sensors	S, T, D	Proficient, expert and layman	Public and restricted	Standard, specialised, and bespoke	Medium and large	Cyberattacks can be controlled by anomaly detection algorithms or by an IDS installed in the infrastructure
Processing models	S, T	Proficient	Public and critical	Specialised and bespoke	Large	Redundant sensors can ensure that this attack is controllable
Infrastructure edge server	E, D	Proficient and layman	Restricted	Standard and specialised	Medium and large	Unauthorised physical access can be blocked, and a CCTV system can be installed for monitoring the infrastructure

use the physical access to the server to gain the administer permission, exploit further threats, and violate the authorisations from the cyber security requirements. In the other columns, we have assessed the required level of expertise, knowledge, equipment, and window-of-opportunity to be able to attack the target surface. Finally, in the rightmost column, we have described potential methods to detect and control each threat. All these features will be used by TARA+ for quantifying the risk of each attack.

5.2 *TARA+*

Threat Analysis and Risk Assessment Plus (TARA+) [11] has been developed to analyse cyber risks related to the society of automotive engineers (SAE) Level 3 automated driving functions and beyond, in order to account for shared system/driver responsibility in the CAV's control. TARA+ is an enhanced version of TARA which integrates the system's/driver's controllability factor on cyberattacks that renders the risk analysis more realistic. In this paper, we focus on Level 4 autonomy that still has an option for the driver to take control, despite this is not necessary [22]. Therefore, only the system's controllability factor is considered hereafter. The output of TARA+ is a value that indicates the severity of the risk. Its calculation depends on multiple factors that are detailed next.

Attack Potential The attack potential P_o is a linear function of the threat actor's expertise E_x, the required equipment E_m, the knowledge regarding the target system K_t, and the window-of-opportunity W_o. If we assign equal weights to these parameters we get

$$P_o = E_x + E_m + K_t + W_o. \tag{1}$$

In Table 3, lower values in the rightmost column are associated with higher attack potentials. For instance, in case a threat actor with layman's expertise on cyber security, public information about the target system, standard equipment, and unlimited window-of-opportunity can successfully apply an attack, we will regard this attack as very possible. After calculating P_o, we can also quantify the attack probability ranking, P_r, of the identified attack, see [11, Table IV].

Table 3 Attack potential factors and ranking [11, Table I]

Expertise (E_x)	Knowledge of the target (K_t)	Equipment (E_m)	Window-of-opportunity (W_o)	Value
Layman	Public Information	Standard	Unlimited	0
Proficient	Restricted Information	Specialised	Large	1
Expert	Sensitive Information	Bespoke	Medium	2
Multiple Experts	Critical Information	Multiple Bespoke	Small	3

Table 4 Risk ranking matrix [11]

Risk ranking (R^*)	$P_r = 0$	$P_r = 1$	$P_r = 2$	$P_r = 3$	$P_r = 4$
$MI = 0$	QM	QM	QM	QM	Low
$MI = 1$	QM	Low	Low	Low	Medium
$MI = 2$	QM	Low	Medium	Medium	High
$MI = 3$	QM	Low	Medium	High	High
$MI = 4$	Low	Medium	High	High	Critical

Impact Factor The impact factor I_f quantifies the cost incurred by an attack to the system. It is a linear function of the attack severity, S_v, the operational malfunction, O_f, the financial cost, F_c and the privacy/legislative cost, P_c, where the scaling weights are adjustable.

$$I_f = 3\,S_v + F_c + 2\,O_f + P_c. \tag{2}$$

To assign values to the four parameters, we have used TARA+ [11, Table II]. One may already deduce that although an attack might be very probable, it might still be possible to ignore, in case it is associated with a low impact factor.

Modified Impact and Risk Ranking The modified impact, MI_f, depends on the impact factor and the controllability of the attack. The controllability factor $C \in \{0, 1, 2, 3, 4\}$ quantifies the resilience of the system against attacks. If the system can detect the attack, and it continues to be fully operational with a sufficient level of redundancy, the attack can be safely controlled, and the controllability factor is defined to be zero ($C = 0$). When the system can detect the attack, but the operation level is compromised due to safety reasons, the controllability factor is equal to three ($C = 3$). Finally, if the system cannot detect the attack, the controllability factor is set equal to four ($C = 4$). See [11, Table III] for more details. Finally, the modified impact MI_f can be read as

$$MI_f = \frac{I_f \cdot C}{4}. \tag{3}$$

The modified impact ranking, MI, results from the quantisation of MI_f, see [11, Table V]. With the parameters MI and P_r at hand, one can determine the risk ranking of the threat R^* based on Table 4. The risk ranking ranges from “QM” (Quality Management) at the lowest level to “Critical” at the highest level.

Table 5 Risk assessment of the abuse cases by TARA+

Attack scenario	Attack surface	Attack potential (P_o) and probability (P_r)	Controllability factor (C)	Impact ranking (MI)	Risk ranking (R^*)
Jamming the radar detector	Infrastructure sensors	$Ex = 1$, $E_m = 2$, $K_t = 0$, $W_o = 1$ $P_o = 4$, $Pr = 3$ (*Possible*)	$C = 3$	$S_v = 2$, $F_c = 2$, $O_f = 3$, $P_c = 0$ $I_f = 14$, $MI = 2$ (*Medium*)	*Medium*
Tampering with the image object detector	Processing models	$E_x = 1$, $E_m = 1$, $K_t = 3$, $W_o = 1$ $P_o = 6$, $P_r = 3$ (*Possible*)	$C = 3$	$S_v = 2$, $F_c = 2$, $O_f = 3$, $P_c = 0$ $I_f = 14$, $MI = 2$ (*Medium*)	*Medium*
Blinding the V2X infras-tructure	Infrastructure sensors and processing models	$E_x = 1$, $E_m = 2$, $K_t = 3$, $W_o = 1$ $P_o = 7$, $P_r = 2$ (*Unlikely*)	$C = 4$	$S_v = 4$, $F_c = 4$, $O_f = 4$, $P_c = 0$ $I_f = 24$, $MI = 4$ (*Critical*)	*High*

6 Risk Assessment and Mitigation Schemes

In this section, we will rank the risk of the abuse cases presented in Sects. 4.2 and 4.3 using TARA+ and also discuss potential mitigation schemes.

Jamming the Radar Detector In order to rank the risk of radar jamming, we first need to model the threat actor and evaluate the attack potential P_o, see Eq. (1) and Table 3. We have selected: (i) *Proficient* expertise, $E_x = 1$, because a general security knowledge about popular attacks, like jamming, is only required. (ii) *Bespoke* equipment, $E_m = 2$, because the threat actor does not only need a signal analyser and a signal generator but also need a frequency multiplier to enhance the maximum frequency range of the generated waveforms and ensure the jamming of the radar. (iii) *Public information* for the knowledge of the target, $K_t = 0$, because the threat actors do not require the detailed specifications as long as they can get the basic and public knowledge about the location of the radar. (iv) *Large* window-of-opportunity, $W_o = 1$, as the radar jamming can be executed remotely with minimal time constraints. After substituting these values into (1), we end up with $P_o = 4$. According to [11, Table IV], the attack probability, $P_r = 3$, is classified as *possible*.

Next, we use Eq. (2) and [11, Table II] to assess the impact factor I_f of the attack. We have selected its severity $S_v = 2$ and the financial cost $F_c = 2$ to be medium, and the operational malfunction of the vehicle $O_f = 3$ to be high, as it affects a primary function of the CAV. In addition, $P_c = 0$ because there are no privacy considerations for the cyber security requirements listed in Table 1. These values yield $I_f = 14$. Since only the radar is jammed, anomaly detection algorithms and/or intrusion detection systems (IDS) can identify the attack by monitoring the difference between the output of the radar signal's processor and the output of other sensors like the camera [23]. As a result, we have selected the controllability factor $C=3$. After substituting $C = 3$ and $I_f = 14$ into (3), we get $MI_f = 10.5$. Subsequently, the modified impact ranking is $MI = 2$ or *medium* according to [11, Table V]. Finally, from Table 4 we conclude that the risk ranking R^* of this abuse case is *medium*.

Tampering with the Image Object Detector The risk ranking follows the same procedure to that of radar jamming. The calculation of the values for the parameters, P_o, P_r, I_f, etc., is presented in the second row of Table 5. Training an adversarial image to tamper with the image object detector requires just a GPU to train the machine learning model and a colour printer. Therefore, specialised equipment, $Em = 1$, is sufficient. Also, recall from Sect. 4.2 that the threat actors require full knowledge of the targeted image object detector to carry out this attack, hence, $K_t = 3$. In Table 5, we see that the ranking for the attack probability $P_r = 3$ and the modified impact ranking $MI = 2$ are equal to those of radar jamming and thus, the risk ranking R^* is *medium* too.

Blinding the Road Infrastructure If both attacks are executed simultaneously, the RSU becomes completely blind. To evaluate the attack potential P_o of the combined attack, we have selected the maximum value (over the previous abuse cases) for each of the four parameters involved in its calculation, yielding $P_o = 7$ and $P_r = 2$ *(Unlikely)*, see the last row of Table 5. At the same time, the combined attack has the highest controllability factor $C = 4$, because it is generally undetectable by the system. It is also likely to cause devastating effects on other traffic objects. For instance, if the CAV uses an incorrect trajectory to merge based on the received inaccurate perception knowledge from the RSU, it is probable to cause serious safety issues along the highway. Accordingly, we have assigned the highest severity, financial, and operational cost to the combined attack yielding $I_f = 24$. Finally, based on Eq. (3) and [11, Table V], we have calculated the modified impact ranking MI as *critical*, and the risk ranking R^* as *High*.

Mitigation Schemes We will describe mitigation schemes for each of the three abuse cases. Attacking the object detector requires that the threat actor has the complete knowledge of the machine learning model running in the RSU, e.g., the architecture of the neural network, the trained weights, etc. Therefore, the safety recommendation is to guarantee the model's confidentiality, which is related to the threat class, "information disclosure" in STRIDE. Furthermore, one way to mitigate the impact of a jamming attack is to deploy a frequency-agile radar, which can switch its operation among different frequencies. It is noted that including both a radar and a camera at the RSU is a mitigation mechanism per se. When the attack aims at an individual sensor, the risk decreases, if the system can detect the inconsistencies between the outputs of different sensors. In that case, the RSU can go into safe-mode operation, for instance, it can stop broadcasting CPMs. Finally, if both sensors are under attack, we should combine the mitigation schemes of the previous abuse cases. It is expected that soon there will be further redundancy at the RSU, equipping it with more sensors, e.g., lidar and ultrasound devices, which will mitigate further the cyber security risks.

7 Comparative Study

In this section, we will present a comparative study between our focused risk assessment model, Tara+, and two previous models, see the studies in [5] and [13], to discuss the validation of the selected parameters in Table 5. These studies utilised the defined threat actor profiles on Table 6, and based on these profiles, the values of various parameters like expertise, motivation, knowledge about the target system, equipment in use and financial availability are specified. Afterwards, the attack capability can be calculated as a function of these parameters.

Specifically, the risk assessment model of [5] calculates the attack potential from the difference between the threat actor's capability to execute a successful attack and the system's resistance to the attack. The system's resistance is the minimum required capability to realise a successful attack. The required capability can be determined by doing an attack surface analysis in a similar manner as we did earlier for Tara+, see Table 2. Likewise, the SARA method [13] requires to identify the minimal required threat actor profile which is able to execute a successful attack to the system, and based on that it evaluates the attack potential. The minimal threat actor profile to the attack scenarios of our system's RA (jamming of the radar and/or image object detection in the camera) is *Organised Crime* from Table 6. Unlike [5] and [13], the parameters of our risk assessment using Tara+ have been determined by considering the minimal requirements of a successful attack from the system perspective and not from the perspective of the threat actor.

The results of our comparative study are presented on Table 7. Unlike Tara+, the risk assessment methodology in [5] does not categorise the risk as low, medium, or high, but it visualises the three characteristics of a cyber risk (i.e., potential, motivation, and impact) using contour plots in the two-dimensional space. Thus, it will help security analysts to understand which countermeasures to prioritise. Although trends of the results are consistent with our analysis using Tara+, the methodology could not assess that the attack probability of *Blinding the V2X infrastructure* should be a lot less than the other two. This is because the risk assessment methodology in [5] does not consider the system's controllability. On the other hand, SARA [13] includes a metric called observation which is defined as the system's ability to detect errors and operates on risk calculation. The attack probabilities of our scenarios are quantified as *Moderate* according to the SARA's attack mapping table, because the threat actor has been identified as *Organised Crime*, and any attack executed by *Organised Crime* cannot be lower than *Moderate*. However, *Blinding the V2X infrastructure* in our RA is a lot less likely as compared to the other two individual scenarios, due to the system's redundancy. Sara methodology is not able to capture the reduced probability (the likelihood) of the combined attack. Nevertheless, it quantifies a higher risk, R7 instead of R4, for the combined attack.

Table 6 Threat actor profiles [5]

Threat agent	Motivations	Finances	Expertise	Knowledge	Equipment
Thief	Financial	Low	Layman	Public	Standard
Owner	Financial	Low	Layman	Public	Standard
Organised Crime	Financial	High	Proficient	Restricted	Bespoke
Mechanic	Financial	Low	Expert	Critical	Specialised
Hacktivist	Ideology, Passion	Low	Multiple Experts	Sensitive	Multiple Bespoke
Terrorist	Ideology	Low	Layman	Public	Standard
Foreign Government	Financial, Ideology	High	Multiple Experts	Restricted	Multiple Bespoke

Table 7 Comparison study of risk assessment models

Attack scenario	TARA+	Risk ranking by [5]	SARA risk ranking by [13]
Jamming the radar detector	Attack probability: Possible Risk: Medium	Attack probability: $\Delta Ex = 0$, $\Delta E_m = 1$, $\Delta K_t = 2$, $\Delta W_o = 0$ **Pr=3** Impact: $S_v = 2$, $F_c = 2$, $O_f = 3$, $P_c = 0$ **I=9** Motivation: $m_f = 3$, $m_r = 1$ **M=2**	Attack probability: $C_a = 16$, $T = 2$, $W_o = 1$ **Pr=3 (Moderate)** Severity:(2, 2, 3, 0) **S=3** Risk: **R4**
Tampering with the image object detector	Attack probability: Possible Risk: Medium	Attack probability: $\Delta Ex = 0$, $\Delta E_m = 2$, $\Delta K_t = 0$, $\Delta W_o = 1$ **Pr=3** Impact: $S_v = 2$, $F_c = 2$, $O_f = 3$, $P_c = 0$ **I=9** Motivation: $m_f = 3$, $m_r = 1$ **M=2**	Attack probability: $C_a = 16$, $T = 2$, $W_o = 1$ **Pr=3 (Moderate)** Severity:(2, 2, 3, 0) **S=3** Risk: **R4**
Blinding the V2X infrastructure	Attack probability: Unlikely Risk: High	Attack probability: $\Delta Ex = 0$, $\Delta E_m = 1$, $\Delta K_t = 0$, $\Delta W_o = 1$ **Pr=2** Impact: $S_v = 4$, $F_c = 4$, $O_f = 4$, $P_c = 0$ **I=20** Motivation: $m_f = 3$, $m_r = 2$ **M=1**	Attack probability: $C_a = 16$, $T = 1$, $W_o = 1$ **Pr=3 (Moderate)** Severity:(4, 4, 4, 4) **S=4** Risk: **R7**

8 Conclusion

The highway merging functionality with Level 4 autonomous driving capability from a slip road will improve traffic efficiency and safety and help increase the driver's comfort. In this paper, we have considered a merging scenario where the autonomous vehicle does not rely only on the on-board sensor knowledge to perform the merge, but it also receives perception messages from the road infrastructure (off-board radar and camera) over V2X. Although the enhanced perception brings clear benefits, it also creates a challenge from the cyber security perspective. We have identified the attack surfaces in the system's reference architecture and applied STRIDE to classify the threats. We have used TARA+ to evaluate the cyber security risks for three abuse cases, where the threat actors attack the off-board sensors by jamming the infrastructure radar and/or tampering with the input image to the object detector at the camera. The abuse case that combines both attacks and blinds completely the road infrastructure is, to the best of our knowledge, new. It turns out that the associated risk is medium in case a single sensor is under attack and high when both sensors are under attack. Apart from deploying more sensors like lidar and ultrasound devices at the infrastructure, we have suggested that the confidentiality of the image object detector and the frequency agility of the radar are crucial to avoid security breaches. We believe that this work can help the automotive stakeholders towards the understanding and mitigation of cyber security threats before the commercial rollout of autonomous vehicles. In the future, it is important to conduct simulation-based studies of the attack scenarios discussed in this paper. By doing so, we will measure the impact of the identified cyber security attacks on the V2X autonomous driving systems and be able to select cost-effective countermeasures to improve the resilience of these systems.

Acknowledgments This work was partly funded by UK Research and Innovation through INNOVATE UK in project AutopleX (project reference 104272) and the European Union's Horizon 2020 research and innovation programme in project L3Pilot under grant agreement No 723051. The sole responsibility of this publication lies with the authors. The authors would like to thank all partners within AutopleX and L3Pilot for their cooperation and valuable contribution.

References

1. A. Stevens et al., Cooperative automation through the cloud: The CARMA project, in *12th ITS European Congr.*, 2017
2. D. Bevly et al., Lane change and merge maneuvers for connected and automated vehicles: A survey. IEEE Trans. Intell. Veh. **1**(1), 105–120 (2016)
3. C. Maple, Security and privacy in the internet of things. J. Cyber Policy **2**(2), 155–184 (2017)
4. J. Petit, S.E. Shladover, Potential cyberattacks on automated vehicles. IEEE Trans. Intell. Transp. Syst. **16**(2), 546–556 (2014)
5. D. Dominic et al., Risk assessment for cooperative automated driving, in *Proc. 2nd ACM Workshop Cyber-Phys. Syst. Security Privacy (CPS-SPC)*, 2016, pp. 47–58

6. M.M. Islam et al., A risk assessment framework for automotive embedded systems, in *Proc. 2nd ACM Workshop Cyber-Phys. Syst. Secur.*, 2016, pp. 3–14
7. K. Kim, J.S. Kim, S. Jeong, J.-H. Park, H.K. Kim, Cybersecurity for autonomous vehicles: Review of attacks and defense. Comput. Secur. **103**, 102150 (2021)
8. Z. El-Rewini et. al., Cybersecurity challenges in vehicular communications. Veh. Commun., **23** (2020). https://doi.org/10.1016/j.vehcom.2019.100214
9. L. Sequeira et al., A lane merge coordination model for a v2x scenario, in *Proc. Eur. Conf. Netw. and Commun. (EuCNC)*, 2019, pp. 198–203
10. J. Ziegler et al., Making bertha drive—an autonomous journey on a historic route. IEEE Intell. Transp. Syst. Mag. **6**(2), 8–20 (2014)
11. A. Bolovinou et al., Tara+: Controllability-aware threat analysis and risk assessment for l3 automated driving systems, in *IEEE Intell. Vehicles Symp.*, June 2019, pp. 8–13
12. N. Vignard et al., Deliverable 4.2 Legal requirements to AD piloting and cyber security analysis, L3 Pilot Driving Automation, Tech. Rep., 04 2019. [Online]. Available: https://bit.ly/3o7FzBr
13. J.-P. Monteuuis et al., Sara: Security automotive risk analysis method, in *Proc. 4th ACM Workshop on Cyber-Phys. Syst. Secur.*, 2018, pp. 3–14
14. C. Yan, W. Xu, J. Liu, Can you trust autonomous vehicles: Contactless attacks against sensors of self-driving vehicle, in *DEF CON*, vol. 24, 2016
15. S. Thys, W. Van Ranst, T. Goedemé, Fooling automated surveillance cameras: adversarial patches to attack person detection, in *Proc. IEEE Conf. Comput. Vis. Pattern Recognit. (CVPR)*, 2019, pp. 1105–1112
16. D. Song et al., Physical adversarial examples for object detectors, in *12th {USENIX} Workshop on Offensive Technologies*, 2018
17. S.-T. Chen et al., Shapeshifter: Robust physical adversarial attack on faster R-CNN object detector, in *Proc. Joint European Conf, on Machine Learning and Knowledge Discovery in Databases* (Springer, 2018), pp. 52–68
18. ETSI, Intelligent Transport Systems (ITS); Vehicular Communications; Basic Set of Applications; Analysis of the Collective Perception Service (CPS) (2017). [Online]. Available: https://bit.ly/35hiGBC
19. J. Redmon, A. Farhadi, Yolo9000: Better, faster, stronger, in *Proc. IEEE Conf. Comput. Vis. Pattern Recognit. (CVPR)*, 2017, pp. 7263–7271
20. M. Sharif et al., Accessorize to a crime: Real and stealthy attacks on state-of-the-art face recognition, in *Proc. ACM SIGSAC Conf. on Comput. and Commun. Secur.*, 2016, pp. 1528–40
21. K. Eykholt et al., Robust physical-world attacks on deep learning visual classification, in *Proc. IEEE Conf. Comput. Vis. Pattern Recognit (CVPR)*, 2018, pp. 1625–1634
22. SAE On-Road Automated Driving Committee, *Taxonomy and Definitions for Terms Related to Driving Automation Systems for On-Road Motor Vehicles*, jun 2021
23. O. Y. Al-Jarrah et al., Intrusion detection systems for intra-vehicle networks: A review. IEEE Access **7**, 21266—21289 (2019)

A Machine Learning Framework for Intrusion Detection in VANET Communications

Nourhene Ben Rabah and Hanen Idoudi

1 Introduction

Vehicular ad hoc networks (VANET) stand for different communication schemas that can be performed between connected vehicles and anything (V2X). This includes vehicle-to-vehicle communications, vehicle-to-roadside infrastructure components, or intra-vehicle communications.

A VANET system relies on two main components: Roadside Unit (RSU) and On-Board Unit (OBU). RSU is the roadside communication equipment. It provides Internet access to vehicles and ensures exchanging data between vehicles. The OBU is the mobile treatment and communication unit embedded on the vehicle. It allows communication with other vehicles or with the infrastructure's equipment. VANET communication can be deployed according to different architectures, such as vehicle-to-vehicle (V2V), vehicle-to-infrastructure (V2I), infrastructure-to-vehicle (I2V), infrastructure-to-infrastructure (I2I), and hybrid [1]. Furthermore, a VANET system is composed of three planes: vehicular plane, RSU plane, and services plane. In the vehicular plane, each vehicle is equipped with OBU. The latter allows V2V communication. The RSU plane facilitates V2I, I2V, and I2I communications. In the service plane, different types of services can be deployed such as safety, infotainment, payment, Internet, and cloud-based services. A VANET has some similar features of MANET (mobile ad hoc networks) such as omnidirectional broadcast, short

N. Ben Rabah
Centre de Recherche en Informatique, Université Paris 1 Panthéon Sorbonne, Paris, France

ESIEE-IT, Pontoise, France
e-mail: nbenrabah@esiee-it.fr

H. Idoudi (✉)
National School of Computer Science, University of Manouba, Manouba, Tunisia
e-mail: hanen.idoudi@ensi-uma.tn

K. Daimi et al. (eds.), *Emerging Trends in Cybersecurity Applications*,
https://doi.org/10.1007/978-3-031-09640-2_10

transmission range, and low bandwidth. In contrast, it has particular characteristics. First, a VANET has a highly dynamic topology due to the high mobility of vehicles. This leads also to frequent disconnections. Secondly, target vehicles can be reached upon their geographical location. Thirdly, signal propagation is affected by the environment such as buildings, trees, etc. [1]. Finally, energy, storage failure, and computing capacity are less critical for VANETs as for MANET. Despite that, the serious challenge for VANET is processing huge amount of data in a real-time manner.

This diversity of communication schemas and the inherent characteristics of wireless communications make VANETs vulnerable to many security attacks and vulnerabilities. This is emphasized by the critical aspect of some exchanged information that is used for road safety purposes. Security breaches are several and can affect all network layers and all communication aspects in VANET. Moreover, VANETs suffer from traditional vulnerabilities that affect any wireless environment but are also subject to new and specific attacks exploiting inherent vehicular characteristics [1]. Most of the security solutions defined for traditional networks are not suitable for vehicular networks. Subsequently, researchers are looking for appropriate systems that support vehicular network characteristics and provide robust security mechanisms.

Different security countermeasures have been proposed such as key management systems, anonymity, traceability techniques, cryptographic algorithms, trust management methods, etc. [2]. Recently, many researchers showed that integrating artificial intelligence (AI) in intrusion detection systems increases their effectiveness in detecting attacks on V2X networks. IDS are a widely used approach that analyzes the traffic for indicators of security breaches and creates an alert for any observed security anomaly. Moreover, machine learning (ML) can realize anomaly-based detection systems capable of detecting unknown and zero-day attacks, learning, and training itself by analyzing network activity and increasing its detection accuracy over time.

Applying ML techniques for intrusion detection in VANET is of particular interest due to the huge amount of exchanged data and the diversity of attacks that can occur. In recent years, many published datasets, describing real traces of VANET communication, have allowed the assessment of ML techniques performances for intrusion detection.

This work intends to define a novel comprehensive framework to design an IDS for V2X communications. Furthermore, unlike most existing works, we use a very recent dataset to evaluate and compare both ensemble and standalone learning techniques to detect various types of DOS and DDOS attacks in VANET.

We define first a novel framework for applying ML techniques to detect anomalies in VANET communication. Then, we use a very recent dataset, VDOS-LRS dataset, that describes urban vehicular traffic to assess and compare the performances of well-known standalone ML methods and ensemble ML methods to detect DOS and DDOS attacks in urban environment.

The rest of this chapter is structured as follows.

In Sect. 2, we review related works related to security issues in VANET, and we review most important works on ML-based IDS for VANETs. In Sect. 3, we expose our framework for designing ML-based IDS for VANET communication. Main results are discussed in Sect. 4 where we study the performances of several ML techniques, both standalone and ensemble learning techniques, on detecting DOS and DDOS attacks in urban traffic using a very recent VANET dataset, namely, VDOS-LRS dataset.

Finally, Sect. 5 gives the conclusion of the study.

2 Security of VANET Communications

In this section, we discuss the security issue in VANET communication; then, we focus on the most important works that considered the use of machine learning-based intrusion detection systems for VANET.

2.1 Security Attacks and Vulnerabilities in VANET

In-vehicle communications involve embedded units mainly interacting via CAN-Bus, Ethernet, or WiFi standards whereas inter-vehicle networks refer to different kind of interactions between vehicles and other components of the ITS system. These latter can be vehicle-to-infrastructure (V2I), vehicle-to-cloud (V2C), vehicle-to-vehicle (V2V), and vehicle-to-device (V2D) communications [1]. This diversity of architectures and communication schemes led to the inception of the vehicle-to-anything or V2X paradigm.

Many security attacks are targeting VANET communications taking profit from the highly heterogeneity of such environments, the highly dynamic topology induced by mobility, and the lack of standard security so far [1, 2]. Security requirements such as availability, data integrity, confidentiality, authenticity, and non-repudiation can be compromised.

Denial of Service (DoS) and Distributed DoS (DDOS) attacks aim to disrupt network service's availability by flooding the OBU (On-Board Unit) and/or RSU (Roadside Unit) communication channels with an unhandled huge amount of requests, resulting in network out of service [3]. In black hole and gray hole attacks, an attacker can capture illegitimate traffic, then drops, retains, or forwards them to erroneous destinations [4]. In Sybil attacks, malicious nodes may create several virtual cars with the same identity to mislead some functionalities. Node impersonation attack tries to impersonate legitimate node's identity. Additionally, GPS spoofing or position faking attacks, also known as hidden vehicle attacks, generate fake position alarms [5].

Different attacks can also threaten the integrity and/or the confidentiality of data such as tampering attacks and spoofing [1, 2].

In-vehicle communications are equally vulnerable as the inter-vehicle communications and can also suffer from all kinds of attacks following the illegitimate intrusion of malicious data [6].

We compare the characteristics of some notable security attacks in Table 1 with regard to the targeted environment and the compromised security requirement.

2.2 *Security Countermeasures*

Many security mechanisms are considered to secure vehicular communication while taking into account their inherent characteristics. Most important cover the following categories.

– *Cryptography*

Its aim is to ensure confidentiality of data while being transmitted from one source node to a destination node. Moreover, they involve encryption algorithms, hash functions, and digital signature algorithm and can provide solutions for diverse types of threats at different levels in VANET. New lightweight solutions for data encryption are more considered to tackle the limited computation capacities of different VANET equipment. The Elliptic Curve Digital Signature Algorithm (ECDSA) is one of the most widely used digital signatures algorithms in IoT in general and in securing VANET communications [7] [8].

– *Key Management Systems*

PKI are core ITS component for identity and key management and can be implemented as centralized, decentralized, or distributed systems. Many enhanced solutions based on PKI are proposed to secure authentication and revocation [9]. For instance, in [33], authors define Enhanced Secure Authentication and Revocation (ESAR) scheme for VANETs which is responsible for revocation checking, processing, and PKI key pair updating.

– *Anonymity, Unlinkability, and Traceability Techniques*

These strategies intend to ensure the privacy of users' data by means of data suppression, randomization, or cloaking to prevent unauthorized access. They offer a countermeasure against several attacks such as eavesdropping, trajectory tracking, or location disclosure.

For instance, anonymity techniques are based on the use of pseudonyms by Group Signature and Pseudonymous Authentication schemes. In a group signature approach, a group private key will be used by all vehicles, whereas in pseudonymous authentication schemes, each vehicle is assigned a set of identities that it stores locally. Hybrid approaches that combine both group signature and pseudonymous authentication schemes are also considered [10, 11].

To achieve traceability, unique electronic license plate (ELP) should be used. Pseudonyms could be linked with a specific ELP identity. This would allow authori-

Table 1 Characteristics of some VANET security attacks

Attack	Internal	External	Inter-veh.	In-veh.	Targeted security requirement				
					Availability	Integrity	Confidentiality	Authenticity	Non-repudiation
DOS/DDOS	X	X	X	X	X			X	
Black hole/grey hole		X	X		X				
Hidden vehicle	X		X				X	X	
Node impersonation	X		X		X			X	
Spoofing	X	X	X	X				X	
Position falsification		X	X			X			
Sybil	X		X			X		X	X
Replay	X		X	X		X		X	X
Fuzzy	X			X					

ties to trace a misbehaved user whenever it is needed. Moreover, in group signatures, a tracing manager can revoke the malicious vehicles by analyzing their signatures [12].

- *Security Protocols*

Standard communication and routing protocols need to be secured, hence the need for integrating with security protocols at network, transport, or application level. Several security protocols are proposed or adapted to the context of IoT communications in general such as TLS and DTLS [13].

- *Intrusion Detection Systems*

Intrusion detection systems (IDS) are an efficient way to detect and prevent malicious or abnormal activities. A typical IDS relies on three main components:

- Information collector: It relies on sensors commonly deployed at different sensitive locations.
- Analysis engine: Its main purpose is to analyze information collected via sensors.
- Reporting engine: This component is responsible for logging and raising alarms when a malicious node or an abnormal event is detected.

In VANET networks, IDS sensors are generally located at RSU and on vehicles. First, these sensors collect nodes' communication information. Second, the data collected is sent to the analysis engine. Third, the analysis engine analyzes the received data using different methods which depend on the IDS type. If an abnormal event or a malicious node is detected, a report is sent to the reporting engine. Finally, the reporting engine informs the appropriate nodes about the attack.

IDS for VANET are mainly classified into four categories. This classification is based on the techniques used to detect threats. These classes are signature based, watchdog based, behavior based, and hybrid IDS [2].

2.3 ML-Based Intrusion Detection Systems for VANETs

Behavior-based IDS, also known as anomaly-based, use AI and ML as well as other statistical methods to analyze data on a network to detect malicious behavior patterns as well as specific behaviors that may be linked to an attack.

ML-based IDS are part of behavior-based IDS. This approach assumes that intrusive activities are a subclass of abnormal activities. In ML-based IDS, different ML techniques can be used to recognize the misbehavior pattern. In fact, it extracts relations between different attributes and builds attack models [7]. This mechanism allows the RSU or OBU to detect any misbehavior in the network by analyzing received messages and network information. The main advantage of this approach is its ability to detect zero-day attacks and anomalies.

So far, many works adopted ML techniques to build efficient IDS.

In [14], Fuad A. Ghaleb et al. proposed a misbehavior detection model based on ML techniques. Authors used real-world traffic dataset, namely, Next Generation Simulation (NGSIM) to train and evaluate the model. They used artificial neural network.

In [3], authors aimed at detecting wormhole attacks in VANET using ML-based IDS. Firstly, they generated a dataset by using both the traffic simulator Simulation of Urban Mobility Model (SUMO) and NS3. Secondly, two different ML algorithms were applied on the generated dataset to train the model, namely, k-nearest neighbors (kNN) and support vector machines (SVM). Finally, to evaluate the different models, the authors used the accuracy rate and four different alarms which are true positive (TP), false positive (FP), true negative (TN), and false negative (FN). As a result, authors pointed out that both the SVM and kNN performed well on detecting wormhole attacks.

In [15], authors proposed a ML-based IDS to detect position falsification attack in VANET. To train and evaluate ML models, the authors used Vehicular Reference Misbehavior Dataset (VeReMi dataset). Authors used logistic regression (LR) and SVM models. To evaluate the work, they used F-measure. As a result, they proved that SVM performed better than LR.

In [16], authors developed an intrusion detection system based on gradient boosting decision tree (GBDT) for CAN-Bus and proposed a new feature based on entropy as the feature construction of GBDT and used a dataset from a real domestic car to evaluate the model.

Authors in [17] showed that tree-based and ensemble learning models show more performance in detection compared to other models. Random forest, bagging, and AdaBoosting methods are trained and tested on the Can-hacking dataset, and the DT-based model results in yield performance.

Vuong et al. [18] proposed a decision tree-based method for detecting DoS and command injection attacks on robotic vehicles using cyber and physical features to show the importance of incorporating the physical features in improving the performance of the model. They tested their model in a collected dataset. In addition to DoS and command injection attack detection, they also provide in [19] a lightweight intrusion detection mechanism that can detect malware against network and malware against CPU using both cyber and physical input features using the decision tree model.

A tree-based intelligent IDS for the internet of vehicles (IoV) that detects DoS and fuzzy attacks is proposed by Li Yang et al. [20]. Firstly, they tested the performance of decision tree (DT), random forest (RF), extra trees (ET), and XGradient Boost (XGB) methods and applied multi-threading to get a lower execution time. Then, they selected three models that generate the lowest execution time as a meta-classifier in the second layer of the stacking ensemble model. Besides, they used an ensemble feature selection (FS) technique to improve the confidence of the selected features. Finally, the authors tested the model on the car-hacking dataset.

In [34], authors define a novel machine learning model using random forest and a posterior detection based on coresets. Their model showed high accuracy for DOS attack detection.

The use of ML techniques is undoubtedly efficient, but due to the numerous opportunities that ML techniques offer, more works are still needed to investigate the design of the best ML framework for VANET IDS.

In our work, we intend to define a comprehensive framework to design VANET IDS. Furthermore, unlike most existing works, we use a very recent dataset to evaluate and compare both ensemble learning and standalone learning techniques to detect various types of DOS attack.

Our contribution is exposed in the next section.

3 Proposed ML Framework

In this section, we introduce a novel machine learning framework for intrusion detection in V2X communications. The elaboration process comprises three major phases: dataset description, data preprocessing, and the application of standalone and ensemble learning methods, as shown in Fig. 1.

3.1 First Phase: Dataset Description

One of the challenges of building efficient V2X ML-based IDS is the lack of public datasets with a big collection of network traffic logs depicting both normal and abnormal activities. More recent works that tried to tackle IDS design using ML or DL (deep learning) techniques to mitigate more complex or new attacks have pointed out this problem, and some tried to build simulated datasets at that end [21–23]. A survey of the most important and most recent datasets dedicated to VANET communication and involving some well-known security attacks is given in [24].

To evaluate the proposed framework, we used the Vehicular Denial of Service Networks and Systems Laboratory (VDOS-LRS) dataset [25]. It is one of the most

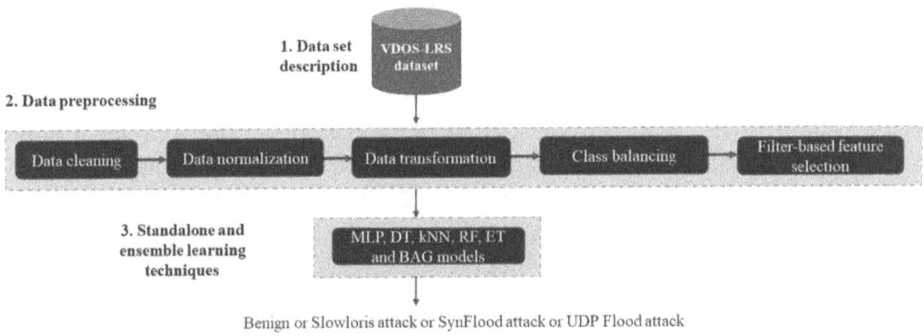

Fig. 1 Proposed ML framework

recently published dataset that incorporate real network traffic collected in different environments (urban, rural, and highway). This dataset involves traces of three DoS attacks:

- SYN flood attack is based on sending a huge number of TCP-SYN requests to a vehicle to make it out of service.
- UDP flood overloads random ports on the targeted host with UDP datagrams.
- Slowloris attack is an application layer DDoS attack that uses partial HTTP requests to open multiple connections towards a target.

For this study, we focused on the urban environment. It is initially presented as a PCAP file. For this purpose, we used the network traffic flow generator and analyzer, CICFlowMeter [26], which allowed us to generate bidirectional flows described through 84 statistical features such as duration, number of packets, number of bytes, packet length, etc. These flows are then saved as a csv file, representing our dataset. It includes 26,334 normal instances, 124,968 SYN flood attack instances, 122,457 UDP flood attack instances, and 650 Slowloris attack instances.

3.2 Second Phase: Data Preprocessing

These different steps are used to improve the data quality and, consequently, the performance of the machine learning models. It includes data cleaning, data normalization, data transformation, and class balancing.

1. *Data Cleaning*

   It is used to handle erroneous, inaccurate, and irrelevant data to improve the dataset quality. Indeed, we do not consider source and destination IP addresses and ports, as attackers can easily modify them [22]. Therefore, we removed these five features: "Flow ID," "Src IP," "Src Port," "DST IP," and "DST Port." Thus, we replaced the missing values of some features with the mean values of these features.

2. *Data Normalization*

   It is performed to avoid bias when feature values belong to very different scales. Some features in our dataset vary between 0 and 1, while others can reach infinite values. Therefore, we normalized these features according to Eq. 1, defined as follows:

$$X_{\text{normalized}} = \frac{X - X_{\text{min}}}{X_{\text{max}} - X_{\text{min}}} \tag{1}$$

where $X_{\text{normalized}}$ is the normalization result and X is the initial value. Here, X_{max} and X_{min} represent the maximum and the minimum values of each feature, respectively.

3. *Data Transformation*

It is used to modify data to fit the input of any ML model. Indeed, some ML models can work with qualitative data (i.e., non-numerical data) such as k-nearest neighbors (kNN), naive Bayes (NB), and decision trees (DT). However, most of them require numerical inputs and outputs to work properly. Therefore, it is important to convert qualitative data to numerical data. In our dataset, each instance is represented by 77 numerical features and one object feature ("Timestamp") that represents the date and time values of the flow. In this step, we propose to replace this feature by six features of numerical type: "Timestamp_year," "Timestamp_month," "Timestamp_day," "Timestamp_hour," "Timestamp_minute," and "Timestamp_second."

4. *Class Balancing*

Class imbalance is a major problem in supervised ML methods. It usually occurs when the dataset traces are collected from a real environment. Indeed, in such an environment, the data is usually unbalanced, and the models learned from the data may have better accuracy on the majority class but very poor accuracy on the other classes. There are three main ways to deal with this problem: modifying the ML algorithm, introducing a misclassification cost, and data sampling [27]. Data sampling is the only solution that can be done independently of the classification algorithm, since the other two require direct or indirect modifications to the algorithm. Data sampling is performed using two methods: undersampling the majority class or oversampling the minority class.

Since the classes in our dataset are unbalanced (see Fig. 2), we use the Synthetic Minority Oversampling Technique (SMOTE) [28, 29] to solve this problem. SMOTE involves synthesizing new examples of the minority classes so that the number of examples of the minority class gets closer to or matches the number of examples of the majority class. After performing it, we get 124,968 instances of each class (see Fig. 3).

5. *Filter-Based Feature Selection*

Feature selection [30] is a very important step that consists in selecting from the initial dataset the most relevant features. Indeed, if there are too many features, or if most of them are not relevant, the models will consume more resources and be difficult to train. On the other hand, if there are not enough informative features, the models will not be able to perform their ultimate tasks.

To achieve such a goal, we propose to use a filter-based feature selection method that consists of selecting the most relevant subsets of features according to their relationship with the target variable. We, therefore, use statistical tests that consist in (a) evaluating the relationship between each input feature and the output feature and (b) discarding input variables that have a weak relationship with the target variable. In other hand, we keep the input features that have strong statistical relationship with the output variable.

There are many statistical tests such as chi-squared, Pearson correlation, permutation feature importance, ANOVA F-value test, and others. The choice of statistical

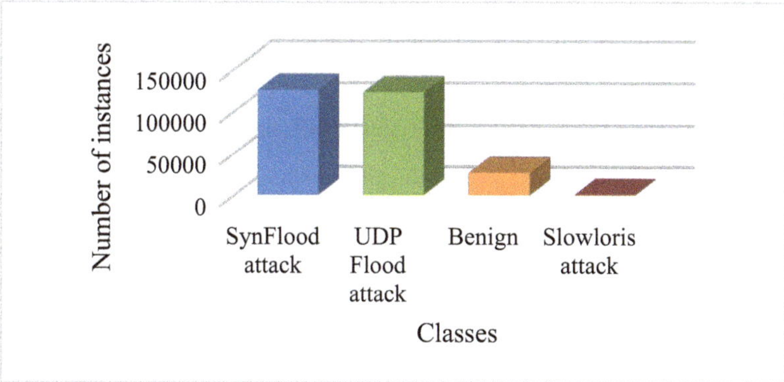

Fig. 2 Number of instances of each class before SMOTE: 124,968 instances of "SYN flood attack" (majority class), 122,457 instances of "UDP flood attack," 26,334 instances of "Benign," and 650 instances of "Slowloris attack" (minority class)

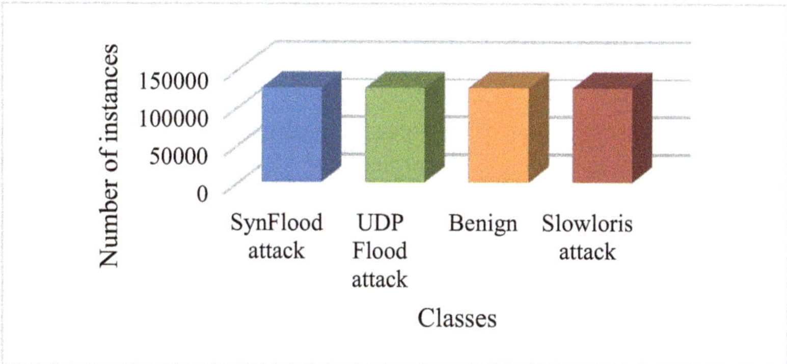

Fig. 3 Number of instances of each class after SMOTE: 124,968 instances of each class

measures depends strongly on the types of input variables and the output variable (numerical or categorical). In our dataset, the input variables are of numerical type, and the output variable is of categorical type (the class), hence the interest to use two statistical measures which are:

- ANOVA F-value test that estimates the degree of linear dependence between an input variable and an output variable while giving a high score to strongly correlated features and a low score to weakly correlated features.
- Mutual Information (MI) that measures the reduction in uncertainty between each input variable and the target variable. The features in the set are classified according to their MI value. A feature with a low MI value implies that it does not have much effect on the classification. Therefore, features with a low MI value can be discarded without affecting the performance of the models [31].

3.3 Third Phase: Standalone and Ensemble Learning Techniques

To validate our framework, we use two types of ML algorithms:

- Standalone algorithms such as multilayer perceptron (MLP), decision tree (DT), and k-nearest neighbors (kNN).
- Ensemble algorithms such as random forest (RF), extra tree (ET), and bagging (BAG).

We used the Scikit-learn library implementation of these algorithms [32]. The choice of these algorithms' hyperparameters has an impact on their performance. For this study, we have used the default values specified by Scikit-learn as they work reasonably well. It should be noted that the hyperparameters may be set using grid search or randomized search, but these methods are slow and costly.

4 Experimental Results

This section presents two strategies to check the results obtained by our proposed framework. First, we evaluate the performance of ML algorithms presented above before and after using the SMOTE method. Then, we outline the most relevant features according to the two filter-based feature selection methods: the ANOVA F-value test and the Mutual Information. All experiments were performed using ten-fold cross-validation.

4.1 Performance Metrics

To measure the performance of ML models, we used different metrics, such accuracy and F-measure. These metrics are calculated from four basic measures assessed for each class:

- True positive of the class C_i (TP_i)
- True negative of the class $C_i(TN_i)$
- False negative of the class $C_i(FN_i)$
- False positive of the class $C_i(FP_i)$

with $i \in$ {Benign, SYN flood attack, UDP flood attack, and Slowloris attack}

In the following, we present these metrics calculated according to these outcomes:

- Accuracy represents the ratio of correctly recognized records to the entire test dataset. It is measured as follows:

Table 2 Multi-class confusion matrix to illustrate TP_{Benign}, TN_{Benign}, FN_{Benign}, FP_{Benign}, and MS_{Benign}

	Benign	**SynFlood**	**UDP Flood**	**Slowloris**
Benign	TP_{Benign}	FN_{Benign}	FN_{Benign}	FN_{Benign}
Slowloris	FP_{Benign}	TN_{Benign}	MS_{Benign}	MS_{Benign}
SynFlood	FP_{Benign}	MS_{Benign}	TN_{Benign}	MS_{Benign}
SynFlood	FP_{Benign}	MS_{Benign}	MS_{Benign}	TN_{Benign}

$$\text{Accuracy} = \frac{\sum_{i=1}^{l} \frac{TP_i+TN_i}{TP_i+FN_i+FP_i+TN_i}}{l} \tag{2}$$

- F-score (Eq. 3) is used to measure precision (Eq. 4) and recall (Eq. 5) at the same time. The F-score is the harmonic mean of precision and recall values and reaches its best value at 1 and worst value at 0. It is calculated as follows:

$$F-\text{score} = \frac{2 * \text{Recall} * \text{Precision}}{\text{Recall} + \text{Precision}} \tag{3}$$

$$\text{Precision} = \frac{\sum_{i=1}^{l} \frac{TP_i}{TP_i+FP_i}}{l} \tag{4}$$

$$\text{Recall} = \frac{\sum_{i=1}^{l} \frac{TP_i}{TP_i+FN}}{l} \tag{5}$$

l is the number of classes.

We also propose to use the confusion matrix (CM) as it is representing performance results in an intuitive way for non-experts in ML. Each column of the matrix represents the instances in a predicted class, while each row represents the instances in a real class. For example, we present in Table 2 a multi-class confusion matrix to illustrate TP_{Benign}, TN_{Benign}, FN_{Benign}, and FP_{Benign}. TP_{Benign} refers to the normal instances that are correctly classified, TN_{Benign} means attack instances (SYN flood, UDP flood, and Slowloris) that are correctly predicted, FN_{Benign} refers to the normal instances that are classified as attacks (i.e., false alarms that are triggered without a real attack), and FP_{Benign} means attack instances that are predicted as normal traffic. The diagonal of the matrix represents the well-classified instances (TP_{Benign} and TN_{Benign}). MS_{Benign} means attacks that are classified as other attacks.

4.2 Evaluation of ML Models Before and After SMOTE

Tables 3 and 4 show the detection performance of the standalone and ensemble models before and after oversampling with the SMOTE method, respectively. Looking at these results, we can see that, whatever the used algorithm, the accuracy is high in the original dataset. It exceeds 98% for all models. For kNN and MLP, the accuracy of the original dataset is even higher than that of SMOTE. Therefore, these results are incorrect because when the classes are not balanced, the minor classes have a negative effect on the accuracy. Therefore, the *F*-score is the best metric when working with an unbalanced dataset.

By analyzing these tables, we can see also that *F*-score values of DT, MLP, BAG, RF, and ET models are improved after oversampling by the SMOTE method. On the other hand, the *F*-score value of the kNN model decreased after oversampling by the SMOTE method, and this shows that the algorithm is not influenced by the class distribution. The model gave better results on the unbalanced dataset. Further observations show that DT, BAG, ET, and RF have the best accuracy using SMOTE (no significant difference). That's why we focus on those classifiers in the following.

To help non-experts in ML understand the performance of models after using SMOTE method, we present in Tables 5, 6, 7, and 8 the confusion matrices of the DT, BAG, RF, and ET models, respectively.

These confusion matrices show that the different models globally correctly classify "Benign" instances and instances of different attacks. In other words, BAG and ET contain less false alarms than DT and RF (see orange columns). We get 3 false alarms for BAG and ET, 5 false alarms for DT and 35 for RF. We thus observe that the models classify very well Slowloris and SYN flood attacks but less for SYN flood attacks.

Table 3 Evaluation of standalone models before and after SMOTE

	Standalone models					
	DT		kNN		MLP	
Methods	Accuracy	*F*-score	Accuracy	*F*-score	Accuracy	*F*-score
None	99.998	0.99960	99.814	0.99704	98.113	0.72191
SMOTE	99.998	**0.99998**	99.672	0.99672	94.176	**0.94150**

Table 4 Evaluation of ensemble models before and after SMOTE

	Ensemble models					
	BAG		RF		ET	
Methods	Accuracy	*F*-score	Accuracy	*F*-score	Accuracy	*F*-score
None	99.998	0.99978	99.991	0.99985	99.999	0.99977
SMOTE	99.998	**0.99998**	99.991	**0.99991**	99.999	**0.99999**

Table 5 Multi-class confusion matrix after SMOTE for DT

	Benign	SynFlood	UDP Flood	Slowloris
Benign	124963	2	2	1
Slowloris	0	124968	0	0
SynFlood	2	0	124966	0
SynFlood	0	0	0	124968

Table 6 Multi-class confusion matrix after SMOTE for BAG

	Benign	SynFlood	UDP Flood	Slowloris
Benign	124965	1	2	0
Slowloris	0	124968	0	0
SynFlood	4	0	124964	0
SynFlood	1	0	0	124967

Table 7 Multi-class confusion matrix after SMOTE for RF

	Benign	SynFlood	UDP Flood	Slowloris
Benign	124933	0	35	0
Slowloris	0	124968	0	0
SynFlood	5	0	124963	0
SynFlood	0	0	0	124968

Table 8 Multi-class confusion matrix after SMOTE for ET

	Benign	SynFlood	UDP Flood	Slowloris
Benign	124933	0	35	0
Slowloris	0	124968	0	0
SynFlood	5	0	124963	0
SynFlood	0	0	0	124968

4.3 Feature Selection and Analysis

In Table 9, we present the performance of the different ML models incorporating the two feature selection methods, ANOVA F-value and Mutual Information, while varying the number of selected features.

The results analysis can be concluded in the following points:

- Of the two feature selection methods implemented, mutual information is comparatively the better performing.
- Among the 4 classifiers implemented, RF and ET give the best accuracies by varying the number of features from 10 to 45.
- Feature selection method using Mutual Information identifies features that have the strongest impact on the prediction. As an example, we can see in Table 10, 10, 12, and 25 features selected by Mutual Information.

Table 9 Comparison between the performance of ANOVA F-value and Mutual Information

		Accuracy			
Feature selection method	Number of features	DT	BAG	RF	ET
ANOVA F-value	**10**	95.746	95.757	95.757	96.969
	12	95.743	95.820	95.760	96.970
	25	99.612	99.618	99.497	99.604
	30	99.709	99.728	99.612	99.723
	35	99.717	99.729	99.597	99.710
	40	99.708	99.728	99.600	99.715
	45	99.709	99.734	99.584	99.709
Mutual Information	**10**	99.979	99.986	**99.990**	99.988
	12	99.992	99.993	99.994	**99.995**
	25	99.993	99.993	99.995	**99.996**
	30	99.993	99.994	**99.996**	99.995
	35	99.994	99.995	**99.996**	**99.996**
	40	**99.998**	99.997	**99.998**	**99.998**
	45	**99.998**	99.998	99.993	**99.999**

Table 10 The selected features using mutual information

Number of features	Features
10	Flow Duration, Flow Pkts/s, Flow IAT Mean, Flow IAT Max, Flow IAT Min, Fwd Header Len, Bwd Header Len, Fwd Pkts/s, Bwd Pkts/s, Timestamp_hour
12	Flow Duration, Flow Pkts/s, Flow IAT Mean, Flow IAT Std, Flow IAT Max, Flow IAT Min, Fwd Header Len, Bwd Header Len, Fwd Pkts/s, Bwd Pkts/s, Init Bwd Win Byts, Timestamp_hour
25	Protocol, Flow Duration, Flow Pkts/s, Flow IAT Mean, Flow IAT Std, Flow IAT Max, Flow IAT Min, Fwd IAT Tot, Fwd IAT Mean, Fwd IAT Max, Fwd IAT Min, Bwd IAT Tot, Bwd IAT Mean, Bwd IAT Std, Bwd IAT Max, Fwd Header Len, Bwd Header Len, Fwd Pkts/s, Bwd Pkts/s, SYN Flag Cnt, Init Bwd Win Byts, Idle Mean, Idle Max, Idle Min, Timestamp_hour

5 Conclusion

VANETs suffer from several vulnerabilities due to the inherent characteristics of vehicles and the open radio environment. Security of VANET communications is hence a critical issue due to the diversity of VANET applications, architectures, and characteristics. Many works have been done to study security attacks and countermeasures that can tackle VANET vulnerabilities. Intrusion detection systems (IDS) are an efficient way to detect and prevent malicious activities; hence, they are necessary before triggering the appropriate countermeasure. The use of machine learning techniques is particularly interesting to tackle unknown and zero-day attacks.

In our work, we introduced a novel comprehensive framework to design VANET IDS. Furthermore, unlike most existing works, we use a very recent dataset to evaluate and compare both ensemble learning and standalone learning techniques to detect various types of DOS and DDOS attacks.

For data preprocessing phase, and after data cleaning, normalization, and transformation, we adopted the Synthetic Minority Oversampling Technique (SMOTE) for class balancing; then, we used ANOVA F and Mutual Information for selecting the most relevant features. Afterward, we applied several standalone ML techniques and ensemble ML techniques.

Experiments showed that using SMOTE improves F-score for both standalone and ensemble ML methods. When comparing the two considered feature selection methods, ANOVA F-value and Mutual Information, while varying the number of selected features, we noticed that Mutual Information performs better and is able to identify features that have the strongest impact on the prediction. Moreover, among the four classifiers implemented, RF and ET give the best accuracies by varying the number of features from 10 to 45.

Incorporating ML techniques when designing IDS is undoubtedly efficient, but due to the numerous opportunities that ML techniques offer, more works are still needed to investigate the design of the best ML framework for VANET IDS. For instance, federated learning is a promising approach that can adapt better to the distributed nature of VANET communication by alleviating the vehicle from a big amount of data processing.

Acknowledgment We would like to thank the research team of the Networks and Systems Laboratory-LRS, Department of Computer Science, Badji Mokhtar University, Annaba, Algeria, for sharing with us their work on the VDOS-LR security dataset.

References

1. A. Ghosal, M. Conti, Security issues and challenges in V2X: a survey. Comput. Netw. **169**, 107093, ISSN:1389-1286 (2020)
2. A. Alnasser, H. Sun, J. Jiang, Cyber security challenges and solutions for V2X communications: a survey. Comput. Netw. **151**, 52–67 (2019)
3. N.A. Alsulaim, R. Abdullah Alolaqi, R.Y. Alhumaidan, Proposed solutions to detect and prevent DoS attacks on VANETs system, in *3rd International Conference on Computer Applications & Information Security (ICCAIS)*, (2020), pp. 1–6
4. K. Stępień, A. Poniszewska-Marańda, Security methods against Black Hole attacks in Vehicular Ad-Hoc Network, in *IEEE 19th International Symposium on Network Computing and Applications (NCA)*, (2020), pp. 1–4
5. J. Montenegro, C. Iza, M.A. Igartua, Detection of position falsification attacks in VANETs applying trust model and machine learning, in *PE-WASUN '20: Proceedings of the 17th ACM Symposium on Performance Evaluation of Wireless Ad Hoc, Sensor, & Ubiquitous Networks*, (2020), pp. 9–16
6. A. Alshammari, M.A. Zohdy, D. Debnath, G. Corser, Classification approach for intrusion detection in vehicle systems. Wirel. Eng. Technol. **9**(4), 79–94 (2018)
7. M.A. Al-Shareeda, M. Anbar, S. Manickam, A. Khalil, I.H. Hasbullah, Security and privacy schemes in vehicular Ad-Hoc network with identity-based cryptography approach: a survey. IEEE Access **9**, 121522–121531 (2021)
8. D. Koo, Y. Shin, J. Yun, J. Hur, An online data-oriented authentication based on Merkle tree with improved reliability, in *2017 IEEE International Conference on Web Services (ICWS)*, (2017), pp. 840–843
9. R. Barskar, M. Ahirwar, R. Vishwakarma, Secure key management in vehicular ad-hoc network: a review, in *International Conference on Signal Processing, Communication, Power and Embedded System (SCOPES)*, (2016), pp. 1688–1694
10. D. Manivannan, S.S. Moni, S. Zeadally, Secure authentication and privacy-preserving techniques in Vehicular Ad-hoc NETworks (VANETs). Veh. Commun. **25**, 100247 (2020) ISSN:2214-2096
11. N. Parikh, M.L. Das, Privacy-preserving services in VANET with misbehavior detection, in *IEEE International Conference on Advanced Networks and Telecommunications Systems (ANTS)*, (2017), pp. 1–6
12. L. Chen, S. Ng, G. Wang, Threshold anonymous announcement in VANETs. IEEE J. Sel. Areas Commun. **29**, 605–615 (2011)
13. S.S.L. André Perez, *TLS and DTLS Protocols, Network Security* (Wiley). ISBN:9781848217584
14. F.A. Ghaleb, A. Zainal, M.A. Rassam, F. Mohammed, An effective misbehavior detection model using artificial neural network for vehicular Ad hoc network applications, in *IEEE Conference on Application, Information and Network Security (AINS)*, (2017), pp. 13–18
15. P.K. Singh, R.R. Gupta, S.K. Nandi, S. Nandi, Machine learning based approach to detect wormhole attack in VANETs, in *Workshops of the International Conference on Advanced Information Networking and Applications*, (Springer, 2019), pp. 651–661
16. D. Tian, Y. Li, Y. Wang, X. Duan, C. Wang, W. Wang, R. Hui, P. Guo, An intrusion detection system based on machine learning for can-bus, in *International Conference on Industrial Networks and Intelligent Systems*, (Springer, 2017), pp. 285–294
17. S.C. Kalkan, O.K. Sahingoz, In-vehicle intrusion detection system on controller area network with machine learning models, in *11th International Conference on Computing, Communication and Networking Technologies (ICCCNT)*, (2020), pp. 1–6
18. T.P. Vuong, G. Loukas, D. Gan, A. Bezemskij, Decision tree-based detection of denial of service and command injection attacks on robotic vehicles, in *IEEE International Workshop on Information Forensics and Security (WIFS)*, (2015), pp. 1–6

19. T.P. Vuong, G. Loukas, D. Gan, Performance evaluation of cyber-physical intrusion detection on a robotic vehicle, in *IEEE International Conference on Computer and Information Technology; Ubiquitous Computing and Communications; Dependable, Autonomic and Secure Computing; Pervasive Intelligence and Computing*, (2015)
20. L. Yang, A. Moubayed, I. Hamieh, A. Shami, Tree-based intelligent intrusion detection system in internet of vehicles, in *IEEE Global Communications Conference (GLOBECOM)*, (2019)
21. S. Iranmanesh, F. S. Abkenar, A. Jamalipour and R. Raad. A heuristic distributed scheme to detect falsification of mobility patterns in internet of vehicles.. IEEE Internet Things J., 2021.
22. A.R. Gad, A.A. Nashat, T.M. Barkat, Intrusion detection system using machine learning for vehicular Ad hoc networks based on ToN-IoT dataset. IEEE Access **9**, 142206–142217 (2021)
23. D.M. Kang, S.H. Yoon, D.K. Shin, Y. Yoon, H.M. Kim, S.H. Jang, A study on attack pattern generation and hybrid MR-IDS for in-vehicle network, in *International Conference on Artificial Intelligence in Information and Communication (ICAIIC)*, (2021), pp. 291–294
24. D. Swessi, H. Idoudi, A comparative review of security threats datasets for vehicular networks, in *International Conference on Innovation and Intelligence for Informatics, Computing, and Technologies (3ICT)*, (2021), pp. 746–751
25. R. Rahal, A. Amara Korba, N. Ghoualmi-Zine, Towards the development of realistic DoS dataset for intelligent transportation systems. Wirel. Pers. Commun. **115**, 1415–1444 (2020)
26. A. Habibi Lashkari, CICFlowMeter (formerly known as ISCXFlowMeter): a network traffic Bi-flow generator and analyzer for anomaly detection 2018. https://github.com/ahlashkari/CICFlowMeter
27. P.D. Gutiérrez, M. Lastra, J.M. Benítez, F. Herrera, Smote-gpu: big data preprocessing on commodity hardware for imbalanced classification. Prog. Artif. Intell. **6**(4), 347–354 (2017)
28. N.V. Chawla, K.W. Bowyer, L.O. Hall, W.P. Kegelmeyer, SMOTE: synthetic minority over-sampling technique. J. Artif. Intell. Res. **16**, 321–357 (2002)
29. R. Alshamy, M. Ghurab, S. Othman, F. Alshami, Intrusion detection model for imbalanced dataset using SMOTE and random forest algorithm, in *International Conference on Advances in Cyber Security*, (Springer, Singapore, 2021), pp. 361–378
30. J. Cai, J. Luo, S. Wang, S. Yang, Feature selection in machine learning: a new perspective. Neurocomputing **300**, 70–79 (2018)
31. A. Thakkar, R. Lohiya, Attack classification using feature selection techniques: a comparative study. J. Ambient Intell. Human. Comput. **1**, 1249–1266 (2021)
32. F. Pedregosa, G. Varoquaux, A. Gramfort, V. Michel, B. Thirion, O. Grisel, E. Duchesnay, Scikit-learn: machine learning in Python. J. Mach. Learn. Res. **12**, 2825–2830 (2011)
33. U. Coruh, O. Bayat, ESAR: enhanced secure authentication and revocation scheme for vehicular Ad Hoc networks. J. Inf. Secur. Appl. **64** (2022). Elsevier
34. H. Bangui, M. Ge, B. Buhnova, A hybrid machine learning model for intrusion detection in VANET. Computing, Springer (2021)

Part IV
Mobile Applications Security

The Implementation of Uncertainty Models for Fraud Detection on Mobile Advertising

Jinming Ma, Tianbing Xia, and Janusz Getta

1 The Competition Between Fraud and Anti-fraud on Mobile Advertisements

Mobile advertisement fraud has been growing ever since mobile Internet became popular. From click redirections to user APPs, many different fraud methods are developed in the last decade. Some of these methods are easy to identify, while some are not. Besides these, some methods are specially designed to avoid fraud detection. These methods are making mobile anti-fraud become a more and more challenging task for mobile Internet companies.

Crussell, J., R. Stevens, and H. Chen [1] identified two types of fraud methods and developed a tool targeting to analyze the fraudsters. Song, L et al. [2] focused on predicting and detecting click fraud on a large scale. Tian T., et al. [4] described a specially designed anti-fraud method against crowd fraud (device-based fraud in this article), and Oentaryo R., et al. [5] used a method of data mining approach to detect click fraud on online advertisement. Besides these references, there are also a few websites that offer suggestions related to fraud detection. In [3], Pooranian Z., et al. introduced different fraud and anti-fraud methods. Tian T., et al. [4] described a specially designed anti-fraud method against device-based crowd fraud. Oentaryo R., et al. [5] showed a method of data mining approach to detect click fraud on online advertisements.

Mobile Internet advertisement fraud is a series of "user" actions. Usually, it aims to gain advertising expenses from advertisers. Mobile advertisement fraud is not only harmful but also attractive. Gaining much profit by doing only a few simple technical

J. Ma · T. Xia (✉) · J. Getta
University of Wollongong, Wollongong, NSW, Australia
e-mail: jm662@uow.edu.au; txia@uow.edu.au; jrg@uow.edu.au

K. Daimi et al. (eds.), *Emerging Trends in Cybersecurity Applications*,
https://doi.org/10.1007/978-3-031-09640-2_11

tricks without worrying about traffic at all is much easier than managing a real honest APP. More importantly, such technical tricks are easy to run, but hard to identify.

The most important difference between fraud actions and real user actions is that real user actions on advertisements usually show attention and interest in the advertisement. That means there's a probability that advertisers may gain profit from the users. On the contrary, fraud actions do not bring any user's attention to the advertisers and are nearly impossible for advertisers to gain profit. It can damage the interests of advertisers and the industry environment of mobile Internet at the same time.

Giving a proper definition of a mobile Internet advertisement fraudster is never an easy job. This is not only because of the difference in fraud methods but also the difference in the intention of different fraudsters. Professional fraudsters aim to gain profit from advertisers, and competitors try to consume competitive advertiser's advertisement budget. Some advertisers may even choose to click their advertisements so that the click rate is increased and their advertisements may win more traffic. Besides all the above, using inducing apps makes the definition and detection of mobile Internet fraud much more complicated.

Rough set theory shows great value in describing uncertainty. Suraj, Z. [6] explained the rough set theory with easy-understanding examples. A survey about the development of rough set theory has been given by Q. Zhang et al. and Tsumoto and Shusaku. [7, 8] showed the history and possibilities in the future of rough set theory. Cornelis, Chris et al. [9] combined fuzzy set theory and rough set theory.

Competitions between mobile advertisement fraud and anti-fraud are two competitors trying to suppress each other. Each side of this competition tries their best to create new methods and defeat the competitor. In this research, we implement methods by using fuzzy set theory to demonstrate how to detect cheaters and developed applications using rough set theory to detect fraudsters. The methods for mobile anti-fraud advertisements have the potential of ending this circle. The analysis in this research and the implementation of these two uncertainty models could be the solutions for mobile Internet advertisement anti-fraud systems.

One of the main problems of mobile anti-fraud is the lack of evidence to prove a user to be a fraudster. In this research, we implement uncertainty of fuzzy theory to demonstrate how to detect cheaters. The advantage of this method is that the hardship in detecting fraudsters in small data samples can be avoided. We achieve this by giving each user a suspicious degree showing how likely the user is cheating and deciding whether a group of users (like all users of a certain APP) together could be fraudsters according to the average suspicious degree. This makes the process more accurate as the data of a single user is too small to be predictable. We implement the uncertainty method of rough set theory. The advantage of this anti-fraud method proposed in this work is that the method is hard to counter. It means that avoiding the detection of this method is very difficult for fraudsters.

Ever since smartphone and mobile Internet became popular, mobile advertisement fraud and anti-fraud have become two competitors both trying to suppress the other. Every time a new fraud method is developed, a specially designed anti-fraud technique will come out soon. After that new fraud methods will keep coming out to avoid being detected. The method does not target any fraud attempts, but it observes the differences between user groups. If the fraudsters do not own related data of real user groups, it is almost impossible for fraudsters to avoid being detected by this method. The implementation of the uncertainty models in this chapter has the potential of ending this circle. In this chapter, we will introduce the methods for mobile fraud detection in the following sections.

2 Analysis Fraud and Risk with Fuzzy and Rough Sets on Mobile Advertising Fraud Detection

Unlike analyzing data problems, such as identifying the users' gender or age, the most serious problem in the mobile anti-fraud area is the hardship in gaining a data set with exactly marked fraudsters. It is almost impossible to get a data sample like that, as even if one gets such a data sample, there is no way of proving the data is correct. This problem blocks the application of supervised learning in the anti-fraud area since there is no training data for a model to learn. In another word, one of the most serious problems in the mobile anti-fraud area is the uncertainty of fraudsters' identification. Thus, to solve the problem of uncertainty, we are applying fuzzy and rough set theory to mobile fraud detection. Fuzzy sets and rough sets address two important and mutually orthogonal characteristics of imperfect data and knowledge. While the former allows that objects belong to a set or relation to a given degree, the latter provides approximations of concepts in the presence of incomplete information [10, 11]. In this work, we demonstrate how we implement these two theories to analyze and detect fraud and risk on mobile advertisements.

2.1 The Implementation of Fuzzy Set Theory to Mobile Fraud Detection

The reason is the hardship in the exact detection of fraudsters. As the behavior of normal users is unpredictable, there is no way to distinguish fraudsters very precisely from honest users. No number could categorically distinguish fraudsters from normal users. To solve this problem, we propose to apply a method based on fuzzy set theory. If it is hard to identify one fraud user, then we can calculate a suspicious degree of how likely a user is cheating. It can be done for every user instead of dividing the users into two categories.

2.1.1 Fuzzy Set and Uncertainty Define

When dealing with uncertainty problems, probability is probably the first method that comes to people's minds. Sometimes, however, even probability itself is uncertain. In this case, fuzzy set theory is a useful tool to describe uncertainty. The definitions of fuzzy subset [12] are below.

Fuzzy subset: Given a set A, a fuzzy subset B of A is defined by its membership function $B(x)$ with values in [0,1] for all x in A.

If $B(x_0) = 1$, then x_0 belongs to B.
If $B(x_0) = 0$, then x_0 does not belong to B.
If $B(x_0) = h$, where $0 < h < 1$, then the membership degree of x_0 in B is h.

Triangular fuzzy number: Given three real numbers $a < b < c$, then a triangular fuzzy number $N = (a/b/c)$ can be defined as a fuzzy subset of R with membership function defined as below.

$$f(n) = \begin{cases} 0 \ (x \leq a \text{ or } x > c) \\ \frac{b-x}{b-a} \ (a < x \leq b) \\ 1 - \frac{c-x}{c-b} \ (b < x \leq c) \end{cases}$$

In this chapter, other types of fuzzy numbers will not be discussed. Thus, when referring to a fuzzy number, we assume that it is always a triangular fuzzy number.

$\alpha-$Cuts: Given a fuzzy number N, an α–cut of N is defined as $N(\alpha) = \{x \in R | N(x) \geq \alpha\}$ where $0 < \alpha \leq 1$.

2.1.2 Fuzzy Statistics on Anti-fraud Methods

Consider the overactive check method. Let U be a set of all users recorded in an hour. Let set O be the fuzzy subset of all overactive users in U. The membership function of a user in a certain hour can be defined as

$$O(u) = \max\ (\min(1, \max(0, 10\text{s} - \text{MT D})),\\ \min(1, \text{costeventdensity} \times \text{costeventrate})),$$

where MTD is the minimum time difference between two cost events (if exist). This membership function is called suspicious degree and is denoted as *susp-degree*. The suspicious degree of a user in an hour is used to define the suspicious degree of an APP on a day. The suspicious degree of an APP on a day is an α −cut of a fuzzy number: (min(susp−degree)/avg(susp−degree)/max(susp−degree)). This is not typically a real α −cut of a fuzzy number, as the fuzzy number and the α are both undefined. We can always assume such a number and α satisfying the suspicious degree in the given circumstance always exist. This suspicious degree of an APP on a day is the result that determines the validity of a user. The avg (susp − degree) will show how likely the whole APP is cheating, while the min(susp−degree) and max(susp−degree) will show the behavior of normal users and fraudsters in the APP.

2.1.3 Anti-fraud Data Analyzing Processes

We focus on using fuzzy set theory to measure how suspicious the activities of a user or app can be. The measurement is called "suspicious degree." We can avoid concluding a small data sample, e.g., a single user with the calculation of suspicious degrees. We use the suspicious degrees to represent the most important feature of one log or user and make a conclusion after the collection of all suspicious degrees of a certain app or source.

We proposed three main processes to analyze different types of fraudsters. The processes include the *origin check process, overactive monitor process, and new user monitor process*. All data obtained from the recording of the users' activities are saved in seven relational tables: logs, origin-check, all-users, new-users, new-users-monitor, and sups-result.

Process Origin Check

To identify the attribution of misleading fraudsters, we designed the *origin check process*. Each time a download type logs with *event* = *cost_event*, this process will check if a show event log with the same ads_id exists in the logs table. If not, the log is very suspicious to be cheating. Otherwise, the process will calculate the sups_degree accordingly to the time difference between the show log and cost event log, as it is also suspicious if the time difference is rather small. This process starts each time a log is updated to the logs table (see Fig. 1).

Process Overactive Monitor

This process analyzes if any user is too active to be normal. Usually, users will not click or download advertisements frequently. On the contrary, fraudsters would report cost event logs in a much higher density. That would result in higher cost event rate, density, and smaller cost event time difference. The overactive monitor process will analyze these indexes to identify the fraudsters. The overactive monitor processes data from the logs every hour to analyze if users are overactive in an hour. Related detail can be seen in Fig. 2 (all operations are applied to the data of a certain hour).

The following is an example algorithm for overactive monitor process:

```
query the database
                let sel_1 be a tulple of cost event rate and
                density of every user in the logs table
                 each element of this tulple represents a user
for each tulple i in sel_1 do
                    timestamp=0, min_timedif=1000000
if i[costEvent_density]>=2 then
query the database
```

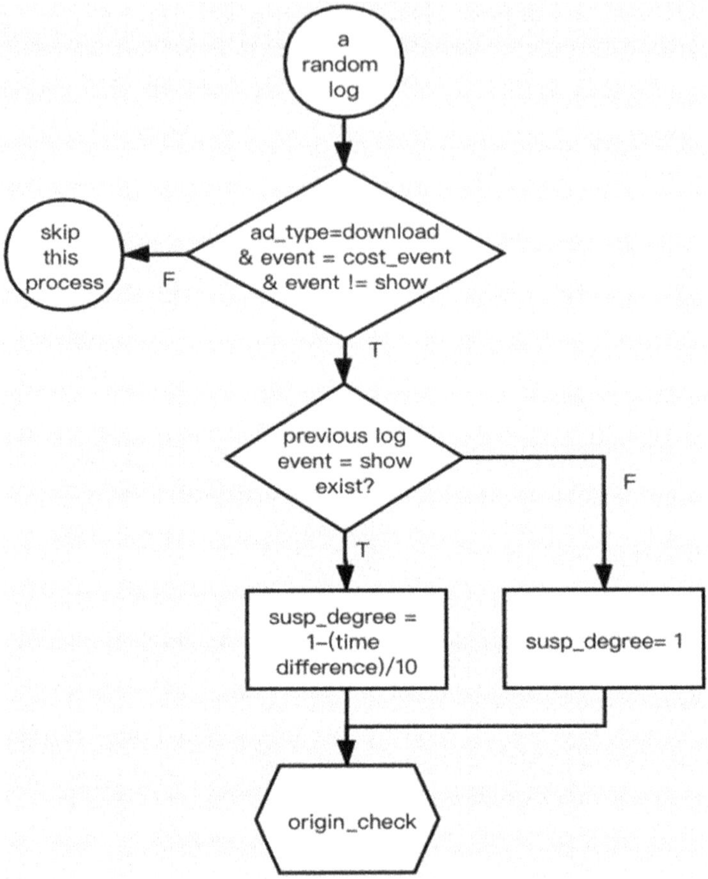

Fig. 1 Process origin check

```
                    let sel_2 be the tulple of timestamp of
                    the given user's all cost event logs
                     whose cost event is not show
                each element of this tuple represents a cost
                event
for each tulple j in sel_2 do
                        min_timedif  ←
                   min (min_timedif, j[timestamp] -
                   timestamp)),
                   timestamp ← j[timestamp]
end for
                   sups_degree ←
                     max (min (1, max (0, 1 - min timedif/
                     10000)),
                     min (1, i[costEvent_rate] * i[costEvent_
                     density]))
```

Fig. 2 Overactive monitor process

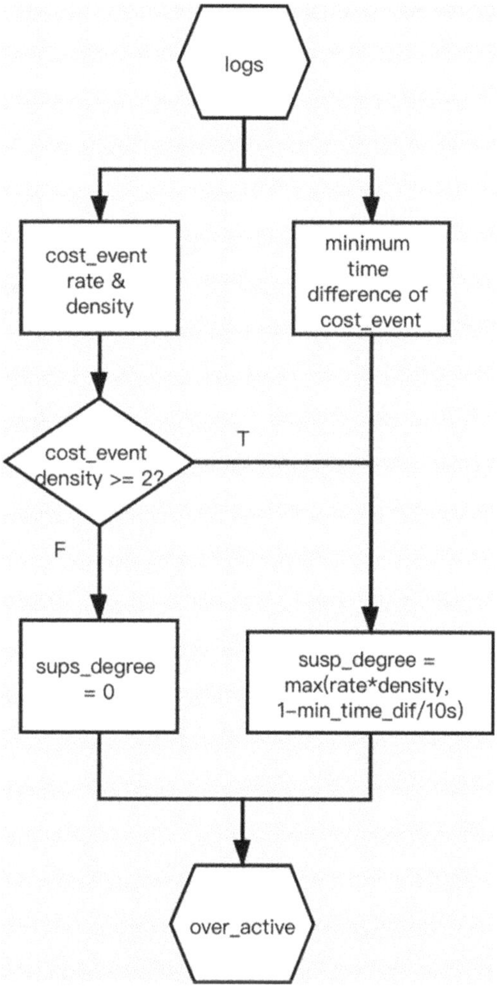


```
else
        sups_degree ← 0, min_timedif ← 1000000
end if
        insert into overactive values
        (i + (min_timedif, sups_degree))
end for
```

Process New User Monitor

The overactive process is effective against many fraudsters. But if the fraudsters, like server-based fraudsters, change their ID frequently, it will be hard for the overactive process to collect enough data for analysis. In this case monitoring the percentage of new users and cost events of new users, namely, NU_density and NU_action_density in the process, can easily identify cheating APPs. Every hour, the new user monitor

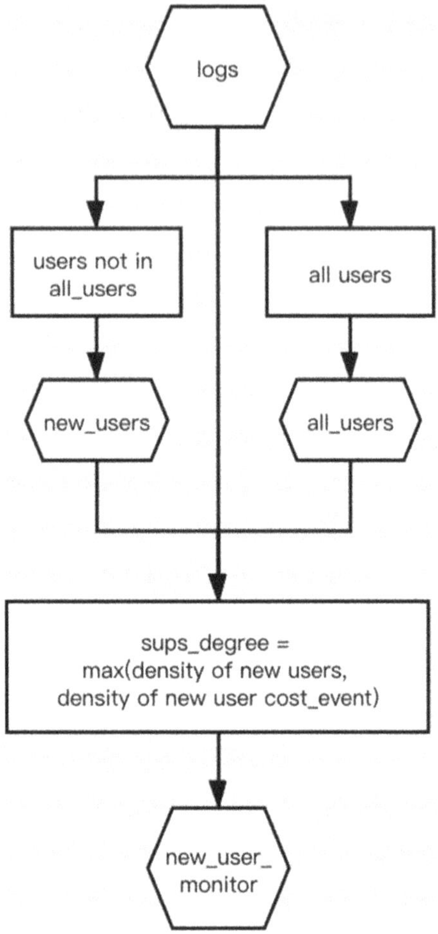

Fig. 3 Process new user monitor

process queries data from logs to analyze if any source or APP contains too many new users. Related detail can be seen in Fig. 3 (all operations are only applied to data of a certain hour). The data flow diagram that overviews the entire anti-fraud data analysis process can be seen in Fig. 4.

2.1.4 Testing Log Generator and Test Result

Testing log generator to test if the anti-fraud process is functional in identifying fraudsters, we designed a generator that could simulate normal users and fraudsters.

The generator will create logs from four different apps, namely, A, B, C, and D. Of the four apps, A is the only normal APP, with only real human behavior like users. B is the attribution misleading fraudster, which has a lot of common users as A and will always generate a cost event log if any of these common users reported

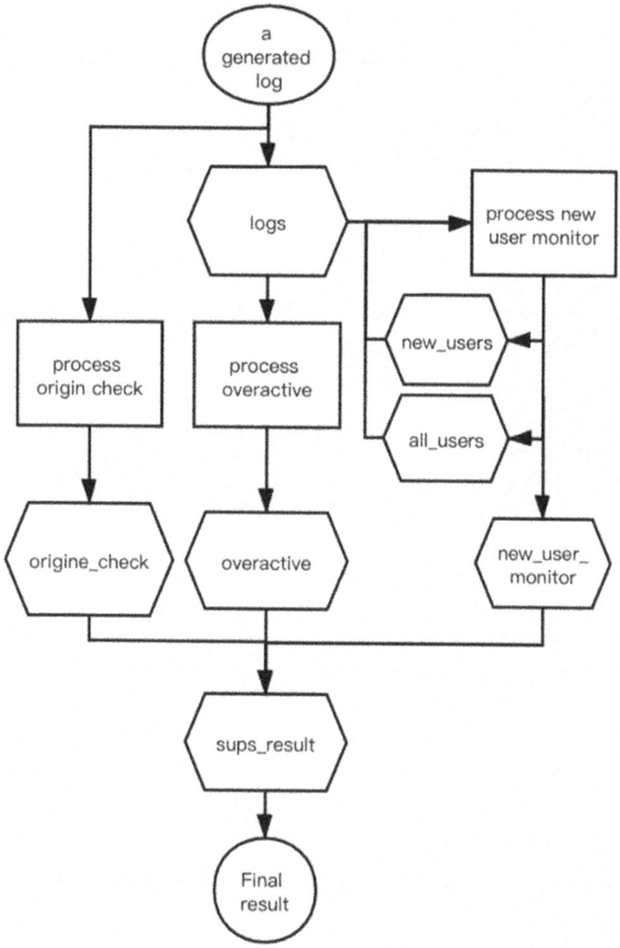

Fig. 4 Entire analyzing process

a cost event log of a download type advertisement on A. Apart from this, B is all the same as A. C is a sever-based fraudster, with the rate of both cost event and new user generation being much higher than A and B. D is a device-based fraudster, with the new user generation rate a little bit higher and cost event generation rate much higher than A and B. There's another difference between the four apps. Since A and B contain mostly human-like users, the time difference between different events of the two apps is a little bit longer than C and D. Also, as C is a fully automatic server-based fraudster, and D is supposed to be operated by a human, the time difference between events of C is even shorter than D, being the shortest of the four APPs. See the two test results on Table 1. A_1, B_1, C_1, and D_1 are for test result 1, and A_2, B_2,

Table 1 Test result for analyzing processes from testing log generator

Unique identity of APP	A_1	B_1	C_1	D_1	A_2	B_2	C_2	D_2
The minimum sups_degree in overactive	0	0	0	0	0	0	0	0
The average sups_degree in overactive.	0	15.6%	4.9%	64.6%	0	11.0%	4.3%	66.8%
The maximum sups_degree in overactive	0	1	1	1	0	1	1	1
The minimum sups_degree in new_user_monitor	4.9%	7.2%	98.3%	19.5%	10.5%	8.5%	97.9%	25.6%
The average sups_degree in new_user_monitor	4.9%	7.2%	98.3%	19.5%	10.5%	8.5%	97.9%	25.6%
The maximum sups_degree in new_user_monitor	4.9%	7.2%	98.3%	19.5%	10.5%	8.5%	97.9%	25.6%
The minimum sups_degree in origin_check	0	0	0	0	0	0	0	0
The average sups_degree in origin_check	41.0%	94.1%	2.2%	1.6%	0	89.0%	2.4%	4.2%
The maximum sups_degree in origin_check	81.0%	1	76.5%	72.4%	0	1	87.1%	85.2%
The number of logs in the origin_check	2	70	206	147	7	51	199	185

C_2, and D_2 are for test result 2. According to the test, the process is functioning as intended.

This work aims to solve the problem using the application of fuzzy statistics. By measuring how likely a log, user, or app is cheating, identifying fraudsters no longer needs to be 100% accurate. Instead, the suspicious degree could offer a numerical view for advertisers to analyze fraudsters. This method can be a solution for mobile Internet advertisement anti-fraud systems.

2.2 *Implement of Rough Set Theory on Mobile Fraud Detection*

It is hard to analyze whether a user group is a fraudster because many other attributes may influence the click rate. The best way is to analyze how the user actions are dependent on attributes. The definition of "dependent" requires knowledge of rough set theory. In rough set theory, an index called dependency can express how one certain group of attributes is dependent on another. The dependency of click actions on attributes like age, gender, platform, or even price and brand of user's device can measure the user's action. If the dependency of two user groups on one same advertisement is different, then it is very likely that one of the groups is a fraudster.

2.2.1 Rough Set Theory and Dependency

The advantage of using dependency on mobile anti-fraud is that this method focuses on the action of groups of people. One of the most serious hardships in anti-fraud is that fraudsters change their ID and IP frequently, which makes it hard for an anti-fraud process to track them down. However, if the new IDs and IPs are classified in the same group (like all users of a certain APP), changing ID or IP would be meaningless for our method. Another advantage of this method is that even if fraudsters are aware of this method, it is still hard for them to avoid being detected as both using and avoiding this method require sample data of real users, which is hard for fraudsters to gain.

Rough set theory is a mathematical theory that is designed to show the relationship between attributes in an information system. This work aims to use the index dependency of attributes in rough set theory to detect fraudsters. Thus, we will introduce how dependency is defined.

In the real world, data are usually stored in relational tables. To apply rough set theory to real-world data analysis, defining relational tables, namely, information systems, in mathematical language is very useful.

So that we can apply rough set theory to an information system, we need to first build an approximation space structure on it. The definition of indiscernibility relation on information systems [13] is below.

Given an information system $S = (U, A)$ and B, a subset of A, the *B-indiscernibility relation* (written as $\mathrm{IND}_S(B)$) can be defined as

$$\mathrm{IND}_S(B) = \left\{\left(x, x'\right) \in U^2 | \forall a \in B, a(x) = a\left(x'\right)\right\}$$

where $a(x)$ is the value of attribute a on element x.

As $\mathrm{IND}_S(B)$ is a subset of U^2, $\mathrm{IND}_S(B)$ suits the definition of indiscernibility relation. More importantly, it is obvious that $\mathrm{IND}_S(B)$ is an equivalence relation. Thus, we have successfully built an approximation space $(U, \mathrm{IND}_S(B))$ on the information system $S = (U, A)$.

Given an information system $S = (U, A)$ and a subset B of A, the following concepts can be easily defined.

Equivalent Class. For any x in U, the equivalent class x of *B-indiscernibility relation* is

$$[x]_B = \left\{x' \in U | \left(x, x'\right) \in \mathrm{IND}_S\,(\mathrm{B})\right\}$$

Lower Approximation. $\underline{B}(X) = \{x | [x]_B \subseteq X\}$

Upper Approximation. $\overline{B}(X) = \{x | [x]_B \cap X \neq \varnothing\}$

Accuracy of Approximation. $\alpha_B(X) = \dfrac{|\underline{B}(X)|}{|\overline{B}\ (X)|}$

Membership Degree. $\mu_X^B(x) = \frac{|[x]_B \cap X|}{|[x]_B|}$

The major goal of this fraud detection is to detect "unusual people." Theoretically, the behavior of different user groups should be similar if the user groups are large enough. Thus, if the behavior between two user groups is quite different, at least one of the two user groups is "unusual," which means very likely to be cheating.

This was only a theoretical method of anti-fraud in the past, as there was no index qualified for this job. In rough set theory, there is an index suitable to express user behavior for fraud detection. This index is called the dependency of attributes.

Dependency of Attributes. Given an information system $S = (U, A)$ and C, D being subset of A, the dependency of attributes C on attributes D in the universe of U is defined as

$$k_U (C, D) = \frac{| \cup_{X \in U/D} C(X) |}{| U |}$$

The dependency of attributes shows the accuracy when using one set of attributes to represent another set of attributes. In the real world, dependency is used to describe how strong one set of attributes can influence another set of attributes. If the dependency of D on C equals one, then the value of D is fully determined by the value of C. If the dependency equals zero, then C does not influence the value of D.

The behavior of different user groups is supposed to be similar. It is safe to assume that the influence of some attributes, like age or gender of users, on the behavior of users, whether they will click on an advertisement, is stable in different user groups. With this assumption, we can use the dependency to detect fraudsters. If the dependency of a click event on a group of attributes is quite different in different user groups, then one of the user groups must contain several fraudsters.

2.2.2 Calculate the Dependency Metric

To make the result more accurate, we use a dependency metric for fraud detection given a data set with user action of click or not and attributes of user's age, gender, platform, and location. Now we give the diagram to show how to calculate the dependency metric:

Given a user group A and an ad ID, the process first inquires the database for logs of the user group and advertisement. After obtaining the logs, the following algorithm computes the dependency metric (Fig. 5).

When dealing with real data, the number of logs with click = 1 will be much smaller than logs with click = 0. That would cause the dependency to be too small to study. The solution to this problem is easy: putting a higher weight on the logs with click = 1 will fix it.

Mobile fraud actions are the actions that intend to gain profit. It is easy to see from this definition. The main difference between normal user action and fraud action is the intention. The normal user actions come from the user's interests. Their actions are influenced by the attributes of users, such as age or gender. On the contrary, the fraud actions are not dependent on the users' attributes as the normal user actions. In other words, the dependency of a click event on attributes, such as age, gender, or location, should be different in the universe of normal users and the universe of fraudsters.

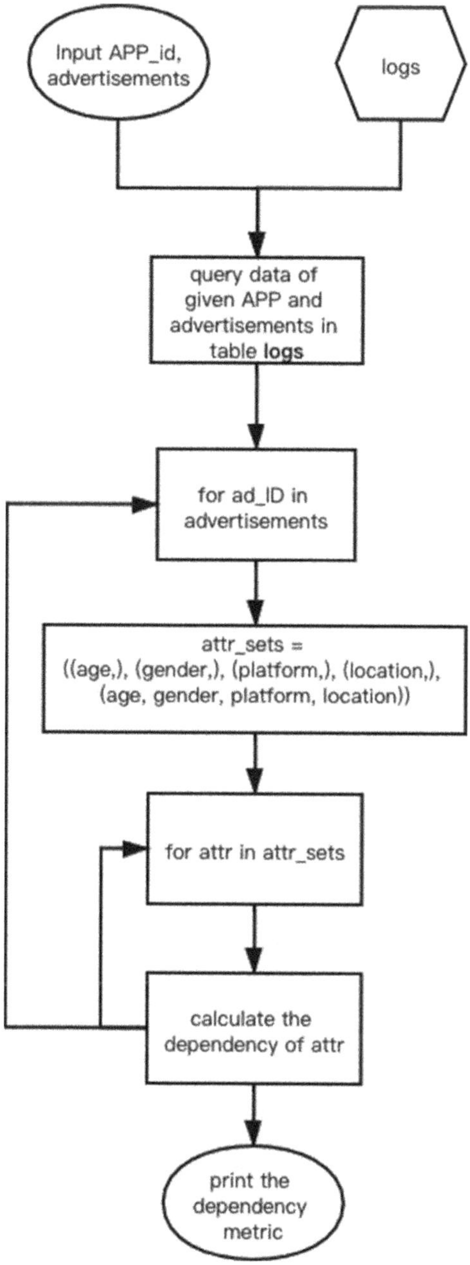

Fig. 5 Calculating the dependency metric

2.2.3 Test Result

As mentioned previously, identifying suspicious APPs will be enough for fraud detection. Thus, we designed five different APPs in our test, each representing one different APP. There are normal APPs and fraudsters' APPs. Users in different APPs have different possibilities to "click" on advertisements. In the real world, different advertisements have different target users. Different attributes also have different influences on whether target users will click or not. Such actions are not easy to simulate for fraudsters. The sever-based fraudster may identify the target users of an advertisement, but it is almost impossible for fraudsters to know the difference between different attributes.

To simulate this phenomenon, we design a series of columns for advertisements correspondence to the power of attributes. For example, we define a column power of age for an attribute age of an advertisement. The power of age describes how strong this attribute will influence the normal user's possibility of clicking an advertisement. We designed three advertisements for the test. The information on the power of attributes of each advertisement is shown below. To verify the fraud detecting process, we design five different APPs that represent normal APPs or fraudsters. Related information on each APP and the test result is written below. The APP1 is a normal APP. Thus, users in APP1 are more likely to click if they are target users of an advertisement. Also, different attributes have a different power to influence whether target users will click. These rules are both applied in APP1. See Table 2 for the result of dependency metric of APP1.

APP2 is a sever-based fraudster. The users in APP2 are false users; thus their actions are different from normal users like APP1. Target users in APP2 are more likely to click, but there's no difference in the power of different attributes. See Table 3 for the result of dependency metric of APP2.

APP3 is a real device-based fraudster. Since all cheating actions are operated by people, all the actions of target users in APP3 are the same as the other users. See Table 4 for the result of dependency metric of APP3.

APP4 is a user-inducing APP. Half of the users in APP4 behaves like users in APP1, while the other half behaves like users in APP3. See Table 5 for the result of dependency metric of APP4.

APP5 is also a normal APP. All the behaviors are set to be the same as APP1. See Table 6 for the result of dependency metric of APP5.

According to the test result, APP1 and APP5 are similar. This suits our assumption, as APP1 and APP5 are both normal APPs. In the result of APP2, as it is a

Table 2 Dependency metric of APP1

ad ID	D_age	D_gender	D_platform	D_location	main_Dependency
0001	0.388	0.324	0.042	0.055	0.435
0002	0.331	0.345	0.316	0.390	0.560
0003	0.056	0.058	0.027	0.053	0.141

Table 3 Dependency metric of APP2

ad ID	D_age	D_gender	D_platform	D_location	main_Dependency
0001	0.212	0.241	0.207	0.204	0.373
0002	0.213	0.226	0.237	0.190	0.375
0003	0.246	0.218	0.218	0.177	0.368

Table 4 Dependency metric of APP3

ad ID	D_age	D_gender	D_platform	D_location	main_Dependency
0001	0.021	0.012	0.018	0.018	0.057
0002	0.006	0.001	0.010	0.010	0.042
0003	0.013	0.001	0.009	0.009	0.047

Table 5 Dependency metric of APP4

ad ID	D_age	D_gender	D_platform	D_location	main_Dependency
0001	0.065	0.047	0.006	0.003	0.095
0002	0.052	0.064	0.074	0.055	0.113
0003	0.007	0.006	0.007	0.013	0.063

Table 6 Dependency metric of APP5

ad ID	D_age	D_gender	D_platform	D_location	main_Dependency
0001	0.338	0.339	0.039	0.062	0.404
0002	0.344	0.340	0.348	0.336	0.562
0003	0.017	0.073	0.066	0.057	0.149

server-based fraudster, APP2 can identify the target users for each app; thus the dependency of each attribute is not as small as APP3 or APP4, but as APP2 is not aware of the power of each attribute, different attributes have little difference in the result. APP3 and APP4 are similar; as they both contain real device-based fraudster, the influence of attributes is so small that the dependencies are all nearly zero.

Thus, the test result shows that our process has the potential of detecting fraudsters. We say potential because the process is tested only by generated date. Besides, as the difference between the power of different attributes and different advertisements is too idealized, it remains a question whether one can find advertisements as good as they are in the test. In conclusion, the process is useful in mobile anti-fraud, but not a practical method yet. To be applied in the real world, there are much more work and tests to be done.

3 Potential Impact on Future Applications for Online Media Anti-fraud Detection

In this chapter we implement uncertainty model theories for online advertising fraud detection. The fuzzy set and rough set theories can be applied not only on mobile advertisement anti-fraud but also in many other areas, such as big data analysis and online media fraud detection [14]. As it is designed specifically for big data analysis, there are strong potential advantage for the future media cybersecurity.

3.1 Reduce or Endow Weight to the Dimensions of Parameters for Machine Learning

Machine learning technology is used in the mobile Internet industry in almost every area. For example, the click-through rate (*CTR*) model is the ratio of users who click on a specific link to the number of total users who view a page, email, or advertisement and usually requires supervised learning technology. The function of the CTR model is to predict the possibility of a user clicking on an advertisement, an article, or anything that shows on an APP. This model is essential for mobile APP companies, as it is the core of personalized push systems. A good CTR model can always send users what they are interested in, which brings a good user experience. In the commercial area, CTR model is not only important for the user experience but also related to the profit of advertisements. One of the most common types of billing for mobile advertising is the cost-per-click type [15]. This means that the mobile APP company will not gain any income unless a user clicks an advertisement. In this case, sending users advertisements that do not catch their interest is not profitable. Thus, increasing the accuracy of the CTR model would benefit both user experience and advertising profit. As suggested in the previous sections, the part dependencies of an advertisement's click rate are different with different attributes. For example, the dependency of a makeup advertisement's click rate on age is probably smaller than the dependency on gender. Such differences for different advertisements have not been applied. Thus, if the dependency on an attribute is high, we endow a higher weight on the attribute. On the other hand, if the dependency is very low, we may consider removing the attribute.

3.2 Data Analysis Support for Advertisers

In the mobile advertisement area, data analysis is very important for advertisers to find data flow is advantaged or disadvantaged. Advertisers need to decide which group of users they prefer to advertise. If an advertiser decides to show an advertisement to every user on an APP [16], then the advertiser will probably pay for a lot of traffic that does not belong to their target users. If an advertiser aims to a small group of users, the advertiser may miss many potential customers. The part dependency of attributes may be of great help for it. Nowadays, when advertisers are analyzing data, they can learn whether elder people are more interested in their advertisement, but it is hard to learn whether gender has a stronger influence on the users' attitude than age. Part dependency is a good fit for this role.

3.3 Membership Degree for Fuzzy Tagging

Mobile Internet personalized push system requires giving tags for users. Usually, many users value their privacy and would share their personal information on the Internet. In this case, some companies would use algorithms to tag the users with different attributes like age or gender. These attributes are valued by algorithms. They are not always accurate. Thus, giving the tags an index showing how likely the tag is true would be helpful for further data analysis. With the data that comes from real users who agree to share their personal information, we can use the membership degree from part rough set theory to describe how likely the user belongs to a certain user group, namely, how likely a certain tag is to be true. If a user is not 100% likely to be a male, then a lower weight should be put on the user when advertising an advertisement targeting only males. This process can be called fuzzy tagging. The membership degree can be applied to the CTR model with fuzzy tagging.

4 Conclusion and Future Work

In the current online media world, so much advertising is running on the Internet for different purposes. Unfortunately, many people are using fraudulent advertising applications to gain profits. In this work, we presented uncertainty models for mobile advertisement fraud detection. One of the most severe hardships is the little evidence to prove whether a user is a fraudster. This method aims to solve the problem using the application of fuzzy set theory. By measuring how likely a log, user, or app is cheating, identifying fraudsters no longer need to be 100% accurate. Instead, the suspicious degree could offer a numerical view for advertisers to analyze fraudsters. This method can be an effective solution for mobile Internet advertisement anti-fraud systems. We designed an algorithm using fuzzy rough set theory to detect mobile fraudsters. In our test, the algorithm proved to have the potential for fraud detection. In this research, we also developed an anti-fraud method with the rough set theory. The method calculates and compares the part dependency metric of different user groups on different advertisements. By testing different user groups with different advertisements, the dependency metric can identify fraudsters without knowing cheating methods. There is also another advantage of this method. As the major job of this method is to compare the dependency metric, fraudsters can't pretend to be a normal user group if they do not have the dependency metric data of real user groups. Since fraudsters usually have little real traffic, it is safe to assume that the fraudsters will not be able to gain the data. Besides, anti-fraud programmers can always use new advertisements for the dependency metric, which means the anti-fraud programmers will become the initial ones with the help of the method in this research. The method has advantages. However, it is still not practical enough for the application. Since the testing data are generated by the computer randomly, we are not sure the dependency metrics between the testing data and the real data are different or not. If the difference of dependency metrics between the testing data and the real data is small, it is not strong enough for fraud detection. In this case, adjustments like using advertisements with higher dependency or adding more attributes to the dependency metric may fix the problem. In conclusion, anti-fraud detection by using uncertainty models has revolutionary potential property as it not only detects fraudsters without knowing their methods but also helps anti-fraud programmers to get an initiative position against fraudsters. But it needs real data for testing and modification to suit the real-world application.

References

1. J. Crussell, R. Stevens, H. Chen, MAdFraud: investigating ad fraud in android applications, in *Proceedings of 12th Annual International Conference on Mobile System Applications and Services*, (2014), pp. 123–134
2. L. Song, X. Gong, X. He, Z. Rong, A. Zhou, Multi-stage malicious click detection on large scale web advertising data, in *Proceedings of CEU Workshop, Italy*, (2014), pp. 67–72
3. Z. Pooranian, M. Conti, H. Hadaddi, Online advertising security: issues, taxonomy, and future directions. IEEE Commun. Surv. Tutor. (2020)
4. T. Tian, J. Zhu, F. Xia, Z. Xin, Z. Tong, Crowd fraud detection in internet advertising, in *The 24th International Conference on World Wide Web*, (2015), pp. 1100–1110
5. R. Oentaryo, E. Lim, M. Finegold, D. Lo, F. Zhu, C. Phua, E. Cheu, G. Yap, K. Sim, M. Nguyen, et al., Detecting click fraud in online advertising: a data mining approach. J. Mach. Learn. Res. **15**(1), 99–140 (2014)
6. Z. Suraj, An introduction to rough set theory and its applications a tutorial, in *Proceedings of ICENCO'2004, Egypt*, (2004), pp. 35–45
7. A.B. Qinghua Zhang, A. Qin Xie, W.A. Guoyin, A survey on rough set theory and its application. Control Theory Appl. **1**(4), 323–333 (2016)
8. Tsumoto, Shusaku, Rough sets: past, present and future. J. Jpn. Soc. Fuzzy Theory Syst. **13**(6), 552–561 (2001)
9. C. Cornelis, M. Cock, A. Radzikowska, in *Fuzzy Rough Sets: From Theory into Practice*, ed. by Handbook of Granular Computing, (2008)
10. A. Kandel, W.J. Byatt, Fuzzy sets, fuzzy algebra, and fuzzy statistics. Proc. IEEE **66**(12), 1619–1639 (1978)
11. J.N. Mordeson, P.S. Nair, *Fuzzy Mathematics: An Introduction for Engineers and Scientists* (Physica Velag Press, 2010)
12. J.J. Buckley, Fuzzy statistics: regression and prediction. Soft Comput. **9**(10), 769–775 (2005)
13. L.A. Zadeh, Fuzzy sets. Inf. Control **8**(3), 338–353 (1965)
14. U. Oegs, G. Kueova, Specific features of descriptive statistics with fuzzy random variables. Inf. Technol. Manag. Sci. **21**, 104–110 (2018)

15. A. Jb, B. Atdaf, D. Amc, A. Ms, E. Rsa, Auto loan fraud detection using dominance-based rough set approach versus machine learning methods. Sciencedirect. Expert Syst. Appl. **11**, 163–190 (2020)
16. N. Vratonjic, J. Freudiger, M. Felegyhazi, J.P. Hubaux, Securing online advertising. Technical Repot LCA, epfl (2008)

Improving Android Application Quality Through Extendable, Automated Security Testing

Nuno Realista, Francisco Palma, Carlos Serrão, Luís Nunes, and Ana Almeida

1 Introduction

The mobile ecosystem popularity continues to grow as more users and organisations are embracing it. Some studies estimate 6.1 billion smartphones by the end of 2020 [1], a market dominated by two major mobile platforms (Android and iOS). Users embrace mobile platforms for messaging, e-commerce, productivity tools, health and fitness, home banking, payments and many more. Moreover, this mobile platform handles many of the end users' personal and critical data [2], which makes them extremely attractive to malicious adversaries that will try to obtain unauthorised access to mobile devices and users' data [3]. These adversaries are continuously targeting both mobile ecosystems – iOS and Android [4]; however, Android, due to its market penetration (around 80% of all mobile devices in the World) and openness (Android is free and open, and smart device manufacturers use it as the basis for their systems) make Android ecosystem more attractive to attackers [5].

The increasing demand for this mobile ecosystem leads to a substantial boost in the number of existing developers and available applications. Unfortunately, this makes these platforms more exposed to development mistakes that are often translated into security vulnerabilities that malicious adversaries can abuse to target end users [6]. Therefore, developers need to be alert when developing their mobile applications to guarantee that security vulnerabilities are not introduced. Most of the time, developers have trouble understanding mobile software development risks and how to address and mitigate them to increase the security of the developed mobile applications. In terms of security, lowering the number of development-introduced

N. Realista · F. Palma · C. Serrão (✉) · L. Nunes · A. Almeida
ISCTE – Instituto Universitário de Lisboa, Information Sciences, Technologies and Architecture Research Center (ISTAR-IUL), Lisbon, Portugal
e-mail: Nuno_Realista@iscte-iul.pt; Francisco_Livramento@iscte-iul.pt; carlos.serrao@iscte-iul.pt; luis.nunes@iscte-iul.pt; ana.almeida@iscte-iul.pt

K. Daimi et al. (eds.), *Emerging Trends in Cybersecurity Applications*,
https://doi.org/10.1007/978-3-031-09640-2_12

security vulnerabilities will result in applications that pose lower security risks for both the end users and the mobile ecosystem. Nevertheless, developers need to be aware of these security risks to avoid threats and attacks to the end user devices and data.

Currently, developers can take advantage of multiple sophisticated tools to analyse potential security vulnerabilities on developed software. However, these tools are not simple to use and provide ambiguous results, and numerous tools often generate different results, making the developers task a nightmare. Additionally, the process needs to be repeated multiple times each time newer application versions are released. All this toolset should ideally be integrated into the mobile application release management and software development processes, taking advantage of DevSecOps paradigm. Therefore, there is the need for a system capable of producing security-related feedback that simplifies mobile application secure development, which can also help developers discover and correct security vulnerabilities when submitting applications to App Stores, providing feedback and helping them understand the risks and correct those vulnerabilities. This feedback will contribute to the overall mobile ecosystem security increase assisting developers in improving the app's security before they are made available on the App Store, downloaded and installed on the end user mobile device.

Mobile application stores, usually known as App Stores, are the traditional way users download and install apps on their devices. These App Stores act as a trusted gateway between the mobile application developers and the end users, ensuring that the application originates from a trustworthy developer and can be installed on the device. When developers submit their applications on App Stores, they usually undergo a set of scrutiny processes before the app gets accepted and is published on the App Store. These review processes depend on the App Store. The most popular Android mobile App Stores currently have quality assurance processes that evaluate the applications' security with signature-based mobile antivirus and antimalware tools to reduce malware prevalence but are not looking specifically into insecure mobile application development practices. This is particularly relevant because a non-malware application with security vulnerabilities may pose an equal higher security risk to the end user as a rogue mobile app.

Part of the work described in this chapter was developed, considering one of the most popular existing App Stores – Aptoide (www.aptoide.com). Aptoide is an Android App Store with over 220 million unique active users and partnerships with over 15,000 app developers, having almost one million apps available. The App Store has the means for analysing the received apps according to security and quality parameters before making them available to consumers. Aptoide strives to face the moving-target nature of mobile malware and poorly developed applications. Fostering safe mobile software development practices is considered a prevention strategy. By reducing the security vulnerabilities in the distributed apps, the App Store makes it hard for malicious applications to exploit or steal personal data/assets from their users. The system described in this chapter was developed in a project called AppSentinel that ran between 2018 and 2020. The problem addressed in this work is not only a problem for the Aptoide App Store, and in this chapter, we also considered Google Play Store and APKPure stores to test the developed system.

The objective of having a better and secure mobile ecosystem can be achieved by analysing apps vulnerabilities as presented in this chapter and providing developers with action-oriented feedback about how to mitigate the security of their app breaches and make their user-base safe. On the other hand, it is also relevant for App Store managers to have pertinent information about the security risk of a submitted application before it is made available on the store. This analysis is conducted in a way that does not require the availability of the application source code and therefore may be used by App Stores when receiving app submissions. This work presents the following contributions:

- The specification and development of a free integrated system, capable of analysing security vulnerabilities on mobile applications (statically analysing the APK of the Android application) and providing feedback and awareness to the software developers about the discovered vulnerabilities and how to address such vulnerabilities [7]. The system is made available as open source Python software in the GitHub platform (https://github.com/pontocom/appsentinel) and is publicly available for anyone to use and contribute to its further development.
- The proposal of an App Security Risk Score (ASRS) that can be used to automatically compute the security risk associated with a smart device application, based on the number of security vulnerabilities discovered.

The remainder of this chapter is organised according to the following structure. In the next section, we provide an overview of some related work and consider secure mobile development. The third part of the chapter is dedicated to the description of the developed system and how it can analyse Android application development vulnerabilities through the integration of multiple scanning tools and provide detailed feedback to the developers and estimate the security risk of an application. The following section demonstrates the system's applicability through its usage with different existing App Stores (Aptoide, Google Play Store and APKPure), collecting and comparing the obtained results. Finally, at the end of the chapter, we present several conclusions from this work and introduce some challenges requiring further research.

2 Related Work

The following section of this chapter approaches the mobile ecosystem and secure mobile application development. As referred previously, developers play a relevant part in the mobile ecosystem cybersecurity by improving the application security execution on the end users' devices and reducing vulnerabilities that might compromise the user data or conduct other obnoxious activities. Mobile application developers need to realise the risks that affect mobile applications and know the appropriate mitigation actions [8]. Modern mobile applications result from the intelligent assemblage of multiple elements and components providing the needed functionalities to end users: Application, Device, Network and Server. The application results from a developer activity to produce the necessary source code and compile it to the appropriate format (an APK, in the case of Android). They are made available for installation on the end user device through an App Store (in Android, under certain conditions, users can install applications on their devices directly without requiring a proper application store). Mobile apps running on the device may access the underlying OS services, access and store data, access hardware sensors, communicate with other applications or services and communicate through the available network. The device corresponds to the physical device that executes the client-side part of the mobile application, composed of all the hardware, sensors, operating system (Android or iOS) and specific operating services and applications. The network represents several means of communication that allow direct interaction with other nearby devices. Also, they can communicate with external services (servers or cloud) to exchange information. Finally, the server represents many external services (cloud) on distributed remote locations accessed through the Internet. These remote services provide extra functionality to the mobile application by providing APIs to exchange information with the application.

From a security perspective, security risks are affecting all the different mobile application components. One of the primary sources for the identification and classification of risks on mobile applications development is OWASP. OWASP has various mobile application security-related projects integrated on the OWASP Mobile Security Project [9]. The OWASP Top 10 Mobile Risks [10] compile the most prevalent risks affecting the security of mobile applications. These risks are (M1) Improper Platform Usage, (M2) Insecure Data Storage, (M3) Insecure Communication, (M4) Insecure Authentication, (M5) Insufficient Cryptography, (M6) Insecure Authorisation, (M7) Client Code Quality, (M8) Code Tampering, (M9) Reverse Engineering and (M10) Extraneous Functionality. Besides this, OWASP also provides two other relevant sources of information related to mobile application security development. The first one is the Mobile Security Testing Guide [8], a comprehensive manual for mobile application security testing and reverse engineering devoted to the iOS and Android mobile platforms. The second one is the OWASP Mobile Application Security Verification Standard [10] used by software architects and developers seeking to develop secure mobile applications and security testers to ensure completeness and consistency of the security test results. ENISA has also been producing some work in this field of mobile application security development. One of the most important and relevant initiatives is the ENISA report on "Privacy and data protection in mobile applications" [11], which presents a study on the mobile application development ecosystem and the technical implementation of the GDPR (EU regulation on privacy). ENISA also published "Smartphone Secure Development Guidelines" [12], a technology-neutral document produced for developers of smartphone applications as a guide for developing secure mobile applications. Another organisation concerned with mobile application security is NIST. NIST published the "Vetting the Security of Mobile Applications" report [13] to help organisations understand the "process for vetting the security of mobile applications, plan for the implementation of an application vetting process, develop application security requirements, understand the types of application vulnerabilities and the testing methods used to detect those vulnerabilities, and determine if an application is acceptable for deployment on the organisation's mobile devices".

The mobile development risks identified by the above entities need to be addressed by applying secure software development methodologies. Secure software development methodologies add to the traditional software development engineering processes the necessary steps to enable the design and implementation of secure software through a holistic and integrated security vision, including security requirements, threat modelling, secure coding, static and dynamic code reviews and specific security tests. For example, Microsoft developed their own Secure Development Lifecycle (SLD) [14] that consists of a set of practices that support security assurance and compliance requirements, helping developers to build more secure software by reducing the number and severity of vulnerabilities in software while reducing development cost [15]. BSIMM (Building Security In Maturity Model) [16] is also another initiative in this field, describing a set of activities grouped into four domains, including the "Secure Software Development Lifecycle Touchpoints" that characterise the practices associated with analysis and assurance of particular software development artefacts and processes, such as architecture analysis, code review and security testing [17]. OWASP also has a relevant initiative in this field that is SAMM (Security Assurance Maturity Model), a framework from OWASP that can help organisations to assess, formulate and implement a strategy for software security [16, 18].

On what relates to the specific topic of Android automated testing, some authors did an intensive literature review and created a taxonomy [19] about Android testing on what concerns test objectives, targets, levels and techniques. One of the test objectives analysed is related to the security of Android applications, where several approaches to test the security of Android applications were considered [20]. However, from the long list of the literature surveyed, it was possible to conclude that most of the works are too focused on specific types of Android application security problems and do not provide a broad overview of all the potential security risks an application might face. Also, in some cases, the security analysis performed followed a "grey box" or "white box" approach, requiring partial (or full) access to the app source code. Finally, to the best of our knowledge, none of the works considered the developer secure coding learning process and a way to provide helpful feedback to improve the security of developed mobile applications.

Considering all of the related work, we propose a system that aims to improve the final quality of the mobile applications through extendable and automated security testing using a "black box" approach. This "black box" approach enables a more realistic analysis scenario since attackers will also have similar access to applications when they are publicly available. Moreover, this system will provide helpful feedback for developers to mitigate security risks identified and improve the quality of their apps in terms of security. The following sections describe the developed system and also the tests and validation conducted to assess its usefulness.

3 Android Application Vulnerability Analysis and Feedback System

In this part of the chapter, we present the architecture and implementation of the system capable of identifying security vulnerabilities and provide feedback for developers. Using a "black box" testing approach, the system identifies potential security vulnerabilities on Android applications (APKs). Then, it offers the necessary input for developers on how to mitigate or eliminate such vulnerabilities using information from different sources collected on a knowledge base. The system is mainly composed of two major components. The first one is an API that allows external entities to use the system to send new APKs to be analysed and to get feedback for the developers about analysed APKs. The second one is a collection of software components, written in Python, properly orchestrated that regularly executes to perform a security analysis on the Android APKs that require it and provide feedback to developers.

The primary objective of the developed system is integration with existing third-party App Stores and their security assurance processes. These processes require the system to handle hundreds of new mobile applications or existing application update submissions per day. Therefore, our approach should test each of the new app submissions, determine the existence of security vulnerabilities (classified according to the OWASP Mobile Top 10 security risks) and return the appropriate feedback to the developers. Moreover, the system should either operate directly over the APK file uploaded to the system, use a remote APK location or use the unique App Store identifier (depending on the App Store itself). As a result, the developer receives appropriate educational feedback containing detailed information about each potential vulnerability discovered and how to conduct the proper measures to mitigate them. The system also computes the overall security risk of the APK itself, returning it together with a detailed explanation about how the risk score was obtained.

(A) System Architecture and Major Components.

The system receives either APK identifiers or APK files from external systems (either developers or App Stores), perform heavy testing on those APKs and provide feedback about the security-related findings to the developers or App Store QA managers. These functionalities are exposed to external clients through a REST API that completely isolates the information exchange processes from the internal operations. Figure 1 illustrates the overall system.

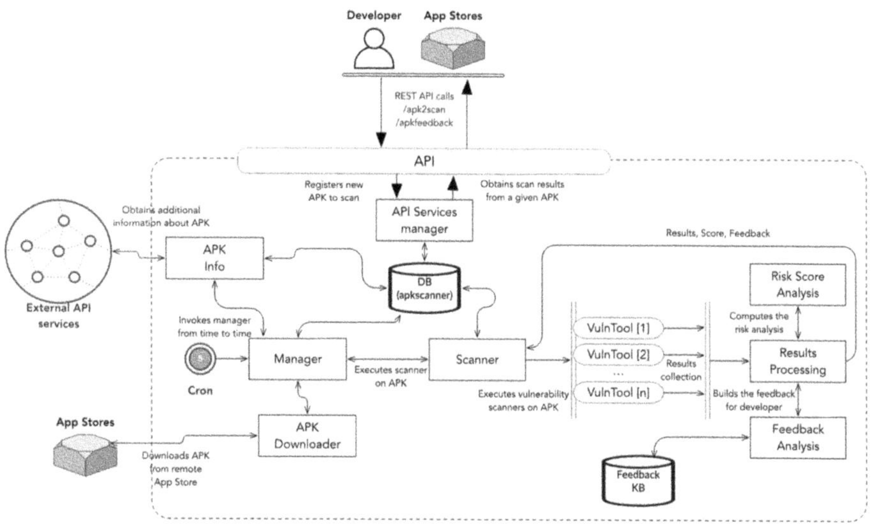

Fig. 1 The overall architecture of the Android Vulnerability Analysis and Feedback System

(B) Information Storage.

The system stores all the necessary information and the results from the vulnerability analysis process in an internal MySQL database. The database holds the following primary information:

- `apk`: Contains all the information about a downloaded APK, so the system analyses all the metadata information about the APK.
- `apk2scan`: Contains the information about the APKs that have been submitted to the system but have not yet been processed (scanned). It is a pool of APKs to be analysed.
- `apkresults`: Contains information about the results obtained after applying the appropriate scanning tools to a given APK. This structure contains information about the different collected analysis results over time.

These database tables aggregate all the required information used to analyse the submitted APKs and store the analysis results (security vulnerabilities, developers' feedback and security risk score).

(C) System REST API.

The REST API allows external systems to interact with the vulnerability system internal operations. Furthermore, it offers the functionalities of analysing APKs and collecting results and feedback from the analysed APKs. In resume, this system REST API possesses two main capabilities: (a) the ability to receive the indication of a new APK to be scanned (`apkscan`) and (b) the capability of providing integrated scanning results and feedback about an analysed APK (`apkfeedback`).

The “`apkscan`” API entry is used by external entities to send a new APK to the system for analysis. The client should pass (POST) an APK unique identifier (that depends on the AppStore) or the URI of the APK file to be scanned. This API entry operates in the following manner:

(1) The client sends a POST request for the “`apkscan`” API entry, passing either the APK identifier or the URI of the APK file to download and scan.
(2) API Services manager receives the request and checks for the existence of the APK identifier.
(3) API Services manager stores the APK identifier on the internal system database (`apkscanner`).
(4) API Services manager returns the result of the operation to the client. This result is a JSON-formatted messaging containing a Boolean “`status`” with the operation result and a string “`message`” field with further information.

An external client entity uses the “`apkfeedback`” API entry to request the results for a previously submitted APK to the system. The client should request (GET) the results passing the APK unique identifier to retrieve the results and feedback. This API entry operates in the following manner:

(1) The client sends a GET request for the “`apkfeedback`” API entry, passing the APK identifier that was given previously for scanning.
(2) API Services manager receives the request and checks for the existence of the APK identifier.
(3) API Services manager requests to the database the existing results and feedback for the given APK.
(4) API Services manager returns the results to the client. The result is a JSON-formatted message with a different structure according to the situation:
    (a) Suppose an error occurs or the requested APK is not yet processed. In that case, a JSON-formatted message is received containing a status field indicating the success or failure of the API call and a message field with some further details of the result of the operation.
    (b) A JSON-formatted message is received if the system has already processed the APK and the available results. This message contains a specific structure that is detailed in the following section.

(D) Vulnerability Analysis and Feedback System.

This section describes the system’s core and integrates with the previously described API that allows external entities to communicate with the system to perform APK security analysis and receive detailed feedback. The internal operations of the different system components and the different interactions are detailed to perform the services exposed by the REST API. The process of the internal components to analyse the APKs is utterly independent of the API invocation allowing for multiple clients to invoke the API endpoints without having to wait for the vulnerability scanning operations to complete (some of these tasks may take a long time, depending on the type of APK, its dimension and the number of security scanning tools to apply). The vulnerability analysis system checks for new APKs to scan, batch processes them and stores the results on a database.

The system operates using available vulnerability scanning tools over the requested Android APKs, collecting the results while returning the appropriate feedback to developers and App Store QA managers. Feedback will allow developers to improve their application quality by mitigating development vulnerabilities that might represent security threats for the end users. Also, the App Store QA managers have a new metric that can be used as decision criteria to accept or reject an app submission to the App Store. The sequence of operations conducted by the system sums up in the following:

- To avoid exhausting the resources on the server, the developed system uses the Unix/Linux “cron” daemon to execute the necessary operations over a group of APKs. Cron is set up to run according to a specific time interval and executes the “Manager”. The “Manager” is responsible for the central operations over the APK to analyse, for the collection of the results and for providing developers feedback.
- When the “Manager” starts, it connects to the “`apkscanner`” database and checks for the existence of newer APKs to test – these APKs have been added previously by some external system using the REST API. For example, suppose there are new APKs to scan on the database. In that case, the “Manager” will request the “APK Downloader” to download the APK file from some remote location (from an App Store or some other site) to initiate the vulnerability analysis scanning.
- If necessary, the “Manager” may require some more additional information about the APK and may use the “APK Info” to connect to some extra remote services to get the needed information (such as version, number of downloads and other).
- The “APK Info” stores APK metadata information on the database.
- The “Manager” invokes the “Scanner” component.
- The “Scanner” checks for the availability of existing and integrated security vulnerability scanning tools on the system (a pluggable system is used to enable the integration of multiple specific security vulnerability analysis tools on the system, allowing for the extensibility of the system).
- For each of the installed security scanning tools discovered on the system, it is executed the appropriated “plugin” (scanning tool).
- Each scanning “plugin” executes its specific tests over the selected APK file.
- Each scanning “plugin” temporarily writes its results to the file system.
- After the “Scanner” finalises its work, it normalises the findings obtained, using the “Results Processing” module. The normalisation process involves going through each of the vulnerabilities identified by the multiple scanning “plugins” and eliminating the duplicate findings and providing adequate feedback for each of the unique identified vulnerabilities.
- The “Feedback Analysis” module looks at the different discovered vulnerabilities and uses the “Feedback KB” knowledge database to build the appropriate feedback report to return to the developers and App Store QA managers.

- At the same time, the "Risk Score Analysis" module computes the security risk of the APK, based on the different vulnerabilities identified.
- Finally, the "Scanner" writes the "normalised" results and the feedback on the "`apkscanner`" database.
- The work is completed until the next round of execution requested by the "cron" daemon.

This core part of the system is independent of the REST API, allowing the system to receive multiple APKs to analyse without depending on the actual analysis's time. Furthermore, one of the objectives is to integrate this system with external App Stores, where thousands of new Android apps (or new versions of already existing apps) are submitted every day. Therefore, it is essential to make the REST API and the system core independent for extensibility and performance issues.

(E) System Vulnerability Analysis Extensibility.

One of the objectives in the design and implementation of the system was the ability to allow the system to extend and adapt to newer security vulnerabilities and analysis tools that might be released in the future. There are already a plethora of multiple Android application static and dynamic vulnerability analysis tools. Tools such as AndroBugs [21, 22], DroidstatX [23], SUPER (Secure, Unified, Powerful and Extensible Rust) Android Analyzer, AndroWarn [24], CuckooDroid [25, 26], EviCheck [27], Quick Android Review Kit [28], StaConAn [27], Mobile Security Framework [29] and others are used to detect different types of security vulnerabilities. The system can support those and any other new scanning tool that might appear in the future. Therefore, any developer can select a new scanning tool, write a simple extension plugin for the tool and make it available for the system to use and scan APKs. To do that, the developer must create a Python plugin file using a specific supplied integration template file (this file is provided on the GitHub repository) and adapt it to the particular scanning tool. Also, the developer must define a result integration dictionary that maps the exact results from the scanning tool into general system results and feedback to be provided to the developer (Fig. 2).

To prove the extensibility of the system to support multiple scanning tools, we have already developed four different plugins that offer support for four various Android static vulnerability analysis tools – AndroBugs, DroidstatX, SUPER Android Analyzer and AndroWarn. These plugins are executed side-by-side over the Android application file (APK) and produce a set of specific results integrated and managed by the "Results Processing" module of the system. As more plugins (and vulnerability scanning tools) get integrated, better will be the obtained results about the existence of vulnerabilities and better will be the feedback returned to the developers about the vulnerabilities detected on the applications and how to mitigate them.

Any developer willing to develop a new plugin for the existing system can accomplish that easily by using the source code of the provided template plugin (Fig. 3). With that template, he can implement the specificities of the scanning tool inside that template. Each scanning tool has its internal mechanisms, options and

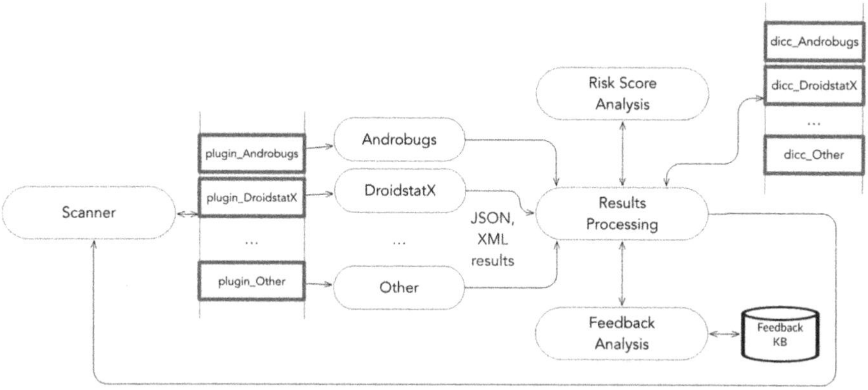

Fig. 2 Security vulnerability analysis plugin architecture on the system

```python
# Plugin to handle the tools capable of setting the input for "TEMPLATE" and handle the output
import os
import configparser
import logging as log


config = configparser.ConfigParser()
config.read('config.ini')

log.basicConfig(filename=config['GENERAL']['logDir'] + "appsentinel.log", filemode='a', format='%(asctime)s,%(msecs)d | %(name)s | %(levelname)s | %(funcName)s:%(lineno)d | %(message)s', datefmt='%H:%M:%S', level=log.DEBUG)

pluginName = "Template"
enable = False

# Define any specific configuration directives here
TOOLSPECIFICLocation = config['TOOLSPECIFIC']['TOOLSPECIFICLocation']

# this one is mandatory -> where to place the results of the tool
jsonResultsLocation = config['SCANNER']['jsonResultsLocation'] + "/" + pluginName + "/"


class PluginClass:
    def __init__(self):
        ''' constructor '''

    def run(self, apk_file, md5):
        print("Running the TEMPLATE plugin!...")
        # test the existence of the results directory
        if not os.path.exists(jsonResultsLocation):
            os.system("mkdir " + jsonResultsLocation)

        # everything bellow this is specific of your plugin
```

Fig. 3 Python template for integration of new scanning tools on the system

behaviours, and the developed plugin should adopt that to allow the scanning task to accomplish its objectives and provide results. Also, another critical requirement for the plugin developer is to provide a specific dictionary (a simple file in JSON format) (Fig. 4). This dictionary maps the results discovered by the particular scanning tool (developed for the plugin) into the generic, integrated and categorised results format provided by the Android Vulnerability Analysis and Feedback System. The dictionary uses the OWASP Mobile Top 10 risk classification methodology, aggregating the tools results in the "M1..M10" OWASP risk methodology. A detailed description of the results and feedback to return to the developer and App Store QA manager is provided in the following section.

```
{
    "results": [
        {
            "name": "Possibility of SQL Injection in exported ContentProvider",
            "category": "M1",
            "level": "Warning",
            "cvssV3": 8.4,
            "vectorStringV3": "AV:L/AC:L/PR:N/UI:N/S:U/C:H/I:H/A:H",
            "keywords": ["SQLInjection"]
        },
        {
            "name": "Possibility of Path Traversal in exported ContentProvider",
            "category": "M1",
            "level": "Warning",
            "cvssV3": 6.8,
            "vectorStringV3": "AV:L/AC:L/PR:N/UI:N/S:U/C:H/I:L/A:N",
            "keywords":["pathTransversal"]
        },
        {
            "name": "Application is debuggable",
            "category": "M1",
            "level": "Critical",
            "cvssV3":9.8,
            "vectorStringV3": "AV:N/AC:L/PR:N/UI:N/S:U/C:H/I:H/A:H",
            "keywords":["debuggable"]
        },
        {
            "name": "Activities vulnerable to Fragment Injection",
            "category": "M1",
            "level": "Critical",
            "cvssV3": 8.1,
            "vectorStringV3": "AV:L/AC:H/PR:N/UI:N/S:C/C:H/I:H/A:L",
            "keywords":["fragmentInjection"]
        },
        {
```

Fig. 4 Example of a dictionary file that maps the specific results of the tool in a common classification

(F) Results and Feedback.

The objective of the presented system is to automate APK testing and produce appropriate feedback for Android mobile application developers and App Store QA managers about existing vulnerabilities on submitted applications and ultimately improve their quality from a security standpoint. The system also provides appropriate feedback that can help developers to learn how to mitigate those vulnerabilities. Also, it gives security risk metrics that help App Store QA managers decide whether to accept or reject an app on the store. Properly integrated on an App Store, mobile application acceptance pipeline can reduce insecure Android applications. It can accomplish this objective by warning the App Store QA managers and application developers about possible security vulnerabilities that require proper mitigation.

Therefore, the system produces integrated security vulnerability analysis using specific existing mobile application security static or dynamic analysis tools. The system collects the findings of each one, aggregates, classifies and categorises them into a single result object that follows the OWASP Mobile Top 10 risks methodology (the risks range from "M1" to "M10"). These aggregated results are returned to the client (via the "`apkfeedback`" REST API entry, as previously presented) using a JSON-formatted response composed of several different structures. It is possible to interpret these structures by various means (other tools, a mobile or web application or others). Another important aspect is the valuable educational feedback provided to the developers for each of the identified vulnerabilities. Developers receive additional help (URLs for websites, videos, books and other resources) that will allow them to understand better the reported vulnerabilities, how they work, how attackers exploit them and how to correct or mitigate them. Also, the answer returns to the user an application risk score computed by the system, which will be helpful

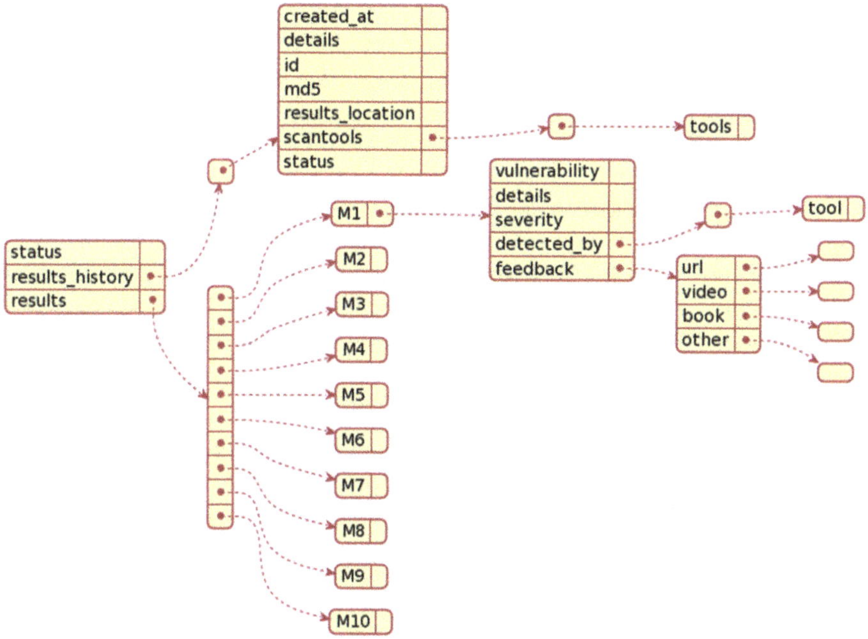

Fig. 5 Feedback JSON-formatted message structure

for App Store QA managers to make informed decisions about the acceptance of a submitted APK.

The "`response`" that contains the appropriate vulnerability scanning results and the developer's feedback is a JSON-formatted message that is composed of multiple structures – "`status`", "`results_history`" and "`results`". The first one, "`status`", is a simple JSON object that contains a Boolean value that indicates if the API request has succeeded or not and an optional "`message`" that includes some additional information in the event of an error (Fig. 5).

The second element on the result is the "results history" element composed of an array of JSON objects representing the history of the different vulnerability analyses conducted over time on a given Android mobile application or on the different versions of that application. The results history is essential because it allows tracking the security maturity of a given application over time. Each of the "`results_history`" elements has the following structure:

- "`created_at`": Represents the time and date of the performed scanning activity.
- "`details`": A string of data that provides extra details about the performed scanning activity. These additional details may include details about some specificities of the tools used or some particularities about the execution conditions that were present.

- "`id`": Each of the scanning activities is uniquely represented and identified by the system. This identifier is part of the JSON object.
- "`apkid`": This is the unique representation of the Android mobile application APK in the App Store – this is dependent on the App Store.
- "`results_location`": The server stores the different results produced by the multiple vulnerability scanning tools used. This field is the server storage location containing the specific scanning results with further detailed information.
- "`scantools`": This represents the identification of the multiple specific scanning tools used to perform the security analysis and vulnerability identification on this run.
- "`status`": This represents the success or not of the specific scanning operation that took place at this moment in time.

Finally, the last element on the result object is "`results`", composed of a specific subset of JSON objects and named according to the OWASP Mobile Top 10 terminology, ranging from "`M1`" up to "`M10`". The system gathers the specific results from the multiple vulnerability scanning tools used, categorising and classifying them according to the OWASP naming. This approach will help organise and enumerate the various vulnerabilities into general risk classification metrics allowing developers to quickly understand the security risks of their Android applications and learn how to take the appropriate actions to mitigate them. Each of the "`M1..M10`" structures is composed of an array of multiple vulnerabilities (if they exist) that have the following format:

- "`vulnerability`": This corresponds to a single sentence that clearly describes the vulnerability identified in the conducted analysis. This vulnerability might be identified by a single tool or by a different set of tools.
- "`details`": In this field, a more detailed description of the vulnerability is presented, providing a more lengthy explanation of the detected vulnerability with all the relevant details as well as some possible CVE or CWE enumeration, if it exists.
- "`severity`": Not all the vulnerabilities detected share the same severity classification. Vulnerabilities are grouped according to their criticality. The higher the severity level, the more critical they are, ranging from a highly exploitable vulnerability that can compromise the end user easily to less exploitable and let an attacker get some information about the execution environment. If more than one scanning tool reports the same vulnerability with different severity levels, then the average level of those severities is considered.
- "`detectedby`": this field indicates which scanning tool has detected this vulnerability – if different scanning tools detected this vulnerability, then they are enumerated here.
- "`feedback`": the feedback includes not only information about the existing development security vulnerabilities that are present on their Android applications but also information about the way the detected vulnerabilities operate and how they affect the security of the end users and how they can correct or mitigate them. Thus, this feedback mechanism intends also to educate mobile application

developers not only about development risks but also about good development practices that need to be employed to correct these development security risks. The system returns a plethora of additional information in a specific format. First, an "`url`" field with a list of URLs points to online resources containing relevant information about the vulnerability and how to correct it. Second, a "`video`" field contains links to videos with information pertinent about the vulnerability. Third, a "`book`" field indicates books that refer to vulnerability. Fourth, an "`other`" field contains any other relevant information related to the vulnerability.

This structure returns upon invoking the REST API by any client application that submits some Android application for analysis that can process it and present to the developer. Currently, the system is integrated with a specific Android App Store (Aptoide) security and quality assurance system. A web-based system offers the results allowing further analysis and configuration to personalise how to deliver the end user feedback.

(G) Computing the App Security Risk Score (ASRS).

To give a quantified idea of how secure an app is in terms of implementation, this system is also composed of a module that calculates a risk associated with the vulnerabilities found during the static analysis. The App Security Risk Score (ASRS) corresponds to a computed metric between 0 and 1 – the higher the metric, the greater is the risk. It is calculated depending on what type of discovered vulnerabilities and on which tools detect those vulnerabilities. Not every vulnerability has the same impact on compromising the user's data and not all the vulnerabilities can be easily exploited. Therefore, those that can be easily manipulated and have a higher impact on user data will have a more profound effect on the calculation of the ASRS. A similar approach is also considered for the scanning tools; in other words, not all tools have the same impact on the ASRS calculation – several tools might detect the same vulnerability, but only one will add more weight to the final score. The weight considered for a specific tool depends on its reliability, specifically if it has scientific research supporting it, high popularity among the community and still receiving support from the developers. Each of the App Stores' security analysts may adjust the different weights conducting the audits to the apps and considering the various vulnerability scanning tools. This affects the scoring values of the applications to a given App Store, because there might be App Stores that might accept riskier applications while others might not.

Furthermore, to calculate the ASRS, we decided to use a weighted arithmetical mean that considers all the discovered vulnerabilities and group them by each scanning tool that could detect them. The formula for calculating the ASRS is the following:

$$\text{ASRS} = \frac{\sum_{t=1}^{\infty} \left(\frac{\sum_{v=1}^{\infty} \text{Critical}_v \ x \ P_c + \text{Warning}_v \ x \ P_w + \text{Notice}_v \ x \ P_n}{v} \right) x \ P_t}{t}$$

- P_c – Critical vulnerability category weight
- P_w – Warning vulnerability category weight
- P_n – Notice vulnerability category weight
- P_t – Scanning tool weight
- v – Total number of detected vulnerabilities in specific category
- t – Total number of scanning tools.

The severity categories are indicated by the vulnerability scanning tools used on the system. There are three different categories: critical, warning and notice. Critical vulnerabilities represent a confirmed security vulnerability, highly exploitable by an attacker, that should be solved immediately. Warning is suspicious vulnerability that might not be automatically confirmed by the scanning tool – this requires manual confirmation by the developers. Notice is a low priority issue detected by the scanning tool that may not represent a direct security issue to the application but may provide information for a potential attacker.

First, each tool detects v vulnerabilities for each APK. Then, these vulnerabilities are grouped according to their severity category and multiplied by their associated weight. For critical vulnerabilities, it is multiplied a weight of P_c; for warnings, a weight of P_w; and for notices, a weight of P_n. In the end, everything is summed and divided by the total number of vulnerabilities. Therefore, it is also essential that not all the vulnerabilities have the same weight; for example, not all critical vulnerabilities have the same weight. Some of them have more impact than others. The same applies to the other severity groups. So, for each severity group, we have:

$$\text{Severity}_v \; x \; P_s = \left(\text{Severity}_{v1} \; x \; P_{s1} + \text{Severity}_{v2} \; x \; P_{s2} + \cdots + \text{Severity}_{vn} \; x \; P_{sn}\right)$$

where $Severity_{vn}$ corresponds to the number of vulnerabilities of that severity group for a specific weight, P_{sn}. After all tools have calculated their mean with all vulnerabilities detected, a weight associated with each tool, Pt, will be multiplied by their corresponding mean. With this, we have the impact of each tool that will increase or decrease the weight of the detected vulnerabilities. Knowing that all scanning tools detect a vulnerability might mean that it is not a false positive, and its weight in the final risk score will be higher. Finally, it will be divided by t, and the sum of all tool weights must equal the total number of scanning tools. If a tool fails its analysis, the weights of the remaining tools will be adjusted, and the total number of tools, t, will decrease.

To add weights to the vulnerabilities, we considered the type of attacks performed by exploiting the detected vulnerabilities. We use the CVSS v3.0 (Common Vulnerability Scoring System) metric as a weight which is a quantitative metric and uses simple formulas that return a value as the severity of the vulnerability. The CVSS is being widely used and accepted by the community, and it has been growing in popularity. This scoring system is composed of three metric groups: Base, Temporal and Environmental. This research only used the base score because it represents the innate characteristic of each vulnerability that is not affected by time, place or system since we are always in the Android environment. To apply the CVSS to the vulnerabilities in this system, we read the dictionaries of each tool

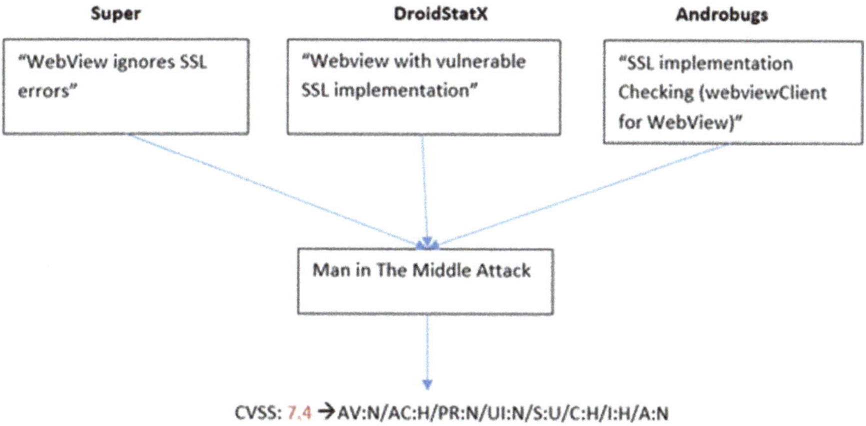

Fig. 6 Example of CVSS score and vector string attribution to vulnerabilities that suffer from the same attack

to find the characteristics of the vulnerabilities and compare them. Frequently we saw the same vulnerabilities in the different scanning tools but with other names and descriptions. To work around this problem, we used keywords to define those types of vulnerabilities and to apply the same CVSS score. In Fig. 6 it is possible to visualise an example of how this method was performed. Three different tools (SUPER Android Analyzer, DroidstatX and AndroBugs) identify the same vulnerability with other names. Despite those different names, we can set keywords to unify them. In this case, we used the words "WebView" and "SSL"; those words point to one type of attack, MITM (man in the middle). Knowing which type of attack can exploit those vulnerabilities, it is easy to set a CVSS score to quantify their severity. Also, whenever it is possible, we provide a vector string that explains how the score was achieved. The vector string represents the defined base metrics of the standard calculation method of CVSS.

Consequently, for future tools added to our system, it will be easier to give vulnerability scores with the set of keywords if they match the names given by the third-party tools. An expert review will always be necessary because it is possible to find vulnerabilities with different names, and they might not match any of the keywords, but they point to the same issue.

In conclusion, to guide developers and App Store QA managers, we defined thresholds to classify the security level of an application. There are three main groups, low risk, moderate risk and high risk.

- Low-risk values between 0 and 0.5; usually with a low number of detected vulnerabilities and especially a very low number of critical vulnerabilities. Applications with these risk scores are secure and will not compromise the user's data. Still, a review of the code is always a good practice.
- Moderate-risk values between 0.5 and 0.7; usually, with important notices and warnings, critical vulnerabilities can be also significant but not that often. Appli-

cations with these risk scores might not compromise the users, but reviewing the code before publishing an app is advisable.

- High-risk values between 0.7 and 1; usually, these apps have many critical vulnerabilities or critical vulnerabilities detected by many high weighted tools. Therefore, these apps should not be published, and a review of the code must be performed.

4 Validating and Testing the System

We have conducted several tests to evaluate the developed system and assure that the automated APK testing and developer feedback mechanisms effectively detected Android mobile application security vulnerabilities. These tests were conducted on three different App Stores: Aptoide, Google Play Store and APKPure. The testing procedure was entirely automated without any manual intervention and consisted of two phases. In the first phase, we have selected the top popular applications from each of the categories existing on the different App Stores tested (42 categories in Aptoide, 33 categories in Google Play Store and 32 categories in APKPure) in terms of downloads, by the end users. The selection process resulted in a total of 1065 Android applications tested by our system and the appropriate identification of possible security vulnerabilities. The selected APKs had an average size of 27 MB, with the larger APK having 1.37 GB and the smallest one having 27 KB in size. In terms of downloads from the Aptoide App Store, the average number of downloads of the selected apps was 263 million (most downloaded app: 376 million downloads; least downloaded app: only 40 downloads).

The tests were executed on a single server equipped with an AMD Ryzen 7 Pro 3700 running at 4.4GHz with 64GB of DDR4 ECC 2666 MHz RAM. The server was also equipped with 2x SSD NVMe 960GB and was running Ubuntu Server 18.04 with the 4.15.0–109-generic kernel as the operating system. The system was configured to use three different vulnerability analysis plugins – AndroBugs, DroidstatX and SUPER Android Analyzer. These plugins are executed against all the selected Android applications on the various App Stores. The execution took a total of 7 h and 47 min to complete on the different App Stores (different stores with a different number of applications to test), with a global average analysis time for each app of 29 seconds (the most laborious application took 4 min, while the easiest one took just 1 s). Figure 7 displays the total time for each of the App Stores to complete the analysis and the maximum, minimum and average time to analyse the APKs.

After the system concluded the security analysis of the 1065 Android apps selected, it was possible to observe that some of the security analysis plugins were unable to analyse some of the APKs; however, the fact that three plugins were used allowed the system to offer a 100% coverage of all the testing targets – all the APKs were adequately tested, and results were collected.

The Android application vulnerability testing system has produced evidence that resulted in the identification of a total of 25,106 security vulnerabilities (summing

Fig. 7 Execution times of the system on the different App Stores and globally

	Aptoide	Google Play Store	APKPure	Global
Total	03:11:31	02:28:53	02:07:08	07:47:32
Max	00:04:24	00:04:10	00:03:56	00:04:10
Min	00:00:01	00:00:02	00:00:01	00:00:01
Average	00:00:28	00:00:36	00:00:24	00:00:29

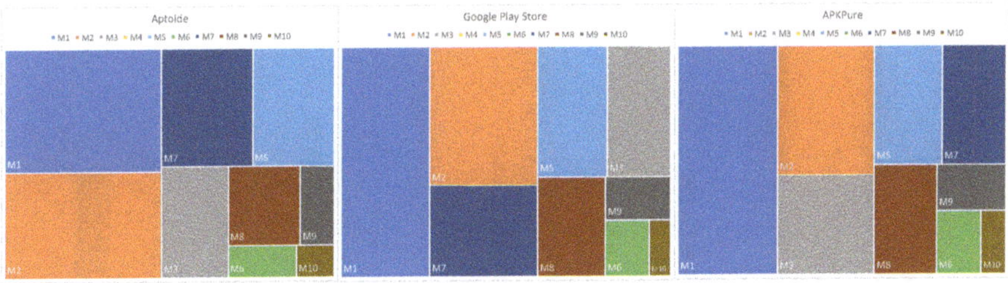

Fig. 8 Visual distribution of the detected vulnerabilities according to OWASP Mobile Top 10 (Aptoide, Google Play Store and APKPure)

the vulnerabilities from the three App Stores). These vulnerabilities were mapped according to the OWASP Mobile Top 10 risk identification methodology, aggregating the test results from the different vulnerability scanning plugins used on the system to conduct the APK security audit (Fig. 8). From this analysis, it is possible to identify (M1) Improper Platform Usage, (M2) Insecure Data Storage, (M7) Client Code Quality and (M5) Insufficient Cryptography as the most prevalent vulnerabilities identified on the tested mobile apps. These four types of OWASP risks account for 72% of the detected vulnerabilities (Fig. 9). It is also interesting that the automated tests conducted detected no vulnerabilities related to insecure authentication (M4). This metric reveals that the analysed applications are not using any type of authentication mechanism or that most mobile application developers are putting determined efforts into this specific aspect of their mobile apps.

While comparing the different App Stores analysed by our system, the total of 25,106 security vulnerabilities was divided into 5663 critical vulnerabilities, 11,087 warnings and 8356 notices. The critical vulnerabilities are the most important vulnerabilities and the ones that need more attention from the developers and App Store QA managers' side. It is also relevant to notice that the average number of vulnerabilities per app was 23.6, and from those, 5.3 were critical (Aptoide with 6.9, Google Play Store with 4.6 and APKPure with 4.0) (Fig. 10).

Another relevant aspect of the security analysis offered insight into which app categories on the App Stores contained more security vulnerabilities. The analysed App Stores include different app categories, so it is hard to make correspondences. Nevertheless, it is possible to notice that in the case of the Aptoide, the three most vulnerable app categories are "BEAUTY", "WEATHER" and "ENTERTAINMENT"; in the case of Google Play Store, are "COMMUNICATION", "GAME"

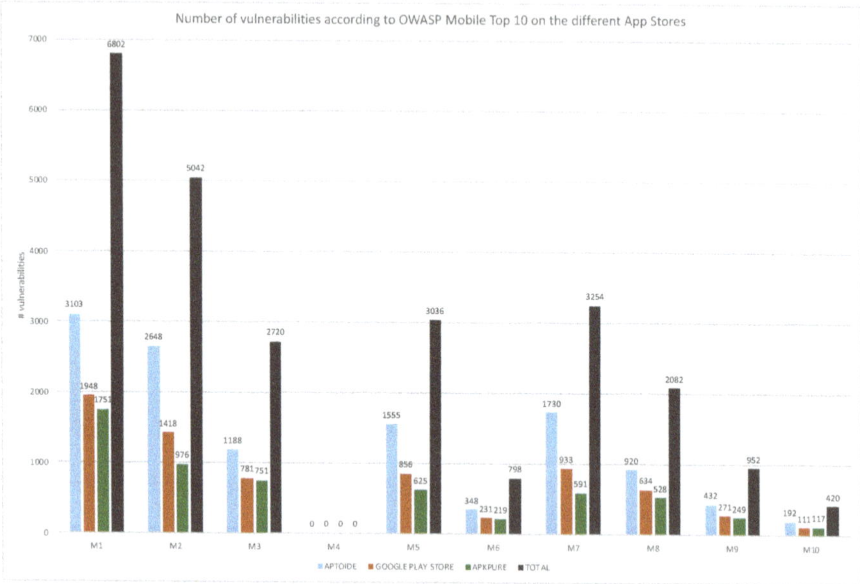

Fig. 9 Number of identified vulnerabilities according to OWASP Mobile Top 10 and per App Store

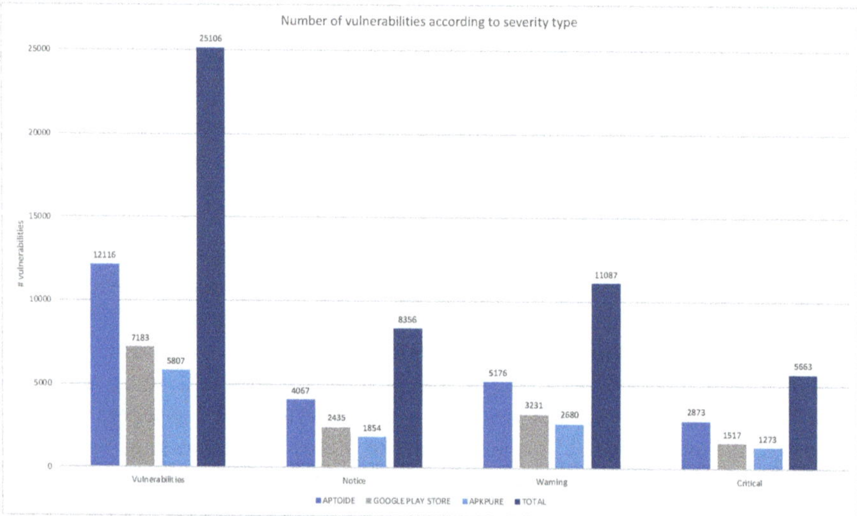

Fig. 10 Number of vulnerabilities according to their severity per App Store

and "FAMILY"; and finally in the case of APKPure, are "COMMUNICATION", "MAPS_AND_NAVIGATION" and "PRODUCTIVITY" (Fig. 11).

An important metric that the system produces that is of most relevance for App Store QA managers is the App Security Risk Score (ASRS). The ASRS provides

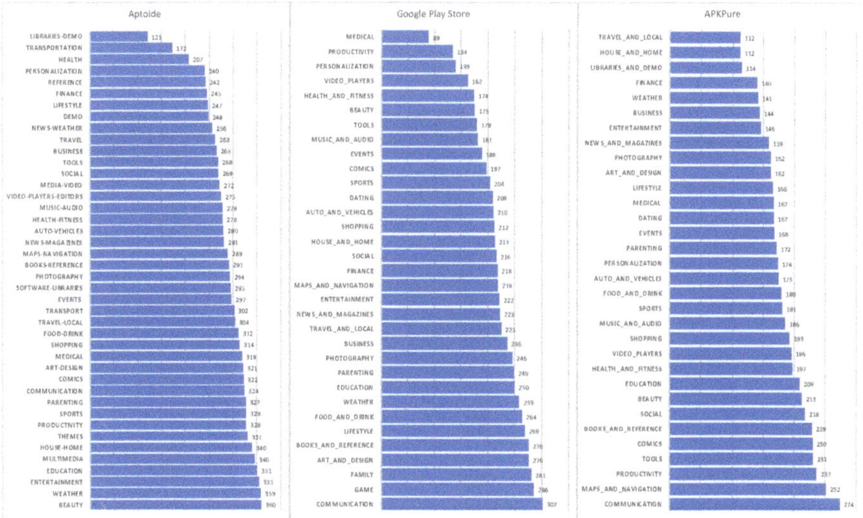

Fig. 11 Total number of vulnerabilities according to the different App Store categories

to the App Store QA manager a critical metric to determine the security of the submitted app, and that can be used to either accept or reject it on the App Store. For example, from the analysis conducted on three selected App Stores (Aptoide, Google Play Store and APKPure) (Fig. 12), it was possible to determine that according to our ASRS computing criteria, the App Store that presents a slightly higher average ASRS level is APKPure (53%) when compared to Aptoide (both with 52.37%).

The tests resulted in 1065 security feedback reports sent to the developers – one per APK submitted for analysis. These reports included information about each of the security vulnerabilities discovered by the system. They also contained specific feedback that allowed developers to understand better the security vulnerability and its risks and some advice on how to further investigate the vulnerability and correct it.

From the tests, it was possible to conclude two main aspects. First, the required average time to conduct the proper security tests and analysis on a specific App Store-submitted Android app is around 29 sec. Depending on the complexity and size of the app, this time can grow up to 4.5 min (on the conducted test hardware configuration). In addition, the system used three different scanning plugins, so the average time to perform these types of security tests would grow with the increase in the number of plugins. Second, the system detected an average of 23.6 security vulnerabilities per Android app and an average of 223 security vulnerabilities per app category. This number sheds light on the security of the applications that end users install on their own devices. Also, it reveals how these vulnerable Android applications evade the quality assurance processes implemented on the App Stores and how they are made available for download and installation on millions of end user Android devices.

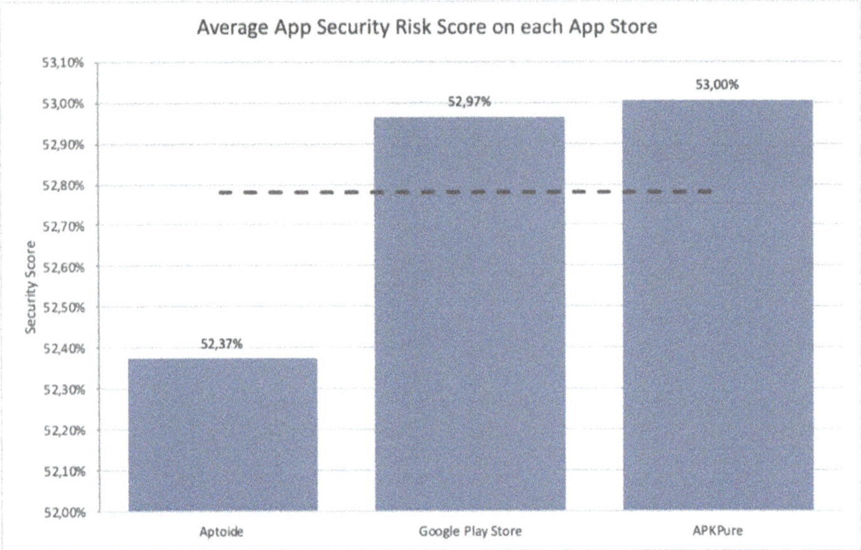

Fig. 12 App security risk score per App Store

5 Conclusions and Future Work

As the number of users using smartphones and mobile applications continues to grow, the security risks exposed to user data are climbing. These mobile platforms are already a choice for attackers/criminals exploring several attacks to compromise the users.

Developers of mobile applications have an essential role in the security of mobile applications because they are responsible for their design and implementation. However, some implementation errors often lead to security vulnerabilities. Most of the time, developers have in-depth knowledge of the technologies used to develop mobile applications but are not aware of good development security practices. This chapter proposes an automated system aiming to improve this panorama. It provides Android mobile application developers with the opportunity to automate security testing and evaluate the development of security vulnerabilities on their apps and receive adequate feedback to help developers correct their apps and mitigate vulnerabilities. At the same time, the system can integrate with the Android App Store quality assurance processes offering QA managers specific information and valuable metrics to improve the acceptance or rejection of newly submitted apps. The system tests and analyses the multiple submitted applications and reports the security problems to developers and App Store QA managers before the applications are accepted on the App Store and made available to the millions of end users and their devices.

Although the proposed system contributes to the mobile application security improvement and developer's education, it uses an automated test process that detects security vulnerabilities. As with any fully automated process, it has some limitations. The most important limitation is the existence of false positives, i.e. the erroneous detection of security problems that need manual confirmation. Therefore, as with any security analysis, any specific finding needs to be adequately verified. The proposed system can help pinpoint potential security problems, but they need to be further confirmed by the developers. The specific confirmation of the detected vulnerabilities by the system is out of the scope of this chapter.

Another important future development for the system includes the usage of machine learning techniques to provide developers better feedback about the security vulnerabilities found and enhance the vulnerability identification process by reducing the number of false positives. Moreover, we are working to improve the process of normalising the detected vulnerabilities by different scanning tools through intelligent text matching patterns. In addition, the system is also being improved to analyse the Android applications' vulnerability history to determine their security maturity.

Acknowledgements This work is part of the AppSentinel project, co-funded by Lisboa2020/Portugal2020/EU in the context of the Portuguese Sistema de Incentivos à I&DT - Projetos em Copromoção (project 33953).

References

1. J. Clement, Mobile app usage – statistics & facts. Statista (2019). https://www.statista.com/topics/1002/mobile-app-usage/
2. A. Ahmad, K. Li, C. Feng, S.M. Asim, A. Yousif, S. Ge, An empirical study of investigating mobile applications development challenges. IEEE Access **6**, 17711–17728 (2018)
3. J. Khan, H. Abbas, J. Al-Muhtadi, Survey on mobile user's data privacy threats and defense mechanisms. Procedia Comput. Sci. **56**, 376–383 (2015)
4. P. Faruki, V. Laxmi, A. Bharmal, M.S. Gaur, V. Ganmoor, AndroSimilar: Robust signature for detecting variants of android malware. J. Inf. Secur. Appl. **22**, 66–80 (2015)
5. I. Mohamed D. Patel, Android vs iOS security: A comparative study, in *2015 12th International Conference on Information Technology-New Generations* (2015), pp. 725–730
6. T. Petsas, A. Papadogiannakis, M. Polychronakis, E. P. Markatos, T. Karagiannis, Rise of the planet of the apps: A systematic study of the mobile app ecosystem, in *Proceedings of the 2013 conference on Internet measurement conference* (2013), pp. 277–290
7. F. Palma, N. Realista, C. Serrão, L. Nunes, J. Oliveira, A. Almeida, Automated security testing of android applications for secure mobile development, in *2020 IEEE International Conference on Software Testing, Verification and Validation Workshops (ICSTW)* (2020), pp. 222–231
8. R. Mahmood, N. Esfahani, T. Kacem, N. Mirzaei, S. Malek, A. Stavrou, A whitebox approach for automated security testing of android applications on the cloud, in *2012 7th International Workshop on Automation of Software Test (AST)* (2012), pp. 22–28
9. OWASP, OWASP Mobile Security Project. https://www.owasp.org/index.php/OWASP_Mobile_Security_Project. Accessed 11 Dec 2019
10. OWASP, OWASP Mobile Mobile Top 10 (2016). https://www.owasp.org/index.php/Mobile_Top_10_2016-Top_10. Accessed 11 Dec 2019
11. ENISA, *Privacy and Data Protection in Mobile Applications* (2018)
12. ENISA, Smartphone Secure Development Guidelines (2017)

13. S. Quirolgico, J. Voas, T. Karygiannis, C. Michael, K. Scarfone, Vetting the Security of Mobile Applications (2015).. https://doi.org/10.6028/NIST.SP.800-163
14. M. Howard, S. Lipner, *The Security Development Lifecycle*. O'Reilly Media, Incorporated (2009)
15. M. Howard, Building more secure software with improved development processes. IEEE Secur. Priv. **2**(6), 63–65 (2004). https://doi.org/10.1109/MSP.2004.95
16. G. McGraw, Software security and the building security in maturity model (BSIMM). J. Comput. Sci. Coll. **30**(3), 7–8 (2015)
17. B. Chess, B. Arkin, Software security in practice. IEEE Secur. Priv. **9**(2), 89–92 (2011)
18. G. McGraw, Building secure software: Better than protecting bad software. IEEE Softw. **19**(6), 57–58 (2002). https://doi.org/10.1109/MS.2002.1049391
19. P. Kong, L. Li, J. Gao, K. Liu, T.F. Bissyandé, J. Klein, Automated testing of android apps: A systematic literature review. IEEE Trans. Reliab. **68**(1), 45–66 (2018)
20. A. Amin, A. Eldessouki, M.T. Magdy, N. Abdeen, H. Hindy, I. Hegazy, Androshield: Automated android applications vulnerability detection, a hybrid static and dynamic analysis approach. Information **10**(10), 326 (2019)
21. Androbugs, AndroBugs Framework (2015), [Online]. Available: https://github.com/AndroBugs/AndroBugs_Framework
22. N. Drong, J. Van Thuijl, Upgrading and Extending the AndroBugs Framework (2020)
23. C. André, DroidstatX (2019), [Online]. Available: https://github.com/clviper/droidstatx
24. D. Thomas, AndroWarn (2019), [Online]. Available: https://github.com/maaaaz/androwarn
25. I. Revivo, O. Caspi, Cuckoo-Droid (2017), [Online]. Available: https://github.com/idanr1986/cuckoo-droid
26. G. Suciu, C.-I. Istrate, R. I. Ruaducanu, M.-C. Dictu, O. Fratu, A. Vulpe, Mobile devices forensic platform for malware detection, in *6th International Symposium for ICS \& SCADA Cyber Security Research 2019 6* (2019), pp. 59–66
27. M. N. Seghir, D. Aspinall, Evicheck: Digital evidence for android, in *International Symposium on Automated Technology for Verification and Analysis* (2015), pp. 221–227
28. Linkedin, Quick Android Review Kit (2017), [Online]. Available: https://github.com/linkedin/qark/
29. MobSF, Mobile Security Framework – MobSF (2019), [Online]. Available: https://github.com/MobSF/Mobile-Security-Framework-MobSF

Part V
Energy Applications Security

A Provably Secure Data Sharing Scheme for Smart Gas Grid in Fog Computing Environment

Rachana Patil, Yogesh H. Patil, Aparna Bannore, Arijit Karati, Renu Kachhoriya, and Manjiri Ranjanikar

1 Introduction

Smart devices and sensors connected to the Internet are part of the Internet of Things (IoT). Statistic predicts that by 2025, there will be 30.9 billion active IoT-connected devices, including sensors, nodes, and gateways, all across the world. The Internet of Things (IoT) is playing an increasingly important role in the development and implementation of smart city infrastructure. Water management, irrigation, garbage management, parking, and intelligent street lighting are a few of the smart city's vital domains. It is possible to find all the datasets used in the design and simulation of these domains on the Internet.

Increased demand for gas energy is driving the development and deployment of smart natural gas meters as efficiency in energy management is becoming increasingly important. Nearly every energy industry has looked into the idea of a "smart grid" in the last few years [1, 2], with the goal of using big data and networking expertise to better manage energy usage.

R. Patil (✉) · R. Kachhoriya · M. Ranjanikar
Pimpri Chinchwad College of Engineering, Pune, Maharashtra, India
e-mail: rachana.patil@pccoepune.org; manjiri.ranjanikar@pccoepune.org

Y. H. Patil
Dr. D. Y. Patil Institute of Technology, Pune, India
e-mail: meetyog@gmal.com

A. Bannore
SIES Graduate School of Engineering, Navi Mumbai, India
e-mail: bannore.aparna@siesgst.ac.in

A. Karati
National Sun Yat-sen University, Kaohsiung, Taiwan
e-mail: arijit.karati@mail.cse.nsysu.edu.tw

K. Daimi et al. (eds.), *Emerging Trends in Cybersecurity Applications*,
https://doi.org/10.1007/978-3-031-09640-2_13

The progress of smart gas metering, on the other hand, is much slower than the development of the smart power grid [3]. By utilizing IoT (Internet of Things) technologies, it's now possible to install a smart gas metering system along gas pipelines to access and transmit real-time data on gas consumption across an extended network [4]. A smart gas grid technology [5] can be divided into three stages, one for sensing, and another for networking, and a third for applications, much like a traditional Internet of Things (IoT) system [6, 7].

In the sensing stage, smart gas sensors are installed to collect real-time gas consumption data and shared it to the application layer via the network layer [8, 9]. The use of cloud computing is common in the application or networking layer to process and stock up actual gas consumption data [10] and deliver the data to the needy end users. A successful smart gas grid relies on precise and consistent energy metering data that can be collected in real time, transmitted to multiple users, and used to tackle the information problem that exists in existing energy grids [11].

The sensors of smart gas metering systems, which are installed along gas pipelines, are a key component of the system. Furthermore, in smart gas grids, users and customers may place high demands on the meter's intellect, expediency, and data security in addition to the essential requirements of gas utilization measurement.

Our Smart Gas Grid in Fog Computing Environment (SGG-FCE) aims to achieve these objectives by integrating the Internet of Things (IoT) into the gas distribution system and improving gas infrastructure management in terms of hazard minimization. When it comes to building such critical systems, security and privacy are often neglected. We have devised a proxy signcryption scheme based on hyperelliptic curves as a solution.

The significant contributions of this paper are as follows:

- IBPSC-SGG-FCE is a new secure hyperelliptic curve identity-based proxy signcryption scheme for the smart gas grid network in a fog computing environment.
- Security analysis of the proposed IBPSC-SGG-FCE shows that it is resistant to HYEC-DLP and HYEC-DHP.
- Resilience against replay and man-in-the-middle attacks is demonstrated by AVISPA performance analysis of the proposed IBPSC-SGG-FCE system.
- In the end, the comparison with the current scheme shows that IBPSC-SGG-FCE is more cost-effective in terms of calculation and communication than the current designs.

1.1 Smart Gas Networks

Natural gas is utilized in homes and for businesses around the world for heating and fuelling. Several research and development firms are working on natural gas smart grids based on the Internet of Things (IoT). Detecting gas theft, leakage, corrosion in the pipelines, and remote cutoff in emergency scenarios are all possible, thanks to the real-time data that gas companies collect from their pipelines [12]. Communication

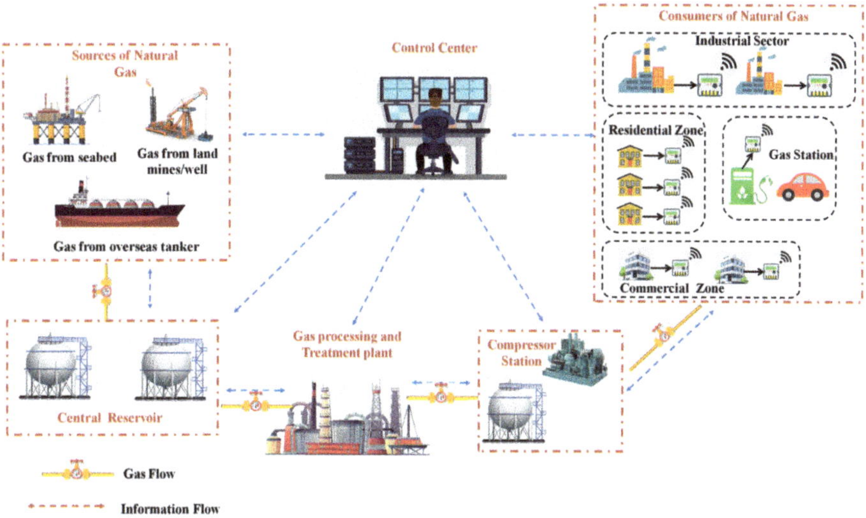

Fig. 1 Basic architecture of SGG system

networks with modern infrastructure are needed to accomplish this; smart meters and sensors transmit data over these networks. A smart grid is made up of all the components. Data can be obtained from non-smart gas meters using additional devices and sensors once the network is created. Then, using artificial intelligence software, the data is examined for subsequent actions.

Figure 1 depicts basic architecture of SGG system and its functionality. It shows that the main sources of gas are seabed, land mines or wells, and overseas tankers. It is stored at high-capacity central reservoir tank and then processed and treated for usable form that is further distributed to commercial, residential, and various applications through compressor stations. While implementing this smart system, a number of instruments and devices are connected to central control center for controlling, monitoring, and smooth functioning of SGG system.

1.2 Smart Gas Meters

Sensors, wireless communication modules, real-time clocks, electric motors to operate valves, a display unit, and a battery make up a smart gas meter that is an IoT device. By connecting wirelessly via wide area networks to an intelligent gas grid, it can not only measure gas flow but also provide data access, remote location monitoring and monitoring of infrastructure, and billing [13]. A gas cutoff can be requested from a distance if it detects an emergency, such as an earthquake or a gas leak. When it comes to powering smart meters, nothing beats batteries. To extend the life of the battery in a smart meter, low-power modules are used instead of high-power ones. Using several wireless interface standards in a smart meter design allows it to simultaneously connect to different heterogeneous and homogeneous networks.

1.3 Chapter Contribution

The significant contributions of this paper are as follows:

- IBPSC-SGG-FCE is a new secure hyperelliptic curve identity-based proxy signcryption scheme for the smart gas grid network in a fog computing environment.
- Security analysis of the proposed IBPSC-SGG-FCE shows that it is resistant to HYEC-DLP and HYEC-DHP.
- Resilience against replay and man-in-the-middle attacks is demonstrated by AVISPA performance analysis of the proposed IBPSC-SGG-FCE system.
- In the end, the comparison with the current scheme shows that IBPSC-SGG-FCE is more cost-effective in terms of calculation and communication than the current designs.

1.4 Chapter Organization

The organization of the chapter is explained as follows: Sect.2 consists of the related work about IBPSC schemes. In Sect. 3, we discuss the formal model of the proposed IBPSC-SGG-FCE scheme and the security assumptions. Section 4 describes the proposed IBPSC-SGG-FCE scheme. Section 5 consists of security analysis of the proposed scheme and also performance analysis using AVISPA and computation and communication cost comparison of proposed scheme with existing IBPSC schemes. Section 6 concludes our proposed scheme.

2 Related Work

An identity-based proxy signcryption was proposed by the authors of [14] in 2004. (ID-BPS). A security requirement of unforgeability is not met, nor is the concept of proxy protections met, by the currently proposed scheme.

Several years later, in 2005, [15] proposed an ID-BPS that was based on the theory of [14]. When compared to [14], the scheme is more competent in terms of computation costs. An efficient ID-BPS with public verifiability and forward secrecy was developed in [16].

ID-BPS was first proposed in 2008 by [17], allowing for the delegated signing rights of a proxy agent. It is possible to distinguish between the signatures of a proxy signer and those of the principle signer in the proposed scheme.

As far back as 2009, [18] improved upon [19]'s projected ID-BPS. Security is a major concern in this project's design (i.e., confidentiality, public authenticity, acceptance, verified, and forward confidentiality).

For forward secrecy and public verifiability, [20] proposed a proficient ID-BPS in 2012. It is claimed by the authors that the projected work is more computationally resourceful than [15, 16]. Another ID-BPS was later proposed by [20, 21] using the discrete logarithm problem for elliptic curves.

The proposed scheme was improved by [22] in 2014. Under the random oracle model (ROM), [23] proposed a safe ID-BPS system. In terms of computational power, the method is efficient. ID-BPGS was introduced in ROM in 2018 by [24]. ID-BPS with universal ROM compensability was developed in 2018 by [25]. The idea was to keep computations as low as possible. A well-organized ID-BPGS for cloud data sharing was developed by [26] in 2020. It is possible to delegate the rights to sign encryption to a manager's proxy agents in the proposed scheme. In place of their respective manager, the proxy agent generates encrypted data. An authorized user can download and decrypt the data generated by the proxy agent by uploading it to the cloud server. [27] Presented an ID-BPGS in the ROM in the year of 2020. According to computation costs, the proposed method is an efficient one.

However, the utilization of bilinear pairing (BP) and elliptic curve cryptosystem (ECC) builds the aforementioned schemes to be capable to maintain high computation and communication powers. Devices with limited processing power may be unable to handle such computationally intensive tasks. The existing schemes also lack realistic application scenarios such as smart gas grid, water grids, etc. In comparison to BP, ECC is believed to be the most compact and powerful cryptographic mechanism because of its higher efficiency and shorter key length, as we discussed in the introduction section. It is better suited to the smart gas grid in fog computing environment because the HCC utilizes the key size of 80 bits. Because of this, we came up with IBPS-SGG-FCE, which is both lightweight and secure.

3 Complexity Assumptions

1. *Hyperelliptic curve discrete logarithm problem (HYEC-DLP) assumption* [27–29]

Consider there is a divisor D_h of order p from the group of Jacobian (f_p). Also, there is an equation $\eta_1 * D_h = \eta_2$ where $\eta_1 \in f_p$; therefore, computation of η_1 is having a negligible advantage.

2. *Hyperelliptic curve computational Diffie-Hellman (HYEC-CDH) assumption*

Consider there is a divisor D_h of order p from the group of Jacobian (f_p). Also, there is an equation $\mathscr{L} = \eta_1 * \eta_2 * D_h =$ where η_1 and $\eta_2 \in f_p$; therefore computation of η_1 and η_2 is having a negligible advantage.

3.1 Formal Model of Proposed IBPSC-SGG-FCE Scheme

The IBPSC-SGG-FCE scheme is divided into six phases as follows.

1. Setup phase: This phase is accountable for generating public parameters P_{pub} which are openly accessible to all the participating entities and master secret M_{SK} which is a secret of the trusted third party.
2. Key extraction phase: Every individual user sends his/her unique identity $\mathfrak{U}_A$ to the trusted third party. The secret key for the user Sk_A is generated and returned via the secret channel.
3. Warrant generation and delegation phase: The original signer shall make a message warrant Ω_w which contains the information about the type of delegation and time of delegation; it also defines the type of documents to be signcrypted by proxy signcryptor. This phase is accountable for generating the signing warrant S_w and delegating it to proxy signer.
4. Warrant verification phase: This phase is accountable for the verification of signing warrant received from original signer. If the warrant is verified correctly, then the proxy signer executes the next algorithm.
5. Proxy signcryption phase: This phase takes the message to be sent m, proxy signers identity $\mathfrak{U}_{pr}$, proxy signers private key Sk_{pr}, identity of receiver $\mathfrak{U}_B$, and public parameters P_{pub} as input and generates the signcrypted message and sends to the receiver via a secure channel.
6. Unsigncryption phase: This phase takes received signcrypted message, receivers secret key Sk_B and the identity of both sender and receiver, $\mathfrak{U}_A$ and $\mathfrak{U}_B$, and generates the original message M if the signcrypted message has not been tampered with, else it returns $\perp$.

3.2 Security Definition

The proposed IBPSC-SGG-FCE scheme must satisfy confidentiality and unforgeability of original message. Let us consider that there exists an adversary $\mathcal{A}_d$ for the proposed scheme and $\mathbb{C}_h$ is a challenger [30]. For indistinguishability against adaptive chosen cipher text attack (IND-IBPSC-SGG-FCE-CCA) the following interaction between adversary $\mathcal{A}_d$ and challenger $\mathbb{C}_h$ is defined.

Definition 1 If adversary $\mathcal{A}_d$ with no polynomial time and having non-negligible advantage wins the following game, then the proposed scheme IBPSC-SGG-FCE can achieve the security requirements of IND-IBPSC-SGG-FCE-CCA.

Initialize The challenger $\mathbb{C}_h$ executes the setup algorithm to get the public parameters and a master secret λ. Then $\mathbb{C}_h$ sends the public parameters to adversary $\mathcal{A}_d$ and keeps λ with itself.

Phase 1 Adversary $\mathcal{A}_d$ executes the following queries which are interdependent.

1. Key extraction query: Adversary $\mathcal{A}_d$ selects the unique identity as $\mathfrak{U}_{ID}$. The challenger $\mathbb{C}_h$ runs key extraction algorithm and returns the Sk_{ID} to adversary $\mathcal{A}_d$.
2. Warrant generation and delegation query: The adversary $\mathcal{A}_d$ sends the request for signing warrant. The challenger $\mathbb{C}_h$ returns the warrant Ω_w and signing warrant S_w.
3. Warrant verification query: The adversary $\mathcal{A}_d$ verifies the signing warrant received from challenger $\mathbb{C}_h$.
4. Proxy signcryption query: The adversary $\mathcal{A}_d$ selects message m and the identities $\mathfrak{U}_A$, $\mathfrak{U}_B$, and $\mathfrak{U}_C$. The challenger $\mathbb{C}_h$ executes Key Extraction and Warrant generation and Delegation to get secret keys of Sk_A and Sk_B and the signing warrant S_w and then executes Proxy Secret key Generation to get SK_{Pr}. Finally the challenger $\mathbb{C}_h$ runs Proxy Signcryption and sends the signcrypted ciphertext ρ to $\mathcal{A}_d$.
5. Unsigncryption query: The adversary $\mathcal{A}_d$ selects the signcrypted ciphertext ρ and the identities $\mathfrak{U}_A$, $\mathfrak{U}_B$, and $\mathfrak{U}_C$. The challenger $\mathbb{C}_h$ runs key extraction algorithm to get the Sk_c, then executes unsigncryption algorithm, and sends result to $\mathcal{A}_d$.

Challenge The adversary $\mathcal{A}_d$ wishes to be challenged on the two messages M_0 and M_1 and identities $\mathfrak{U}_i$ and $\mathfrak{U}_j$. In the first stage, $\mathcal{A}_d$ cannot query for secret key of any of the identity. The challenger $\mathbb{C}_h$ produces the random bit $\mathfrak{b} \in_R \{0,1\}$ for which the $\rho = signcrypt\left(M, Sk_b, \mathfrak{U}_C\right)$ and sends to $\mathcal{A}_d$.

Phase 2 The adversary $\mathcal{A}_d$ executes the queries like phase 1. Except Key Extraction query for identities $\mathfrak{U}_i$ and $\mathfrak{U}_j$ and unsigncrypted text for ρ.

Guess The adversary $\mathcal{A}_d$ produces the random bit $\mathfrak{b}' \in_R \{0,1\}$. If $\mathfrak{b} = \mathfrak{b}'$ the adversary $\mathcal{A}_d$ wins the game. We have the following advantage of $\mathcal{A}_d$.

$$\text{Adv}\left(\mathcal{A}_d\right) = \left|Pr[\mathfrak{b} = \mathfrak{b}'] - \frac{1}{2}\right|$$

Definition 2 The proposed scheme IBPSC-SGG-FCE can achieve the existential unforgeability against adaptive chosen messages attack (EUF-IBPSC-SGG-FCE-SPA) if adversary $\mathcal{A}_d$ with no polynomial time and having non-negligible advantage in the following game.

Initial: The challenger $\mathbb{C}_h$ executes the setup algorithm to get the public parameters and a master secret λ. Then $\mathbb{C}_h$ sends the public parameters to adversary $\mathcal{A}_d$. Then $\mathcal{A}_d$ performs polynomial limited number of queries like in IND- IBPSC-SGG-FCE -CCA. Finally, adversary $\mathcal{A}_d$ generates $(\rho, \mathfrak{U}_i, \mathfrak{U}_j)$. In phase 2 the private key for $\mathfrak{U}_i$ was not asked, and the adversary $\mathcal{A}_d$ wins the game if the output of Unsigncryption $(\rho,\ Sk_i, \mathfrak{U}_j)$ is not $\perp$.

4 Proposed IBPSC-SGG-FCE Scheme

The system model for the proposed IBPSC-SGG-FCE consists of five entities as described in Fig. 2. The trusted third party is the PKG, which is the trusted authority specially used for the purpose to generate the private keys for data owners, data users, and proxy signcryptor, respectively. At fog layer, the proxy signcryptor is an entity that generates the signcrypted messages on behalf of data owners and later sends it to the cloud server. As a storage service, the cloud server acts as an intermediary between (data owners and data users) and facilitates secure delivery of signcrypted ciphertext to the authorized recipient. To decrypt the message, the data user must first validate the message's content and then decrypt the signcrypted ciphertext.

4.1 Construction of IBPSC-SGG-FCE

The proposed IBPSC-SGG-FCE scheme is described in this section. The proposed scheme consists of six phases as setup, key extraction, warrant generation and delegation, warrant verification, proxy signcryption, and unsigncryption. The description of notations used is mentioned in Table 1.

1. *Setup Phase*

   Input: Security parameters l of hyperelliptic curve
   Output: Public system parameters P_{pub}

   1. Identity of each participating entity is represented as $\mathfrak{U}_i$.
   2. Select $\in_R \mathbb{Z}_p$, where λ is a master secret key (M_{SK}).
   3. Master public key $M_{Pub} = \lambda * D_h$.
   4. $\mathcal{H}_1$, $\mathcal{H}_2$, $\mathcal{H}_3$, and $\mathcal{H}_4$ are secured one-way hash functions.
   5. The PKG publishes the public system parameters as $P_{pub} = \{M_{Pub}, D_h, \mathcal{H}_1, \mathcal{H}_2, \mathcal{H}_3, \mathcal{H}_4\}$. The λ is kept secret.

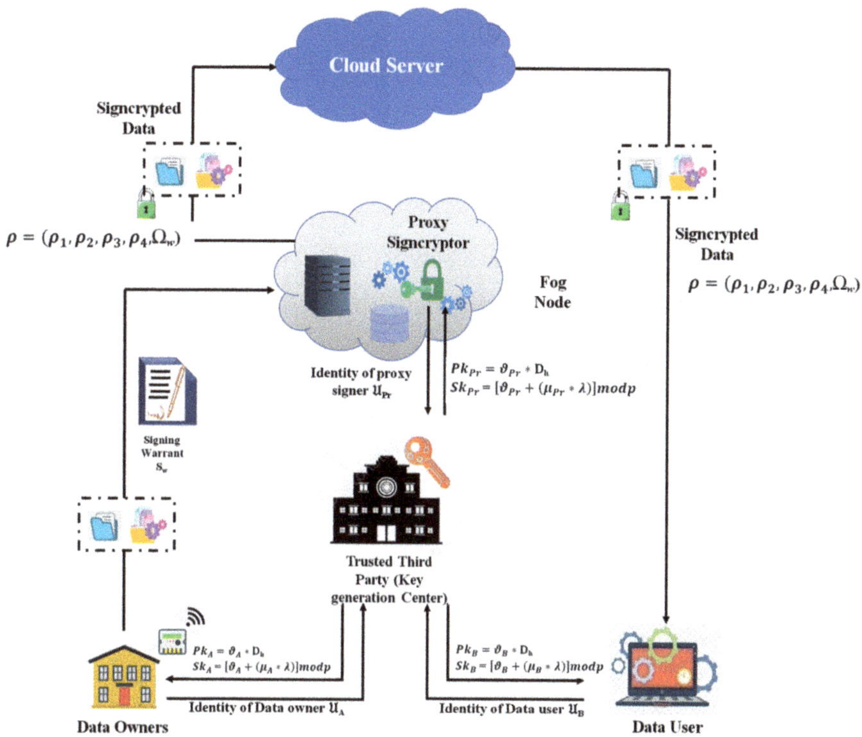

Fig. 2 The system model for the proposed IBPSC-SGG-FCE scheme

Table 1 Notations used in this work

Notation	Meaning
l	Security parameter of HYEC
M_{Pub}, λ	Master public and private key
$\mathscr{H}_1$, $\mathscr{H}_2$, $\mathscr{H}_3$, $\mathscr{H}_4$	Irreversible cryptographic hash function
D_h	Divisor of HYEC
$\mathfrak{U}_{\mathrm{A}}$	Identity of data owner
$\mathfrak{U}_{\mathrm{pr}}$	Proxy signer
$\mathfrak{U}_{\mathrm{B}}$	Data requester
Sk_{A}, Pk_{A}	Secret key and public key of data owner
Sk_{pr}, Pk_{pr}	Secret key and public key of proxy signer
Sk_{B}, Pk_{B}	Secret key and public key of data requester
$\mathrm{S_W}$	Signcrypting warrant
Ω_{W}	Message warrant
ρ	Signcrypted message
$\oplus$	Bitwise XOR operation

2. *Key Extraction Phase*

   Input: Identity of participating entities $\mathfrak{U}_i$
   Output: Public and secret keys for $\mathfrak{U}_i$

   1. Compute $\vartheta_A = \mathscr{H}_1(\mathfrak{U}_A, \lambda)$.
   2. The public key of user A with identity $\mathfrak{U}_A$ is $Pk_A = \vartheta_A * D_h$.
   3. Compute $\mu_A = \mathscr{H}_1(\mathfrak{U}_A, Pk_A)$.
   4. The secret key of user A with identity $\mathfrak{U}_A$ is $Sk_A = [\vartheta_A + (\mu_A * \lambda)] mod p$.

3. *Warrant Generation and Delegation Phase*

   Input: P_{pub}, Sk_A, Ω_w
   Output: Signcrypting warrant S_w
   The original signer shall make a warrant Ω_w which contains the information about the type of delegation and time of delegation (t_1); it also defines the type of documents to be signcrypted by proxy signcryptor.
   By using warrant Ω_w the original signer generates signcrypting warrant S_w by using original signer's private key Sk_A.

   1. Select $\varphi \in_R \mathbb{Z}_p$.
   2. $\gamma = \varphi * D_h$.
   3. $S_w = (\varphi - Sk_A . \mathscr{H}_2(\gamma, \Omega_w, pr, t_1))$ mod p.

   The original signer sends $(\gamma, S_w, \Omega_w, t_1)$ to proxy signcryptor.

4. *Warrant Verification Phase*

   Input: (γ, S_w, Ω_w).
   Output: Accept or reject the signing warrant.

   1. The proxy signer checks the freshness of $(\gamma, S_w, \Omega_w, t_1)$.
   2. On successful event, it verifies the received delegation by computing

   $\gamma' = S_w * D_h + \mathscr{H}_2(\gamma, \Omega_w, \mathfrak{U}_{pr}, t_1) Pk_A$.

5. *Proxy Signcryption Phase*

   Input: P_{pub}, Pk_B, Sk_{pr}, S_w, M
   Output: Signcrypted message ρ

   1. Select $\alpha, \beta \in_R \mathbb{Z}_p$.
   2. Compute $\rho_1 = \alpha * D_h$.
   3. Compute $\rho_2 = \beta * D_h$.
   4. Compute $\mu_B = \mathscr{H}_0(\mathfrak{U}_B, Pk_B)$.

$$5. \quad \text{Compute } \aleph = \alpha\,(Pk_B + \mu_B * M_{Pub}) \quad (1)$$

6. Compute $\acute{K} = \mathscr{H}_3(\aleph)$.
7. Compute $\rho_3 = M \oplus \acute{K}$.
8. Compute $£ = \mathscr{H}_4(M)$.
9. Compute $\rho_4 = [S_w + \beta + (£ * Sk_{pr})] \bmod p$.
10. $\rho = (\rho_1, \rho_2, \rho_3, \rho_4, \Omega_w)$.

The proxy signcryptor uploads the signcrypted ciphertext ρ on cloud.

6. *Unsigncryption Phase*

Input: P_{pub}, Sk_B, ρ
Output: Original message M or $\perp$
The receiver with identity $\mathfrak{U}_B$ will download the signcrypted ciphertext ρ from cloud and perform the following operations to compute the original message M.

1. Compute $\aleph' = Sk_B * \rho_1$.
2. Compute $\acute{K}' = \mathscr{H}_3(\aleph')$.
3. Compute $M' = \acute{K}' \oplus \rho_3$.
4. Compute $£' = \mathscr{H}_4(M')$.
5. The receiver with identity $\mathfrak{U}_B$ will then decrypt the cipher text, and the message M is accepted only if the following condition holds

$$\rho_4 * D_h = \gamma - \mathcal{H}_2\ (\gamma, \Omega_w)\ \left(Pk_A + \mu_A * M_{pub}\right) + \rho_2 + £ * \left(Pk_{pr} + \mu_{pr} * M_{Pub}\right) \tag{2}$$

Otherwise it returns $\perp$.

5 Security Analysis of the Proposed Scheme

Proposition 1 Correctness for the generation of $\acute{K}'$. The proxy unsigncryptor first computes $\aleph'$.

Let us consider Eq. (1).
$\aleph = \alpha(Pk_B + \mu_B * M_{Pub})$
Substitute $Pk_B = \vartheta_B * D_h$ and $M_{Pub} = \lambda * D_h$.
$\aleph = \alpha(\vartheta_B * D_h + \mu_B * \lambda * D_h)$
$\aleph = \alpha D_h(\vartheta_B + \mu_B * \lambda)$
Substitute $Sk_B = \vartheta_B + \mu_B * \lambda$ and $\rho_1 = \alpha * D_h$.
$\aleph = Sk_B * \rho_1$
$\aleph = \aleph'$
Then compute $\acute{K}' = \mathscr{H}_3(\aleph')$.

Proposition 2 Correctness for the signature. The proxy unsigncryptor checks the correctness of received signature by verifying Eq. 2.

$$\rho_4 * D_h = \gamma - \mathcal{H}_2\ (\gamma,\ \Omega_w)\ \left(Pk_A + \mu_A * M_{pub}\right) + \rho_2 \\ + £ * \left(Pk_{pr} + \mu_{pr} * M_{Pub}\right) \tag{2}$$

For the correctness of Eq. 2, first we prove the correctness of $S_w * D_h = \gamma - \mathscr{H}_2\ (\gamma, \Omega_w)\ (Pk_A + \mu_A * M_{pub})$.

Let us consider LHS $S_w * D_h$.

Substitute $S_w = (\varphi - Sk_A.\ \mathscr{H}_2\ (\gamma, \Omega_w))$.

$= (\varphi - Sk_A.\ \mathscr{H}_2\ (\gamma, \Omega_w))*D_h$

$= \varphi * D_h - Sk_A.\ \mathscr{H}_2\ (\gamma, \Omega_w)*D_h$

Substitute $Sk_A = \vartheta_A + (\mu_A * \lambda)$.

$=\varphi * D_h - \vartheta_A + (\mu_A * \lambda).\ \mathscr{H}_2\ (\gamma, \Omega_w)*D_h$

$=\varphi * D_h - \mathscr{H}_2\ (\gamma, \Omega_w)\ (\vartheta_A * D_h + \mu_A * \lambda * D_h)$

Substitute $Pk_A = \vartheta_A * D_h$ and $M_{Pub} = \lambda * D_h$.

$=\varphi * D_h - \mathscr{H}_2\ (\gamma, \Omega_w)\ (Pk_A + \mu_A * M_{Pub})$

Substitute $\gamma = \varphi * D_h$.

$$S_w * D_h = \gamma - \mathcal{H}_2\ (\gamma, \Omega_w)\ \left(Pk_A + \mu_A * M_{pub}\right) \tag{3}$$

=RHS

Now, let us consider Eq. 2.

$\rho_4 * D_h = \gamma - \mathscr{H}_2\ (\gamma, \Omega_w)\ (Pk_A + \mu_A * M_{pub}) + \rho_2 + £ * (Pk_{pr} + \mu_{pr} * M_{Pub})$

Let us consider LHS $\rho_4 * D_h$.

Substitute $\rho_4 = [S_w + \beta + (£ * Sk_{pr})]$.

$=[S_w + \beta + (£ * Sk_{pr})] * D_h$

$=(S_w * D_h) + (\beta * D_h) + (£ * Sk_{pr} * D_h)$

Substitute $\beta * D_h = \rho_2$ and $Sk_{pr} = [\vartheta_{pr} + (\mu_{pr} * \lambda)]$.

$=(S_w * D_h) + \rho_2 + £ * [\vartheta_{pr} + (\mu_{pr} * \lambda)] * D_h$

$=(S_w * D_h) + \rho_2 + £ * [\vartheta_{pr} * D_h + (\mu_{pr} * \lambda) * D_h]$

Substitute $Pk_{pr} = \vartheta_{Pr} * D_h$ and $M_{Pub} = \lambda * D_h$.

$=(S_w * D_h) + \rho_2 + £ * (Pk_{pr} + \mu_{pr} * M_{Pub})$

Substitute Eq. 3 $S_w * D_h = \gamma - \mathscr{H}_2\ (\gamma, \Omega_w)\ (Pk_A + \mu_A * M_{pub})$.

$= \gamma - \mathscr{H}_2\ (\gamma, \Omega_w)\ (Pk_A + \mu_A * M_{pub}) + \rho_2 + £ * (Pk_{pr} + \mu_{pr} * M_{Pub})$

=RHS

Hence proved.

Theorem 1 The proposed scheme is indistinguishable against adaptive chosen ciphertext attack (IND-IBPSC-SGG-FCE-CCA), if the solution for HYEC-DHP is infeasible for adversary $\mathcal{A}_d$.

Proof

Let us consider that there exists an adversary $\mathcal{A}_d$ for the proposed scheme and $\mathbb{C}_\mathrm{h}$ is a challenger. Suppose the two randomly selected numbers are η_1 and η_2. The challenge for challenger $\mathbb{C}_\mathrm{h}$ is to solve $\eta_1 * D_h = \eta_2 * D_h = \eta_1 * \eta_2 * D_h$.

Initialization

The challenger $\mathbb{C}_\mathrm{h}$ takes *security parameters l of hyperelliptic curve*. Select $\in_R \mathbb{Z}_p$, where λ is a master secret key (M_{SK}) selected randomly, and then compute master public key $M_{Pub} = \lambda * D_h$. The challenger $\mathbb{C}_\mathrm{h}$ generates the public system parameters as $P_{pub} = \{M_{Pub}, D_h, \mathscr{H}_1, \mathscr{H}_2, \mathscr{H}_3, \mathscr{H}_4\}$ and sends it to adversary $\mathcal{A}_d$.

Phase I

After completing the successful initialization phase, the adversary $\mathcal{A}_d$ executes the following queries which are interdependent.

$\mathscr{H}_1$ *queries*: The challenger $\mathbb{C}_\mathrm{h}$ maintains the list $\mathscr{LH}_1$ of triplet ($\mathfrak{U}$i, Pk_i, ϑ_i); when adversary $\mathcal{A}_d$ asks for the ($\mathfrak{U}$i, Pk_i, ϑ_i), the challenger $\mathbb{C}_\mathrm{h}$ selects ϑ_i randomly and computes $\vartheta_i = \mathscr{H}_1(\mathfrak{U}_\mathrm{i}, Pk_\mathrm{i})$, and add ($\mathfrak{U}_\mathrm{i}$, Pk_i, ϑ_i) to list $\mathscr{LH}_1$.

$\mathscr{H}_2$ *queries*: The challenger $\mathbb{C}_\mathrm{h}$ maintains the list $\mathscr{LH}_2$ of (γ, Ω_w, $\mathfrak{U}_\mathrm{i}$, $\mathrm{t_i}$, $\mathrm{S_w}$); when adversary $\mathcal{A}_d$ asks for the (γ, Ω_w, $\mathfrak{U}_\mathrm{i}$, $\mathrm{t_i}$, $\mathrm{S_w}$), the challenger $\mathbb{C}_\mathrm{h}$ selects $\mathrm{S_w}$ randomly and computes $\mathrm{S_w} = \mathscr{H}_1(\gamma, \Omega_w, \mathfrak{U}_\mathrm{i}, \mathrm{t_i})$, and add ($\gamma$, Ω_w, $\mathfrak{U}_\mathrm{i}$, $\mathrm{t_i}$, $\mathrm{S_w}$) to list $\mathscr{LH}_2$.

$\mathscr{H}_3$ *queries*: The challenger $\mathbb{C}_\mathrm{h}$ maintains the list $\mathscr{LH}_3$ of ($\aleph$, $\acute{K}$); when adversary $\mathcal{A}_d$ asks for the ($\aleph$), the challenger $\mathbb{C}_\mathrm{h}$ selects $\acute{K}$ randomly and computes $\acute{K} = \mathscr{H}_3(\aleph)$, and add ($\aleph$, $\acute{K}$) to list $\mathscr{LH}_3$.

$\mathscr{H}_4$ *queries*: The challenger $\mathbb{C}_\mathrm{h}$ maintains the list $\mathscr{LH}_4$ of (M, £); when adversary $\mathcal{A}_d$ asks for the (M), the challenger $\mathbb{C}_\mathrm{h}$ selects £ randomly and computes £$= \mathscr{H}_4(M)$, and add (M, £) to list $\mathscr{LH}_4$.

Public key extraction query: The challenger $\mathbb{C}_\mathrm{h}$ maintains the list $\mathscr{LH}_\mathrm{pk}$ of ($\mathfrak{U}_\mathrm{i}$, Pk_i). When adversary $\mathcal{A}_d$ asks for the public key, the challenger $\mathbb{C}_\mathrm{h}$ chooses $\vartheta_i \in_R \mathbb{Z}_p$ and sets $Pk_i = \vartheta_i * D_h$. Then add ($\mathfrak{U}_\mathrm{i}$, Pk_i, ϑ_i) to list $\mathscr{LH}_\mathrm{pk}$ and send Pk_i to adversary $\mathcal{A}_d$.

Secret key extraction query: The challenger $\mathbb{C}_\mathrm{h}$ maintains the list $\mathscr{LH}_\mathrm{sk}$ of ($\mathfrak{U}_\mathrm{i}$, Pk_i, ϑ_i). When adversary $\mathcal{A}_d$ asks for the secret key, the challenger $\mathbb{C}_\mathrm{h}$ checks if $\mathfrak{U}_\mathrm{i} = \mathfrak{U}^*$; it terminates the game. Else, it takes a triple ($\mathfrak{U}_\mathrm{i}$, Pk_i, ϑ_i). From $\mathscr{LH}_1$, set $Sk_A = [\vartheta_\mathrm{A} + (\mu_\mathrm{A} * \lambda)]$ mod p. Then add ($\mathfrak{U}_\mathrm{i}$, Pk_i, Sk_i) to list $\mathscr{LH}_\mathrm{sk}$ and send Sk_i to adversary $\mathcal{A}_d$.

Warrant generation and delegation query: The adversary $\mathcal{A}_d$ asks for the warrant generation and delegation query; if $\mathfrak{U}_\mathrm{A} = \mathfrak{U}^*$, then the challenger $\mathbb{C}_\mathrm{h}$ replies with Ω_w using warrant generation and delegation algorithm to the adversary $\mathcal{A}_d$, or else it calculates $\gamma = \varphi * D_h$ where $\varphi \in_R \mathbb{Z}_p$; then set warrant as ($\gamma$, $\mathrm{S_w}$, Ω_w, $\mathrm{t_1}$) and reply to adversary $\mathcal{A}_d$.

Proxy signcryption query: Using the identity of the proxy $\mathfrak{U}_\mathrm{pr}$ and identity of receiver $\mathfrak{U}_\mathrm{B}$, the challenger $\mathbb{C}_\mathrm{h}$ performs the following computations with message M, after receiving proxy signcryption query.

- It checks if $\mathfrak{U}_{pr} = \mathfrak{U}^*$; the challenger $\mathbb{C}_h$ calls proxy signcryption algorithm and sends ρ to adversary $\mathcal{A}_d$.
- Else select α, β, $£ \in_R \mathbb{Z}_p$. Compute $\rho_1 = \alpha * D_h$ and $\rho_2 = \beta * D_h$. calculate $\aleph = \alpha(Pk_B + \mu_B * M_{Pub})$, set $\acute{K} = \mathscr{H}_3(\aleph)$, compute $\rho_3 = M \oplus \acute{K}$, compute $\rho_4 = [S_w + \beta + (£ * Sk_{pr})] \bmod p$, and send ρ to adversary $\mathcal{A}_d$.

Unsigncryption query: If the adversary $\mathcal{A}_d$ asked, if $\mathfrak{U}_B \neq \mathfrak{U}^*$, then the challenger $\mathbb{C}_h$ replied by calling unsigncryption algorithm.

Challenge: An adversary $\mathcal{A}_d$ may output two messages M_0 and M_1 and two identities $\mathfrak{U}_{pr}$ and $\mathfrak{U}_B$ and send it to challenger $\mathbb{C}_h$. Then $\mathbb{C}_h$ compares if $\mathfrak{U}_{pr} \neq \mathfrak{U}^*$; then it terminates. If not then the challenger $\mathbb{C}_h$ selects b ϵ {0,1} and completes the following operations to generate ρ^*. It selects two random numbers α, $\beta \in_R \mathbb{Z}_p$ and computes $\rho_1 = \alpha * D_h$, $\rho_2 = \beta * D_h$. Compute $\mu_B = \mathscr{H}_0$ ($\mathfrak{U}_B$, Pk_B) and $\aleph = \alpha(Pk_B + \mu_B * M_{Pub})$. Set $\acute{K} = \mathscr{H}_3(\aleph)$, compute $\rho_3 = M \oplus \acute{K}$ and $\rho_4 = [S_w + \beta + (£ * Sk_{pr})]$, and send $\rho^* = (\rho_1, \rho_2, \rho_3, \rho_4, \Omega_w)$ to adversary $\mathcal{A}_d$. Then the adversary $\mathcal{A}_d$ continues with $\mathscr{H}$queries, public key extraction query $\left(\mathcal{Q}_{pk}\right)$, secret key extraction query $\left(\mathcal{Q}_{sk}\right)$, warrant generation and delegation query $(\mathcal{Q}_{gd})$, proxy signcryption queries $\left(\mathcal{Q}_{psc}\right)$, and unsigncryption query $\left(\mathcal{Q}_{usc}\right)$.

Guess: An adversary $\mathcal{A}_d$ may output $b' = b$; then adversary $\mathcal{A}_d$ is successful and identifies the solution for HYEC-DHP instance, or else an adversary $\mathcal{A}_d$ terminates. Then challenger $\mathbb{C}_h$ can solve HYEC-DHP and be successful in challenge phase and its probability as $\frac{1}{\mathcal{Q}_{pk} - \mathcal{Q}_{sk}}$ so we have the probability as

$$\text{Adv}\ (\mathcal{A}_d)^* \geq \text{Adv}\ (\mathcal{A}_d)\left(1 - \frac{\mathcal{Q}_{sk}}{\mathcal{Q}_{pk}}\right)\left(1 - \frac{1}{2^{\lambda}}\right)\left(\frac{1}{\mathcal{Q}_{pk} - \mathcal{Q}_{sk}}\right).$$

Theorem 2 The proposed IBPSC-SGG-FCE scheme is unforgeable, if an adversary $\mathcal{A}_d$ has the capability of existential forgery for (EUF-IBPSC-SGG-FCE-SPA) selected plaintext attack.

Proof

Suppose the challenger $\mathbb{C}_h$ receives a challenge to extract the randomly selected number η_1 from $\eta_1 * D_h = \eta_2$ for the adversary $\mathcal{A}_d$ that is called hyperelliptic curve discrete logarithm problem (HYCDLP).

Initialization

The challenger $\mathbb{C}_{\mathrm{h}}$ takes *security parameters l of hyperelliptic curve.* Select $\in_R \mathbb{Z}_p$, where λ is a master secret key (M_{SK}) selected randomly, and then compute master public key $M_{Pub} = \lambda * D_h$. The challenger $\mathbb{C}_{\mathrm{h}}$ generates the public system parameters as $P_{pub} = \{M_{Pub}, D_h, \mathscr{H}_1, \mathscr{H}_2, \mathscr{H}_3, \mathscr{H}_4\}$ and sends P_{pub} to adversary $\mathcal{A}_d$.

Queries

After initialization phase the adversary $\mathcal{A}_d$ executes $\mathscr{H}$queries, public key extraction query $\left(\mathcal{Q}_{pk}\right)$, secret key extraction query $\left(\mathcal{Q}_{sk}\right)$, warrant generation and delegation query $(\mathcal{Q}_{gd})$, proxy signcryption queries $\left(\mathcal{Q}_{psc}\right)$, and unsigncryption query $\left(\mathcal{Q}_{usc}\right)$, similar to theorem 1.

Forgery

The adversary $\mathcal{A}_d$ wins the game if the following cases hold.

Case 1: Here we focus on outputs of a valid delegation as $\{(\gamma, \mathrm{S_w}, \Omega_{\mathrm{w}}), {}_{\mathrm{A}}\}$. By using the forking concept, the challenger $\mathbb{C}_{\mathrm{h}}$ generates a valid delegation signature $\{(\gamma', \mathrm{S_w}, \Omega_{\mathrm{w}}), {}_{\mathrm{A}}\}$. Here $\sigma = \mathcal{H}_2\left(\gamma, \Omega_{\mathrm{w}}, \mathfrak{U}_{\mathrm{pr}}, \mathrm{t}_1\right)$, and $\sigma' = \mathcal{H}_2'\left(\gamma', \Omega_{\mathrm{w}}, \mathfrak{U}_{\mathrm{pr}}, \mathrm{t}_1\right)$, and $\sigma \neq \sigma'$. If ${}_{\mathrm{A}} = \mathfrak{U}_*$ then the challenger $\mathbb{C}_{\mathrm{h}}$ can get the solution for HYCDLP by calculating $\eta_1 = \frac{\gamma - \gamma'}{(\sigma - \sigma') - \mu_A * \lambda}$.

Case 2: Here we focus on outputs of a valid delegation as $\rho = (\rho_1, \rho_2, \rho_3, \rho_4, (\gamma, \mathrm{S_w}, \Omega_{\mathrm{w}}), \mathfrak{U}_{\mathrm{A}})$. By using the forking concept, the challenger $\mathbb{C}_{\mathrm{h}}$ generates a valid delegation signature as $\rho = \left(\rho_1, \rho_2, \rho_3, \rho_4, \left(\gamma', \mathrm{S_w}, \Omega_{\mathrm{w}}\right), \mathfrak{U}_{\mathrm{A}}\right)$. Here $\sigma = \mathcal{H}_2\left(\gamma, \Omega_{\mathrm{w}}, \mathfrak{U}_{\mathrm{pr}}, \mathrm{t}_1\right)$, and $\sigma' = \mathcal{H}_2'\left(\gamma', \Omega_{\mathrm{w}}, \mathfrak{U}_{\mathrm{pr}}, \mathrm{t}_1\right)$, and $\sigma \neq \sigma'$ and $£' = £ = \mathscr{H}_4(\mathrm{M})$. If $\mathfrak{U}_A = \mathfrak{U}_*$ then the challenger $\mathbb{C}_{\mathrm{h}}$ can get the solution for HYCDLP by calculating $\eta_1 = \frac{\gamma - \gamma'}{(\sigma - \sigma') - \mu_A * \lambda}$.

Case 3: Here we focus on outputs of a valid delegation as $\rho = (\rho_1, \rho_2, \rho_3, \rho_4, \Omega_{\mathrm{w}})$. By using the forking concept, the challenger $\mathbb{C}_{\mathrm{h}}$ generates a valid delegation signature as $\rho = \left(\rho_1, \rho_2, \rho_3, \rho_4', \Omega_{\mathrm{w}}\right)$. Here $£ = \mathscr{H}_4(\mathrm{M})$, $£' = \mathscr{H}_4(M')$, and $£' \neq £$. *If* ${}_{\mathrm{A}} = \mathfrak{U}_*$ then the challenger $\mathbb{C}_{\mathrm{h}}$ can get the solution for HYCDLP by calculating $\eta_1 = \frac{£' - £}{(\rho_4' - \rho_4) - \mu_{\mathrm{A}} * \lambda}$.

5.1 *Performance Analysis*

In this section, the performance analysis of the proposed IBPSC-SGG-FCE scheme is discussed. We use the well-known AVISPA tool [28, 29] to discuss the security proof and demonstrate that the proposed scheme is not susceptible to replay and man-in-the-middle attack.

Figure 3 describes the AVISPA architecture. SPAN receives the CAS+ protocol specification in Alice and Bob notation and converts it to HLPSL specification script.

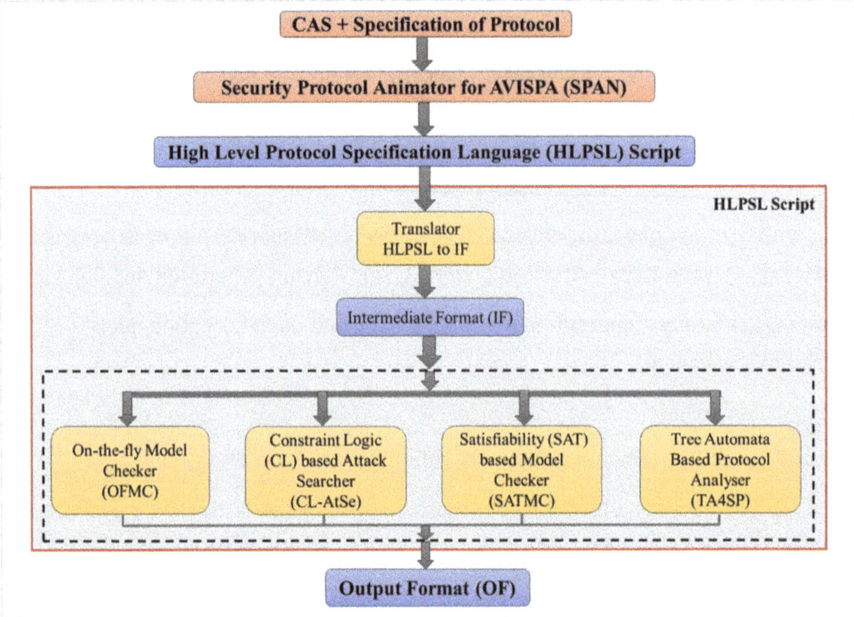

Fig. 3 Architecture of AVISPA

HLPSL script is fed into IF translator, which converts it to IF and analyses it using AVISPA's OFMC and CL-AtSe backends. The backend will execute the protocol indefinitely until it is deemed safe for the given number of sessions or an attack is discovered.

The HLPSL [30] code is written for the proposed scheme with the different roles like data owner, proxy signcryptor, data user, and trusted third party. This code is then executed using SPAN and AVISPA with the backends OFMC and CL-AtSe. We can see that no attacks were discovered by OFMC. In other words, for a limited number of sessions as specified in the role of the environment, the stated security goals were achieved. The proposed protocol is also executed with CL-AtSe backend for bounded number of sessions. The output shows that the protocol is safe under CL-AtSe also. The software resources such as Oracle VM VirtualBox and security protocol animator (SPAN) are used. The output of AVISPA shows that the proposed IBPSC-SGG-FCE is safe under OFMC and CL-AtSe backends. Figures 4 and 5 describe that the search time is 0.24 s, and the visited number of nodes is 208, and 11 states are analyzed, of which 6 states are reachable by the proposed technique.

The performance proposed IBPS-SGG-FCE scheme is analyzed in terms of both computation and communication costs, by comparing with the existing proxy signcryption schemes proposed by authors of [20, 24, 26, 27, 32, 33].

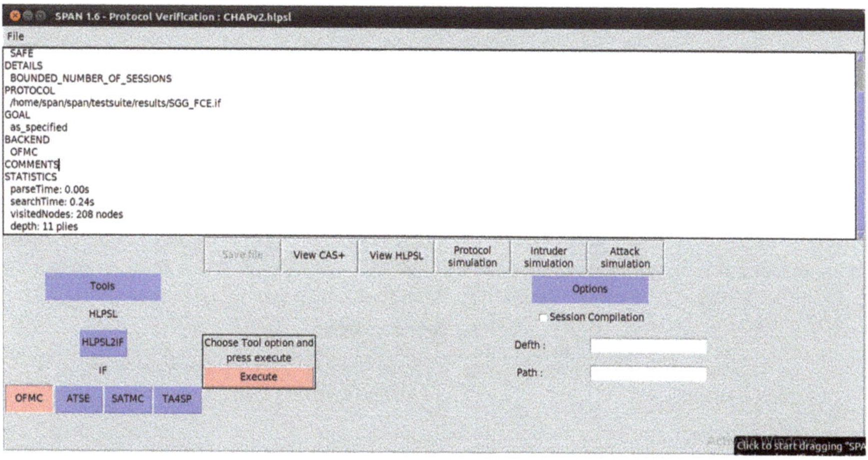

Fig. 4 OFMC output

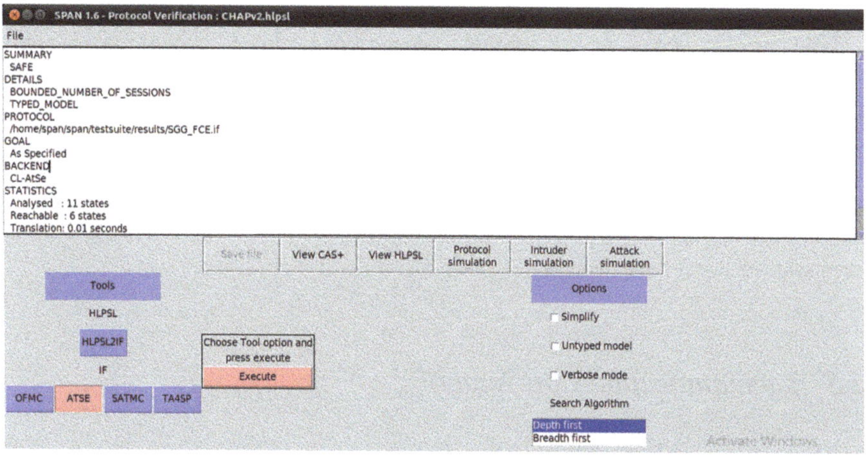

Fig. 5 CL-ATSE output

5.2 *Computational Cost Analysis*

Based on expensive arithmetic computations such as pairing ($\wp$), pairing-based point multiplication ($\mathfrak{B}$PM), elliptic curve point multiplication ($\mathscr{E}$PM), and hyperelliptic curve divisor multiplication ($\mathscr{H}$DM), we compare our proposed work to the work proposed in [20, 24, 26, 27, 32, 33] in terms of performance efficiency. However, operations such as division, encryption, addition, decryption, and hashing have been overlooked due to their minimal processing time. As a result, we conducted a comparison of our technique with other proxy signcryption schemes proposed in [20, 24, 26, 27, 32, 33] using the aforementioned mathematical procedure.

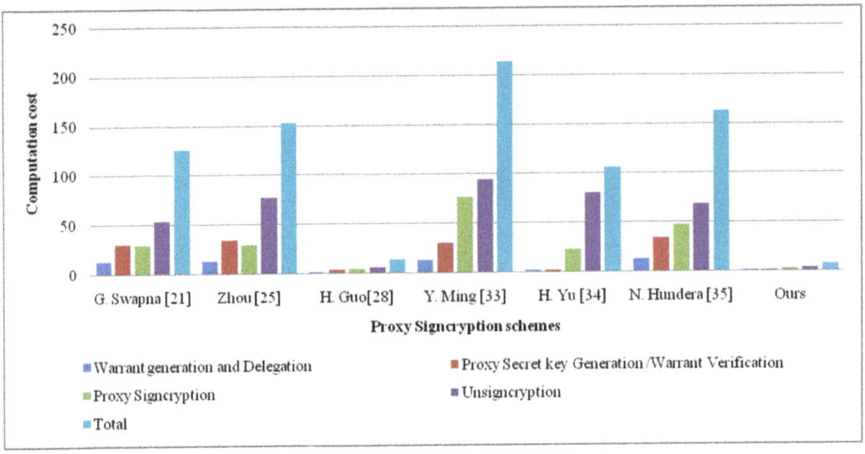

Fig. 6 Comparison of computation cost in milliseconds

The performance of the cryptographic operations evaluated by authors of [31] is 4.31 ms for $\mathfrak{B}$PM, 14.90 ms for $\wp$, 1.25 ms for $\mathscr{E}$, 0.97 ms for $\mathscr{E}$PM, and 0.48 ms for $\mathscr{H}$DM. The simulated results were analyzed on a computer running Windows 10 operating system 64-bit and powered by an Intel Core i7 processor operating at 2.0 GHz with 8 GB of RAM. According to Table 2, it is noticeable that the proposed scheme outperforms the other systems proposed in [20, 24, 26, 27, 32, 33] when it comes to computation power in millisecond (ms). Figure 6 further shows that computation costs have decreased (Table 3).

5.3 *Communication Cost Analysis*

The term "communication overhead" refers to the number of additional bits that will be sent along with the original message. We evaluated our schemes' communication overhead to that of [31]'s work to see how efficient it is. $|G| = 1024$ bits for bilinear pairings, $|q| = 160$ bits for ECC, $|n| = 80$ bits for HCC, and $|m| = 100$ bits for the message (m)e are assumed to be in the range of our comparison. The hypothesis is that the proposed IBPS-SGG-FCE communication cost is $3|m| + 7|n|$. That means the proposed scheme is more cost-effective than other schemes proposed in [20, 24, 26, 27, 32, 33], which is clear from Tables 4 and 5. In addition, a reduction in communication costs is clearly apparent in Fig. 7.

Table 2 Computation cost comparison

Scheme	Warrant generation and delegation	Proxy secret key generation/warrant verification	Proxy signcryption	Unsigncryption	Total
G. Swapna [20]	$3\mathfrak{B}$PM	$2\wp$	$3\mathfrak{B}\text{PM} + 1\mathscr{E} + 1\wp$	$2\ \mathfrak{B}\text{PM} + 3\ \wp$	$8\mathfrak{B}\text{PM} + 1\mathscr{E} + 4\wp$
Zhou [24]	$3\mathfrak{B}$PM	$2\wp + 1\mathfrak{B}\text{PM}$	$3\mathfrak{B}\text{PM} + 1\wp + 1\mathscr{E}$	$4\mathfrak{B}\text{PM} + 4\wp$	$11\mathfrak{B}\text{PM} + 1\mathscr{E} + 7\wp$
H. Guo[27]	$1\mathscr{E}$PM	$3\mathscr{E}$PM	$4\mathscr{E}$PM	$6\mathscr{E}$PM	$14\mathscr{E}$PM
Y. Ming [32]	$3\mathfrak{B}$PM	$2\wp$	$3\mathfrak{B}\text{PM} + 4\wp + 3\mathscr{E}$	$4\text{M} + 6\wp$	$6\mathfrak{B}\text{PM} + 7\mathscr{E} + 12\wp$
H. Yu [33]	1M	1M	$1\mathfrak{B}\text{PM} + 3\mathscr{E} + 1\wp$	$1\mathfrak{B}\text{PM} + 1\mathscr{E} + 5\wp$	$2\mathfrak{B}\text{PM} + 6\mathscr{E} + 6\wp$
N. Hundera [26]	$3\mathfrak{B}$PM	$1\mathfrak{B}\text{PM} + 2\wp$	$4\mathfrak{B}\text{PM} + 2\wp$	$2\mathfrak{B}\text{PM} + 4\wp$	$10\mathfrak{B}\text{PM} + 8\wp$
Ours	$2\mathscr{H}$DM	$2\mathscr{H}$DM	$5\mathscr{H}$DM	$6\mathscr{H}$DM	$15\mathscr{H}$DM

Table 3 Computation cost in milliseconds

Scheme	Warrant generation and delegation	Proxy secret key generation/warrant verification	Proxy signcryption	Unsigncryption	Total
G. Swapna [20]	12.93	29.8	29.08	53.32	125.13
Zhou [24]	12.93	34.11	29.08	76.84	152.96
H. Guo [27]	0.97	2.91	3.88	5.82	13.58
Y. Ming [32]	12.93	29.8	76.28	94.4	213.41
H. Yu [33]	1.25	1.25	22.96	80.06	105.52
N. Hundera [26]	12.93	34.11	47.04	68.22	162.3
Ours	0.96	0.96	2.4	2.88	7.2

Table 4 Communication cost comparison

Scheme	Proxy delegation	Proxy signcryption	Total
G. Swapna [20]	$1\mathcal{M} + 2\mathfrak{C}$	$2\mathcal{M} + 3\mathfrak{C}$	$3\mathcal{M} + 5\mathfrak{C}$
Zhou [24]	$3\mathcal{M} + 1\mathscr{H}$	$1\mathcal{M} + 1\mathfrak{C} + 2\mathscr{H}$	$4\mathcal{M} + 1\mathfrak{C} + 3\mathscr{H}$
H. Guo [27]	$1\mathcal{M} + 2\mathcal{Q}$	$2\mathcal{M} + 5\mathcal{Q}$	$3\mathcal{M} + 7\mathcal{Q}$
Y. Ming [32]	$1\mathcal{M}$	$1\mathfrak{C} + 2\mathscr{H}$	$1\mathcal{M} + 1\mathfrak{C} + 2\mathscr{H}$
H. Yu [33]	$1\mathcal{M} + 1\mathscr{H}$	$1\mathfrak{C} + 2\mathscr{H}$	$1\mathcal{M} + 1\mathfrak{C} + 3\mathscr{H}$
N. Hundera [26]	$\mathcal{M} + 2\mathfrak{C}$	$2\mathcal{M} + 1\mathfrak{C} + 1\mathscr{H}$	$3\mathcal{M} + 3\mathfrak{C} + 1\mathscr{H}$
Ours	$1\mathcal{M} + 2N$	$2\mathcal{M} + 5N$	$3\mathcal{M} + 7N$

Table 5 Communication cost in bits

Scheme	Proxy delegation	Proxy signcryption	Total
G. Swapna [20]	4096	7168	11,264
Zhou [24]	6656	3232	9888
H. Guo[27]	2368	6656	9024
Y. Ming [32]	2048	1184	3232
H. Yu [33]	2560	1184	3744
N. Hundera [26]	4096	5200	9296
Ours	2208	4896	7104

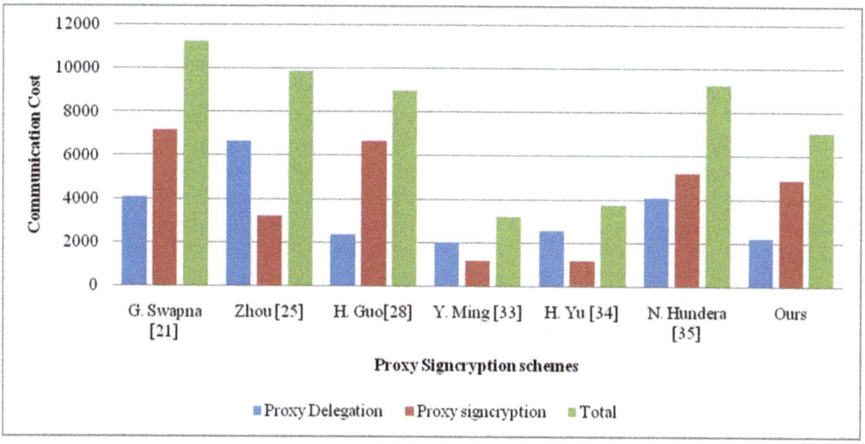

Fig. 7 Comparison of communication cost in bits

6 Conclusion

Despite the supremacy of SGG, the insecure and high-latency links between cloud data centers and smart gas metering devices are a source of concern in the practical usage of the SGG. Fog computing and SGG integration promise to be a possible way to solve this issue. This chapter explains how to design a safe identity-based proxy signcryption and apply it to fog-based SGG. Under the hardness of the

HYEC-DLP and HYEC-DHP assumption, the proposed method meets sufficient security criteria. The revised protocol is safe in practice, according to the simulation research conducted with AVISPA tools. Furthermore, a thorough analysis of its performance reveals its efficiency in terms of processing and transmission expenses. The development of an attribute-based proxy signcryption method with PRE for fine-grained access control will be the focus of our future study.

References

1. M. Sheha, K. Mohammadi, K. Powell, Techno-economic analysis of the impact of dynamic electricity prices on solar penetration in a smart grid environment with distributed energy storage. Appl. Energy **282**, 116168 (2021)
2. O.M. Butt, M. Zulqarnain, T.M. Butt, Recent advancement in smart grid technology: future prospects in the electrical power network. Ain Shams Eng. J. **12**(1), 687–695 (2021)
3. A. Panayiotou, N.P. Stavrou, E. Stergiou, Applying the industry 4.0 in a smart gas grid: the Greek Gas Distribution Network case, in *2021 International Symposium on Electrical, Electronics and Information Engineering*, (2021), pp. 180–184
4. A.R. Al-Ali, T. Landolsi, M.H. Hassan, M. Ezzeddine, M. Abdelsalam, M. Baseet, An IoT-based smart utility meter, in *2018 2nd International Conference on Smart Grid and Smart Cities (ICSGSC)*, (IEEE, 2018), pp. 80–83
5. H. Lund, Renewable heating strategies and their consequences for storage and grid infrastructures comparing a smart grid to a smart energy systems approach. Energy **151**, 94–102 (2018)
6. F. Dababneh, L. Li, Integrated electricity and natural gas demand response for manufacturers in the smart grid. IEEE Trans. Smart Grid **10**(4), 4164–4174 (2018)
7. Kamrani, F., Fattaheian-Dehkordi, Sajjad; Gholami, Mohammad; Abbaspour, Ali; Fotuhi-Firuzabad, Mahmud; Lehtonen, Matti 2021. A Two-Stage Flexibility-Oriented Stochastic Energy Management Strategy for Multi-Microgrids Considering Interaction With Gas Grid
8. S. Colombo, Y. Lim, F. Casalegno, Deep vision shield: assessing the use of hmd and wearable sensors in a smart safety device, in *Proceedings of the 12th ACM International Conference on PErvasive Technologies Related to Assistive Environments*, (2019), pp. 402–410
9. A. Elmrini, A.G. Amrani, Wireless sensors network for traffic surveillance and management in smart cities, in *2018 IEEE 5th International Congress on Information Science and Technology (CiSt)*, (IEEE, 2018), pp. 562–566
10. S. Surnov, I. Bychkovskiy, G. Surnov, S. Krasnov, Smart Monitoring: remote-monitoring technology of power, gas, and water consumption in Smart Cities. arXiv preprint arXiv:1910.08759 (2019)
11. J. Qiu, J. Zhao, H. Yang, D. Wang, Z.Y. Dong, Planning of solar photovoltaics, battery energy storage system and gas micro turbine for coupled micro energy grids. Appl. Energy **219**, 361–369 (2018)
12. E.A.M. Ceseña, P. Mancarella, Energy systems integration in smart districts: robust optimisation of multi-energy flows in integrated electricity, heat and gas networks. IEEE Trans. Smart Grid **10**(1), 1122–1131 (2018)
13. J. Savickis, L. Zemite, I. Bode, L. Jansons, Natural gas metering and its accuracy in the smart gas supply systems. Latv. J. Phys. Tech. Sci. **57**(5), 39–50 (2020)
14. X. Li, K. Chen, Identity based proxy-signcryption scheme from pairings, in *IEEE International Conference on Services Computing, 2004 (SCC 2004). Proceedings*, vol. 2004, (IEEE, 2004), pp. 494–497
15. Q. Wang, Z. Cao, Efficient ID-based proxy signature and proxy signcryption form bilinear pairings, in *International Conference on Computational and Information Science*, (Springer, Berlin/Heidelberg, 2005), pp. 167–172

16. M. Wang, H. Li, Z. Liu, Efficient identity based proxy-signcryption schemes with forward security and public verifiability, in *International Conference on Networking and Mobile Computing*, (Springer, Berlin/Heidelberg, 2005), pp. 982–991
17. H.M. Elkamchouchi, Y. Abouelseoud, A new proxy identity-based signcryption scheme for partial delegation of signing rights. IACR Cryptol. ePrint Arch. **2008**, 41 (2008)
18. X.X. Tian, J.P. Xu, H.J. Li, Y. Peng, A.Q. Zhang, Secure ID-based proxy signcryption scheme with designated proxy signcrypter, in *2009 International Conference on Multimedia Information Networking and Security*, vol. 1, (IEEE, 2009), pp. 351–355
19. C.C. Tsai, K.C. Liao, T.H. Chen, W.B. Lee, Security enhancement of a novel proxy key generation protocol, in *31st Annual International Computer Software and Applications Conference (COMPSAC 2007)*, vol. 2, (IEEE, 2007), pp. 681–688
20. G. Swapna, P.V.S.S.N. Gopal, T. Gowri, P.V. Reddy, An efficient ID-based proxy signcryption scheme. Int. J. Inf. Netw. Secur. **1**(3), 200 (2012)
21. H.M. Elkamchouchi, Y. Abouelseoud, E.F.A. Elkhair, K. Elsheikh, An efficient ID-based proxy signcryption scheme without bilinear pairings. Int. J. Comput. Appl. **975**, 8887 (2013)
22. N.W. Lo, J.L. Tsai, A provably secure proxy signcryption scheme using bilinear pairings. J. Appl. Math., 2014 (2014)
23. H.Y. Lin, T.S. Wu, S.K. Huang, Y.S. Yeh, Efficient proxy signcryption scheme with provable CCA and CMA security. Comput. Math. Appl. **60**(7), 1850–1858 (2010)
24. C. Zhou, Y. Zhang, L. Wang, A provable secure identity-based generalized proxy signcryption scheme. Int. J. Netw. Secur. **20**(6), 1183–1193 (2018)
25. H. Yu, Z. Wang, J. Li, X. Gao, Identity-based proxy signcryption protocol with universal composability. Secur. Commun. Netw. **2018** (2018)
26. N.W. Hundera, Q. Mei, H. Xiong, D.M. Geressu, A secure and efficient identity-based proxy signcryption in cloud data sharing. KSII Trans. Internet Inf. Syst. (TIIS) **14**(1), 455–472 (2020)
27. H. Guo, L. Deng, An identity based proxy signcryption scheme without pairings. Int. J. Netw. Secur. **22**(4), 561–568 (2020)
28. P.R. Yogesh, Formal verification of secure evidence collection protocol using BAN logic and AVISPA. Proc. Comput. Sci. **167**, 1334–1344 (2020)
29. P.R. Yogesh, Backtracking tool root-tracker to identify true source of cyber crime. Proc. Comput. Sci. **171**, 1120–1128 (2020)
30. R.Y. Patil, S.R. Devane, Network forensic investigation protocol to identify true origin of cyber crime. J. King Saud Univ. Comput. Inf. Sci. (2019)
31. C. Zhou, Z. Zhao, W. Zhou, Y. Mei, Certificateless key-insulated generalized signcryption scheme without bilinear pairings. Secur. Commun. Netw. **2017** (2017)
32. Y. Ming, Y. Wang, Proxy signcryption scheme in the standard model. Secur. Commun. Netw. **8**(8), 1431–1446 (2015)
33. H. Yu, Z. Wang, Construction of certificateless proxy signcryption scheme from CMGs. IEEE Access **7**, 141910–141919 (2019)

Countering Cybersecurity Threats in Smart Grid Systems Using Machine Learning

Mais Nijim, Hisham Albataineh, Viswas Kanumuri, Ayush Goyal, Avdesh Mishra, and David Hicks

1 Introduction

The traditional electrical power grid and network are rapidly evolving into smart grid systems. Initially, electricity is produced when the energy needs were simple (more than 100 years ago). At that time, it helped a lot to manage the home needs, but when the industry evolution started, the power consumption has increased more and more, which created an imbalance between demand and supply. Power is built when the electric needs are simple (more than 100 years ago). It helped a lot to manage the utilities and work when the industries less use the power, then the power necessity has been increased, and it failed in demand and supply. Then, the smart grid came into existence. A grid is a chain connected to everything like power generation, transmission lines, distribution networks, charging stations, etc. These new developments integrate the traditional grid with (ICT) solutions, thus empowering power needs and making them available to every consumer who needs them [16]. Smart grid network constantly monitors, controls, and manages clients' demands. Smart grid systems allow overall seamless management of the grid network at reduced costs. In addition, smart grid systems provide a two-way communication channel between clients and electric power companies, making it possible for service providers to optimize their services based on information received from clients.

The original version of the chapter has been revised. A correction to this chapter can be found at https://doi.org/10.1007/978-3-031-09640-2_21.

M. Nijim (✉) · H. Albataineh · V. Kanumuri · A. Goyal · A. Mishra · D. Hicks
Texas A&M University, Kingsville, TX, USA
e-mail: mais.nijim@tamuk.edu; hisham.albataineh@tamuk.edu; ayush.goyal@tamuk.edu; avdesh.mishra@tamuk.edu; David.hicks@tamuk.edu

K. Daimi et al. (eds.), *Emerging Trends in Cybersecurity Applications*,
https://doi.org/10.1007/978-3-031-09640-2_14

Smart grid systems stand out as the most basic and challenging artificial intelligence systems in the current growing world. The system combines such functionalities and features as monitoring, sensing, controlling, and two-way communication between service providers and clients. The evolution of smart grid systems aimed to put into place self-healing, resilient, sustainable, and efficient systems. This development was (and continues to be) primarily supported by increased penetration and demand for renewable and distributed energy resources. The National Institute of Standards and Technology (NIST) briefs grid systems as an inbuilt of the old century grid system with ongoing modern-day information, communication, and technology solutions [18]. Through this integration, utility firms, consumers, and ICT developers can handle the grid by placing and connecting various energy sources and consumer devices. In addition, the integration of the legacy network system with the cyberinfrastructure enables the collection of massive and voluminous data from the lack of connected devices such as meters, measurement units of Phasor, and circuit breakers, among others.

Compared to the traditional power grid, the Smart grid systems are designed to provide multiple communication mechanisms considered their heart. However, their extensive integration with various hardware devices will give an edge to open up loopholes for cyber attackers to launch a wide range of attacks, leading to loss of property and life and disruption of services, among other effects. The integration and over-dependency on ICT infrastructure exponentially increase cybersecurity risks, threats, and vulnerabilities. On the contrary, most smart grid system services and critical control processes such as load aggregation, on-demand responses, and state estimation rely on fast, secured, and robust grid systems. Therefore, secure and stable grid systems become indispensable aspects of the entire system. System weaknesses cause or make it possible for common attacks to occur in the system. For example, devices like smart meters, electronic gadgets, automatic robots, car electric charging points, etc., can cause vulnerabilities that may help system adversaries to manipulate meter box readings, measurements, system parameters, and price information. In extreme cases, attackers may interrupt and get direct access to the carping system regularly and unsettle the whole system in inconstant ways.

The Department for Homeland Security and Department of Energy acknowledge that control energy systems are at risk of cyberattacks. The two US-based agencies point out how attackers have progressively run after cunning means to explore the flaws in smart grid systems' units, telecoms methods, protocols, and frequent operating systems to access more critical system infrastructure [20]. Additionally, technology advancements have made available highly knowledgeable cyberattack tools that demand less tech knowledge or experience to apply the technology. As a result, they could run the malware without core technical knowledge. Security, therefore, remains the most dominant issue in the grid system due to the risk inconveniences that attacks could cause on customers, service providers, and nation(s) in general. Thus, most security solutions are geared towards providing and ensuring that smart grid systems' services run uninterrupted and according to user demands, maintaining the integrity of the information exchanged, and that users' data is confidential.

Efforts to secure smart grid systems have also heightened and born considerable successes. Various reactive and proactive security solutions and methodologies have been proposed or designed to reduce security incidences. The solutions also aim to increase systems' aptness to hunt out and timely spot unusual actions, which should quickly initiate all the possible countermeasures. In case of a security threat, restore the system to normalcy the soonest possible. "Since the nature of threats and vulnerabilities are constantly changing, the applications of current best security practices are necessary but not sufficient" [22]. Newer technologies can complement previous security solutions and help protect data, both at rest and in transit. This chapter primarily focuses on machine learning as a potential solution to smart grid cybersecurity problems. The report is centered on safety issues for the entire grid system lifecycle phases, from data preprocessing to storage. The paper is organized into the following sections: smart grid background and technology, literature review, security risk in smart grid, machine learning background and types of machine learning applications in smart grid safety solutions, and conclusion sections.

2 Smart Grid Background and Technology

The smart grid has its advantages. Solar energy and wind energy are very complicated because of maintaining a steady power balance. However, it is even more complicated as the power from their sources is more inconsistent. Here comes the smart grid, which balances these small changes to constant. It uses sensors to constantly monitor a series of devices that control the flowing current through different points.

Electricity was first started in 1882 by Thomas Edison, but he introduced direct current, which has many limitations because of the voltage. At first, the power can only move half a mile. Tesla thought Edison was wrong, and he proved it by bringing the power into alternate current. In this, power flows back and front. Tesla uses electric transformers to convert alternate current to higher voltages for more efficient transport over long distances, which is harder to do in the case of direct current. Renewable energy tends to generate direct current, then different forms of power come into existence.

A smart grid is a level data preprocessing framework in the factors of edge computing and cloud computing. Here the nano grids are also known as microgrids, which are nothing but individual connections in a grid. Nano grids were situated at the edge of the IoT that will straightly communicate with the database. The grid seated on the cloud dispenses the current to each microgrid focused on the consumer supply needs.

2.1 The Need for Smart Grid

It is rated that worldwide energy claims would hit 44% by 2030 as per the article by Reuters (https://www.reuters.com/article/us-eia-global-demand/global-energy-demand-seen-up-44-percent-by-2030-idUSN2719528620090527). To touch this request, the continuous cradle of power is the rapidly spreading origin of world power, with expenditure growth by an estimated 3.0% [14]. The shift from non-continuous to endless sources of power has taken shape. Similar efforts have resulted in replacing the older power delivery systems with acclaim and service networks. However, old power distribution methods are inadequate and cannot touch the high demand, and power consumption patterns shift. Therefore, an efficient structure in which individuals, frameworks, arrangements, and business processes are dynamic and adaptable in answering innovation changes, client solicitations, guidelines, and strategies is needed. This kind of mechanism is achieved along with the support of the notion of the grid system – "an electricity network that can intelligently integrate the actions of all users connected to it to efficiently deliver sustainable, economic, and secure electricity supplies [14]."

By definition, a smart power grid system consults power-sharing networks that use digital communication technologies to detect changes in local power consumption and automatically respond to these changes without human intervention. The system's efficiency is improved by using interconnected smart devices, among them smart meters and appliances and renewable and efficient energy resources. The system allows clients to interact with the grid through its two-way communication feature and overcomes many challenges associated with traditional electrical grid systems. Notably, the smart grid reduces energy consumption and costs to consumers by using smart means. Power supply companies also make efficient use of energy as they can meet the varying energy demands of the clients.

The development and consequent transition from legacy to smart grid systems is market-based. The traditional grid is mainly inadequate, outdated, and unable to address the twenty-first-century power supply needs and challenges. On the other hand, smart grids are advantageous in terms of reliability, efficiency, security, safety, environment, and economics. Smart grid systems reduce the costs of power interruptions and disturbances, thus reducing the chances and consequences of widespread blackouts. A smart grid system is termed a reliable network because it delivers electricity to consumers only when they need it. They are also associated with reduced outages (both in number and time). The systems use smart metering infrastructure that instantly detects power quality issues and power loss, thus enabling system operators to respond and diagnose the problems rapidly. Whereas the demand response helps in reducing the stress on the system's resources during peak conditions, thus decreasing their chance of failing. Further, the reliability of smart grid systems is enhanced by ubiquitous sensors and intelligent controls that provide service providers with situational awareness of the system.

Smart grid systems provide new market opportunities for distributed power generation and storage. Smart grid systems help improve economics by lowering power bills and creating opportunities to develop new product services. Power generating companies are presented with opportunities to leverage their resources and reduce operating and maintenance cost at baseload generating plants. Smart grid systems also increase efficiency through improvements that reduce the costs associated with power generation, distribution, and consumption [3]. Efficiency can be measured in terms of how smart grid networks reduce transmission congestion while giving power companies greater access to markets. Finally, smart grid power systems help improve the environment by reducing harmful emissions and discharges.

Nevertheless, smart grid systems could be disadvantageous due to the long run's high initial and maintenance costs. Most of the smart devices needed for the system are more expensive than legacy network devices. Furthermore, the system requires that continuous communication be available in all situations, a service that may be harbored during abnormal weather conditions such as windstorms, heavy rains, and lightning conditions. As a result, performance is a massive challenge during emergency and network congestion periods. Additionally, smart meters and the entire grid system could be hacked, leading to disastrous effects such as disruption of services and increasing and decreasing electric power demand.

2.2 The NIST Smart Grid Conceptual Model and Components

The NIST proposes a conceptual model that describes smart grid systems and appliances' overall composition and architectural design. The conceptual model aims to provide a high-level view that all stakeholders can understand. The model was introduced in 2010 through the NIST's first-ever publication titled Smart Grid Interoperability Framework. With every framework, NIST requires that the conceptual model be upgraded to reflect increases in the number and kinds of sharing power resources (DERs) and be used throughout the power grid that shows the increasing automation of distributed systems.

The roles and responsibilities of most actors and stakeholders have remained unchanged for most conceptual models. Therefore, it becomes understandable how the grid's functions and equipment will change contextually or according to the domain in which it is used. Moreover, the conceptual model is used to reinforce the contrast between the complexities of information exchanges that take place on the grid. For example, a few physical connections are used to produce and consume electrical energy. In these days the type of energy generations like (solar power, nuclear power plant, Thermal power, wind mils etc.) are existed and also new power sources are identified and distribution technologies are becoming more diversified across the grid as dynamics lessen even more.

On the contrary, smart grid systems collect large amounts of data through grid communication channels and are exploding as consumers leverage the use of low-cost smart devices. The NIST conceptual model is divided into seven types: the client, market, specialist organization, activities, age, transmission, and appropriation spaces. In addition, the model is divided into sub-domains that conceptualize roles and services for the intelligent grid system at the lowest level [7]. These sub-domains define the types of services, stakeholders, interactions, and information exchanges that take place in the course of service delivery. Roles of the domain and sub-domains in the grid interact, thus enabling the system's functioning, as shown in Fig. 1.

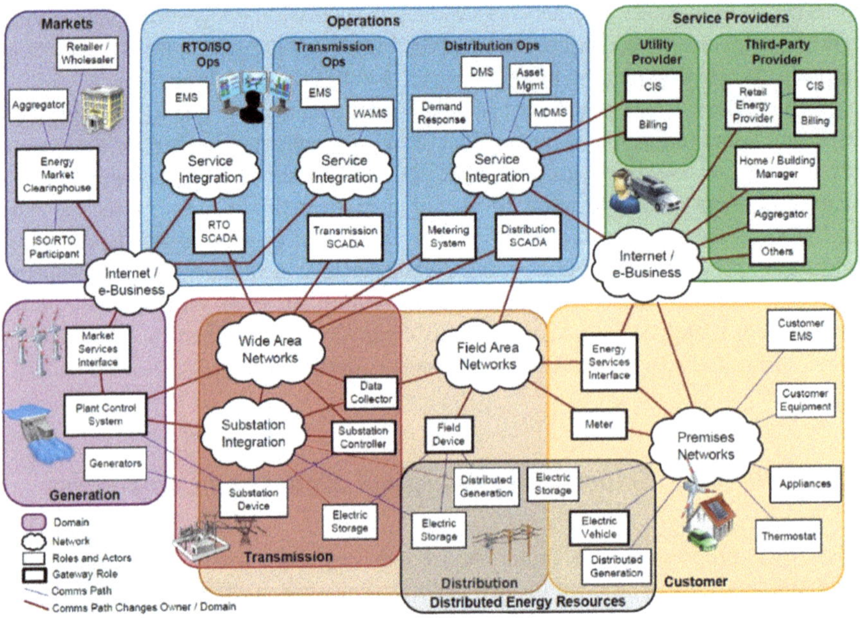

Fig. 1 System functionality smart grid reference model

The customer domain was created to support a significant stakeholder in the entire power lifecycle – the customer. It describes the domain from which power is consumed, actively managed, and generated. The customer domain enables customers to manage their energy generation and consumption, providing secure interfaces for customer interactions with utility service providers. Additionally, the energy service interface (ESI) is used as a link, connecting grid methods such as structure mechanization systems or management systems installed at customers' local premises. Functions and applications supported by the customer domain include storage, micro-generation, industrial automation, and building/home automation.

Smart grid system's assets and services are bought and sold at the market domain. Markets are instrumental in defining the future of smart grid systems and DERs. At this level, entities exchange pricing information in their attempts to equal the demand and supply curves in the grid system. Information flow from the market state to other states is considered critical because it leads to the efficient matching of production and consumption, depending on markets and their proxies. Boundaries within this domain include the edge of operations domain where controlling takes place, the repair provider state, and the consumer domain. The market domain presents suppliers with emerging opportunities for interactions between traditional and nontraditional grid actors.

Further, technological developments have significantly reduced the cost of coordinating and facilitating trades. ICT, therefore, uncovers new market ventures and economic opportunities since reducing transaction costs leads to improved value propositions in the market. Typical functions and applications defined in the market domain include market management, trading and retailing, DER aggregation, market operations, ancillary operations, and platforms for connecting diverse organizations and actors.

The service provider domain is the third domain used to encourage the business operation of power method investors, customers, and distributors. It supports such processes as billing, clients' accounts management, and customer service enhancement. Service providers create fresh and unconventional products and services to meet the ever-changing market needs. Once created, they are performed by the grid, service provider, or contracted third parties. Functioning at the service provider level calls for the development of standards and interfaces to support all the critical power infrastructure besides supporting the entire business ecosystem. In addition, the interfaces so created must be able to work over an assortment of system administration environments and technologies without distorting message semantics or compromising system security, reliability, stability, integrity, and safety. Typical applications in the service provider domain include customer management, billing, accounting, energy management, installation, and maintenance.

The operations domain deals with the smooth performance of the power grid. Actors in the operations domain are responsible for monitoring roles and supervising network connectivity, loading conditions, and equipment status. Other functions include state assessment, client assistance, augmentation arranging, upkeep and development, training, network calculations, and fault management. The domain uses a network calculation algorithm in real time using measured parameters to produce the information necessary to optimize and operate the entire grid system. Actors also foster long-haul plans for the power framework's unwavering quality and timetable development and maintenance.

The generation and DER domain are where actual power generation takes place. Electricity generation is referred to as the process of creating electricity from different forms of energy. Usually, it involves a variety of primary energy sources, including solar radiation, nuclear fission, flowing water, wind and geothermal heat, and conversions. After generating and series of energy conversions, the generated power is ready for transmission. Transmission is the colossal exchange of electrical power from different age sources to dispersion through various substations. A transmission network is worked by the dispatch-owning utility that is mandated to keep balance the solidity of the entire grid by managing an equal balance of supply and clients' demands. After transmission of power to the clients, the electricity need to be tracked for that the power station unit and customer domain are interconnected using the distribution domain at various meters focuses on utilization, stockpiling, and age. The brilliant framework dissemination framework is organized in an assortment of designs that collaboratively guarantee the system's reliability.

2.3 Smart Grid System's Technology

In this smart grid system technology, clients both consume and create energy and will return that to the grid. This two-way communication is the most important feature of smart grid networks since both parties synergistically manage power costs, delivery, and environmental impacts of power generation. To achieve operational efficiency, more robust mechanisms that contribute to the overall system intelligence are needed [19]. The added smartness differs depending on the levels at which it is being evaluated – whether on generation, transmission, distribution, operations, or customer domain level. From the power generation and DER domain, this intelligence is looked to as the ability to automatically adjust prices. In contrast, from a customer's perspective, the intelligence may be considered to reduce consumption costs and provide efficient energy utilization in both industries and home levels. ISLAND: It is the state where the microgrid is interlinked among the grid, but it could resolve itself from the problem; this kind is considered as an island; this action happens when there is a grid mistake or system error or any other dangers; smart grid system can be perceived as a layered system whose different layers play specific roles and interact from generation to distribution and pricing at the customer's interface. The physical layer consists of generations, transmissions, distribution, consumption, renewing, and storing electrical energy. The communication layer uses different types of communication protocols and networks – guided and unguided to support the smart grid's two-way communication platform. These could be home area networks, neighborhood networks, and office and core networks, among others [18]. The system integration platform integrates computing infrastructure, networks, security management, data, and applications for seamless service delivery. Finally, the software layer, the topmost layer in the model, consists of meter data analysis, generation of billing information, load control, customer service provider interfaces, and consumer information systems, among others. Figure 2 shows the functional model of a smart grid system.

3 The Cybersecurity Aspects of the Smart Grid

"Machine learning techniques have been applied in many areas of science due to their unique properties like adaptability, scalability, and potential to rapidly adjust to new and unknown challenges" [6]. Cybersecurity is the field that advances every time concerning demands. It has great attention and much greater advancements in the term of cloud and also in mobile computing, web technologies, online banking, smart grid, and some social networks. Assault on smart grid methods is divided according to the microgrid and location of the microgrid that is under trouble. This diversification classification tells us to arise to system and component microgrid. Components microgrid involves smart meters, measurement of phasors, and other smart devices, while framework hubs incorporate the high-level metering foundation

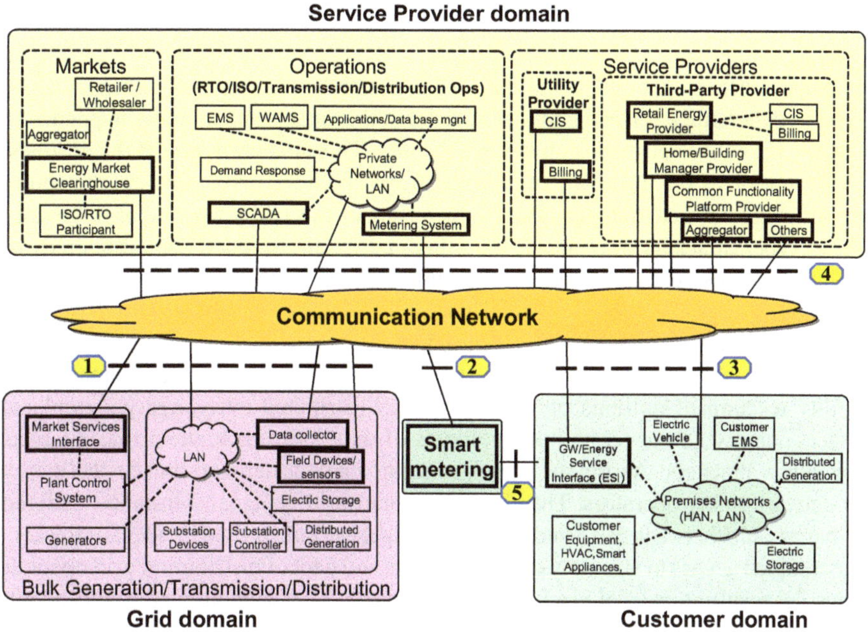

Fig. 2 Functional model of smart grid system [11]

as well as the conveyed administration framework. Smart grid methods are weaker to cyber threats due to various reasons:

Increased installations of intelligent electronic devices imply increased attack sites.

Installation of third-party components in the grid increases vulnerability since they may be infected with Trojans, which can spread to the network.

Inadequate workforce preparing legitimate preparation is essential for individuals to use the network lest they fall victims to phishing attacks or unintentionally publicize confidential data.

Use of Internet protocols: Some Internet protocols are not secure since they transfer data in unencrypted formats and may therefore be a contender for information extraction by means of man-in-the-center assaults. Upkeep essential maintenance: Our aim is to make the system working normally but may itself become a vector for cyberattacks.

Operators may unintentionally disable security mechanisms while conducting tests, thus creating a loophole in the lattice. The National Institute of Standards and Technology system identifies requirements for power grid methods, which include access control, audit, and accountability, configuration management, incidence reporting, identification and authentication, personnel security, protection of smart grid communication systems and their integrity, risk management, and assessment and continuity of operations, among others.

3.1 History of Attacks on Smart Grid Systems

In August 2017, a petrochemical industry in Saudi Arabia was badly hit with a new type of cyberattack. Investigators believed that this attack intentionally happened to destroy the entire data and to shut down the unit. This attack was one of the deadliest cyberattacks, as the unknown enemies destroyed both the drive data and the ability to inflict big physical damage. The US government was really worried of it, because these cyber attackers may attack the USA and its allies, since some hundreds of industries across the globe rely on the same software systems that was attacked by the cyber attackers in Saudi Arabia. Up to now, the culprits were not found and even the country origin was unable to be trace [1].

Critical infrastructure such as smart grid systems is under constant threat. Their ability to connect millions of smart devices (distributed over large geographical spaces) and which generate huge volumes of data have made them major targets for cyber attackers. As threats increase, smart grid systems are put at the mercy of network safety operators. The Ukrainian smart grid system, for instance, suffered cyber assaults between 2015 and 2016. "During these unexpected incidents, attackers have gained access to remotely closed breakers and a lot of distribution grid operator consoles resulting in local blackouts. In this attack, 30 substations were switched off, and around 230,000 people were affected by the blackout [23]." This is the first super successful cyberattack of the smart grid system. Hackers breached communication channels, altered data, and overwhelmed the highly interlinked communication with traffic, ultimately limiting administrators' ability to screen and deal with the lattice. This attack was brought about by an outsider's unapproved access into the organization's SCADA framework (a later survey involved the Russian government) in an assault that impacted seven 110 kV and 23–35 kV substations for 3 hours. Although the administration reestablished administrations, it is accounted for to have encountered facilitated cyberattacks that were executed for 30 minutes.

The Iran nuclear facility also faced similar attacks in 2010 [12]. The Stuxnet attack caused many rotators at the uranium enhancement plant to wear out. This malware was invented to cause service disruption and sabotage the nuclear power generation. It is also could inflict physical damage to the system's critical infrastructure by focusing on machine regulators and SCADA frameworks. The USA also has had its share of smart grid disruptions. In 2003, a huge-voltage current line in Northern Ohio slammed into congested trees, resulting in one of the major country's widespread blackouts. Although the issue could have set off a caution in the control room, the framework fizzled as a strong current flew through the line, gradually weakening it. The transmission line cut off and triggered several failures throughout the system. Eleven people were killed while over 500 thousand people missed power for 2 days in the event.

3.2 Cyberattack Detection and Mitigation Techniques

Diverse machine learning-based solutions have been used in different environments to address computer security problems. The transition from traditional cybersecurity solutions to machine learning-based ones is informed by the fact that assault procedures are turning out to be more modern in entering frameworks. Machine learning methods resolve the most challenging and complex security problems by adapting quickly to such circumstances. Notably, smart grid systems involve multiple stakeholders, and this makes smart grid security a daunting task. Kalogridis et al. developed a security system that provides a set of rules for connecting safety and privacy around various domains to enhance this security. This framework divides smart grid security into three classes: communication, system control, and secured computing controls. Communication security is aimed at securing communication in the smart grid system using network privacy, routing security, and cryptosystem. This is mainly achieved through end-to-end encryption, multiple hop routing, and key management system.

Various machine learning approaches are used concerning the different domains in the smart grids; the most widely used ML algorithms in different approaches of smart grids [21] are decision trees, CNN, K-means, KNN, logistic regression, naive Bayes, linear regression, DB scan, etc.

4 Security Risks in Smart Grid

4.1 Security Requirements for Smart Grid Systems

From both the layered and domain-based conceptual model architectures, smart grid systems bring together many operators and actors – homeworkers, service providers, field engineers, and marketing staff – all of whom expect admittance to the working programming bundles and apparatuses through many passageways. Providing all these actors with access to the system raises security concerns and imposes serious cybersecurity threats. Verification and approval to safeguard the framework, therefore, become critical aspects of the network. The shrewd matrix organization method is made of many intelligent and interconnected devices that communicate in two ways – duplex mode [16]. A lot of information data and operational data is generated during this communication and may become a major target for maliciously intended persons. Information, in this case, implies data about power consumption, billing, trending, clients' information, and geographical locations. On the other hand, operational data may include real-time current and voltages, capacitor banks, current loads, transformer feeders, and relay status. This information is very important in the way smart grid systems operate and require a high level of security.

Salsabeel et al. identify various security requirements for smart grid systems. These include availability, authentication, integrity, authorization, and non-repudiation. Availability, integrity, and confidentiality (CIA) are the most common security goals defined by the CIA security triangle.

(1) The availability aspect of the system implies that system services are 99.99% available and that the system is powerful enough to restore itself to normalcy (robustness) in the event of failure. It also implies that information is timely accessible in the grid. Loss of availability largely affects service delivery and communication services. It could be attributed to disavowal of administration assaults that intend to disrupt services and data transfer and end up locking legitimate system users out [15].
(2) Authentication refers to the process of validating the identity of communicating parties. Authenticating smart grid system users and machines is crucial since a lack of it or any weakness may give attackers access to private information or allow a device to illegitimately perform undesired actions on the grid [4].
(3) System or data integrity security goal is aimed at preventing unauthorized modification of data and information by illegitimate users. Without system integrity, sensor and meter information could be modified, creating a mismatch between users' requests and suppliers' responses. The general effect of loss of integrity is poor power management.
(4) Authorization, which is the last security requirement, aims to provide users and connected devices with permissions or grant them access to the system. Authorization is a key security factor for the proper management of information and system resources.
(5) Confidentiality security goal is aimed at preventing unauthorized users from gaining access to information and, by so doing, protects personal privacy and safety. Through smart system communication grid, information of varying privacy and sensitivity levels is exchanged, making it important to secure it against falling into the wrong hands.
(6) The non-repudiation security requirement for the smart grid system implies that actions performed by the system or a given user cannot be denied.

4.2 *Security Risks, Threats, and Vulnerability Concerns*

The influence of cyberattacks on power systems has become a burning issue in recent years. Smart grid systems use a highly complex and extremely integrated architecture that predisposes them to a myriad of cyber threats, risks, and vulnerabilities [16]. Implementing smart grid methods requires the utilization of multiple communication mechanisms and electronic devices that, due to their complex integrations, are more weak and susceptible to cyber threats, risks, and attacks. Among the many vulnerabilities affecting smart grid systems are:

(1) Consumers' lack of awareness – smart grid systems include comprehensive security architecture. Unfortunately, most clients and smart grid users do not understand security features used to detect and analyze attacks. This creates the need for deliberate efforts to adequately educate users about security risks, costs, and security features for smart grid systems.
(2) The emergence of new and unknown technologies – technology is dynamic and is constantly being advanced to provide better services and products. However, this development comes along with newer techniques and environments used to launch attacks on various IT systems. As new technologies and innovations become used in the smart grid, they present with new and unknown vulnerabilities.
(3) Scalability – smart grid systems use artificial intelligence features to scale up and down based on growth in demand. Growth in the amount of data exchanged through the method and the volume of the physical infrastructure influence the system's intricacy. Increased volume of data and intricacy cause amassing of data and hinder efficient data flow.
(4) Absence of norms and guidelines – the brilliant lattice is a relatively new power grid solution that has minimal standards and regulations. Additionally, interoperability, the ability of various systems to work cooperatively and exchange information with each other, may be a serious hurdle. Therefore, standards and regulations are needed to achieve interoperability in smart grid network systems.
(5) Using existing ICT technologies to actualize smart grid systems may lead to inheriting susceptibilities and unresolved security issues such as IP morphing and disapproval of service attacks, among others, from these devices to smart grid systems.
(6) Smart meters collect large volumes of data about clients and transport it to utility companies, service providers, and consumers [2]. The collected data contain personally identifiable information and might infer to their activities, thus compromising their security.
(7) The autonomous and intelligent smart devices used in the system may go about as assault passage focuses. Additionally, the power grid system's massiveness makes network observing and the executives a difficult errand. With the right tools (notwithstanding technical knowledge), attackers could exploit the above and many more system vulnerabilities and cause considerable damage to the network. Flick and Morehouse classified attackers as either being terrorists who find smart grid systems as attractive targets since they affect millions of citizens, or non-malicious attackers who opine that security is a puzzle to be cracked, or disgruntled employees, or clients driven by retaliation and perniciousness towards different shoppers or rival companies attacking each other for financial gains [5]. After gaining access to the system, attackers can cause component-wise, protocol-wise, or topology-wise attacks – they could target physical components of the system, topology, or communication protocols.

    (a) Malware attack – Attackers can develop malware, inject it into the system, and cause it to spread or self-replicate. The malware (malicious software) can infect smart meters, and servers replace or add functions to the system, leading to system malfunction.

(b) Replay assault – This assault happens when attackers send packets and inject bogus data into the organization. When injected into smart meters, such data might have huge financial implications to create false alarms and prices.

(c) Modbus security may also be compromised. Modbus is specifically designed to facilitate information exchange in smart grid systems and other critical systems. Skilled attackers could gain access to the system and launch such attacks as fake broadcast messages, passive reconnaissance, and sending benign messages through the network.

(d) (MITM) – The full form of MITM is a man-in-the-middle attack. This type of threat describes an eavesdropping attack in which adversaries make isolated associations with communicating parties at two edge points and transmit data among themselves. The communicating parties assume that they are slightly communicating with each other utilizing personal connections. MITM attacks are common in systems that rely on user datagram protocol (UDP) to transmit data without other security measures. Additionally, using public communication line (wide area network) creates more loopholes for man-in-the-middle attacks. They are usually used to damage the data, such as control instructions, pricing signals, and values of measurements.

(e) Distributed denial-of-service attacks (DDoS) – While using WAN, accessing the phasor measurement unit (PMU) presents some vulnerabilities. Malicious software can be installed on a substation edge router, or a brute force attack is used to guess access passwords, giving attackers access to the system. Denial-of-service (DoS) attacks are aimed at making important resources and services unavailable to legitimate users. In smart grid systems, DoS affects systems' reliability by making communication channels unavailable to clients, distributors, and service providers. Distributed denial-of-service is a special type of DoS that happens when Trojan programs infest the system and multiply themselves in critical components of the system, generating false traffic (by sending large STA packets that consume the network's bandwidth) that ultimately prevent authorized users from accessing any services. DDoS attacks can also target protocols – exploit protocol vulnerabilities or intentionally hinder steering tables from deteriorating the activity effectiveness of parcel dispersion in the advanced metering infrastructure packet exchange network.

5 Machine Learning and Types of Machine Learning

Machine learning architecture is used to communicate between cloud computing and edge computing. It is also mainly used for load balancing, where reinforcement is the best technique. The smart grid mainly contains two layers, cloud and array, where the cloud acts as a primary grid and the array acts as the array of some edge servers. The first layer is the cloud, as it comprises the essential framework that appropriates the capacity to the microgrid in the nth layer, and the edge layer consists of n edges; each edge again has the m number of nano grids. The nano grid is well briefed as an interconnected subgroup of low-voltage power systems. The machine learning algorithm will calculate the peak usage of the individual customer in each edge area and add the additional power if needed; it will be sent from the other edges.

5.1 Data Analysis in Machine Learning

Machine learning refers to a kind of data understanding which seeks to automate analytical model building. A broad branch of artificial intelligence allows computers to act like humans in terms of their memory and feelings and improve their learning as they encounter more data [17]. Machine learning technology allows computers and information systems to learn to make decisions and predictions without being directly programmed to do so. Machine learning is concerned with designing algorithms that allow machines (computers) to learn. These algorithms are organized into different taxonomies based on their desired outcomes. They include supervised and unsupervised learning, reinforcement, transduction, and semi-supervised learning.

In the smart grid perspective, machine learning is used in various aspects like managing demand and supply. The algorithm will continuously monitor the status of each node to cross-check whether it fulfills the customer demands or not. If the customer demands are not fulfilled, the additional supply will be assigned, or it is managed from the other grids or lanes. It will also be used to manage different types of grids like distribution stations, smart cities, smart buildings, industries, power plants, electric vehicles, and many others, as shown below in Fig. 2. At the same time, machine learning is also helpful to detect the differences that may happen or happen to the smart grid and helps in how to overcome them.

Figure 3 illustrates how machine learning is applied to the smart grid. On the left side of the smart grid are various power generation sources such as nuclear power plants, renewable power generation units, etc. On the right-hand side of the smart grid, we have the power consumption, such as households, industries, etc. The smart grid acts as a mediator between the source and destination. Various machine learning techniques are applied to the smart grid by taking the consumers' previous power consumption data and analyzing the data to predict multiple things such as load forecasting, stability, theft detection, possible attacks on the smart grid, stage patterns, etc. Machine learning is mainly used in the smart grid to increase the grid's efficiency and the high prediction accuracy of the system.

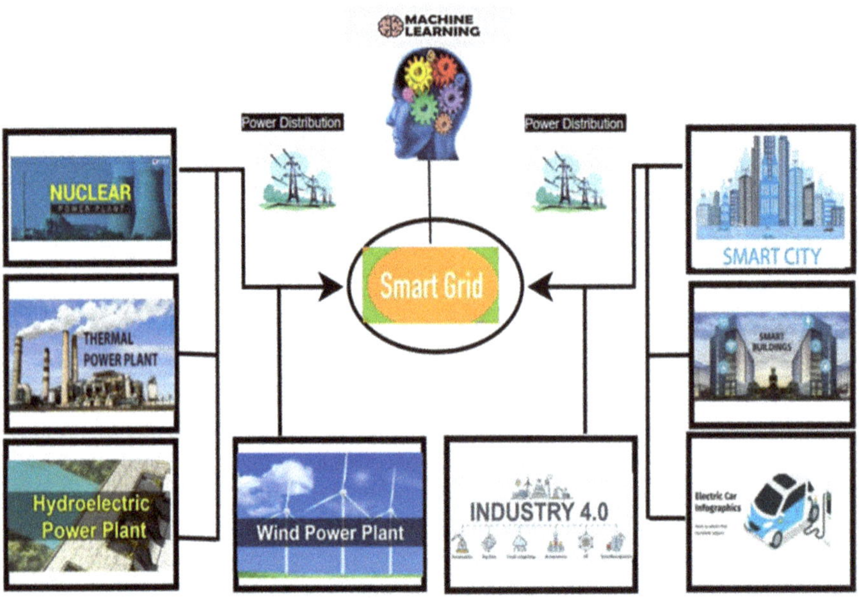

Fig. 3 Machine learning applications in smart grid

5.2 *Supervised Learning*

Supervised learning is a subdivision of machine learning that uses labeled (attributes) datasets to test and train algorithms that classify data or give outcomes. A machine learning paradigm is used to acquire the relationship between inputs and outputs for a given system in light of a given arrangement of matched input-yield preparing tests. The outcome of supervised learning is to create an intelligent framework that can gain proficiency with the planning between inputs and outputs by itself and use its learning experiences to predict outputs for new sets of inputs. Where the result takes a limited arrangement of discrete qualities, the learned mapping leads to data classification. The labeled data sets are fed into the model that automatically adjusts their weights until it is appropriately fitted. This fitting takes place during the cross-validation process. The training (labeled) datasets "teach" the model how to arrive at the desired output, and it takes inputs and corresponding targets. Dataset allows the algorithm to learn by itself over time and to measure its precision using the loss function.

Supervised learning is advantageous since all classes of outputs manipulated using supervised learning are significant to people and can be utilized for both discriminative example arrangement and information relapse. Classification uses the learning model to accurately allot test information into explicit classifications. It does so by identifying how explicit substances in the dataset lead to some given conclusions. Classification algorithms used include support vector machine, random forest, and k-nearest neighbor, among others. Supervised learning uses data regression to establish the relationship between independent and dependent variables. Common regression algorithms include linear and logistical regression.

5.3 *Unsupervised Learning*

It describes a type of machine learning technique that uses minimal human supervision to search for previously undetected patterns within a dataset that has no pre-existing labels. Unlike supervised learning, unsupervised machine learning models probability densities over inputs with no human labels. The self-organization method uses cluster analysis and principal components to study data. These calculations break down and group unlabeled datasets and find stowed away examples without human intercession. The ability of unsupervised learning algorithms to discover similarities and differences in given sets of data makes it the best solution for cross-selling strategies, exploratory data analytics, image recognition, customer segmentation, and other applications.

Clustering refers to an unsupervised data mining technique that is used to group unlabeled data based on their differences and similarities. These calculations are utilized to deal with crude and unclassified information items into different groups with common patterns. Based on the approach or mode of clustering, an algorithm can either be exclusive, overlapping, hierarchical, or probabilistic [10]. The most common clustering algorithm is the k- means clustering method. The method is an exclusive clustering algorithm that assumes that a data point can only exist only in a cluster. Data points are first assigned to a given number of clusters (K) in view of the separation from one another's centroid. The information focuses that are close to a given centroid are then clustered under the same category.

The principal component analysis describes a dimensionally reducing algorithm that is used to reduce redundancies while compressing datasets through feature extraction. Using linear transformation, principal component analysis transforms data into a new representation, thus yielding a new set of principal components. The first of these components seeks to maximize dataset variance, while the second is either orthogonal or perpendicular and uncorrelated to the first.

5.4 *Reinforcement Learning*

IT refers to a machine learning technique that is feedback-based. An agent knows how to act in a climate by performing activities and seeing the consequences of these actions. The agent gets good feedback for every good move and bad feedback or penalty for a negative move that is performed. The algorithm is keen to only know by itself. Since data sets are not labeled. The agent, therefore, collaborates with the climate and investigates it, and its first goal is to further develop execution by way of receiving the highest number of good feedback. From the agent's trial and error method, the agent learns to perform a task better. Reinforcement learning can be implemented using value-based, model-based, or policy-based techniques. The value-based approach focuses on finding the optimal function's amount of value at a given turn and under a policy. The policy-based approach on its side finds the ideal approach for most extreme potential compensations without utilizing a worth capacity. Instead, the agents try applying such policies that the actions performed in every step help in maximizing future rewards. Finally, the model-based technique creates a virtual model for the learning environment from which the agents learn. The approach uses no particular solution or algorithm since different learning environments use different model representations.

5.5 *Semi-supervised Learning*

This is a machine learning problem that uses a small number of label (supervised) datasets and a large number of unlabeled (unsupervised) datasets from which models learn to make future predictions – on unlabeled sets.

6 Applications of Various Machine Learning Techniques in Smart Grids

ML has become a significant a crucial technique for solving many problems, thanks to the development of big data and the steady evolution of computational methods that enhance data collection, management, and analysis. Machine learning is a data analysis technique in which machines and systems are taught how to make decisions based on past experiences. Various machine learning techniques provide efficient ways of analyzing and making decisions and can therefore be implemented in smart grid system applications. These applications include prediction consumption prices, generating power, optimizing power supply and scheduling, fault detection, adaptive controls, sizing, and detecting network intruders. Machine learning analyzes available data sets using given instructions, thus providing data-driven predictions and decisions [9].

Protecting brilliant matrix frameworks against the present digital dangers requires a more noteworthy coordinated effort between engineers, IT managers, clients, and safety managers. Utilities need to consider how cybersecurity solutions evolve and put in place good defense mechanisms. It is important to note that traditional cybersecurity technologies may not adequately counter some threats. As such, leveraging machine learning and big data technologies becomes the solution. Machine learning algorithms generate actionable and predictive insights that help power and utility companies to make better and more informed decisions to protect the grid. Machine learning helps companies and service providers to detect threats and respond to them in time by constantly checking the digital climate and with the accuracy level that no one but machines can. Machine learning and AI technologies, in a broader perspective, offer integrated instruments like endpoint recognition and reaction arrangements, firewalls, and information misfortune avoidance solutions which can naturally answer to assaults by sifting and blocking malicious activities in the system.

While the literature on ML applications in smart grid systems is nascent, there exists a wide scope of exploration that has been led in the view of finding machine learning-dependent safety solutions for power grid methods. Machine learning provides solutions to such attacks as false data injection, convert cyberattacks, electricity theft, and denial-of-service attacks with a lot of accuracies [8]. A core vector machine (CVM) algorithm is used. Other software applications like, power plant model approval instrument to analyze phasor estimation unit and miniature phasor estimation unit information used for framework representation and recurrence recognition.

Support vector machine-based techniques could be used to detect false data injection into smart meters and phasors. The model outperforms the traditional statistical model when trained with sufficient data and can efficiently detect any anomalies in the smart grid system. Additionally, the conditional deep belief network model (CDBN) extracts worldly highlights from dispersed sensor estimations. The model is vigorous against different assaulted estimations and natural clamor levels [8]. CDBN has powerful features that perform even better than SVM and fake neural organization-based identification components. Distributed forswearing administration and disavowal of administration assaults have received the biggest attention in smart grid system security research. One of the many proposed models includes the Vijayan and DoS attack detection framework that utilizes diverse multi-layer deep learning algorithms to precisely identify and detect threats by analyzing traffic on smart meters.

In more specific applications, machine learning algorithms have been redesigned for use in smart grids for the detection of network capacity problems. They aim to reduce network capacity problems, diminish energy misfortunes, and accomplish more prominent effectiveness in service delivery. The benefits of machine learning applications in smart grid systems include better network monitoring, opportune and dependable shortcoming of the executives, constant organization reconfiguration, as well as control of power stream in the organization. In non-intrusion load monitoring, machine learning strategies dominate the use of significant learning and advanced multi-mark gathering systems "where the requirement for earlier extraction is decreased because such techniques' have been programmed in a way to highlight the extraction capacity, which is always one of the difficulties for the conventional concepts of AI." Machine learning is also used in designing digital assault identification techniques. These models can adapt towards versatility and detect network intrusion with preciseness.

Machine learning can be applied throughout the power generation lifecycle in four phases: data collection, feature extraction, classification, and results or outputs. During data collection, data is collected from thousands of distributed smart devices such as relays and protection, distributed energy resources, power and communication networks, among others. Feature extraction focuses on dimensionality reduction and starts from a set of estimated data as it seeks to build perfect values – extracting important features considered to be enlightening and non-excess. Feature extraction also works with ensuing learning and speculation steps. Classification is the third phase in the framework and involves categorizing data sets into multiple classes [13]. Classification leads to fast data solutions for load forecasting and generation, demand response, cybersecurity and fault detection, and protection.

7 Conclusions

Using smart grid network systems revolutionizes the energy sector in a significant way. Unlike the century-old systems, smart grids are intelligent, self-healing, cognitive, and self-monitoring systems that provide a bidirectional flow of data/information and electricity in a widely distributed network. The system increases reliability and efficiency in energy production and distribution. It also connects millions of smart devices that generate or share large volumes of data. However, the interconnectedness and integration make smart grid systems susceptible to major cyberattacks. Machine learning, a relatively new technology, offers exciting solutions to cyberattacks in the energy sector. They allow machines to learn and make decisions and predictions independently with or without minimal human intervention. These algorithms are used to detect and report anomalies in network traffic, filter malicious traffic, and provide secure system access for authorized users. Security is perceived as a framework made of three classes: the communication component, system control, and secured computing components. This study aims to counter the cyber security attacks in smart grids with machine learning approaches. The limitation is that these can only be applied to the datasets instead of ongoing live data. We would like to apply these machine learning techniques to real-time data, giving continuous live prediction every 1 h as a future enhancement.

References

1. S.S. Alhashim, M.M.H. Rahman, *Cybersecurity threats in line with awareness in Saudi Arabia* (International Conference on Information Technology (ICIT), 2021)
2. F. Aloul, A.R. Al-Alia, A.-D. Rami, A.-M. Mamoun, E.-H. Wassim, Smart grid security: Threats, vulnerabilities and solutions. Int. J. Smart Grid Clean Energy **1**(1), 1–6 (2012)
3. K. Dodrill, Understanding the benefits of the smart grid. *NETL Smart Grid Implement* (2010)
4. Z. El Mrabet, K. Naima, E.G. Hassan, E.G. Hamid, Cyber-security in smart grid: Survey and challenges. Comput. Electr. Eng. **67**, 469–482 (2018)
5. T. Flick, M. Justin, *Securing the Smart Grid: Next-Generation Power Grid Security* (Elsevier, 2010)
6. V. Ford, S. Ambareen, Applications of Machine Learning in Cyber Security, in *Proceedings of the 27th International Conference on Computer Applications in Industry and Engineering. 118*, (IEEE Xplore, Kota Kinabalu Malaysia, 2014)
7. A.A. Gopstein, S.B. Danielle, W. Kerry, V. Christopher, *Framework and Roadmap for Smart Grid Interoperability Standards Regional Roundtables Summary Report* (US Department of Commerce, National Institute of Standards and Technology, 2020)
8. N.I. Haque, S. Md Hasan, D. Md Golam, D. Anjan, P. Imtiaz, S. Arif, A.R. Mohammad, Machine learning in generation, detection, and mitigation of cyberattacks in smart grid: A survey. *arXiv preprint arXiv:2010.00661* (2020)
9. E. Hossain, I. Khan, F. Un-Noor, S.S. Sikander, M.S. Sunny, Application of big data and machine learning in smart grid, and associated security concerns: A review. Ieee Access **7**, 13960–13988 (2019)
10. IBM Cloud Education, *Unsupervised Learning*. Retrieved from IBM Cloud Learn Hub (2020, September 21). https://www.ibm.com/cloud/learn/unsupervised-learning
11. N. Katayasha, D. Niyato, W. Ping, H. Ekram, Communication architectures and models for smart grid: An architectural view. *Smart grid communications and networking* (2012)
12. V. Kumar, C.P. Guptha, Cyber Security Issue in Smart Grid IEEE 4th International Conference on Computing, Power and Communication Technologies (GUCON) (2021). https://www.reuters.com/article/us-eia-global-demand/global-energy-demand-seen-up-44-percent-by-2030-idUSN2719528620090527
13. A. Kumbhar, G.D. Pravin, K. Shobha, P. Uday, M. Pravin, A comprehensive review: Machine learning and its application in integrated power system. Energy Rep. (2021)
14. A.S. Omid, H. Omar, D. Tharam, R.T. Azadeh, Intelligent Decision Support System for Including Consumers' Preferences in Residential Energy Consumption in Smart Grid, in *Second International Conference on Computational Intelligence, Modelling and Simulation*, (Curtin University of Technology, 2014), pp. 154–159
15. R.K. Pandey, M. Mohit, Cyber Security Threats – Smart Grid Infrastructure, in *2016 National power systems conference (NPSC)*, (IEEE, 2016), pp. 1–6
16. M.M. Pour, A. Arash, S. Arif, A review on cyber security issues and mitigation methods in smart grid systems, in *SoutheastCon 2017*, (IEEE, 2017), pp. 1–4
17. W. Sanchez-Huertas, G. Víctor, H. Cesar, Machine learning techniques and smart grid applications: A review. Int. J. Appl. Eng. Res. **13**(21), 14876–14885 (2018)
18. S. Shapsough, Q. Fatma, A. Raafat, A. Fadi, A.R. Ali, Smart Grid Cyber Security: Challenges and Solutions, in *2015 international conference on the smart grid and clean energy technologies (ICSGCE)*, (IEEE, 2015), pp. 170–175
19. O.A. Sianaki, H. Omar, D. Tharam, R.T. Azadeh, Intelligent Decision Support System for Including consumers' Preferences in Residential Energy Consumption in Smart Grid, in *2010 Second International Conference on Computational Intelligence, Modelling and Simulation*, (IEEE, 2010), pp. 154–159
20. G.N. Sorebo, C.E. Michael, *Smart Grid Security: An End-to-End View of Security in the New Electrical Grid* (CRC Press, 2011)

21. Tacio Souza Bomfirm. Evolution of machine learning in smart grids. IEEE 8th international conference on smart Energy grid Engineering (SEGE) (Aug 2020)
22. S. Tan, D. Debraj, S. Wen-Zhan, Y. Junjie, K.D. Sajal, Survey of security advances in smart grid: A data driven approach. IEEE Commun. Surv. Tutor. **19**(1), 397–422 (2016)
23. S. Tufail, P. Imtiaz, B. Shanzeh, S. Arif, A survey on cybersecurity challenges, detection, and mitigation techniques for the smart grid. Energies **14**(18), 5894 (2021)

Preserving the Privacy and Cybersecurity of Home Energy Data

Richard Bean, Yanjun Zhang, Ryan K. L. Ko, Xinyu Mao, and Guangdong Bai

1 Introduction

Smart solar inverters are increasingly deployed due to the combination of falling prices of solar panels, batteries and inverters, and government policy [1, 2]. The "smart" aspect of the inverters refers to such features as the ability to monitor energy usage from a smartphone through the Internet connectivity of the inverter and its software. In some cases, the smartphone applications can be used to control devices in the house, and set up schedules for their operation.

Home energy data is also becoming more fine-grained. Since the move to five-minute pricing in the Australian wholesale electricity market, all new home smart meter installations must also record electricity usage on a five-minute basis [3].

With advances in solar technology, solar energy analyses play increasingly important roles in facilitating sustainable smart city and intelligent energy management [4]. As a data-driven study, an energy analysis typically requires a large sample size [5], and sharing solar energy data on a large scale (e.g., at national level) thus becomes essential. For example, the Australian Solar Energy Forecasting System (ASEFS) uses data from small-scale systems as training data for solar forecasting [6].

Although identifiers could be removed with data anonymization techniques, the solar energy data may include covert channels that allow an attacker to infer private

R. Bean (✉)
Centre for Energy Data Innovation, University of Queensland, Brisbane, QLD, Australia
e-mail: r.bean1@uq.edu.au

Y. Zhang · R. K. L. Ko · X. Mao · G. Bai
School of Information Technology and Electricity Engineering, University of Queensland, Brisbane, QLD, Australia
e-mail: yanjun.zhang@deakin.edu.au; ryan.ko@uq.edu.au; xinyu.mao@uq.edu.au; g.bai@uq.edu.au

K. Daimi et al. (eds.), *Emerging Trends in Cybersecurity Applications*,
https://doi.org/10.1007/978-3-031-09640-2_15

information which is not intended to be released by house owners. Such privacy implications have been largely overlooked in both literature and practice.

In this chapter, we elucidate this privacy risk through the location inference attack. We demonstrate that by combining solar energy data with publicly available data, such as climate datasets, an attacker would be able to determine or narrow down the location of particular houses whose temporal occupancy has been learned by the attacker from their fine-grained load associated with the energy data. If the derived location information is combined with the results of an "occupancy detection attack" [7] based on electricity usage data, an attacker can gain insight into the movement patterns and break into houses when occupants are unlikely to be present. This attack suggests that even though the location information has been removed, the correlations between energy data and the public information still put an individual's privacy at risk.

Figure 1 shows a schematic of the system model, with the smart solar inverter at the heart of the system. Energy for the house comes from the solar panel, the grid, or the battery (excess energy may also be exported to the grid or battery) while the data in this process is transferred wirelessly to smartphone apps via a "data cloud".

The attack surfaces in the process are also shown in Fig. 1, marked in red. The data of interest to the attacker is historical or "real-time" generation, which can be intercepted during transmission or obtained from a centralized database, or through vulnerable smartphone apps.

We evaluated our attacks using four large-scale real-world datasets containing over 2300 houses from Australia and New Zealand. Our results demonstrate that the attack can identify the location of households within a range of three grid points of historical weather data. For some subsets of the dataset with higher data accuracy

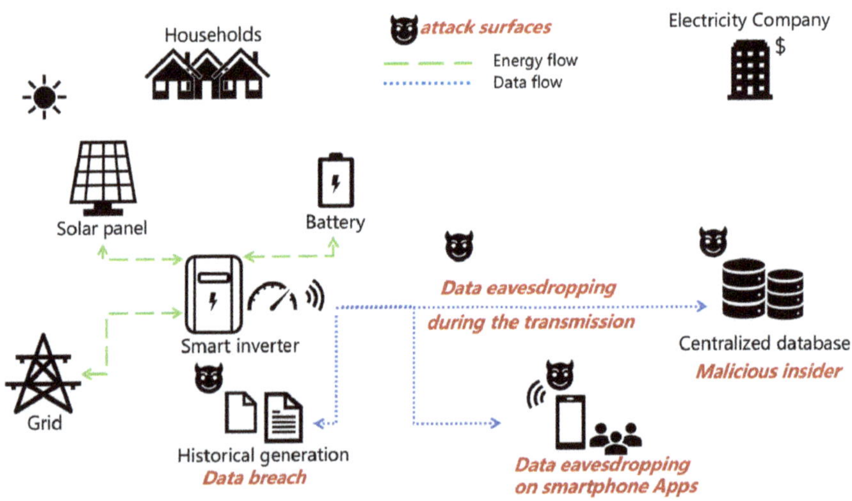

Fig. 1 System model

and quantity, the exact location of particular houses can be successfully determined, where the population density of the area is low.

We summarize our contributions by the following.

- We reveal a covert channel that may leak the location of households through the combined information in solar energy data and publicly available data. We demonstrate this attack is feasible and can scale up in datasets of a whole-country scale. To our best knowledge, our work is the first of its kind.
- We conduct our attack on large-scale real-world datasets. Our results show for high quality data in the house datasets, the houses can be identified to within a mean 39 km radius, with some houses in low-density rural areas able to be precisely located. The quality metrics are described further in Sect. 4.
- We discuss common use cases of home energy data.
- Finally, we responsibly propose defences against the proposed attacks, using privacy and cryptographic techniques.

2 Background and Related Work

The adoption of solar energy is increasing around the world. The penetration of solar panel installation in Australia is one of the highest in the world. For instance, in the state of Queensland, colloquially known as the "Sunshine State", 39.6% of dwellings have a photovoltaic (PV) installation [8]. Depending on local government incentives, local government areas may have even higher penetrations. For instance, the Somerset Regional Council area in Queensland has 54.2% of a total of approximately 11,000 dwellings with solar panels and Adelaide Plains LGA in South Australia has a 56% penetration.

In the case of "smart" solar inverters, a key selling point is the ability of the user to monitor their solar output via a smartphone app remotely so that the homeowner can observe their electricity usage and bills in real time. Retailers and "prosumers" (producers of solar energy and users of grid energy) alike benefit from sharing the household's solar energy production. Retailers can tailor the size of their battery offerings and offer customized tariffs to customers with detailed knowledge of solar outputs. They are also able to remotely identify panel shading and inverter malfunctions to assist the customers. Customers can make informed decisions on sizing and tariffs when more information is shared. Customers can also achieve transparency around these decisions by comparing their inverter and panel performance to similar offerings in their vicinity by sharing their output with such sites as `pv-output.org`.

However, sharing such data is not without risks. For instance, if attackers were able to obtain fine-grained solar data without location data, they may be able to utilize this data in conjunction with historical weather data to estimate the location of the house. If the weather data is granular enough and the density of the surrounding area is sufficiently low, this opens up the possibility of re-identification of the house location. This presents a heightened security threat if the attacker is able to obtain "house load" or electricity usage data of the house and guess when the occupants are likely to be away from home.

These so-called "occupancy detection attacks" have been implemented recently with both a genetic programming approach [7] and LSTM models [9]. Razavi et al. concentrates on determining a household's daily and weekly routine and measures the accuracy of such attacks. The conclusion was that with one week of half-hourly electricity consumption data, time slots where the occupants were "away" could be identified with high precision. Yilmaz et al. proposed a defence method for such attacks called "AMLODA" to simultaneously protect user data and preserve billing integrity.

In the following, we review existing related studies. The "Weatherman" software of Chen and Irwin described in [10], extending on the original [11] "Sunspot" paper, looked at 10 different houses with solar data in the continental United States and attempted to identify the location of each of the houses. The size of the solar installation at each house was not specified. The algorithm used the so-called "DarkSky" API to gather weather data, also implemented in [5]. Unfortunately, this API is no longer publicly accessible, and the solar data and the source code for the program are unavailable as well. Thus it is no longer possible to execute the original algorithm with our data and difficult to verify if the approach generalizes.

It is also no longer clear what the coverage of the DarkSky weather stations in various areas of the world is. In this chapter, we use publicly available weather data and the approach can be generalized to countries worldwide.

3 Location Inference Attack

3.1 Threat Model

A possible attack is the location inference attack. We assume that the attacker has access to an anonymous inverter energy dataset, in which the location information, including any (quasi-)identifiers of a location (e.g., owner's name, zip code), has been removed. The access becomes increasingly feasible for such an attack due to the growing need of sharing solar energy information. One of the selling points of inverters is that house owners can review their live and historical solar energy generation and possibly their house electricity usage. This gives owners the opportunity to see how much money they are making from solar export, or saving from offsetting house electricity usage by solar generation. If a house owner is paying the volatile wholesale electricity price, they can adjust or shift household tasks to minimize their electricity bills.

This setup presents a few opportunities for interception. Risks arise if generation is transmitted to central storage and transmitted or stored without being encrypted;

or if the historical generation is stored on the inverter itself, or on a user phone, tablet or personal computer.

We assume the attacker also has access to publicly available datasets comprising side-channel information that can be combined with the energy dataset to reveal the location information. In this chapter, we use publicly available weather data as our side-channel information.

The adversary's re-identification tactic is to exploit the correlations among energy data and weather data. For example, the intensity of solar radiation can affect the amount of energy generated by solar panels. Due to the existence of the correlations, even though the location information has been removed from the energy datasets, the inherent correlations could still put the location privacy at risk.

3.2 *Our Inverter Dataset Across Two Countries*

In this section, we examine a large dataset of hourly data from 2201 houses in Australia and 116 houses in New Zealand with residential-scale solar. The houses cover all states of Australia and all regions across New Zealand's North and South Islands. The data for each house is available from periods of 169 hours to 13,777 hours, or approximately one week to 19 months. The data is from inverters manufactured by Redback Technologies.

The solar energy was recorded at either 0.1 or 0.01 kWh resolution, depending on the software version of the inverter, and aggregated to hourly resolution from the original one minute data. This aggregated hourly data is obtained from the cloud storage, and is the basis for our investigations as it aligns with hourly weather observations. The location of each site is known to within two decimal places of latitude and longitude. At the studied latitudes, 0.01 degrees corresponds to approximately 900 m.

Prior to the aggregation and storage in the cloud, generation data is calculated with two smart inverter software versions, known as "ROSS 1" and "ROSS 2". The ROSS 1 software calculates energy values on the inverter itself, using instantaneous voltage and current measurements every five seconds. In contrast, ROSS 2 software takes the energy values from the inverter and meters; the ROSS 2 value we use in this chapter is a cumulative PV "all time" reading. If some values are missing in the ROSS 2 data collection process, the inverter performs software interpolation to replace the missing values. The breakdown by country and software version is shown in Table 1.

Table 1 Count of inverters by software version and country

	ROSS-1	ROSS-2
Australia	1336	865
New Zealand	83	33

3.3 ERA5-Land Reanalysis

We analysed the ROSS 1 and ROSS 2 energy data in conjunction with two high temporal and spatial resolution datasets which acted as our "side-channel information". These are the ERA5 and ERA5-Land datasets from the European Centre for Medium-range Weather Forecasting "Copernicus Climate Data Store" (CDS) [12].

Using this data, and making basic assumptions about inverter type, orientation, and tilt, we simulated the solar output at each grid point in the CDS data using algorithms from the PVLib library, a product of the National Renewable Energy Laboratory (NREL) in the US, which has been developed over many years [13].

The ERA5-Land dataset [14] is a reanalysis of world climate available at 0.1 degree grid resolution (latitude and longitude) since 1981. The dataset provides a range of weather variables at single and different pressure levels, at hourly temporal resolution.

In this analysis, we extracted data at a single pressure level, from two bounding boxes. One encapsulates mainland Australia and Tasmania, and the other encapsulates New Zealand. At these latitudes, 0.1 degree resolution is approximately 9 km between grid points.

The ERA5 dataset is similar, but covers ocean as well as land areas, at a 0.25 degree resolution hourly since 1979. This corresponds to a resolution of approximately 22 km between grid points in Australia and New Zealand.

For Australia, these boundaries are 113 to 154 degrees East in longitude (411 points) and 9 to 44 degrees South in latitude (351 points). Of these $411 \times 351 = 144{,}261$ points, 70,079 locations contain data for each weather variable in the ERA5-Land dataset. Similarly, for the ERA5 dataset, this is 165×141 points, or 23,265 total.

As for New Zealand, these boundaries are 166.5 to 178.6 degrees East in longitude (122 points) and 34.4 to 46.7 degrees South in latitude (124 points). Of these $122 \times 124 = 15{,}128$ points, 2887 points have data for each weather variable. Clearly, the shape of Australia is much more efficiently represented by a latitude/longitude bounding box in ERA5-Land, with 48.6% of points on land compared to 19.1% in New Zealand. With ERA5, there are 50×53 points or 2650 points (using 166.5 to 178.75 degrees East and 34 to 47 degrees South).

Five variables were of interest to us in the analysis, as they formed the input to the PVLib algorithm which estimated the inverter energy output for each hour. These were:

- SSRD (surface solar radiation downwards) cumulative hourly value in J/m^2. These were divided by 3600 to obtain the Global Horizontal Irradiance input for PVLIB, which is in W/m^2.
- T2M, the temperature at 2 metre height, in Kelvin adjusted to Celsius for PVLIB
- U10, zonal (east-west) wind speed at 10 metre height, in m/s
- V10, meridional (north-south) wind speed at 10 metre height, in m/s
- TCC, total cloud cover in percent (available in ERA5 data but not ERA5-Land)

Statistics for these values seen in the input data are shown in Table 2.

Table 2 Statistics for ERA5 datasets by country

Variable	ERA5 AU			ERA5-Land AU			ERA5 NZ			ERA5-Land NZ		
	Min	Max	Mean	Min	Max	Mean	Min	Max	Mean	Min	Max	Mean
T2M	−10	49	21	−11	49	22	−16	36	14	−18	35	9
SSRD	0	1165	224	0	1169	236	0	1127	179	−1	1122	160
U10	−30	26	−0.3	−17	19	0.8	−21	23	1.5	−17	19	0.8
V10	−26	24	0.7	−22	24	0.6	−24	21	−0.1	−22	18	−0.4
TCC	0	100	61				0	100	47			
Modelled output (W)	−1.5	4985	972	−1.5	4985	1402	−1.5	4985	719	−1.5	4985	1073
Positive output (W)			2013			2848			1533			2243

We performed decumulation of the SSRD value for the analysis of ERA5-Land in order to compare it against the PV output values observed at each house.

3.4 PV Simulation

Using the PVLib software mentioned, for each hour in the observed data, we simulate the energy output of a solar inverter at each point in the ERA5 and ERA5-Land datasets. As noted, for the ERA5 data, we used SSRD (solar radiation), 2 metre temperature, zonal and meridional wind speeds, and total cloud cover variables. For the ERA5-Land data, we use the same variables except for the cloud cover which is not available. The grid points in the ERA5 data are 0.25 degrees (about 22 km) apart) and in the ERA5-Land data are about 0.1 degrees (about 9 km) apart. In Sect. 4 we use this data to attempt to infer real-world inverter locations.

We make the following assumptions about inverter type, orientation, and tilt. We first assume that the inverter parameters were closely related to the "Outback Power" 5 kW inverter from the "CEC" list in PVLIB [13]. This inverter is denoted as "OutBack Power Technologies Inc SBX5048 120 240 240V" in the PVLIB library. The "SBX5048" model modelled has similar properties to the Redback Technologies inverters being studied here. In particular, the maximum continuous AC output power is the same as the Redback SH5000 model, at 5000 V, and the derating over 45°C is equivalent. Of all the PVLIB library inverters, this is the closest to the Redback Technologies inverters under study (Table 3).

Table 3 Comparison of specifications for Outback and Redback inverters

	Outback	SH4600	SH5000
AC frequency	60 Hz	50 Hz	50 Hz
Max cont AC output power	5000 VA (derate >45C)	4600W AC (derate >45C)	5000W AC (derate >45C)
AC voltage	120/240 V split-phase	230 VAC single-phase	230 VAC single-phase
Battery nominal DC voltage	48 V	48 V	48 V
Operating temp range	−20 to 60°C	−25 to 60°C	−25 to 60°C
MPPT	250–600 V	125–500 V	125–550 V
Typical efficiency	>97%	97%	97%
Max PV Sys Voltage	600 V	500 V	580 V
Max In/Out current	20/24 A	22/20 A	22/21.7 A

We also assume that the solar panels are in a 6.6 kW configuration, comprised of 24,275 W panels from the "Jinko" manufacturer. The panels are assumed to be in a 2 strings per inverter, 12 panels per string configuration.

In terms of orientation, we assume the panels are all north facing (being in the Southern hemisphere as opposed to the Northern in [15]) at an inclination equal to the latitude of the location, at sea level. These assumptions are said to be an "approximation of the necessary conditions for maximum output during a year without any tracking system" by Camargo and Schmidt and are also mentioned as rules of thumb in the survey paper [16].

We then simulate the PV output for each of the 70,079 Australian and 2887 New Zealand locations in the ERA5-Land dataset for March 2019 to August 2020, producing a 70,079 × 13,201 matrix for Australia and a 2887 × 13,201 matrix for New Zealand. For the ERA5 dataset, the Australian matrix is 23,265 × 18,007 and the NZ matrix is 2650 × 18,151.

In the data processing step for ROSS 1, any hours where the same PV value (denoted "PVTotalToday", a cumulative count) occurs for 24 hours or more are removed, along with any hours where the PV value increases by more than 7 kW in any given hour, as this is considered to be beyond the capacity of the inverter.

Our approach uses the PVLIB Python [13] library with simulations. We check the correlation of the vector of PV values with each of the locations in the simulated PV matrix (that is, a vector corresponding to each location) to try to identify the most likely location for the inverter. Then the inverter is assumed to be closest, geographically, to the point with the maximum Pearson correlation found.

4 Evaluation

With the current dataset of 2,201 inverters in Australia and 116 inverters in New Zealand, the algorithm was run to determine the location for each house in the dataset. The geographic point with the highest correlation to the simulated PV trace was recorded. It was demonstrated that the exact address of houses can be identified or inferred with relative ease.

Figures 2a and b show a contour plot of the correlations for an inverter in New South Wales. This inverter has 12,663 hours of data (4938 hours of non-zero data) and maximum correlation coefficient 0.76. The distance between the actual location of the inverter and the calculated grid point location is approximately 1.1 km. (Note that the "actual" location is only known to within approximately 900 metres due to the rounding of the latitude and longitude values.)

In addition, as this inverter is in a rural location with a population density of approximately 1.9 persons per square kilometre, only five houses were within 1.1 km of the actual house location. It is worth noting that, using services such as Google Map, NearMap, or Mechanical Turk, it would be enough to identify the exact address of the house at the moment, when the distance is small, e.g., less than 5 km [10], depending on the population density in the vicinity of the house.

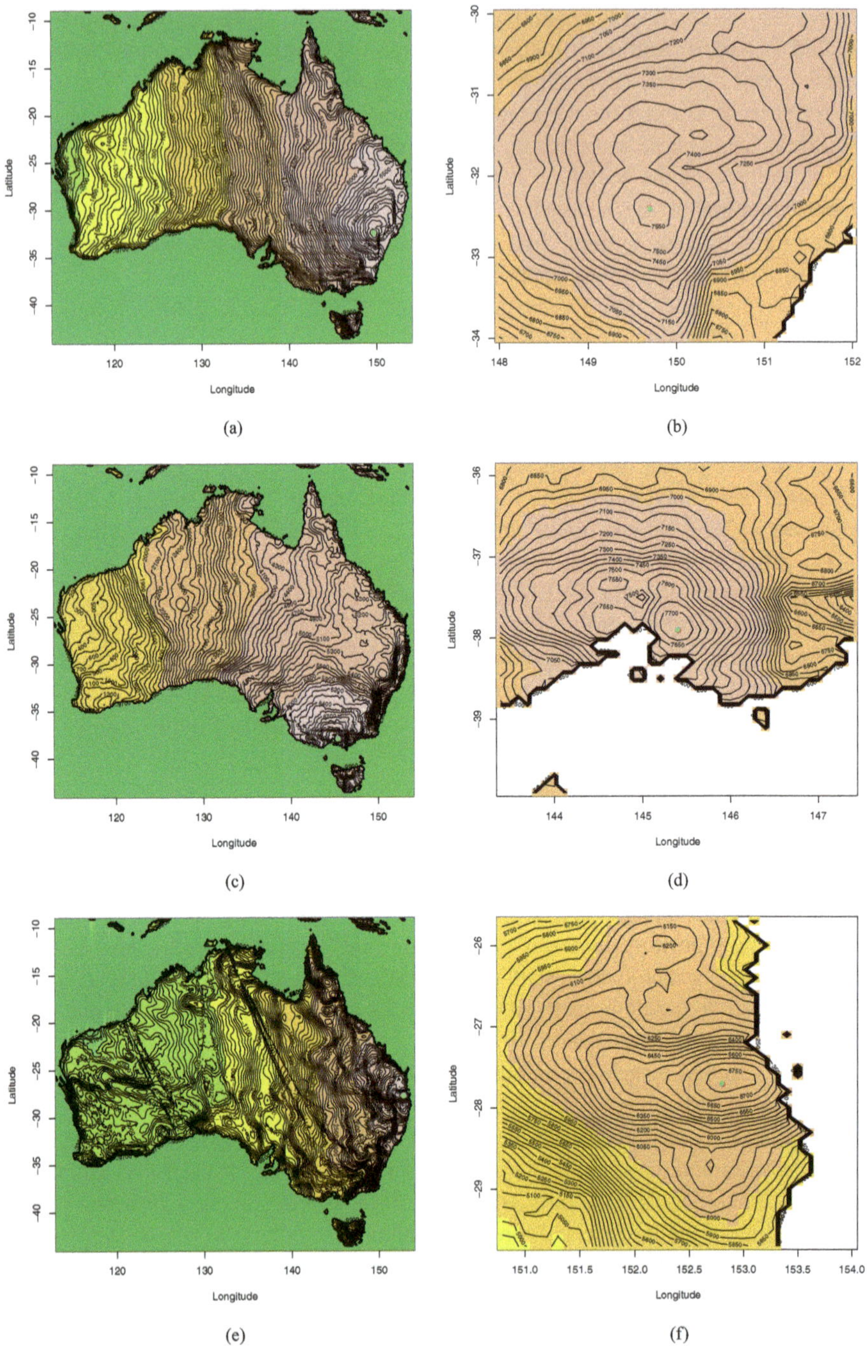

Fig. 2 Correlation map (Australia) between an inverter and its actual location. (**a**) An inverter in New South Wales. (**b**) An inverter in New South Wales (detail). (**c**) An inverter in Victoria. (**d**) An inverter in Victoria (detail). (**e**) An inverter in Queensland. (**f**) An inverter in Queensland (detail)

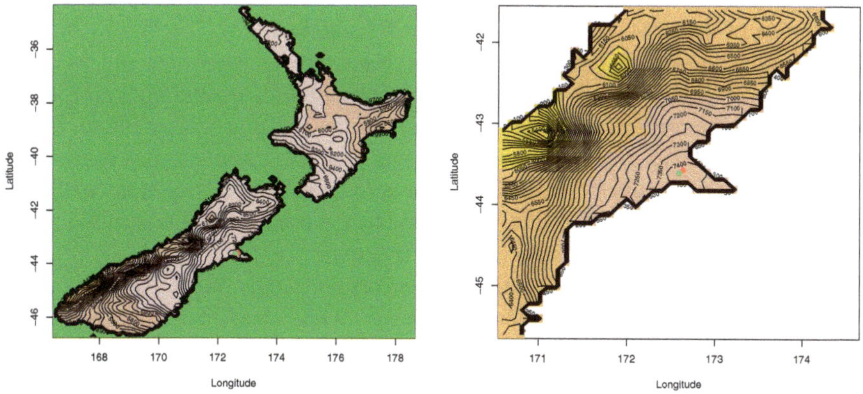

(a) (b)

Fig. 3 Correlation map (New Zealand) between an inverter and its actual location. (**a**) An inverter in Christchurch, New Zealand. (**b**) An inverter in Christchurch, New Zealand (detail)

Table 4 Statistics for distance in points and kilometres, by country and data

	ROSS 1			ROSS 2		
	Median	Median	Mean	Median	Median	Mean
Variable	Points	Km	Km	Points	Km	Km
Australia ERA5	7	57	347	4	76	473
Australia ERA5-Land	6	47	276	9	70	617
NZ ERA5	7	51	90	3	49	95
NZ ERA5-Land	4	29	70	6	46	63

Figures 2c and d show a contour plot of the correlations for an inverter in Victoria near Melbourne. The maximum correlation is 0.78. This inverter has 12,689 hours of data (4017 hours of non-zero data). The distance between the actual location of the inverter and the calculated grid point location is approximately 1.4 km. Figures 2d and e show a contour plot of the correlations for an inverter near Ipswich in Queensland. The maximum correlation is 0.68. This inverter has 12,689 hours of data, but only 532 hours of non-zero data. However, despite the lack of data, the distance between the actual location of the inverter and the calculated location is approximately 1.5 km.

Figures 3a and b show the algorithm placement of an inverter in Christchurch, New Zealand. The inverter has 11,917 hours of data, 2008 hours of non-zero data, and a correlation coefficient of 0.74.

The distance results are shown in Table 4. The values are given in points and kilometres.

In all cases, the ERA5-Land median values in kilometres are lower than the corresponding ERA5 values, despite the lack of cloud cover data in the PVLIB simulation of the solar output using the ERA5-Land data.

Sometimes a large number of hours is necessary to determine the location, while a corresponding large number of non-zero output hours and high correlation coefficient demonstrates greater certainty about the location. For example, taking the Australian ROSS 1 data using ERA5-Land, and restricting it to inverters which have at least 8760 hours, 5000 hours of non-zero data, and a correlation coefficient of at least 0.75, we have 84 inverters. The median and mean distances to the actual inverter locations are then 31 and 39 km, respectively.

Some of the large location identification errors are clearly due to poor data as the ratio of non-zero hours to total hours is very low, while other errors are due to poor assumptions about the inverter. For instance, some inverters in the list are three-phase inverters which can reliably generate over 10 kW instead of being limited to 5 kW as assumed. The other sources of error are due to assumptions about the orientation, tilt and shading of panels and inverter and panel parameters. A higher correlation coefficient and high numbers of non-zero hours, as shown, can provide greater certainty around the localization result for a given inverter.

Further research could investigate quality issues in the data which prevent more accurate re-identification of house locations, as, for example, a circle with a radius of 39 km (the mean distance above) will typically cover many houses even in low-density rural areas; and associating a probability to given predictions. Ideas such as inverse distance weighting could be used to further improve the prediction, instead of providing the nearest grid point as we have done here.

5 Mitigation

Privacy-preserving sharing techniques including cryptographic mechanism and differential privacy can help defeat this attack. Cryptographic solutions, such as homomorphic encryption [17–19] and secure multi-party computation [20, 21], enable computation without disclosing data in plaintext. Differential privacy [22, 23] is another potential mitigation mechanism against the attack.

5.1 Overview of Privacy-Preserving Sharing Techniques

5.1.1 Differential Privacy

Differential privacy (DP) is a methodology that provides a strong standard for privacy guarantees. It ensures the that an observer of an algorithm's output cannot tell if a particular individual's information was used in the computation. It follows that no risk is incurred by joining the database, providing a mathematically rigorous means of minimizing the chances of identifying records of statistical databases while maximizing the accuracy of queries. The common practice for achieving differential privacy is based on additive-noise mechanisms which perturb the original data/models by adding noise.

DP adds sophisticated noise to the data, statistically preserving individual data privacy and general data usability from re-identification. An (ϵ, δ)-DP promises that for two adjacent datasets $\mathcal{D}$ and $\mathcal{D}'$ which differ in at most one record, their results from the domain $\mathcal{S}$ of the randomized mechanism function $\mathcal{M}$ satisfy the following formula:

$$\Pr[\mathcal{M}(D) \in \mathcal{S}] \leq \exp(\epsilon) \cdot \Pr\left[\mathcal{M}\left(D'\right) \in \mathcal{S}\right] + \delta,$$

with the privacy budget $\epsilon > 0$ to control the level of privacy protection, and smaller ϵ providing stronger protection; $\delta \geq 0$ allows uncontrolled privacy breach; and we say the randomized mechanism $\mathcal{M}$ gives (ϵ, δ)-differential privacy. Specifically, when $\delta = 0$, the randomized mechanism $\mathcal{M}$ gives the strictest protection and is called ϵ-differential privacy.

Two principal mechanisms are currently available to realize DP on numerical data and non-numerical data separately: the Laplace mechanism [24] and the Exponential mechanism [25].

The Laplace mechanism $\mathcal{M}(D) = f(D) + n$ adds Laplacian noise $n \sim \text{Lap}(\Delta f/\varepsilon)$ to the result of a query function $f : D \to \mathbb{R}$, according to the sensitivity of the query function: $\Delta f = \max\limits_{(D,D')} \left\| f(D) - f\left(D'\right)\right\|$, which measures the maximal difference between the query results on the adjacent datasets $\mathcal{D}$ and $\mathcal{D}'$. Typical application scenarios are as follows:

1. **Counting queries**. When we want to know how many records satisfy our requirements, the change of one record in the dataset will accordingly cause the difference of at most one record in the query result, therefore a Laplacian mechanism can be realized by directly adding noise $n \sim \text{Lap}(1/\varepsilon)$ to the result.
2. **Histogram queries**. A histogram is composed of bins of data and these bins are independent. Similarly to the first case, we only need to add the $\text{Lap}(1/\varepsilon)$ noise.

The Exponential mechanism $\mathcal{M}(D) = \left\{\text{return } r \text{ with probability} \propto \exp\left(\frac{\epsilon q(D,r)}{2\Delta q}\right)\right\}$ samples discrete results from an exponential distribution, paired with a score function $q(D, r)$ to measure the quality of a result r based on the dataset. The sensitivity is defined upon the adjacent datasets and a set of results $\mathcal{R}$: $\Delta q = \max\limits_{(D,D',r\in\mathcal{R})} \left\|q(D, r) - q\left(D', r\right)\right\|$. Intuitively, a result with a high score of quality is more likely to be sampled, but different from the Laplacian mechanism, the "noise" comes from the randomized sampling process and the results are still the same as the original ones. Based on this property, the Exponential mechanism is also suitable for value-sensitive cases, such as a truthful auction.

The usage of DP can be categorized into data publishing and data analysis [26]. Data publishing aims to share the dataset or the results derived from the data, whereas data analysis refers to building machine learning models on the data and release to the public. DP can provide mathematically rigorous protection for both the two processes: publishing the results in the data publishing process and computing the model parameters in the data analysis process.

5.1.2 Federated Learning

Federated Learning (FL) is a promising solution to reconcile the need for data sharing with the concern for privacy. It is a framework that keeps the sensitive original data locally and instead uses data-derived parameters, such as gradients, to enable training for machine learning tasks. Therefore, it satisfies data protection by design, which complies with privacy laws such as GDPR for data sovereignty, and also facilitates collaborative model training tasks by multiple parties. FL is usually combined with practical privacy-preserving techniques such as DP to enhance its security. Horizontal Federated Leaning (HFL) and Vertical Federated Learning (VFL) are the two commonly discussed categories in FL studies [27].

HFL is a sample-based FL; different parties sharing the same feature space of the data can be federated by their samples. HFL is suitable for companies with the same business but different in groups of customers. It can also be applied in cross-device settings [28], such as a training model with a large amount of mobiles [29].

VTL is a feature-based FL, which is designed for companies who have overlapping customers (samples) but different businesses (features) to train a model jointly. It can be applied in cross-silo settings; for example, a bank and a e-commerce company in the same city having an intersection of customers will benefit from the model trained with more features under VTL.

5.1.3 Homomorphic Encryption

Homomorphic Encryption (HE) is a method to realize secure multi-party computation. It is a cryptographic solution for computing with encrypted data, which means the original data is concealed from parties other than the user during the whole process, while the veracity of the final result is guaranteed. It is suitable for the case when the original data is not intended to be uploaded and published, but corresponding computation is necessary.

HE permits a third party (e.g., cloud service) to perform computation on encrypted data without first decrypting. This property prevents possible privacy breaches under the cloud computing background. HE can be achieved through three schemes: Partially Homomorphic Encryption (PHE), Somewhat Homomorphic Encryption (SHE), and Fully Homomorphic Encryption (FHE).

As HE preserves arithmetic operations on the encrypted data, the three schemes provide different levels of computation: PHE allows unlimited times of addition or multiplication, but not both; SHE can only perform the two operations limited times;

FHE enables unlimited additive and multiplicative operations, but with a cost of computation [30].

5.2 Applying Privacy-Preserving Sharing Techniques to Use Cases of Solar Energy Data

5.2.1 Solar Generation with Scheduling

Solar generation with scheduling is useful in efficiency terms as in a distributed system generation is moved closer to the load, avoiding transmission losses and the subsequent need to perform large scale scheduling of generating units. In a large-scale electricity system, the losses need to be estimated with loss factors periodically and these need to be considered in each dispatch of the whole system.

Loss factors do not need to be estimated as the main factor to consider is the round-trip efficiency of the battery-solar inverter combination chosen. This helps with decarbonising the energy system.

To perform scheduling of a battery inverter combination most effectively, an accurate forecast should be produced incorporating the recent behaviour of the home occupants; that is, a recent load and solar trace. The weather data can be used as training data to predict future solar and house load. For example, hourly Numerical Weather Prediction (NWP) data is available historically from many sources, such as the European Centre for Medium-range Weather Forecasting (ECMWF).

Utilities often keep energy data for each user by periodically performing a virtual or physical meter reading. In states such as Victoria in Australia, all homes have a smart meter which records energy usage periodically. The Victorian smart meters usually collect data at a 30 minute resolution, which allows standard tariffs to be aligned with half-hour boundaries, if necessary.

This data is useful for utilities to understand and plan energy management systems at the macro and micro level but also produces a possible privacy threat, particularly if the energy data resolution is high (e.g., one minute or one second resolution data).

Another reason utilities would like to collect this data is to provide recommendations on the installation of a solar inverter, panel or battery combination system tailored to the energy use of a particular house. This would be highly beneficial for both the customer and the utility. In certain systems, the user may not even have to pay the up-front cost of the inverter system and can own such a system after paying a discounted tariff for some years. In other situations, the user can be provided relative certainty on the "pay-off" period of their system if their energy usage is well understood. This enables more efficient investment decisions on the part of both the homeowner and the utility or retailer.

Since solar energy data can reveal the pattern of customer's residency and activities, and incurs privacy breaches, privacy-preserving techniques should be applied to secure the business. Considering this data is numerical and may be at different levels of resolution, techniques corresponding to the usages of data should be chosen.

1. **Data collecting for computation.** A typical scenario is billing; as it involves financial interest, the value of the actual data is sensitive and should not be distorted. HE-based encryption methods preserve security of the actual data, and allows computation on the encrypted data, therefore is suitable for this case.
2. **Aggregative queries.** If the utility company wants to query the volume of energy in certain area at a particular time, this aggregation can be secured by DP with Laplacian mechanism. The aggregation query is at the macro level, and thus a particular customer's privacy can be hidden statistically with the added noise.

5.2.2 Solar Recommendation System

A utility may wish to provide a recommendation for an existing or potential customer, comprising a recommendation for any combination of inverter, solar panels and batteries.

The solar recommendation system would ideally have as input data a high resolution perfect forecast of usage into the future, so that the battery can be scheduled with perfect accuracy.

The inverter size, solar panel size and battery size (and properties versus temperature, degradation over time versus pay-off period) can be chosen based on this. For example, a smart solar inverter manufacturer may have, say, a 5 kW and 10 kW model inverter, with 6.6 kW and 13.2 kW of solar panels, and then the battery choice can be: no battery, and in incremental blocks of 3.3 kWh up to 13.2 kWh depending on the desired up-front cost and pay-off period.

At a finer level, we may consider the properties of the battery; different battery types (for instance, lithium ion, zinc bromine, or lead acid) have quite different charging, discharging, efficiency, minimum and maximum capacity and degradation properties versus time and temperature which must all be considered in an optimization program [31–33].

In practice, not all of these requirements can be met. For instance, the house may not have a smart meter, and billing may only be available every three months. Battery choice may be limited to lithium ion batteries from a particular manufacturer. Solar panels may be only available in discrete blocks of large size, or there may be only one size of inverter available. Thus the choice of installation can be between just a few kinds of system.

In this case, proxy measures are used to estimate daily usage, based on assigning the house to a cluster, and comparing the house to similar houses in the same geographic location.

For instance, the following factors can be considered in a machine learning algorithm to estimate the load pattern as closely as possible:

- the size of the roof—to develop an upper bound on solar panel capacity
- how many people are living in the home
- how many bedrooms are in the house

- the type of dwelling—townhouse, apartment, or house
- the age of the people living in the home
- the working status of people living in home
- the age of the home
- the area of the land and home
- which direction the roof is facing
- any shading issues
- the material of the roof and the walls
- the air conditioning available and whether it is controllable
- the hot water system and whether it is controllable
- whether the house has gas or electric cooking
- how many garage space, vehicles, and electric vehicles (EVs) are present
- whether any EVs present have bidirectional flow capability

One quick heuristic or rule-of-thumb used in deciding whether a system is suitable for battery scheduling versus weather is the ratio of average solar generation to average load in the house [34].

It is better for a potential user to provide hourly net load over a one year period. In this case, an attacker may be able to estimate solar usage if they know the location of the house, and net this time series out, providing a useful snapshot of home energy usage.

Alternatively, attackers may be able to guess some of the properties of the houses with fine-grained energy data and then map out high-value houses to break into. For instance, the presence and value of EVs may possibly be inferred with NILM (non-intrusive load monitoring).

As the features used to build a machine learning model contain a large amount of private information, we can choose a FL framework to leave these sensitive features locally and apply DP on the gradient transmission of the FL model training. Specifically, customers are distributed in various areas and their data shares the same design of features. Thus, the FL framework is horizontal (sample-based). A HFL framework is not enough to prevent attacks on the parameters, since gradient leakage is possible for image and text data [35], and Byzantine attacks can also poison the model training [36]. DP can be integrated into the FL framework to help build a robust training [37], but need to balance the trade-off between security and data usability.

6 Case Study—Billing with Homomorphic Encryption

In the Australian state of Victoria, the smart meters measure kilowatt hour (kWh) load every half hour. Then, billing is calculated using the measured energy and the tariff in c/kWh. The reference [34] contains example tariffs for Australian houses. The tariffs divide the hours of the day into different price periods. Typically these are one of three divisions: flat, peak/off-peak, or peak/off-peak/shoulder pricing.

Some energy retailers such as Flow Power and Amber Electric allow residential customers to pay the wholesale electricity price allowing for savings depending on the ability of the customer to modify their usage.

As we have demonstrated the possible attack on published anonymous datasets and its hazardousness, applying encryption is critical to protect data privacy from re-identification attacks. This section will give a case based on a typical electricity billing scenario, implementing HE as encryption and showing its security.

The billing scenario is calculating user bills from both the amount of energy consumed and the energy exported to the grid. Under this setting, customers have installed solar panels which can generate energy and the excess part can be sold back to the grid. The bill of one customer at a particular hour h is computed as the cost of importing energy from the power grid minus the revenue of exporting energy to the power grid, referencing to [34]:

$$B_h = t_h I_h - f_h E_h$$

Here, the tariff of imported energy t_h and the revenue of exported energy f_h are public, and customer's energy data I_h and E_h are private. Considering the computation is usually performed on a central server, and the private data will be collected, this process requires an encryption. HE is naturally suitable for this case, because it allows direct arithmetic operations on encrypted data and thus avoids the risk when decrypting the data. The whole process includes HE Encryption, HE Computation and HE Decryption.

HE Encryption The HE algorithm firstly generates a public key p_k for encryption and a private key s_k for decryption. ENC denotes an encryption function which takes the plaintext m and the public key p_k as inputs and outputs the ciphertext c.

Input: public key p_k, energy imported from the grid of all customers in given hours I, energy exported to the grid of all customers in given hours E.
Output: the encrypted data $c_i = \text{ENC}(I, p_k)$, $c_e = \text{ENC}(E, p_k)$

HE Computation HE enables arithmetic operations (for example, addition and multiplication), denoted as OP, on the encrypted data, and evaluates the encrypted computation result. The computation result of the encrypted data should be equivalent to the encrypted result of unencrypted data: $OP(c_i, c_e) = ENC(OP(I,E), p_k)$.

Input: the encrypted data c_i and c_e, tariff t, revenue f.
Output: the encrypted bill of all customers $B_{enc} = t \cdot c_i - f \cdot c_e$

HE Decryption For decryption, a function DEC takes the private key s_k and the ciphertext c, and returns the decrypted final result.

Input: the encrypted bill B_{enc}, the private key s_k.
Output: bill of all customers $B = DEC(B_{enc}, s_k)$

We simulated the situation where 83 customers have 4000 hours usage on average, the tariff is 33.5 c/kWh and the revenue is 11.3 c/kWh. When generating the amount of I_h and E_h for an individual customer, we treat them as average values for simplicity, and randomly sample from reasonable intervals: I_h in [0.29, 1.75], E_h in [0, 3.86]. For the HE implementation, we choose the Paillier scheme which is a PHE algorithm for utility consideration. The simulation consumes 0.33 min for encryption, 0.016 min for computation, and 0.05 min for decryption, which is moderate for the time cost of HE.

7 Concluding Remarks

Our work demonstrated that given a large inverter dataset of over 2300 inverters across both Australia and New Zealand, for hourly resolution data of up to 19 months, we could determine the location of an inverter, simply by using a publicly available weather dataset to within a median of three grid points (0.1 degree resolution data). When the quality of the data was assured through the ratio of non-zero to total hours and the correlation coefficient, the result was greatly improved. For instance, the location can be determined to within a mean of 39 km for a selected subset of inverters, and this result was achieved without using any knowledge of cloud cover. This has immediate implications for the privacy of all rural dwellings using inverter technologies.

Further research could include assessing the quality of the input data further to identify which houses are likely to be reidentified exactly, and also assessing whether integrating other datasets such as MERRA-2 [38] into the investigation can improve the results.

We also investigated defence mechanisms against such privacy attack. In particular, we explored the applicability of the differential privacy mechanism [39], introducing the perturbation of data and also providing a clustered profile. Perturbation reduces the utility of the data for other purposes, and a clustered profile protects privacy since the location can be protected to within an arbitrary distance of the inverter. This provides assurance to inverter customers that even if their profile data was ever obtained by a malicious actor, no further inferences about their residential locations can be made. We also examined federated learning and homomorphic encryption as mechanisms to reduce the efficacy of the proposed attacks.

References

1. Australian Government Clean Energy Regulator, Small-scale renewable energy scheme (2018). http://www.cleanenergyregulator.gov.au/RET/About-the-Renewable-Energy-Target/How-the-scheme-works/Small-scale-Renewable-Energy-Scheme. [Retrieved: December, 2021]
2. R. Best, P.J. Burke, S. Nishitateno, Understanding the determinants of rooftop solar installation: evidence from household surveys in Australia. Aust. J. Agric. Resour. Econ. **63**(4), 922–939 (2019)

3. Australian Energy Market Commission, Five minute settlement (2021). https://www.aemc.gov.au/rule-changes/five-minute-settlement. [Retrieved: December, 2021]
4. A.S. Spanias, Solar energy management as an internet of things (iot) application, in *2017 8th International Conference on Information, Intelligence, Systems & Applications (IISA)* (IEEE, 2017), pp. 1–4
5. D. Syed, H. Abu-Rub, A. Ghrayeb, S.S. Refaat, M. Houchati, O. Bouhali, S. Bañales, Deep learning-based short-term load forecasting approach in smart grid with clustering and consumption pattern recognition. IEEE Access **9**, 54992–55008 (2021)
6. Australia Energy Market Operator, Solar and wind energy forecasting (2016). http://www.aemo.com.au/Electricity/National-Electricity-Market-NEM/Planning-and-forecasting/Solar-and-wind-energy-forecasting. [Retrieved: December, 2021]
7. R. Razavi, A. Gharipour, M. Fleury, I.J. Akpan, Occupancy detection of residential buildings using smart meter data: A large-scale study. Energy Buildings **183**, 195–208 (2019)
8. Australian Photovoltaic Institute, Mapping Australian photovoltaic installations (2021). https://pv-map.apvi.org.au/historical. [Retrieved: December, 2021]
9. I. Yilmaz, A. Siraj, Avoiding occupancy detection from smart meter using adversarial machine learning. IEEE Access **9**, 35411–35430 (2021)
10. D. Chen, D. Irwin, Weatherman: Exposing weather-based privacy threats in big energy data, in *2017 IEEE International Conference on Big Data (Big Data)* (IEEE, 2017), pp. 1079–1086
11. D. Chen, S. Iyengar, D. Irwin, P. Shenoy, Sunspot: Exposing the location of anonymous solar-powered homes, in *Proceedings of the 3rd ACM International Conference on Systems for Energy-Efficient Built Environments*, pp. 85–94 (2016)
12. B. Raoult, C. Bergeron, A.L. Alós, J.-N. Thépaut, D. Dee, Climate service develops user-friendly data store. ECMWF Newsletter **151**, 22–27 (2017)
13. W.F. Holmgren, C.W. Hansen, M.A. Mikofski, pvlib python: A python package for modeling solar energy systems. J. Open Source Softw. **3**(29), 884 (2018)
14. H. Hersbach, B. Bell, P. Berrisford, S. Hirahara, A. Horányi, J. Muñoz-Sabater, J. Nicolas, C. Peubey, R. Radu, D. Schepers, et al., The era5 global reanalysis. Q. J. Roy. Meteorol. Soc. **146**(730), 1999–2049 (2020)
15. L.R. Camargo, J. Schmidt, Simulation of long-term time series of solar photovoltaic power: is the era5-land reanalysis the next big step? Preprint (2020). arXiv:2003.04131
16. A.K. Yadav, S. Chandel, Tilt angle optimization to maximize incident solar radiation: A review. Renew. Sustain. Energy Rev. **23**, 503–513 (2013)
17. M.A. Will, R.K. Ko, A guide to homomorphic encryption, in *The Cloud Security Ecosystem*, ed. by R. Ko, K.-K. R. Choo (Syngress, Boston, 2015), pp. 101–127
18. M.A. Will, B. Nicholson, M. Tiehuis, R.K. Ko, Secure voting in the cloud using homomorphic encryption and mobile agents, in *2015 International Conference on Cloud Computing Research and Innovation (ICCCRI)*, pp. 173–184 (2015)
19. M.A. Will, R.K. Ko, I.H. Witten, Privacy preserving computation by fragmenting individual bits and distributing gates, in *2016 IEEE Trustcom/BigDataSE/ISPA*, pp. 900–908 (2016)
20. Y. Zhang, G. Bai, X. Li, C. Curtis, C. Chen, R.K.L. Ko, Privcoll: Practical privacy-preserving collaborative machine learning, in *Computer Security – ESORICS 2020*, ed. by L. Chen, N. Li, K. Liang, S. Schneider (Springer International Publishing, Cham, 2020), pp. 399–418
21. Y. Zhang, G. Bai, X. Li, C. Curtis, C. Chen, R.K.L. Ko, Privacy-preserving gradient descent for distributed genome-wide analysis, in *Computer Security – ESORICS 2021*, ed. by E. Bertino, H. Shulman, M. Waidner (Springer International Publishing, Cham, 2021), pp. 395–416
22. C. Dwork, Differential privacy: A survey of results, in *International Conference on Theory and Applications of Models of Computation* (Springer, 2008), pp. 1–19
23. M. Abadi, A. Chu, I. Goodfellow, H.B. McMahan, I. Mironov, K. Talwar, L. Zhang, Deep learning with differential privacy, in *Proceedings of the 2016 ACM SIGSAC Conference on Computer and Communications Security*, pp. 308–318 (2016)
24. C. Dwork, F. McSherry, K. Nissim, A. Smith, Calibrating noise to sensitivity in private data analysis, in *Theory of Cryptography Conference* (Springer, 2006), pp. 265–284

25. F. McSherry, K. Talwar, Mechanism design via differential privacy, in *48th Annual IEEE Symposium on Foundations of Computer Science (FOCS'07)* (IEEE, 2007), pp. 94–103
26. T. Zhu, G. Li, W. Zhou, S.Y. Philip, Differentially private data publishing and analysis: A survey. IEEE Trans. Knowl. Data Eng. **29**(8), 1619–1638 (2017)
27. Q. Yang, Y. Liu, T. Chen, Y. Tong, Federated machine learning: Concept and applications. ACM Trans. Intell. Syst. Technol. (TIST) **10**(2), 1–19 (2019)
28. P. Kairouz, H.B. McMahan, B. Avent, A. Bellet, M. Bennis, A.N. Bhagoji, K. Bonawitz, Z. Charles, G. Cormode, R. Cummings, et al., Advances and open problems in federated learning. Preprint (2019). arXiv:1912.04977
29. B. McMahan, E. Moore, D. Ramage, S. Hampson, B. A. y. Arcas, Communication-efficient learning of deep networks from decentralized data, in *Artificial Intelligence and Statistics* (PMLR, 2017), pp. 1273–1282
30. A. Acar, H. Aksu, A.S. Uluagac, M. Conti, A survey on homomorphic encryption schemes: Theory and implementation. ACM Comput. Surv. (CSUR) **51**(4), 1–35 (2018)
31. K. Abdulla, J. De Hoog, V. Muenzel, F. Suits, K. Steer, A. Wirth, S. Halgamuge, Optimal operation of energy storage systems considering forecasts and battery degradation. IEEE Trans. Smart Grid **9**(3), 2086–2096 (2016)
32. B.O. Bilal, V. Sambou, P. Ndiaye, C. Kébé, M. Ndongo, Optimal design of a hybrid solar–wind-battery system using the minimization of the annualized cost system and the minimization of the loss of power supply probability (LPSP). Renewable Energy **35**(10), 2388–2390 (2010)
33. B.S. Borowy, Z.M. Salameh, Methodology for optimally sizing the combination of a battery bank and PV array in a wind/PV hybrid system. IEEE Trans. Energy Convers. **11**(2), 367–375 (1996)
34. R. Bean, H. Khan, Using solar and load predictions in battery scheduling at the residential level, in *Proceedings of the 8th Solar Integration Workshop, Stockholm 2018* (2018)
35. L. Zhu, S. Han, Deep leakage from gradients, in *Federated Learning* (Springer, 2020), pp. 17–31
36. V. Shejwalkar, A. Houmansadr, Manipulating the byzantine: Optimizing model poisoning attacks and defenses for federated learning. Internet Society, 18 (2021)
37. K. Wei, J. Li, M. Ding, C. Ma, H.H. Yang, F. Farokhi, S. Jin, T.Q. Quek, H.V. Poor, Federated learning with differential privacy: Algorithms and performance analysis. IEEE Trans. Inf. Forensics Secur. **15**, 3454–3469 (2020)
38. R. Gelaro, W. McCarty, M.J. Suárez, R. Todling, A. Molod, L. Takacs, C.A. Randles, A. Darmenov, M.G. Bosilovich, R. Reichle, et al., The modern-era retrospective analysis for research and applications, version 2 (merra-2). J. Climate **30**(14), 5419–5454 (2017)
39. C. Dwork, A. Roth, et al., The algorithmic foundations of differential privacy. Found. Trends Theor. Comput. Sci. **9**(3-4), 211–407 (2014)

Part VI
Cyber-Physical Systems, Artificial Intelligence, and Software Applications Security

Non-stationary Watermark-Based Attack Detection to Protect Cyber-Physical Control Systems

Jose Rubio-Hernan, Luca De Cicco, and Joaquin Garcia-Alfaro

1 Introduction

Nowadays, an ever increasing number of companies and industrial facilities require to access critical data from any location, ensuring the control of both data and processes. This need makes the combination of network security and industrial control security a key research topic. Fields involved in this research area are: (1) Information and Communications Technology (ICT), which encompasses the control of computer networks and communication; (2) traditional cybersecurity, focused on creating detection techniques and countermeasures against attacks in the cyber domain; (3) Industrial Control Systems (ICS), focused on designing controllers to make the physical processes behave in such a way that specific static and dynamic performance metrics are met; and (4) safety in industrial processes, focused on methodologies to avoid failures and accidents in the process.

We define the terminology used throughout this chapter: (i) the terms *cyber* and *physical* are used to refer to second upper layers in information and communication technologies and control systems, respectively; (ii) *cyber-physical systems* are the new generation of systems that combine cyber and physical components using data in the digital and continuous domain [33]; (iii) *Networked Control Systems*, which are a subset of cyber-physical systems dedicated to industrial control systems; and (iv) *Control Systems*, defined as an interconnection of components that form a physical system (often referred to as *factories* or *physical environment*) designed to provide a given desired response under a control law. We focus specifically on

J. Rubio-Hernan (✉) · J. Garcia-Alfaro
Télécom SudParis, SAMOVAR, Institut Polytechnique de Paris, Palaiseau, France
e-mail: jose.rubio_hernan@telecom-sudparis.eu; joaquin.garcia_alfaro@telecom-sudparis.eu

L. De Cicco
Dipartimento di Ingegneria Elettrica e dell'Informazione, Bari, Italy
e-mail: luca.decicco@poliba.it

K. Daimi et al. (eds.), *Emerging Trends in Cybersecurity Applications*,
https://doi.org/10.1007/978-3-031-09640-2_16

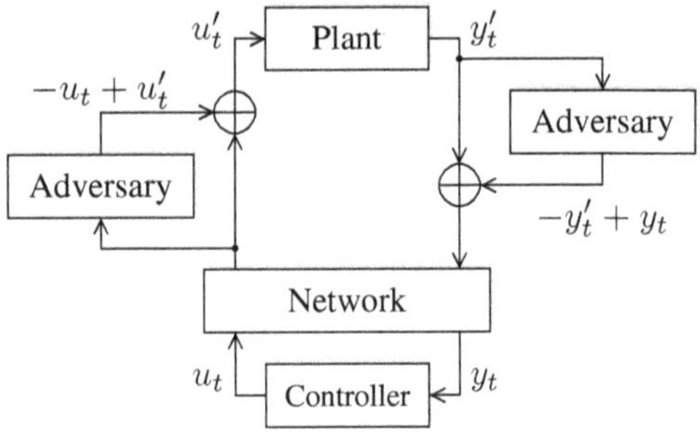

Fig. 1 Representation of a cyber-physical industrial attack against a networked control system. u_t and y_t represent the correct input and output vectors of the system. u'_t and y'_t represent the attack vectors

closed-loop control systems that are control system requiring feedback from sensors to continuously compute the control actions in order to reach the control goals.

The security of cyber-physical systems is attracting the attention of the research community [11], especially after the *StuxNet* malware [14] revealed the potential of security attacks against such systems.

Several authors have investigated the requirements for addressing emerging security issues when designing security mechanisms for cyber-physical systems. In [8], Cardenas et al. define the security issues in these systems by analyzing the problem separately, first from an information security perspective, and then by examining specific control issues. In [9], Cardenas et al. describe for the first time the difference between the security of conventional enterprise networks, and the security of cyber-physical systems. Figure 1 shows how adversaries conducting a cyber-physical attack can be represented as a block diagram scheme, a classical representation employed in the control systems community. The symbol $\bigoplus$ in the figure represents a *summing junction*, i.e., a linear element that outputs the algebraic sum of a number of input signals. The figure represents a closed-loop control system and how the adversaries are able to disrupt the nominal behavior of the system. Specifically, the adversaries modify the control input u_t (by inserting u'_t instead) to affect the state of the system and disrupt the normal operating conditions. Adversaries do not need the knowledge of the system's process model. However, access to all sensors (i.e., all components of the y_t vector) or communication protocols it is required to perform an attack, i.e., to be able to insert the correct y_t vector in place of the y'_t vector generated due to the malicious u'_t data. This type of adversary is then undetectable with a detector that only checks for faulty measurements.

From a cyber perspective, the Supervisory Control and Data Acquisition (SCADA) technologies are used to control industrial environments (such as power distribution, or transportation systems). In addition, protocols based on network-wide control systems need to cover control rules such as delays and anomalies [6]. Indeed, most industrial control protocols (e.g., MODBUS, DNP3, AGA-12, PROFINET, and EtherNet/IP) are designed to provide system safety, but not information security across the network. However, there are protocols with security extensions. AGA-12 uses cryptography to add integrity and confidentiality protection, but with a high deployment cost. DNP3 can be equipped with an extension called DNP3-SA (Secure Authentication, as of the fifth IEEE-1815-2012 release), adding message integrity and authentication to DNP3. However, current cyber-physical systems use these protocols over TCP/IP or UDP/IP (e.g., MODBUS, DNP3 and PROFINET over TCP, EtherNet/IP over TCP or UDP). In this case, there are just security mechanisms up to the application layer, such as TLS and IPSec.

At the application layer, we also find protocols that have evolved. For example, PROFINET can be complemented with a new layer, PROFIsafe, which is designed to provide security, thus protecting the protocol against malfunction (e.g., transmission errors). Unfortunately, this does not provide security against intentional malicious acts [1]. It should be noted that most protocols that run over Ethernet or TCP/IP are modifications of serial protocols that do not provide security. Although the transport and network layers can provide a certain level of security, these mechanisms are not sufficient to provide protection for control data. To fully address the problem of control data protection, it is necessary to add new cyber-physical solutions to these protocols.

In the literature, some authors have proposed the use of a physical attestation at the cyber layer [28], a physical signature sent by the cyber layer to the physical layer in order to verify the correct behavior of the physical processes [24], or a signature on the physical data in order to avoid identifying the real value of the data and secure the communication [4]. In [2], Arvani et al. describe a detection method using discrete wavelet signal transformation. Do et al. [13] investigate strategies for handling cyber-physical attacks using statistical detection methods. These proposals are only valid when adversaries perform a replay or integrity attack without the ability to gain knowledge about physical processes.

1.1 Objectives and Contributions

This chapter focuses on the security between the cyber layer and the physical layer, forming cyber-physical systems. We start with a security analysis based on theoretical detection mechanisms proposed by Mo and Sinopoli [22] and Chabukswar et al. [10], who study the use of stationary signatures to detect attacks to cyber-physical systems. Continuing the approach of signature-based detectors, we propose a new detection mechanism using *non-stationary signatures* to cover a larger number of threats. This new mechanism increases the attack detection rate while keeping the same performance cost as the previous approach. Next, we analyze the limitations of the new proposal. This analysis leads us to improve the detection mechanism, as well as to create a new control and security strategy capable of avoiding the security weaknesses generated by the cyber layer's membership in the physical and control domain.

Current security for cyber-physical systems focuses on either cyber adversaries or physical adversaries, but not both. For this reason, the new security challenges in cyber-physical systems require the analysis of control strategies and security mechanisms to detect attacks. This analysis will allow us to create a new control and security strategy, improving the existing detection mechanisms in the literature, in order to secure the cyber and physical layer against cyber-physical adversaries.

Contributions of the Chapter The mechanisms proposed in this chapter allow us to detect threats conducted by cyber-physical adversaries. In addition, we analyze the different cyber-physical adversaries, and we classify these adversaries according to their ability to obtain the correct behavior—or *dynamical model*—of the system. Based on this classification, we can define two different types of cyber-physical adversaries: *parametric* and *non-parametric* cyber-physical adversaries. We also address the shortcomings of centralized detection mechanisms by proposing a decentralized detection strategy that increases the robustness of the system against attacks. Then, we define a distributed detection mechanism that increases the robustness against cyber-physical attacks. Finally, we build SCADA simulations and testbeds to validate the new detection models.

This chapter is organized as follow. Section 2 provides additional preliminaries on watermark-based detector mechanisms, reports their main limitations and presents our extended approach. Section 3 presents a distributed detection mechanism, as an evolution of existing watermark-based detectors. Section 4 reports a training cyber-physical testbed to validate the mechanisms presented in this chapter. Section 5 outlines some future research lines that we believe one worth being pursued. Section 6 concludes the chapter.

2 Dynamic Challenge-Response Authentication Scheme

In this section, our focus is on integrity issues due to the interconnection between the *cyber* and *physical* domains in control systems across the network. Specifically, we focus on adapting anomaly detection mechanisms, existing in the physical domain, to handle attacks as well.

The authentication scheme proposed by Mo et al. [23] relies on the adaptation of a real-time anomaly detector based on a *linear* and *time-invariant* model of the system. This scheme, built employing *Kalman Filters* as *linear quadratic estimators (LQE)*, and *linear quadratic regulators (LQR)*, generates authentication signatures to protect the integrity of the physical measurements communicated across the network, because if the messages carrying these measurements are not protected, malicious actions can be taken to mislead the system. However, we show that the detection scheme proposed by Mo et al. only works against certain integrity attacks. We present two new models of adversaries that can evade this detector. These adversaries are classified according to the algorithm used to obtain knowledge of the system dynamics to carry out the attack. Then, we revisit the mechanism proposed in [21, 24] and evaluate its performance against the two new adversary models presented in this section. We adapt this detection scheme to handle uncovered limitations, validating the resulting approach with numerical simulations.

2.1 Problem Formulation

This chapter focuses on physical environments of industrial control systems that can be mathematically modeled as discrete linear time-invariant (LTI) systems. It is worth mentioning that a mathematical model provides a rigorous way to describe the dynamic behavior of a given system. One well-known way to describe the dynamics of such a class of systems is the state-space formulation:

$$x_{t+1} = Ax_t + Bu_t + w_t \tag{1}$$

$$y_t = Cx_t + v_t \tag{2}$$

where $x_t \in \mathbb{R}^n$ is the vector of *state variables*, $u_t \in \mathbb{R}^p$ is the control signal, $y_t \in \mathbb{R}^m$ is the output of the system, and $w_t \in \mathbb{R}^n$ and $v_t \in \mathbb{R}^m$ are the *process noise* and the *sensor measurement noise*, respectively. The noises are assumed to be Gaussian white noise with zero mean and co-variance Q, i.e., $w_t \sim N(0, Q)$ and R, i.e., $v_t \sim N(0, R)$. Moreover, $A \in \mathbb{R}^{n\times n}$, $B \in \mathbb{R}^{n\times p}$ and $C \in \mathbb{R}^{m\times n}$ are, respectively, denoted as the state matrix, the input matrix, and the output matrix.

For the class of systems defined above, one of the most widely used control methods is the *Linear Quadratic Gaussian* (LQG) control. This control consists of two components that can be designed independently:

1. A *Kalman filter* that produces an optimal state estimate $\hat{x}_t$ of the state x_t based on the obtained noisy measurements y_t.
2. A *Linear Quadratic Regulator (LQR)* that provides the control law u_t that solves the LQR problem, based on the state estimate $\hat{x}_t$.

2.2 Detector Based on a Stationary Signature

This section briefly describes the detection scheme proposed by Mo et al. [21, 24]. The procedure applies to physical environments that follow a discrete-time LTI model, and are controlled by an LQG controller (cf. Sect. 2.1). Before presenting the detection scheme, we provide a definition of the adversary model considered in [21, 24]:

Definition 2.1 An attacker who has the ability to listen to all messages containing the outputs of the y_t sensor, and inject messages with a y'_t signal to carry out malicious actions, is defined as a *cyber adversary*.

Remark 2.1 It is important to note that the definition given above assumes that the attacker does not possess (or attempt to gather) knowledge of the system model. For this reason, we refer to such an attacker as a cyber adversary.

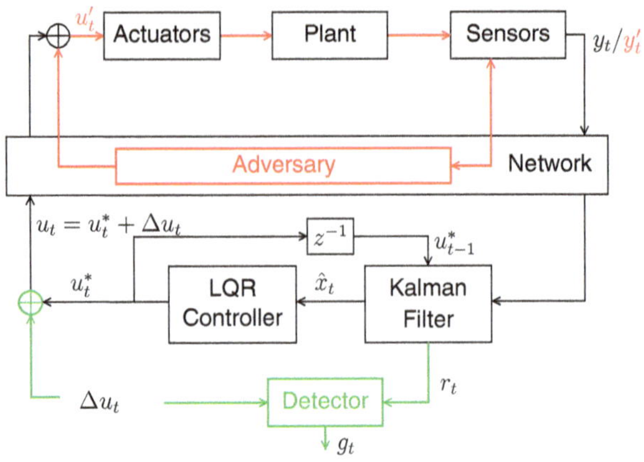

Fig. 2 Signature-based protection in cyber-physical systems [24]

In what follows, we will call u_t^* the output of the LQR controller and u_t the control input that is sent to the physical environment (cf. Eq. (1)). The idea is to superimpose on the optimal control law u_t^* a signature signal $\Delta u_t \in \mathbb{R}^p$ which serves as an authentication signal. Thus, the control input u_t is given by:

$$u_t = u_t^* + \Delta u_t \tag{3}$$

The signature signal is a random Gaussian signal with zero mean, which is independent of both w_t process noise and v_t measurement noise. This authentication signature is expected to detect the repetition and integrity attacks generated by the cyber adversary defined above. Given that the optimal control law u_t^* is equipped with the authentication signal Δu_t, a *detector*—physically co-located with the controller—can be designed with the objective of generating alarms when an attack occurs. For this purpose, Mo et al. [21, 24] propose to use a χ^2 detector, which is a well-known class of real-time anomaly detectors classically used for anomaly detection in control systems [7], with the objective of attack detection. Figure 2 shows the global control system equipped with the attack detector proposed in [21, 24].

An *alarm signal* g_t is computed based on the residuals $r_t = y_t - C\hat{x}_{t|t-1}$ generated by the estimator. Then, g_t is compared to a threshold, γ, to decide if the system is in a normal state. The threshold is set to minimize false alarms [21, 24]. The alarm signal g_t is calculated as follows:

$$g_t = \sum_{i=t-w+1}^{t} (y_i - C\hat{x}_{i|i-1})^T \mathcal{P}^{-1} (y_i - C\hat{x}_{i|i-1}) \tag{4}$$

where w is the detection window size and $\mathcal{P}$ is the co-variance of an independent and identically distributed Gaussian input signal from the sensors.

The system is considered unattacked if $g_t < \gamma$. The system is otherwise considered as attacked and the sensor generates an alarm.

2.3 Cyber-Physical Adversaries

In this section, we introduce an improved adversary that is aware that the system uses the χ^2 detector presented above. Since the detector is based on a stationary signature signal Δu_t, we show that an adversary that is able to extract the model of the system from the control law u_t and the sensor measurement y_t, is able to perform an attack while remaining undetected.

Definition 2.2 An attacker who, in addition to the capabilities of the cyber adversary, is able to listen to messages containing the controller's output (u_t) with the intent of improving his knowledge of the system model using a *parametric or non-parametric identification model*, is defined as a *cyber-physical adversary*.

Depending on how the system behavior is modeled, two different cyber-physical adversaries can be defined as follows:

Definition 2.3 An attacker who uses only the previous input and output of the system to identify the system model is defined as a *non-parametric cyber-physical adversary*.

Remark 2.2 A non-parametric cyber-physical adversary may use, for example, a finite impulse response (FIR) filter-based model identification tool to identify the model of the system [31]. In Fig. 2, the signals u_t' and y_t' are assumed to be the controller output and the sensor output, respectively, when an attack is occurring. We denote by $\Delta u'$ the signature estimated by the non-parametric cyber-physical adversary.

Definition 2.4 An attacker capable of estimating system parameters using input and output data to fool the controller's detector is defined as a parametric cyber-physical adversary.

Remark 2.3 A parametric cyber-physical adversary is capable of estimating system parameters using input and output data to mislead the controller's detector. This adversary can use, for example, an ARX model (autoregressive with exogenous input) or an ARMAX model (autoregressive with dynamic mean and exogenous input) to estimate the dynamics of the system [27].

We assume that the main constraint of this adversary is the energy expended to listen and analyze the communication data, i.e., the number of samples needed to obtain the system model parameters.

2.4 Multi-Signature Based Detector

In the previous sections, we defined three types of adversaries that use different control system vulnerabilities to perform attacks; cyber adversaries, non-parametric cyber-physical adversaries, and parametric cyber-physical adversaries. In this section, we propose a detection scheme that extends the one presented in [21, 24], to detect cyber-physical adversaries. We also study the performance loss of the new detection scheme compared to the one presented in [21, 24].

The goal of our new detection scheme is to increase the difficulty of recovering the authentication signature Δu_t from the control signal u_t, so that the probability of detecting an attack from a non-parametric cyber-physical adversary can be increased. We assume that the attacked control system uses exactly the same type of controllers and detection strategy presented in Sects. 2.1 and 2.2. The only difference in the proposed detection scheme is how the signature signal, Δu_t, is generated. The control input u_t, as in the case of the detection scheme presented in Sect. 2.2, is computed as the superposition of the optimal control signal u_t^* produced by the LQR controller and a signal of several signatures, Δu_t. The idea is to build the authentication signature signal by alternating between N different and independent processes with different co-variance and mean. More precisely, the non-stationary signature, Δu_t, is obtained by changing periodically, with a period T, between N signals $\Delta u^{(i)}$, with $i \in \mathcal{I} = \{0, 1, \ldots, N-1\}$, extracted by different stochastic processes. Therefore, the signature signal Δu_t can be formalized as follows:

$$\Delta u_t = \Delta u_t^{(s(t,T))} \tag{5}$$

where $s : \mathbb{N} \times \mathbb{R} \rightarrow \mathcal{I}$ is a static function that denotes the sample of time, t, and the switching period T to an element of the index set, $\mathcal{I}$, defined as follows:

$$s(t, T) = \left\lfloor \frac{1}{T} \mod (t, NT) \right\rfloor \tag{6}$$

where $mod(x, y)$ is the modulo operator and $\lfloor \cdot \rfloor$ is the default integer function.

Using the proposed signature (cf. Eq. (5)), we now have a suitable adaptive protection mechanism with two main configurable parameters; the number of distributions N and the switching frequency $f = 1/T$. Note that the original signature signal described in Sect. 2.1 is recovered when $f \rightarrow 0$ and when $\Delta u_t^{(0)}$ is a Gaussian and stationary process with zero mean.

2.4.1 Validation Against Non-parametric Cyber-Physical Adversaries

This section validates the previously proposed detection scheme with numerical simulations. In particular, we want to show that the proposed signature signal is able to detect non-parametric cyber-physical adversaries (cf. Sect. 2.3) with a higher detection rate compared to that obtained with the signature proposed in [21, 24]. The simulation is based on Matlab and Simulink models of a plant, as well as the models of the non-parametric cyber-physical adversaries. We use three different (i.e., $N = 3$) randomly switched distributions: a Gaussian distribution, a Rician distribution, and a Rayleigh distribution.

To quantify the effectiveness of the proposed detection scheme, we compute the detection rate as a function of the switching frequency. In particular, for each frequency f considered, we perform 200 Monte Carlo simulations (with randomly generated system parameters) in the case of a non-parametric cyber-physical adversary and a cyber adversary, and compute the cumulative distribution function (CDF) of the detection rate.

We first compare the performance obtained with the non-stationary signature-based detection strategy proposed in this section with that proposed in [21, 24] in the case of a cyber adversary and a non-parametric cyber-physical adversary. We consider here two switching frequencies $f_L = 0.05$Hz (change the signature after 20 steps) and $f_H = 0.14$Hz (change the signature after 7 steps). We verify that the proposed detection strategy in [21, 24], as mentioned before, can detect a cyberattack, but performs poorly when a cyber-physical adversary attacks the system. Nevertheless, the proposed detection strategy based on a non-stationary signature is able to provide a higher detection rate. In particular, we notice that the detector using higher switching frequency f_H provides better performance compared to using the lower switching frequency f_L.

Efficiently Validation Above, we validated the non-stationary signature detector using a static function $\mathcal{I}$ to define the multi-signature. Hereafter, we present the results and validations obtained for a system with the same performance loss between the detector using a stationary signature and the one using a non-stationary signature where this non-stationary signature is generated from a non-static function, $\mathcal{I}_d$. In this simulation, both detectors have a performance loss of 30%, ΔJ, compared to the optimal cost. In addition, the signature uses a dynamic function to define non-stationarity. We find that using the multi-signature (or non-stationary signature) with the same performance loss as the stationary signature, the detection ratio increases as the switching frequency varies in the range $[0, 0.14]$ Hz, where $f = 0$ is the detector of the stationary signature. We confirm that the multi-signature performance increases up to $f = 0.14$ Hz where we observe a peak before the detection ratio stabilizes. This peak before the stability indicates that $f = 0.14$ is the resonance frequency of the system. In the next section, we extend the analysis to the case of parametric cyber-physical adversaries. In addition, we test systems of different order concluding that the detection rate increases with the complexity of the system.

2.4.2 Validation Against Parametric Cyber-Physical Adversaries

Previously, we have seen how the multi-signature detector is able to detect non-parametric cyber and cyber-physical adversaries. Hereafter, we extend the study to the case of parametric cyber-physical adversaries (cf. Definition 2.4). We recall that parametric cyber-physical adversaries are able to identify the parameters of the system model from the input and output signals of the plant (physical environment). A parametric cyber-physical adversary can obtain the system model with high accuracy if control commands and sensor measurements are accessible. For example, using the signature characteristic, a parametric cyber-physics adversary can use an ARX (autoregressive with exogenous input) model to define the system.

Similarly to the previous validation, we analyze the detection ratio for 200 Monte Carlo simulations using 25 order systems, against seven different parametric cyber-physical adversaries. The assumed window size is $\hat{T} = 300$. If the adversaries use a model of the system with the correct order, the detection ratio is about 8%. The set of system orders where the detection ratio does not increase drastically is $[18, 28]$. Otherwise, the probability of detecting the adversary is high. Next, we analyze the detection ratio of the same system, against a parametric cyber-physical adversary with different window sizes (125, 150, 200, 250, and 300), and with the correct system order. We conclude with the results obtained that the window size used by the adversary is inversely proportional to the detection ratio.

Remark 2.4 A parametric cyber-physical adversary is able to obtain the system model, $H(z)$, and mislead the controller by listening to control inputs and sensor measurements. The probability of being detected is equivalent to the probability of obtaining an erroneous model. This probability is directly proportional to the order of the system; and inversely proportional to the size of the window for listening to the data channel.

From Remark 2.4 it follows that, if we consider the real system as a black box, a misidentification of the system depends on the order of the system chosen by the adversaries to recreate the model of the system, as well as the number of samples listened to and the size of the window used by the adversaries to recompute the parameters of the target system. This can be quantified using the Mean Square Error (MSE) [5, 20]. In summary, the probability of obtaining the correct model of the targeted system is directly proportional to the order chosen by the adversaries to generate the model and inversely proportional to the number of samples recovered. The computational cost for adversaries is directly proportional to the order of the system, as such adversaries must increase the order of the model, as well as the window size in order to minimize the MSE. Therefore, the number of samples listened to before performing the attack, and the system order chosen by the adversaries are the two main parameters to escape detection.

2.5 Discussion

We have shown in this section that the detection strategy of classical stationary signature signal (single-watermark approach) is not robust enough from a security perspective. In particular, we have shown that an adversary that learns about the system model is able to separate the watermark from the control signal, to evade detection and successfully attack the system. Indeed, we have shown a quantitative validation that the approach only detects *cyber adversaries*. Then, we have presented an adaptive detection scheme based on multiple signatures with two main configurable parameters: the number of distributions and the switching frequency. The main idea of the new scheme is to use multiple watermark distributions and non-stationary identification signals. The new proposal succeeds in correctly detecting non-parametric cyber-physical adversaries, under the assumption that the signature distributions change frequently. The rationale is that, the non-parametric adversary has little chances of acquiring the necessary information to acquire the watermark and bypass the detector. Moreover, we have confirmed that the multi-watermark detector approach, with the same performance loss as the single-watermark approach, has a higher detection ratio (cf. [29] and citations thereof for further details).

As we have seen, smarter cyber-physical adversaries able to dynamically adapt their behavior can successfully evade detection and disrupt systems whenever the appropriate parameters are met. A more detailed analysis of the strategy used by this new class of adversaries is detailed next. An alternative detection strategy is also proposed. The new detection strategy successfully mitigates the effects of the adversaries uncovered in our analysis.

3 Adaptive Detection Based on Control Theory

As we have shown in Sect. 2, the use of inadequate cyber-physical security mechanisms can have a negative effect in industrial cyber-physical systems [14, 19, 32]. These new systems need the collaboration of a very wide range of disciplines to solve the challenges in terms of autonomy, reliability, usability, functionality, and cybersecurity [3]. Hereafter, we focus on the use of control-theoretic solutions to detect attacks against cyber-physical systems. Traditional literature proposes the use of control strategies to maintain, for example, the closed-loop performance of the system or the safety properties of a communication network connecting the distributed components of a physical system. However, the adaptation of these strategies to manage security incidents is still a challenge.

The monitoring community is actively working on adapting traditional monitoring strategies used to detect accidental flaws and errors, towards detecting malicious attacks [17, 18, 30]. Motivated by the same goals, we present a solution that complements the signature detector to cover these weaknesses. Specifically, the new solution combines the control strategies with the challenge-response strategy analyzed and improved in the previous section. This combination allows handling integrity attacks against cyber-physical systems.

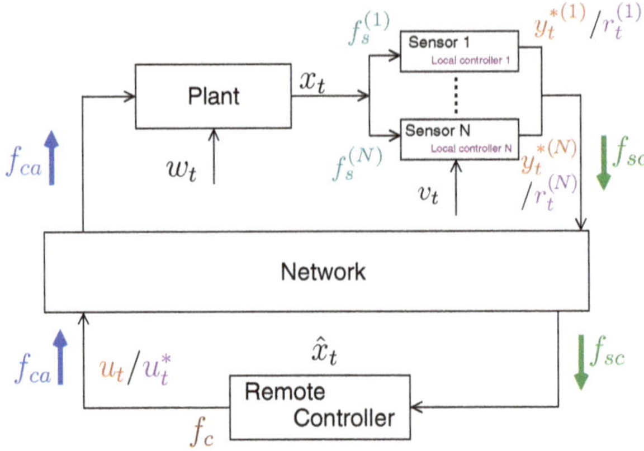

Fig. 3 Diagram of a cyber-physical system with a new security strategy based on system control

3.1 Parametric Cyber-Physical Adversaries Detection

In this section, we present a detection strategy, hereafter referred to as *Periodic and Intermittent Event-Triggered Control Watermark strategy* (PIETC-WD strategy), which aims to detect cyber and cyber-physical adversaries by complementing the strategy proposed in the previous section.

Our strategy consists of a local controller located in each sensor and a common remote controller for the whole system (distributed controller, cf. Fig. 3). The cooperation between the local controllers and the remote controller allows us to create an intrusion detection policy to capture integrity attacks (cf. Definition 3.1). The local controllers manage the dynamics of the physical environment and the remote controller manages the closed loop of the system to ensure the system against integrity attacks. Note that our new system requires an additional controller for each sensor that must have enough computational power to process the data estimates to, among other things, predict the errors between the environmental data and the estimated data. The actuators do not require additional computing power. Nevertheless, during the time between two consecutive events, they must keep the last data received by the remote controller.

Such a system requires defining communication policies among sensors, actuators, and the remote controller. We define two communication policies to ensure the system: (i) *is a periodic communication policy*, with communications between the sensors and the remote controller, with period $T_{sc} = 1/f_{sc}$, and between the remote controller and the actuators, with period $T_{ca} = 1/f_{ca}$; and, (ii) *is an intermittent communication policy*, which allows data to be sent from the sensors to the remote controller if a local controller produces an alarm. Note that T_{sc} cannot be equal to T_{ca} in order to prevent intermittent communication from taking place while periodic communication is being sent.

Definition 3.1 Periodic and Intermittent Event-Triggered Control Watermark Detector (PIETC-WD) is a detection strategy with distributed control tasks. On the one hand, the sensors monitor the system periodically, using their local controllers and signature detectors. On the other hand, the remote controller uses the estimation error received by each sensor to periodically generate the control inputs. This controller also monitors the closed-loop communication with an intermittent signature.

To execute the PIETC-WD strategy, we develop two algorithms. The first one defines the implementation of the remote controller whose input is the data sent by the sensors and the output is the set of control inputs sent to the physical system, and the alarm value. The second one shows the implementation of the local controller, placed in the sensors, whose input is the data obtained from the physical system, y_t^i, with $i \in \mathcal{I} = \{0, 1, \ldots, N-1\}$, and the output is: (i) the residual of the local controllers, r_t^i (with a challenge-response signature), if the alarm is not activated; or (ii) the value obtained by the physical system sensors, y_t^i, if the alarm is activated. In summary, these algorithms define how the remote controller handles data in order to increase the probability of detecting an attack, if the data sent from local controllers is not correct, or if data has been lost. Also, they determine how the sensors change the data sent to the remote controller if an alarm is activated by the sensors.

New Parametric Cyber-Physical Adversary To validate this strategy, we introduce a new parametric cyber-physical adversary that has knowledge of the new detection strategy, in order to evaluate it. This adversary has knowledge of the new communication policies and the existence of different signatures of the data sent from the local controllers or the remote controller. However, it does not know the co-variances of the signatures, the controller parameters used to obtain the correct error between the data, or when the remote controller forces an intermittent communication.

The new adversary can detect the correlation pattern between the inputs and outputs of the physical environment. It can force intermittent communication of sensors with malicious control inputs and deceive the remote controller with read error data to obtain the pattern. Nevertheless, this adversary is not able to know when the communication is periodic or intermittent, since the attacker does not know when the controller adds, to the control inputs, the signature that generates the intermittent communication. The intermittent communication does not change the frequency of communication between the remote controller and the actuators, but produces intermittent communication between the sensors and the remote controller, which is necessary to verify the closed loop.

Using the PIETC-WD strategy, this type of adversary is detected by the localized controllers in the sensors, and by the remote controller when it checks the behavior of the closed loop. These adversaries cannot avoid the alarm in the sensors (local controllers). Nevertheless, the attackers can interrupt the communication between the sensors and the remote controller by misleading the remote controller with the correct residuals (e.g., with replayed residuals). Furthermore, in order to avoid generating an alarm in the remote controller, adversaries can switch between sending the measurement from the sensors or the residuals. However, they then have a high probability of being detected. We validate the PIETC-WD strategy against the new parametric cyber-physical adversaries in the next section.

3.2 Numerical Use Case

This section presents a practical use case where the PIETC-WD strategy proposed in the previous sections could be used in the real world. This use case is based on a chemical plant. This plant has several sensors with local controllers, actuators, and a remote controller that manages all sensor and actuator measurements. The sensors used in this use case send pressure, temperature, and density information. This information is sent when an event generates an alert in a sensor, as well as periodically to indicate system behavior to the remote controller. This installation must be controlled periodically since, if the system receives incorrect or malicious control inputs capable of disturbing the system for ten consecutive periodic samples, it could reach a critical state.

To prevent an adversary from putting the system in a critical state, we use our detector strategy (PIETC-WD) with a remote controller signature management policy defined as follows:

- The controller's watermark uses a policy based on a probability to add the watermark in a specific window of samples. The windows of samples, in this use case, is assumed equal to five. For each window, the probability to add the watermark at each sample is $\beta = 50\%$. The system is able to produce $2^5 = 32$ different sequences with the same probability to be generated, $\theta = 1/2^5$. Nevertheless, if the system send five consecutive samples without watermark, three more samples are used to add a watermark to the control input until a new control sequence starts. These three samples added to the original control sequence add $2^3 = 8$ more sequences when a window of samples has not a watermark. The three last samples have the following probability to add the watermark:
  - The probability to add a watermark in the sixth sample is 60%.
  - The probability to add a watermark in the seventh sample is 50% if a watermark is added in the sixth sample. Otherwise, if a watermark is not added, the probability is 60%.
  - The probability to add a watermark in the eighth sample is 50%, if a watermark is added in the sixth or seventh sample. Otherwise, the probability is 60%.

Figure 4 shows the results of 200 Monte Carlo simulations using the above use case and controller signature policy (local and remote) against the cyber and cyber-physical adversary. These results show that the detection ratio is about 97% against the new parametric cyber-physical adversary and more than 99% against the other

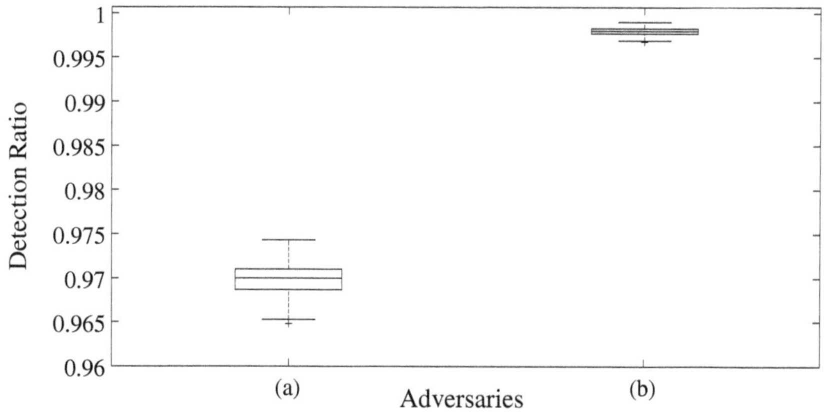

Fig. 4 Detection ratio function with respect to the PIETC-WD strategy with a defined controller's watermark policy; (**a**) against the new parametric cyber-physical adversary; and (**b**) against cyber or other cyber-physical adversaries

cyber and cyber-physical adversaries using the PIETC-WD strategy with a correct policy for the remote controller signature.

4 Experimental Testbed for the Detection of Cyber-Physical Attacks

Experimental testing is essential for the study and analysis of ongoing threats against cyber-physical systems. The research presented in this section addresses some actions to develop a replicable and affordable cyber-physical testbed for training and research. In this framework, our goal is to put into practice the theoretical solutions developed in the previous sections. To achieve this goal, we implement the solutions in realistic scenarios to analyze their effectiveness against intentional attacks. Specifically, we assume cyber-physical environments operated by SCADA technologies and industrial control protocols. We focus on two representative protocols widely used in industry: MODBUS and DNP3 [12, 25]. Both protocols have versions over TCP. This allows us to emulate cyber-physical environments on shared network infrastructures. We assume a Master/Slave design, which primarily dictates that slaves do not initialize any communication unless a master requests it (cf. Fig. 5). One of our goals has been to combine these two protocols, both to allow flexibility and support for multiple devices with MODBUS as well as the security enhancements included in DNP3's functionality. In addition, the cyber-physical detection mechanisms based on the challenge-response strategies proposed in Sect. 2 are included in our SCADA testbed. Similarly, we integrate the control strategy proposed in Sect. 3 to experiment and analyze its actual performance. To

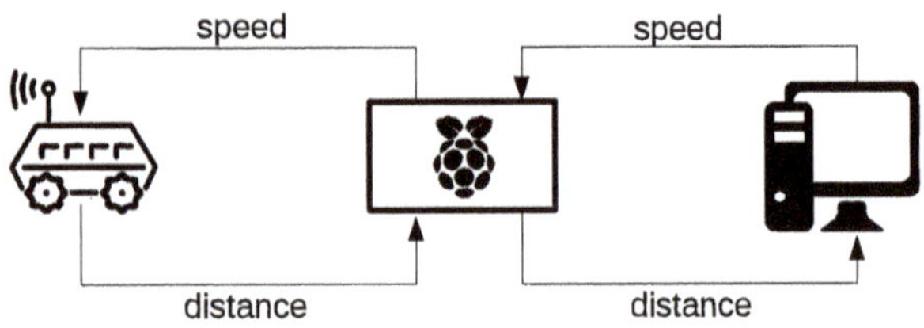

Fig. 5 Test scenario overview

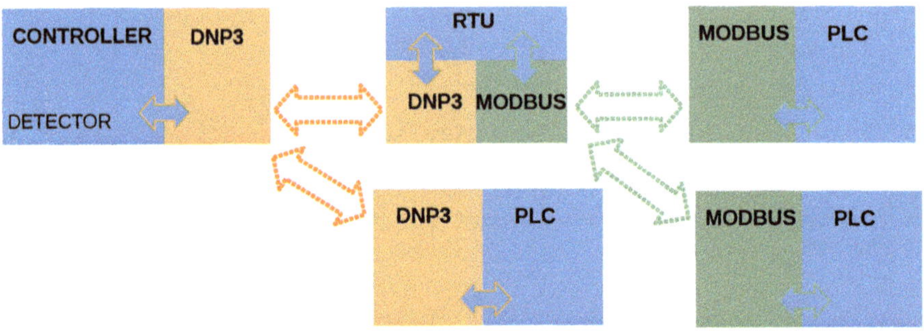

Fig. 6 Abstract architecture overview

complete the testbed, a set of attack scenarios are designed and developed to test attacks against the emulated environment. These scenarios focus on attacking the MODBUS segments of our architecture. The final goal is to analyze the effectiveness of the new security methods implemented on the emulated environment and under the application of some attack models.

4.1 Architecture Design

The proposed architecture of our cyber-physical testbed works as follows. All the elements of the system (controller, sensors, and actuators) can be distributed on several nodes in a shared network combining DNP3 and MODBUS protocols (cf. Fig. 6. Similarly, one or more elements can be integrated into a single device. From a software perspective, the controller never connects directly to the sensors. Instead, it is integrated into the architecture as a Programmable Logic Controller (PLC), with possible connections to other intermediate nodes. Such nodes can translate the controller commands between different protocols (e.g., MODBUS or DNP3). This architecture is capable of handling multiple industrial protocols and connecting to additional SCADA elements, such as PLCs and Remote Terminal Units (RTUs). To evolve the architecture into a complete test bench, new elements can be included in the system, such as additional RTU nodes.

4.2 Adversary Implementation

After getting the architecture up and running, the next requirement is to implement the adversaries reported in the previous sections. To develop these scenarios, we use a common attacker model. It implements most of the underlying capabilities of the adversaries and can be extended to implement the more specific adversaries. For example, we assume that attackers can intercept all communication exchanges between endpoints, and hence modify, store, and analyze what can be replayed to forge false data to and from communication channels. Since this is done using a testbed instead of numerical simulations, all real-life limitations are applied to the attacker. The technique *ARP poisoning* [26] is used by the attacker to intercept the channels and listen to the communications. The attacker has a passive and active mode of operation. During passive mode, the attacker only observes, processes, and analyzes the data without modifying the information contained in the message payload. Ethernet header data, such as MAC addresses, are nevertheless modified due to the compromise of the ARP tables. During the active mode, the attacker starts injecting data into the hijacked communication. This injection, depending on the attacker's model, can be a replayed packet or generated by the attacker:

Replay Attack Attackers use the *ARP poisoning* technique to start listening to the connection (passive mode, from the physical layer perspective). After enough data has been recorded, the *active mode* begins. The attackers inject the old captured data. Before starting to disrupt the physical system, the attacker performs the replay attack between the sensors and the controller. Once the packets are replayed to the controller, the physical system is disrupted by falsifying data between the controller and the automata.

Injection Attack Before launching this attack, the attacker listens to connections using the passive mode *physical layer*, and analyzes the data to determine the system dynamics. This evades the signature-based authentication detector. Once the system model is deduced, the attacker starts injecting correct data into the communication channel to bypass the authentication signature. To evade the detector, the attacker calculates the effect of the signature in the system and attempts to override the detector's ability to detect changes in the return signal. Two different techniques are implemented: (1) a non-parametric adaptive filter, in order to implement the evasion technique presented in Sect. 2, *is a non-parametric cyber-physical attack*; and (2) autoregressive methods, such as ARX and ARMAX, in order to implement the evasion technique presented in Sect. 3, *is a parametric cyber-physical attack.*

The challenge of implementing these two adversaries is to synchronize the output of the adversaries when starting the attack phase. Since the target of the adversaries is to take control of the system, the data sent to the controller must be able to match the current state of the system and the correct signature correlation to avoid detection.

4.3 Attacks and Anomalies Detection

As explained in Sect. 2.2, the metric g_t quantifies the difference between the output of the parametric model and the actual output of the system. An increase in g_t means that the system does not behave or respond to the signature as expected. Therefore, the system is at risk of being attacked. The value of g_t is computed for each iteration and compared to the values of some previous iterations. To eliminate false positives, we implemented an algorithm in the remote controller to separate flaws from serious attacks or failures. Algorithm 1 is used to alert the operator only when actual intervention is required, separating flaws (e.g., latency or inaccuracy events on the sensor) from intentional attacks. For each sample received, the remote controller analyzes g_t. If g_t consecutively exceeds (more than the duration of a pre-defined *window*) a given threshold, it triggers an $alert$. Algorithm 1 is composed of four main functions, to differentiate between faults, accidents, or attacks. They work as follows:

- *alarm_propagation*: this function creates a potential alarm if the detector value, g_t, is over the threshold. This alarm could be a fault peak or an attack[1].
- *alert_propagation*: this function creates an alert if the *DR_value* is below a minor threshold, *min_R*, or above a maximum threshold, *max_R*. The value of *DR_value* is the difference between the detector's value at instant t, and the detector's value at instant $t-1$. The function also verifies if the system has generated faults (alerts or potential alarms), during a precise period of time (denoted by *window size*). This window is the number of samples chosen from the physical system, in order to settle the detector value, g_t. It has to be big enough to minimize the number of false positives and small enough to have enough time to react under critical situations.
- *potential_risk*: this function presents a given system risk level, taking into account alerts and potential alarms; it follows traditional qualitative risk values [16], such as (1) *slow*, (2) *medium*, (3) *high*, (4) *critical*, and (5) *very critical.*
- *real_risk*: this function warns the system under the presence of a real risk, analyzing the conditions received from the other function, i.e., number of potential alarms and alerts.

Using Algorithm 1, the detector can notify potential risks, based on qualitative impact values [16]. Along with these values, it triggers alerts to the operator whenever events are likely to be intentional attacks. Reported alerts, using window size values appropriate to the specific system, are assumed to be triggered early enough, for example, before reaching the *critical level*, to allow security operators to process the information before taking necessary countermeasures, i.e., system safety is assumed to have a higher priority than security.

[1] Notice that we expressly use the term *alarms* to point out towards suspicious events; and *alerts* to point out to events likely to be associated with malicious attacks.

Algorithm 1 Fault and attack differentiation

```
procedure Detector
    alert, alarm ← false
    potential_alarm, potential_attack ← false
    window ← detector_window
    risk, potential_risk ← 0
alarm_propagation:
    if detector_value > threshold then
        potential_alarm ← true
    else
        potential_alarm ← false
    old_detector_value ← detector_value
    goto alert_propagation
alert_propagation:
    DR_value ← detector_value / old_detector_value
    if DR_value < min_R or DR_value > max_R then
        alert ← true
        if 0 < account_fault ⩽ window then
            risk_level ← risk_level + 1
        else
            potential_attack ← true
            alarm_attack ← true
    else
        alert ← false
    goto potential_risk
potential_risk:
    switch risk_level do
        case window/4: potential_risk ← potential_risk + 1
        case 3window/4: potential_risk ← potential_risk + 1
        case window: potential_risk ← potential_risk + 1
    goto real_risk
real_risk:
    if potential_attack = true then
        potential_attack ← false
        alarm_attack ← true
        risk ← risk + potential_risk
    if alarm = true or alert = true then
        account_fault ← account_fault + 1
    else
        account_fault ← 0
    if alarm_attack = true then alarm ← true
    goto alarm_propagation
```

4.4 Experimental Results

Using the previously described testbed, we analyzed the detector with the stationary signature, shown in Table 1, the non-stationary signature, shown in Table 2, and

Table 1 Experimental results using a stationary watermark

	Replay attack	Non-parametric attack	Parametric attack
False negatives	64.06%	85.20%	88.63%
False positives	0.98%	1.66%	1.35%

Table 2 Experimental results using a non-stationary watermark

	Replay attack	Non-parametric attack	Parametric attack
False negatives	62.03%	54.24%	84.61%
False positives	5.10%	3.30%	4.63%

Table 3 Detection performance results using the PIETC-WD detection strategy

	Using only the watermark-based detector	Using as well the PIETC-WD detector
Detection ratio	12.00%	75.25%
Average detection time	6.08s	6.20s
False negatives	88.60%	38.66%
False positives	1.35%	5.23%

Table 4 Detector performance results using a stationary watermark

	Replay attack	Non-parametric attack	Parametric attack
Detection ratio	40.00%	18.00%	12.00%
Average detection time	10.01s	4.89s	6.08s

Table 5 Detector performance results using a non-stationary watermark

	Replay attack	Non-parametric attack	Parametric attack
Detection ratio	60.00%	56.00%	16.00%
Average detection time	9.26s	6.27s	5.63s

the strategy defined in Sect. 3.1, shown in Table 3. Using the stationary signature, we can highlight that the replay attack is the most detectable scenario, with a detection rate of about 40%. The non-parametric attacker has a lower detection rate, of about 18%. This result is expected, as suggested by the theoretical conclusions and the simulation presented (cf. Sect. 2). The parametric attack uses the most robust system identification approach. These attacks can escape the detection process if they succeed in correctly identifying the system attributes. In terms of results, they lead to the lowest detection rate of about 12%.

We should also note that the *average detection time*, shown in Tables 4 and 5, of a replay attack is the slowest of all the analyzed scenarios. This behavior is due to the distribution properties of the signature (cf. Sect. 2.2). At the same time, the injection attacks (parametric or non-parametric version) are detected much faster than the repetition attack. This is due to the transition period required by the attackers to estimate the correct data before deceiving the detector. For this reason, if the attacker does not choose the precise time to launch the attack, the detector implemented at the controller is able to detect the data injected at the beginning of the attack. In addition, the attackers must also synchronize their estimates with the measurements sent by the sensors. In case the synchronization process fails, the detector identifies the uncorrelated data and reports the attack.

Regarding the results of the non-stationary signature detector shown in Table 5, we can verify that the performance obtained with this signature is compatible with the results obtained in the numerical validation (cf. Sect. 2). We show that the replay attack and the non-parametric attackers have a higher detection rate with this strategy, of about 60% and, 56%, respectively. Then, parametric attackers have a small increase in detection rate, from 12% to 16%. Interestingly, the *average detection time* decreases compared to the stationary signature detector. Also, the number of false negatives decreases, which increases the detection accuracy of the strategy against the implemented adversaries. However, the false positives with this strategy increase compared to the stationary signature detector. This means that, in a real testbed, the performance loss of the system is more important, as the number of false positives increases from 1.35% to 4.63%, with the same sensitivity as the previous strategy and a non-stationary signature.

Regarding the results obtained with the PIETC-WD strategy, shown in Table 3, we can highlight that: (1) a system that uses only the signature-based detection mechanism against parametric attackers has a lower detection rate, about 12%. This is possible because attackers can escape the detection process if they manage to correctly identify the system attributes; and (2) a system that uses the strategy proposed in Sect. 3.1 has a higher detection rate, of about 75.25%. In this scenario, the detection rate increases, confirming the theoretical and simulation results reported in Sect. 3. The false negative ratio decreases from 88.60% to 38.66%. In terms of false positives, both scenarios show similar results, but the PIETC-WD strategy generates about 3.9% more. The time between the start of the attack and the time the attack is detected by the remote controller takes longer with the PIETC-WD strategy, since the detection signature handled by the remote controller follows a stochastic law. Therefore, we confirm that the PIETC-WD strategy increases detection performance at the expense of the time used for detection.

5 Future Directions and Research Trends

Critical infrastructures are using cyber-physical systems increasingly and massively. Their complexity also increases their vulnerability to faults and attacks. New defenses are needed, in order to complement standard detection and protection mechanisms. Intrusion and attack detection mechanisms can provide crucial components to build resilient-by-design approaches to handle extreme and complex adversaries. In this book chapter, we have addressed control-theoretic approaches capable of countering unauthorized actions from adversaries. The solutions reported in this chapter aim at identifying and attenuating the impact that adversaries may impose upon the affected elements of a cyber-physical domain. This chapter has also addressed, with limited scope, some challenges on protection in the domain with particular attention to the detection of hidden malicious actions or combined with anomalies and accidents.

In terms of perspectives for future research, several actions remain to be done. Cyber-physical systems encompass many other domains that need to be managed together to improve their resilience to attack and misuse. Perspectives include additional mechanisms allowing the quantification and assurance of resilience of cyber-physical systems, giving some steps further in achieving the security-by-design addressed in this chapter, without losing robustness of the system. Real-time needs must be addressed carefully, in order to develop new resilient systems in which adversarial attacks inflicting a security breach can be managed before detrimental events happen against the system. We envision the use of additional layers, including solutions such as Byzantine fault and intrusion tolerance techniques, as well as self-healing and diversification mechanisms.

With the aforementioned ideas in mind, a first perspective would be to include further analysis about the performance impact of the decentralized protection process presented in this chapter. Likewise, expanding the decentralized model presented in Sect. 3, in order to consolidate the security level of the approach, the impact of the new construction in terms of network performance, as well as the performance of the overall control system. New research in order to fully decentralize the protection strategy that has been initiated in this chapter, as well as an appropriate combination of the cyber and control-physical layers suggested in our work could be developed towards a new generation of cyber-physical SIEM (Security Information and Event Management), capable of correcting cross-layer security incidents.

6 Conclusion

This chapter is based on the premise that, in a cyber-physical system, adversaries can eavesdrop and manipulate information to disrupt the availability and integrity properties of the system. Adversaries can use techniques from both the cyber and physical layers, first to control the network layers and then to disrupt the physical devices. The combination of these techniques can generate stealth attacks, allowing them to escape detection. Attacks against these systems can affect people and physical environments.

In terms of contributions, we started this chapter by reviewing existing technologies on cyber-physical environments from the perspective of traditional ICT security. The state of the art was complemented by three main contributions: (i) A first contribution was to revisit protection approaches related to stationary signatures, transforming them into an adaptive process capable of covering a larger number of adversary models; (ii) We extended the resulting signature detector, used as a physical attestation in the cyber layer, by adding a decentralized strategy to extend the approach to several elements of a cyber-physical environment (not only controllers but also sensors and actuators). The idea is to distribute the detection process over all these elements with sufficient capabilities to identify and manage the system dynamics, in order to identify malicious actions in addition to accidental flaws and errors; and (iii) We validated all our proposals by integrating them into a SCADA technology testbed. The latter was implemented using SCADA protocols used in the industry (e.g., MODBUS and DNP3) and Linux-based embedded devices. It allowed us to test and validate the security performance of our proposals. In addition, several adversaries capable of attacking representative scenarios were provided to complete the numerical simulations. To finish, it is worth noting that the new approach proposed in this chapter does not need synchronization among the different controllers (local or remote controllers). For this reason, it could also allow complementing other techniques such as MTD [15] whose objective is to create randomness in the cyber part of the system in order to detect and make it more resilient.

References

1. J. Åkerberg, M. Björkman, Exploring network security in PROFIsafe, in *Computer Safety, Reliability, and Security: 28th International Conference, SAFECOMP 2009, Hamburg, Germany, September 15–18, 2009. Proceedings* (Springer, Berlin, Heidelberg, 2009), pp. 67–80
2. A. Arvani, V.S. Rao, Detection and protection against intrusions on smart grid systems. Int. J. Cyber Secur. Digit. Forensics (IJCSDF) **3**(1), 38–48 (2014)
3. R. Baheti, H. Gill, Cyber-physical systems. Impact Control Technol. **12**, 161–166 (2011)
4. P. Barbosa, A. Brito, H. Almeida, S. Clauß, Lightweight privacy for smart metering data by adding noise, in *Proceedings of the 29th Annual ACM Symposium on Applied Computing*, SAC '14 (ACM, New York, NY, USA, 2014), pp. 531–538
5. M. Barenthin Syberg, *Complexity Issues, Validation and Input Design for Control in System Identification*. PhD thesis, KTH School of Electrical Engineering, Stockholm, Sweden, 2008
6. S. Brown, Functional safety of electrical/electronic/programmable electronic safety related systems. Comput. Control Eng. J. **11**(11), 14 (2000)
7. B. Brumback, M. Srinath, A chi-square test for fault-detection in Kalman filters. IEEE Trans. Autom. Control **32**(6), 552–554 (1987)
8. A.A. Cardenas, S. Amin, S. Sastry, Secure control: Towards survivable cyber-physical systems, in *The 28th International Conference on Distributed Computing Systems Workshops* (IEEE, 2008), pp. 495–500
9. A.A. Cardenas, S. Amin, B. Sinopoli, A. Giani, A. Perrig, S. Sastry, Challenges for securing cyber physical systems, in *Workshop on Future Directions in Cyber-Physical Systems Security* (DHS, 2009), p. 7
10. R. Chabukswar, *Secure Detection in Cyberphysical Control Systems*. PhD thesis, Department of Electrical and Computer Engineering, Carnegie Mellon University, Pittsburgh, PA, May 2014
11. D. Corman, V. Pillitteri, S. Tousley, M. Tehranipoor, U. Lindqvist, NITRD cyber-physical security panel, in *35th IEEE Symposium on Security and Privacy, IEEE SP 2014*, San Jose, CA, USA, May 18–21
12. K. Curtis, A DNP3 protocol primer. A basic technical overview of the protocol (2005). http://www.dnp.org/AboutUs/DNP3%20Primer%20Rev%20A.pdf, Last access: October 2016
13. V.L. Do, L. Fillatre, I. Nikiforov, A statistical method for detecting cyber/physical attacks on SCADA systems, in *2014 IEEE Conference on Control Applications (CCA)* (Juan Les Antibes, France, 2014), pp. 364–369

14. N. Falliere, L.O. Murchu, E. Chien, W32. Stuxnet Dossier. White Paper Symantec Corp. Secur. Res. **5**, 6 (2011)
15. P. Griffioen, S. Weerakkody, B. Sinopoli, A moving target defense for securing cyber-physical systems. IEEE Trans. Autom. Control **66**(5), 2016–2031 (2021)
16. Group REI-cyber, La Cybersécurité des Réseaux Electriques Intelligents. White book. La Revue de l'Electricité et de l'Electronique (REE), February 2016
17. D. Han, Y. Mo, J. Wu, S. Weerakkody, B. Sinopoli, L. Shi, Stochastic event-triggered sensor schedule for remote state estimation. IEEE Trans. Autom. Control **60**(10), 2661–2675 (2015)
18. W. Heemels, M. Donkers, A.R. Teel, Periodic event-triggered control for linear systems. IEEE Trans. Autom. Control **58**(4), 847–861 (2013)
19. J. Lee, B. Bagheri, H.-A. Kao, A cyber-physical systems architecture for Industry 4.0-based manufacturing systems. Manufacturing Letters **3**, 18–23 (2015)
20. L. Ljung, Perspectives on system identification. Annu. Rev. Control **34**(1), 1–12 (2010)
21. Y. Mo, B. Sinopoli, Secure control against replay attacks, in *47th Annual Allerton Conference on Communication, Control, and Computing* (IEEE, Monticello, IL, USA, 2009), pp. 911–918
22. Y. Mo, T. H.-J. Kim, K. Brancik, D. Dickinson, H. Lee, A. Perrig, B. Sinopoli, Cyber-physical security of a smart grid infrastructure. Proc. IEEE **100**(1), 195–209 (2012)
23. Y. Mo, R. Chabukswar, B. Sinopoli, Detecting integrity attacks on SCADA systems. IEEE Trans. Control Syst. Technol. **22**(4), 1396–1407 (2014)
24. Y. Mo, S. Weerakkody, B. Sinopoli, Physical authentication of control systems: designing watermarked control inputs to detect counterfeit sensor outputs. IEEE Control Syst. **35**(1), 93–109 (2015)
25. Modbus Organization, Official Modbus Specifications (2016). http://www.modbus.org/specs.php, Last access: October 2016
26. S.Y. Nam, D. Kim, J. Kim, et al., Enhanced ARP: preventing ARP poisoning-based man-in-the-middle attacks. IEEE Commun. Lett. **14**(2), 187–189 (2010)
27. H. Natke, System identification: Torsten Söderström and Petre Stoica. Automatica **28**(5), 1069–1071 (1992)
28. T. Roth, B. McMillin, Physical attestation in the smart grid for distributed state verification. IEEE Trans. Dependable Secure Comput., PP(99) (2016)
29. J. Rubio-Hernan, L. De Cicco, J. Garcia-Alfaro, On the use of watermark-based schemes to detect cyber-physical attacks. EURASIP J. Inf. Secur. **2017**(1), 8 (2017)
30. J. Salt, V. Casanova, A. Cuenca, R. Pizá, Sistemas de Control Basados en Red Modelado y Diseño de Estructuras de Control. Revista Iberoamericana de Automática e Informática Industrial RIAI **5**(3), 5–20 (2008)
31. S. Tripathi, M.A. Ikbal, Step size optimization of LMS algorithm using aunt colony optimization & its comparison with particle swarm optimization algorithm in system identification. Int. Res. J. Eng. Technol. (IRJET) **2**, 599–605 (2015)
32. S. Weyer, M. Schmitt, M. Ohmer, D. Gorecky, Towards industry 4.0 - standardization as the crucial challenge for highly modular, multi-vendor production systems. IFAC-PapersOnLine **48**(3), 579–584 (2015)
33. Y. Zhang, F. Xie, Y. Dong, G. Yang, X. Zhou, High fidelity virtualization of cyber-physical systems. Int. J. Model. Simul. Sci. Comput. **4**(2), 1340005 (2013)

Cybersecurity Applications in Software: Data-Driven Software Vulnerability Assessment and Management

Jiao Yin, MingJian Tang, Jinli Cao, Mingshan You, and Hua Wang

1 Software Vulnerability Assessment and Management

With the ongoing adoption of information technology and its impact on national economies and society, software plays a key role in the daily life of both organizations and individuals. However, a growing number of vulnerabilities caused by poor design or overlooked implementation are being disclosed nowadays [1]. The insecurity of information technology is often inevitable, which is a side effect brought by the use of information technology [2].

The scale, type and destructiveness of cyber threats and cyberattacks are increasing year by year, as more and more software vulnerabilities are discovered and publicly disclosed. According to CVE details [3], more than 166,000 software vulnerabilities have been disclosed and archived from 1988 to the end of 2021. More vulnerabilities are available from various channels and venues (e.g., security bulletins, forums, social media and so on). Bilge and Dumitras pointed out that once a vulnerability is disclosed, the chance of being exploited increases by five orders of magnitude[4, 5]. Obviously, unpatched known vulnerabilities impose significant security risks to modern society. Considering the huge number of disclosed

J. Yin · J. Cao (✉)
Department of Computer Science and Information Technology, La Trobe University, Melbourne, VIC, Australia
e-mail: j.yin@latrobe.edu.au; j.cao@latrobe.edu.au

M. Tang
Westpac Banking Corporation, Sydney, NSW, Australia
e-mail: ming.tang@westpac.com.au

M. You · H. Wang
Institute for Sustainable Industries & Liveable Cities, Victoria University, Melbourne, VIC, Australia
e-mail: mingshan.you@live.vu.edu.au; Hua.Wang@vu.edu.au

K. Daimi et al. (eds.), *Emerging Trends in Cybersecurity Applications*,
https://doi.org/10.1007/978-3-031-09640-2_17

vulnerabilities, it is difficult for information system vendors and users to patch each vulnerability in a timely manner. Because of limited budget and resources, vulnerability assessment and management has become critical for both commercial organizations regardless of the size and the entire cybersecurity community to make contingency plans in advance.

This section lays the foundation of software vulnerability assessment and management by introducing the readers to some of the key concepts spanning from vulnerability lifecycle to the entire vulnerability ecosystem.

1.1 Vulnerability and Vulnerability Disclosure

Vulnerability is a term referring to a system flaw that can leave it open to attack. According to the Common Vulnerabilities and Exposures (CVE) consortium, it is formally defined as a weakness in the computational logic (e.g., code) found in software and some hardware components (e.g., firmware) that, when exploited, results in a negative impact on confidentiality, integrity or availability (CIA) [6].

Vulnerability disclosure is the practice of reporting security flaws in computer software or hardware [7]. Vulnerability can be disclosed by multiple parties, including but not limited to third-party or internal software developers, vendors, suppliers, cybersecurity professionals and cybersecurity researchers. Different parties have different preferences for vulnerability disclosure time. Software vendors, suppliers and related developers usually prefer to disclose vulnerabilities after the corresponding patches or remedies are available, while affected end-users, cybersecurity professionals and researchers tend to disclose vulnerabilities as early as possible.

1.2 Exploit and Exploitability

A typical exploit in the cybersecurity domain can be a piece of software, a chunk of data or a sequence of commands, which takes advantage of a bug or vulnerability to cause unintended or unanticipated behaviour [8].

Exploitation is the behaviour of using an exploit to abuse software, hardware or other electronic equipment, including things like gaining control of a computer system, allowing privilege escalation or launching a denial-of-service (DoS) attack [9].

Exploitability is the state or condition of being exploitable. In the cybersecurity domain, a vulnerability is identified as exploitable when the proof-of-concept of the corresponding exploit exists. Exploitability is an important vulnerability assessment metric to reflect the properties of the vulnerability that lead to a successful attack [10].

Table 1 Six vulnerability lifecycle events

Event	Occurred time	Available to public?
Creation	t_{creat}	No
Discovery	t_{disco}	No
Exploit available	t_{explo}	No
Disclosure	t_{discl}	Yes
Patch available	t_{patch}	Yes
Patch installation	t_{insta}	Yes

1.3 Lifecycle of a Vulnerability

Frei et al. described a typical vulnerability lifecycle in [11]. We simplify the main events of the lifecycle in Table 1. A typical vulnerability lifecycle consists of six events, namely creation, discovery, exploit available, disclosure, patch available and patch installation. It should be noted that the order of occurrence of these six events may be slightly different for individual vulnerabilities. For example, vulnerability exploitation may occur before disclosure.

When a vulnerability is disclosed, the vulnerability information has three features, namely free access, independence and validation [11]. Information about disclosed vulnerabilities is available to the public for free. Then, disclosed vulnerability information will be widely accepted and used by the entire cybersecurity community. Finally, the disclosed information undergoes a thorough assessment by a panel of security experts and some assessment results will also be added to the disclosed vulnerability as basic risk ratings.

The time period between vulnerability discovery and disclosure is called the pre-disclosure phase, denoted as $\Delta\ t_{disco}(v)$, where $\Delta t_{disco}(v) = t_{discl}(v) - t_{disco}(v)$; $t_{discl}(v)$ is the disclosure time of v and $t_{disco}(v)$ is the discovery time of v. At this stage, the newly discovered vulnerabilities remain largely private. If they are known by researchers or vendors, they can work to provide patches before they become exploitable or disclosed in public. However, once they are discovered by malicious intruders or cyber-criminals, the potential risk involved can be significantly elevated. However, in this pre-disclosure phase, few things can be done to stop exploitation.

The time period from disclosure to patch available is another important phase, namely the post-disclosure phase, denoted as $\Delta t_{patch}(v)$, where $\Delta t_{patch}(v) = t_{patch}(v) - t_{discl}(v)$ and $t_{patch}(v)$ is the patch available time of v. At this stage, the risks of exploitation soar because more parties, including hackers and cyber-criminals, know of the existence of and have detailed information on the vulnerability. To make matters worse, end-users of the affected products will also be aware of the existence of this vulnerability, which will undoubtedly bring great pressure to vendors and service providers. Therefore, it is crucial for vendors and security information providers (SIPs) to provide a patch or give effective security advice. This research focuses on improving exploitability predictions and analysis to better inform decision-makers to prioritize the most urgent and risky vulnerabilities.

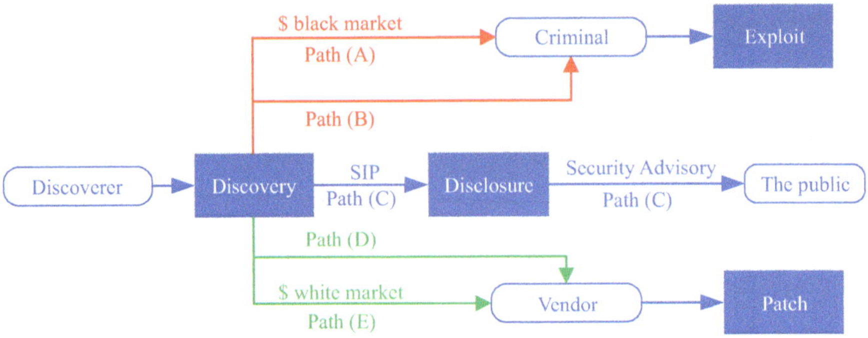

Fig. 1 Cybersecurity ecosystem illustration

Similarly, post-patch phase refers to the time period between vulnerability patch available and patch installation, which is denoted as $\Delta t_{insta}(v)$, where $\Delta t_{insta}(v) = t_{insta}(v) - t_{patch}(v)$ and $t_{insta}(v)$ is the patch instalment time of v. At this stage, if users are able to install the patch of v in a timely manner, the risks of exploitation can be mitigated.

1.4 Cybersecurity Ecosystem

Whenever a new vulnerability is discovered, various parties with different and often conflicting motivations and incentives become involved in a complex way [11]. Participants include but are not limited to discoverers, security advisories, cyber-criminals, traders in white or underground black markets, vendors and the public. The so-called security ecosystem consists of these players and their interactions. Figure 1 provides a high-level view of a cybersecurity ecosystem.

As shown in Fig. 1, Path (A) and (B) in red are at high risk while path (D) and (E) in green have fewer security concerns. Most vulnerabilities go through path (C). Once disclosed, the security of a vulnerability is uncertain, depending on whether it is exploited by attackers or patched by vendors. The risk of exploitation soars after being disclosed, as described in [5], 'after vulnerabilities are disclosed publicly, the volume of targeted attacks increases by five orders of magnitude'.

2 Mainstream Vulnerability and Exploit Databases

Historical vulnerability and exploit records are the most important and valuable digital assets for vulnerability assessment and management. Therefore, many commercial or non-profit organizations are collecting, storing and maintaining their own vulnerabilities and exploit databases. Some of them are available to the public.

This section introduces some well-known and publicly accessible databases and repositories. They are dedicated to comprehensive and credible information on vulnerabilities and potential links to detailed exploits (if exploitable).

2.1 CVE: Common Vulnerabilities and Exposures database

At present, vulnerability disclosure sources mainly include individual vendors, cybersecurity forums and open-source databases. Each disclosed vulnerability will be assigned a unique identification code, CVE-ID. CVE-ID is widely accepted by both local individual information providers/repositories and multiple global vulnerability databases [12]. This unique CVE-ID of each vulnerability facilitates the fast and accurate integration of data across multiple information sources and databases. In other words, it can be used to retrieve and link various information of the same vulnerability from different databases. Apart from CVE-ID, vulnerability disclosure reports may also include disclosure date, the names and corresponding version numbers of affected software products, required permission, the scope of impact and repair suggestions etc. [12].

The CVE database is one of the most well-known vulnerability databases. It stores essential disclosed vulnerability information, such as the CVE-ID, description, one or more public reference links [13]. A vulnerability description is a brief paragraph on each vulnerability, which contains abundant details such as the vulnerability type, names of affected products and vendors, a summary of affected versions, the impact, the access that an attacker requires to exploit the vulnerability and the important code components or inputs that are involved [14]. Depending on the source of disclosure, the description of a software vulnerability is usually written by the party requesting its CVE-ID.

The information in the CVE database serves as the baseline for vulnerability disclosure, and is referenced by many vulnerability databases, security products and services. The vulnerability list in the CVE database organized by year is available for download in several formats, i.e., comma separated format, HTML, text and XML. More than 160,000 vulnerability entries spanning over 20 years from 1999 to the present are included in the CVE database. The CVE database provides multiple attributes of each vulnerability for the public, such as Description, References, Assigning CNA, Date Record Created and Phase. Figure 2 shows a screenshot of vulnerability information listed in the CVE database. For more information, refer to the official website of the CVE database https://cve.mitre.org/index.html.

2.2 NVD: National Vulnerability Database

The NVD database is the U.S. government repository of standards-based vulnerability management data represented using the Security Content Automation Protocol (SCAP) [15]. It provides an analysis on CVE entries that have been published to

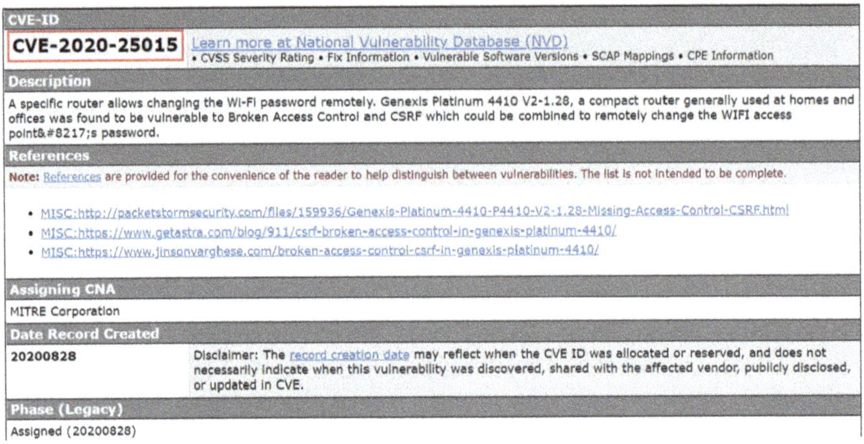

CVE-ID

CVE-2020-25015 Learn more at National Vulnerability Database (NVD)
• CVSS Severity Rating • Fix Information • Vulnerable Software Versions • SCAP Mappings • CPE Information

Description

A specific router allows changing the Wi-Fi password remotely. Genexis Platinum 4410 V2-1.28, a compact router generally used at homes and offices was found to be vulnerable to Broken Access Control and CSRF which could be combined to remotely change the WIFI access point’s password.

References

Note: References are provided for the convenience of the reader to help distinguish between vulnerabilities. The list is not intended to be complete.

- MISC:http://packetstormsecurity.com/files/159936/Genexis-Platinum-4410-P4410-V2-1.28-Missing-Access-Control-CSRF.html
- MISC:https://www.getastra.com/blog/911/csrf-broken-access-control-in-genexis-platinum-4410/
- MISC:https://www.jinsonvarghese.com/broken-access-control-csrf-in-genexis-platinum-4410/

Assigning CNA

MITRE Corporation

Date Record Created

20200828 Disclaimer: The record creation date may reflect when the CVE ID was allocated or reserved, and does not necessarily indicate when this vulnerability was discovered, shared with the affected vendor, publicly disclosed, or updated in CVE.

Phase (Legacy)

Assigned (20200828)

Fig. 2 A vulnerability listed in the CVE database

the CVE database. Based on the descriptions and references provided by the CVE database and other publicly accessed supplemental data, NVD expert panellists conduct an initial vulnerability assessment and give results based on certain standards, such as impact metrics (defined by Common Vulnerability Scoring System (CVSS)), applicability statements (defined by Common Platform Enumeration (CPE)), vulnerability types (defined by Common Weakness Enumeration (CWE)), and also other pertinent metadata [15]. Figure 3 shows the screenshot of information on a vulnerability listed in the NVD database.

Most importantly, the NVD database keeps re-analysing vulnerabilities as time and resources change over time to ensure the information provided by NVD is up to date. The NVD database is updated periodically to maintain the accuracy and real-timeness of vulnerability information and the data feeds in NVD database is available to the public for free [16].

2.3 CVE Details

CVE Details is a website developed by security consultant Serkan Özkan, who wanted to find an easy-to-use list of security vulnerabilities [3]. CVE Details contains information from multiple sources, including NVD XML data feeds, the Exploit Database [17], software vendor statements and additional vendor-supplied data, and Metasploit modules [3]. CVE Details presents each vulnerability entry on a single, easy-to-use web page. Figure 4 shows an example of the vulnerability information listed in CVE Details. Some statistics on vulnerabilities, vendors, products and exploits are also available in tables or figures [3].

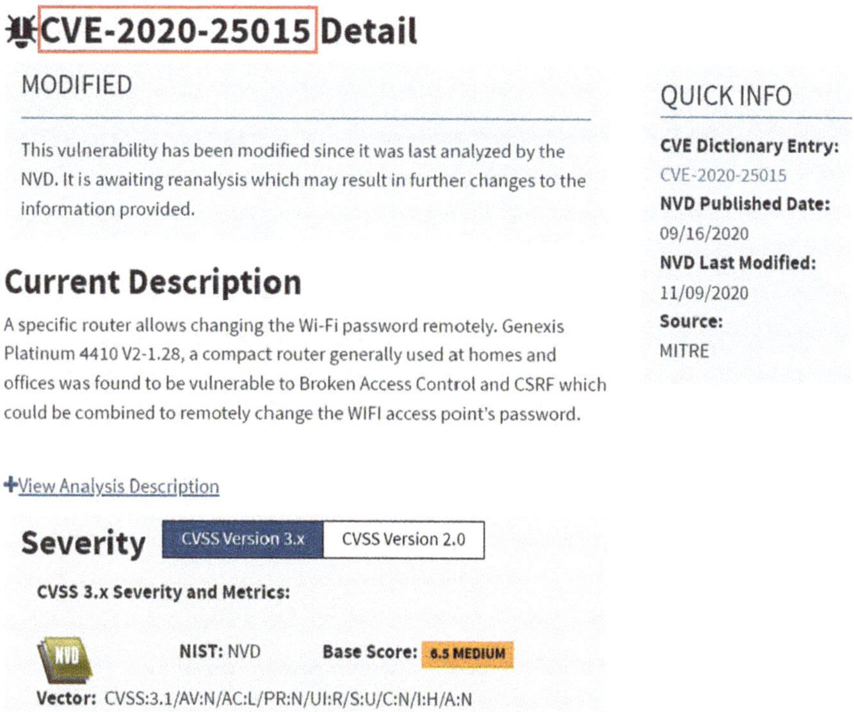

Fig. 3 A vulnerability listed in the NVD database

2.4 EDB: Exploit Database

The Exploit Database is an archive of public exploits and their targeted vulnerabilities, developed for use by penetration testers and vulnerability researchers [17]. The exploits in EDB are gathered from public sources and are freely available to the public. Each exploit in the EDB database has a unique EDB-ID for identification purposes.

The EDB database provides proofs-of-concept rather than advisories for vulnerabilities. Therefore, many researchers use the existence of exploits as the first sign of the exploitability of vulnerabilities [8, 18, 19], although exploitations always appear behind the existence of exploits. Figure 5 shows a screenshot of information on an exploit listed in the EDB database. Apart from the proof-of-concepts' executive codes, other crucial information on an exploit is also provided, such as EDB-ID, CVE-ID, Author, Type and Platform, as shown in Fig. 5. The EDB database is also a CVE-compatible database, making it possible to link the information of vulnerabilities and exploits.

At present, most commercial vulnerability management systems regularly synchronize the vulnerability and exploit information from these mainstream databases.

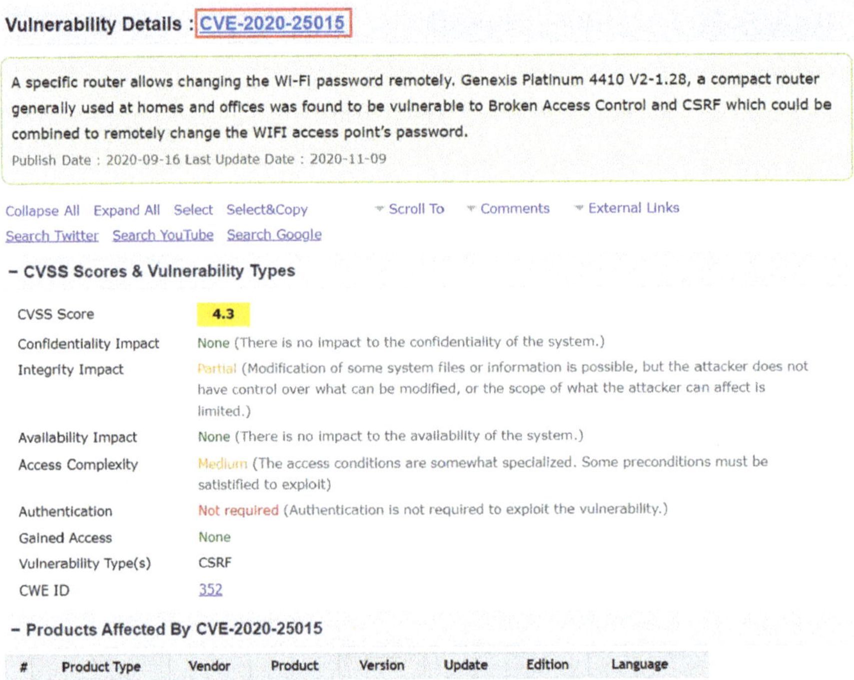

Fig. 4 A vulnerability listed in CVE details

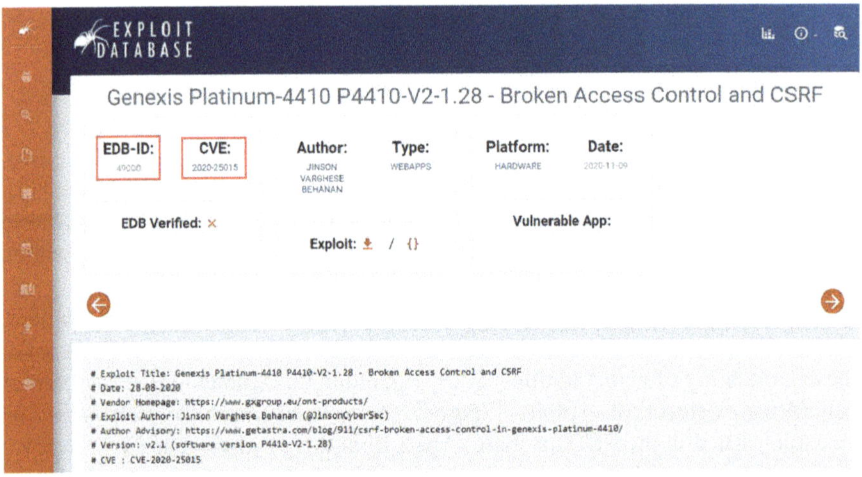

Fig. 5 An exploit listed in the EDB database

Furthermore, the experimental data of most research papers on vulnerability risk assessment come from the integration results of these open-sourced mainstream databases [4, 20–23]. To provide more examples for reference, Table 2 lists the

Table 2 Examples of vulnerabilities and their corresponding exploits listed in CVE, NVD, CVE Details and EDB

CVE-ID/EDB-ID	Database	URL
CVE-2020-25015	CVE	https://cve.mitre.org/cgi-bin/cvename.cgi?name=CVE-2020-25015
	NVD	https://nvd.nist.gov/vuln/detail/CVE-2020-25015
	CVE details	https://www.cvedetails.com/cve-details.php?cve_id=CVE-2020-25015
EDB-ID: 49000	EDB	https://www.exploit-db.com/exploits/49000
CVE-2021-24275	CVE	https://cve.mitre.org/cgi-bin/cvename.cgi?name=CVE-2021-24275
	NVD	https://nvd.nist.gov/vuln/detail/CVE-2021-24275
	CVE details	https://www.cvedetails.com/cve-details.php?cve_id=CVE-2021-24275
EDB-ID: 50346	EDB	https://www.exploit-db.com/exploits/50346
CVE-2021-24287	CVE	https://cve.mitre.org/cgi-bin/cvename.cgi?name=CVE-2021-24287
	NVD	https://nvd.nist.gov/vuln/detail/CVE-2021-24287
	CVE details	https://www.cvedetails.com/cve-details.php?cve_id=CVE-2021-24287
EDB-ID: 50349	EDB	https://www.exploit-db.com/exploits/50349
CVE-2021-24286	CVE	https://cve.mitre.org/cgi-bin/cvename.cgi?name=CVE-2021-24286
	NVD	https://nvd.nist.gov/vuln/detail/CVE-2021-24286
	CVE details	https://www.cvedetails.com/cve-details.php?cve_id=CVE-2021-24286
EDB-ID: 50350	EDB	https://www.exploit-db.com/exploits/50350

Uniform Resource Locator (URL) of examples of more vulnerabilities or exploits contained in the aforementioned four databases. The URL is entered into a browser for detailed information corresponding to that vulnerability or exploit. It is worth mentioning that the exploit listed below each vulnerability in Table 2 is the specific exploit attacking that vulnerability.

3 Common Vulnerability Scoring System

Vulnerability management is a crucial measure for both organizations and the entire cybersecurity community to protect their information systems and networks from cyberattacks, intrusions, malware and various types of data breaches [23]. Since the availability of exploits is much more in quantity than the availability of patches [24], it is important for vulnerability management experts to accurately assess the risk level of existing vulnerabilities. For risk management and vulnerability repair of modern information systems, vulnerability assessment and prioritization are the most basic steps in order to allocate budget and resources efficiently and effectively [25].

To date, various methods have been developed and introduced to assess software vulnerabilities and predict the trends of vulnerability outbreaks [23, 26]. Among them, CVSS plays the role of the de facto standard to assess the severity of software vulnerabilities in industry. It originated from a research project which aimed to promote a common understanding of vulnerabilities and their impact through the development of a common vulnerability scoring system by the National Infrastructure Advisory Council (NIAC) in July 2003 [27]. CVSS is currently at version 3.1 and under the custody of the Forum of Incident Response and Security Teams (FIRST). As a premier organization and recognized global leader in incident response, currently, FIRST has more than 400 members ranging from government, commercial and educational organizations, spread over Africa, the Americas, Asia, Europe and Oceania [28, 29]. Nowadays, CVSS is recommended by a large number of hardware and software vendors, such as Cisco, Oracle and Microsoft [30].

3.1 CVSS Metric Groups

CVSS defines three independent metric groups, namely the base metric group, temporal metric group and environmental metric group, whose detailed metric names are shown in Table 3 [10]. Only the base metric group is mandatory for the calculation of a vulnerability CVSS score.

3.2 CVSS Scores

The values of CVSS metrics shown in Table 3 are either a number between 0–10 or a discrete categorical value, which are given by a cybersecurity experts panel

Table 3 Metrics in CVSS metric groups

Metric group	Metric name (and abbreviated form)
Base metric group	Attack Vector (AV), Attack Complexity (AC), Privileges Required (PR), User Interaction (UI), Scope (S), Confidentiality (C), Integrity (I), Availability
Temporal metric group	Exploit Code Maturity (E), Remediation Level (RL), Report Confidence (RC)
Environmental metric group	Confidentiality Requirement (CR), Integrity Requirement (IR), Availability Requirement (AR), Modified Attack Vector (MAV), Modified Attack Complexity (MAC), Modified Privileges Required (MPR), Modified User Interaction (MUI), Modified Scope (MS), Modified Confidentiality (MC), Modified Integrity (MI), Modified Availability (MA)

according to the basic information of disclosed vulnerabilities [10]. Based on these metric groups, CVSS then calculates an overall score between 0–10.0 as the final CVSS score of a vulnerability according to a specially designed formula, where 10.0 represents the highest risk [4]. The detailed calculation process can be found in [10].

In particular, CVSS includes a formula to calculate the exploitability score of a vulnerability, as shown in Eq. (1) [10],

$$\text{Exploitability} = 8.22 \times \text{AV} \times \text{AC} \times \text{PR} \times \text{UI}, \tag{1}$$

where 8.22 is the coefficient assigned by a panel of CVSS cybersecurity experts; AV, AC, PR and UI are the abbreviated forms of the four base metrics listed in Table 3.

In addition to an overall score between 0 and 10, CVSS also provides a qualitative evaluation method for vulnerabilities by mapping the overall CVSS score to five risk levels, namely none (0.0), low (0.1–3.9), medium (4.0–6.9), high (7.0–8.9) and critical (9.0–10.0).

CVE Details presents the current vulnerability distribution by CVSS scores based on 162,031 vulnerabilities, which shows the weighted average CVSS score for all disclosed vulnerabilities is 6.5 [3].

3.3 Limitations of CVSS

CVSS is a carefully designed scoring system based on expert knowledge and has been accepted by a wide range of organizations. However, it is widely questioned by researchers that an overall score calculated by combining multiple metric groups with fixed weights, such as a CVSS score, can accurately represent the risk level of different software vulnerabilities [4].

Furthermore, CVSS is widely criticized by the academic community for the inconsistency between CVSS scores and the exploitability of vulnerabilities [8, 12, 19]. The overall CVSS scores of existing disclosed vulnerabilities show that there is no significant correlation between the CVSS score of a vulnerability and the possibility of its exploitability.

To further validate the criticism of CVSS in the academic community, the work in [4] visualizes two CVSS metrics, namely base score and exploitability score, that are most relevant to the exploitability of vulnerabilities from two CVSS versions (CVSS V2.0 and V3.0), as shown in Fig. 6. The data samples in Fig. 6 are from all disclosed vulnerabilities recorded in the NVD database from 1988 to 2019. Figure 6 shows exploited vulnerabilities in dark orange and unexploited vulnerabilities in the navy. The Y-axis represents the comparison of the number of these two types of vulnerabilities with the same CVSS metric score. The X-axis indicates the value of the corresponding CVSS metrics and the larger the value, the greater the risk of the corresponding vulnerabilities. Taking the V2 exploitability score shown in subplot (b) as an example, the blue bar with a score between 8 and 10 is very high, indicating that the number of unexploited vulnerabilities in this interval is very

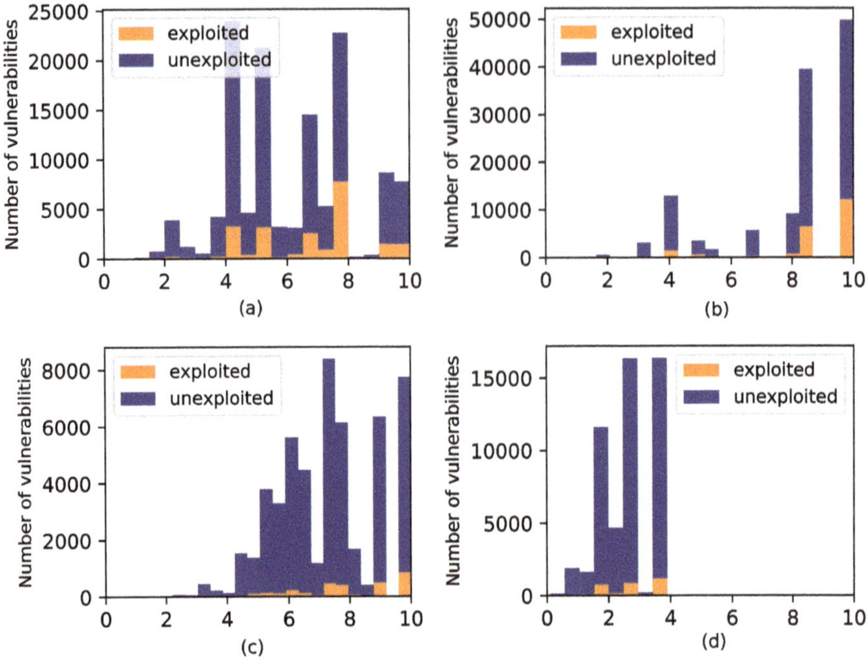

Fig. 6 The CVSS metric score distribution of vulnerabilities disclosed from 1988 to 2019. (**a**) V2 base score. (**b**) V2 exploitability score. (**c**) V3 base score. (**d**) V3 exploitability score

large. Obviously, this contradicts the low probability of unexploited vulnerabilities. Similarly, the contradiction between the CVSS metric score and the exploitability of vulnerabilities is also reflected in subplots (a), (c) and (d). It is worth noting that V3 is an improved CVSS version of V2. However, as can be seen from subplots (c) and (d), the deficiency that CVSS cannot effectively depict the exploitability of vulnerabilities has not been significantly improved.

Other concerns on the application of CVSS as a vulnerability assessment indicator include the following two points. Firstly, the value assignment of CVSS metrics relies on an expert panel, which is costly in time and money. Furthermore, it is difficult to ensure consistency when the personnel changes.

4 Vulnerability Exploitability Prediction and Analysis

Vulnerability exploitability prediction and analysis is one of the most important tasks in vulnerability assessment and management. Considering the inaccuracy of the CVSS Exploitability score calculated by Eq. (1), researchers in the academic community have done a significant amount of work in vulnerability exploitability prediction. This section introduces some representative work in three aspects, see Sects. 4.1, 4.2 and 4.3 for details.

4.1 Vulnerability Exploitability Prediction

The exploitability of a vulnerability indicates if a vulnerability will be exploited or not. With the accumulation of more and more historical data, researchers adopted a variety of machine learning and deep learning models and algorithms to predict the exploitability of vulnerabilities and very promising results are reported. Vulnerability features can be extracted from publicly available information, including descriptions, CVSS metrics, social media streams, etc. [12, 31, 32].

As one of the most influential early works, the work in [12] extracted text features for vulnerabilities using a kind of one-hot representation. Specifically, a dictionary containing important tokens for vulnerability exploitation prediction was formed and if a token in the dictionary appears in the text fields of disclosed vulnerability information, the corresponding position is set to 1 otherwise 0. The achieved classification accuracy in [12] is nearly 85% with the linear support vector machine (LSVM) classifier.

Sabottke, Suciu and Dumitras [31] proposed a Twitter-based exploit detector, predicting real-world vulnerability exploitations. In this work, they manually extract statistical features from Twitter streams and other open-source databases, like NVD and OSVDB.

Along the same line, the work in [32] proposed an exploitability prediction method based on neural language models. Instead of extracting linguistic features using traditional TF-IDF-based representation, it adopts the neural language models to learn word embeddings based on the corpus collected from multiple sources. The experimental results show that the high-dimensional word embedding features extracted by the deep learning language model have better performance on the vulnerability exploitability prediction problem than features extracted by traditional statistical-based text feature extraction methods [32].

The authors in [19] considered two risk factors: (1) the existence of a public proof-of-concept exploit; (2) the existence of an exploit traded in the cybercrime black markets to evaluate the possibility of exploitation using a case-control study methodology.

According to the historical records in the NVD database and EDB database, the work in [33] established a machine learning model to automatically predict the vulnerability of unseen vulnerabilities. The authors compared the impact of different vulnerability feature sources on predictive performance. Results showed that the features extracted from text information such as vulnerability descriptions and external references are the most effective. On the premise that the above-mentioned features have been extracted, features such as CVSS metrics are redundant.

Jacobs, Romanosky et al. proposed an Exploit Prediction Scoring System (EPSS), which has the capability to predict if a vulnerability will be exploited or not in the wild within one year after disclosure [34]. The authors claimed that their system is simple to implement and therefore can be updated in a timely manner when new data becomes available.

The work in [4] proposed a deep neural language model based framework for vulnerability exploitability prediction. They apply the transfer learning technique and fine-tune a widely used pre-trained NLP model, Bidirectional Encoder Representations from Transformers (BERT), on the corpus consisting of vulnerability descriptions to extract domain-specific semantic features from vulnerability descriptions only. The extracted semantic features are fed into a pooling layer and an LSTM classification layer for the final decision-making. The experiments showed that their method achieved 91% in accuracy on a balanced real-world dataset.

4.2 Online Vulnerability Exploitability Prediction

The aforementioned research work treated vulnerability exploitability prediction as an offline machine learning problem. However, the reality is that the features and patterns of vulnerabilities, exploitation and the latent relationship between them are dynamically changing with the development of technology. In practical vulnerability exploitability prediction systems, if the possible concept drift problem is not considered, the performance of the predictive model will get worse and worse along time. Therefore, online learning models for vulnerability exploitability prediction have become a new trend.

The authors in [35] pointed out that exploitability assessment suffers from a class bias because 'not exploitable' labels could be inaccurate over time. Therefore, they proposed a new metric, called Expected Exploitability (EE) to provide a time-varying view of exploitability. In this work, they characterized the noise in exploit prediction as a class- and feature-dependent label noise and developed techniques to incorporate noise robustness into learning EE by capitalizing on domain-specific observations. Furthermore, instead of extracting features by technical analysis on existing metrics, they designed novel feature sets from previously under-utilized artefacts which are published after the disclosure of vulnerabilities, such as technical write-ups, social media discussions and proof-of-concept exploits. Experiment results on a dataset of 103,137 vulnerabilities showed an increase of precision from 49% to 86% was achieved by EE over existing metrics.

The authors in [20] also noticed an 'actual drift' problem existing in vulnerability exploitability labels, which means that the exploitability of vulnerabilities is chronologically variable. An 'unexploitable' vulnerability can become 'exploitable' after several days, months or years. In this work, based on the fact that vulnerability exploitability labels may change from unexploitable to exploitable over time, they proposed an algorithm called class rectification strategy (CRS) to detect the conceptual drift of vulnerability exploitability labels. Furthermore, they improved the real-time performance of the predictive model by updating the model online with vulnerabilities that have experienced label drift.

For the vulnerability exploitability prediction problem, on the whole, unexploitable vulnerabilities are far more than exploitable ones. This class unbalanced state changes dynamically in online learning scenarios. The work in [20] discussed how to improve the performance of vulnerability exploitability prediction under an online learning setting. It proposed a balanced window strategy (BWS) to build a dynamic class-balanced dataset to update the predictive model periodically. The experiment results show that BWS is effective in improving the online exploitability prediction performance of a variety of classifiers.

4.3 Vulnerability Exploitation Time Prediction

A vulnerability can have multiple exploits. The exploitation time of a vulnerability discussed in this section refers to the time difference between the earliest disclosure time of its exploits and the disclosure time of the vulnerability itself [22]. Figure 7 shows the exploitation time distribution of 23,302 vulnerabilities disclosed from 1988 to 2020 [22]. Exploitation time prediction is a more valuable and challenging task than exploitability prediction, which can provide a prediction of how soon a vulnerability will be exploited with an exact exploitation time or an exploitation time range. As shown in Fig. 7, the exploitation time varies in a large range in a biased distribution, which makes it challenging to make a prediction.

As early as 2010, the work in [12] began to study the problem of vulnerability exploitation time prediction. Instead of predicting a specific date, they reported the possible exploitation time in a weekly and monthly manner. An overall cumulative error rate of 15% was reported at the end of online training with a simple linear classier, which is extremely promising.

Sabottke, Suciu and Dumitras [31] proposed a Twitter-based exploit detector, predicting real-world vulnerability exploitations. In this work, they extract features from NVD, OSVDB and Twitter streams and manually select features based on the mutual information between features and labels. One of the contributions of

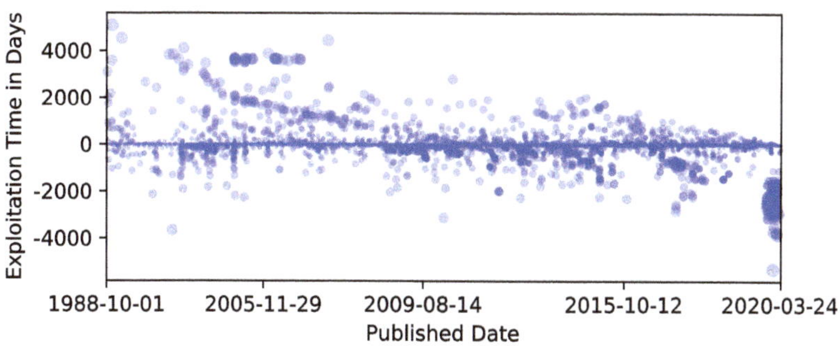

Fig. 7 The distribution of exploitation time

this work is that it not only investigated the vulnerability exploitability prediction problem but can also predict the emergence of exploits on an average of two days in advance.

The work in [21] and [22] investigated the vulnerability exploitation time prediction problem. Specifically, they divided the exploitation time into three classes, Neg, ZeroDay and Pos, based on the time differences between a vulnerability being exploited and being published. Although promising results were reported, more fine-grained exploitation time prediction results are expected.

5 Summary

Rigorous vulnerability evaluation and assessment empowers organizations to make informed and data-driven risk management decisions towards better mitigation and security of their IT environment against malicious exploitations. This chapter touches on cybersecurity applications in software vulnerability assessment and management. Specifically, this chapter includes:

(1) the key concepts, background, significance and foundations of software vulnerability evaluation and assessment;
(2) the valuable digital assets for both industry applications and academic research, the mainstream vulnerability and exploit databases, namely CVE, NVD, CVE Details and EDB;
(3) some of the latest advances as well as open challenges on vulnerability assessment and evaluation in both industry and academia community;
(4) further introduction and review on one of the research hotspots, vulnerability exploitability prediction and analysis.

Previous research works provided some promising solutions for vulnerability assessment and management. However, there are still many unsolved challenges. Some future directions in improving vulnerability assessment and management are listed as follows:

(1) Explore the exploitation time prediction with finer granularity. As the early attempts to predict vulnerability exploitation time, the work in [20–22] divided exploitation time into three classes, Neg, ZeroDay and Pos, based on the time differences between a vulnerability being exploited and being published. Although their prediction results are more detailed than exploitability prediction, a finer-granular exploitation time prediction can be more useful in practice, especially for Pos vulnerabilities. For example, the predicted exploitation time period can be yearly, monthly, weekly or even daily. The main challenge of finer granularity comes from data deficiencies and data imbalance within each granularity. With the increase of available vulnerabilities and exploit data and the development of unsupervised learning techniques, novel solutions will emerge in the future.

(2) Combine the exploitability prediction with other vulnerability assessment metrics to form a more comprehensive vulnerability risk evaluation model. In addition to exploitability, the risk level of a vulnerability is affected by many other aspects, such as the number of devices and users affected, the business process affected and the cost comparison between exploitation and remediation. CVSS is an example of a comprehensive vulnerability risk evaluation model. However, its effectiveness is far from satisfactory. More accurate availability prediction is undoubtedly conducive to vulnerability risk assessment, but how to combine the exploitability prediction results with other risk factors to form a holistic and effective vulnerability risk assessment framework will be grand challenging.

(3) Construct cybersecurity domain specific knowledge graph and explore more knowledge graph powered vulnerability intelligence applications. So far, the major source of vulnerabilities and exploits for research works and industry applications comes from existing well-organized databases, such as NVD, EDB and CVE Details. However, there are still vast amounts of vulnerability raw data from multimodal information sources, such as social media, software vendors, technical forums. This information can be used to build comprehensive cybersecurity knowledge graph. Based on such a domain-specific knowledge graph, novel knowledge-driven applications can be nurtured, including but not limited to subgraph matching to discover multi-stage and highly sophisticated cyberattack tactics and multi-hop question-and-answer systems, which can make highly specialized cyber knowledge more accessible.

References

1. M. Tang, M. Alazab, Y. Luo, Big data for cybersecurity: Vulnerability disclosure trends and dependencies. IEEE Trans. Big Data **5**(3), 317–329 (2017)
2. R. Anderson, T. Moore, The economics of information security. Science **314**(5799), 610–613 (2006)
3. S. Özkan, CVE details, the ultimate security vulnerability database (2021). https://www.cvedetails.com/, [Retrieved: Nov, 2021]
4. J. Yin, M. Tang, J. Cao, H. Wang, Apply transfer learning to cybersecurity: Predicting exploitability of vulnerabilities by description. Knowl. Based Syst., 106529 (2020)
5. L. Bilge, T. Dumitraş, Before we knew it: an empirical study of zero-day attacks in the real world, in *Proceedings of the 2012 ACM Conference on Computer and Communications Security* (Raleigh North Carolina, USA, 2012), pp. 833–844
6. The MITRE Corporation, About CVE - terminology. https://cve.mitre.org/about/terminology.html, [Retrieved: Nov, 2021]
7. L. Rosencrance, Vulnerability disclosure (2017). https://searchsecurity.techtarget.com/definition/vulnerability-disclosure, [Retrieved: Nov, 2021]
8. A. Younis, Y.K. Malaiya, I. Ray, Assessing vulnerability exploitability risk using software properties. Softw. Qual. J. **24**(1), 159–202 (2016)
9. Wikipedia, Exploit (computer security). https://en.wikipedia.org/wiki/Exploit_(computer_security), [Retrieved: Nov, 2021]

10. Forum of Incident Response and Security Teams, Common vulnerability scoring system v3.1: Specification document. https://www.first.org/cvss/v3.1/specification-document, [Retrieved: Nov, 2021]
11. S. Frei, D. Schatzmann, B. Plattner, B. Trammell, Modeling the security ecosystem-the dynamics of (in) security, in *Economics of Information Security and Privacy*, London, England, 2010, pp. 79–106
12. M. Bozorgi, L.K. Saul, S. Savage, G.M. Voelker, Beyond heuristics: learning to classify vulnerabilities and predict exploits, in *Proceedings of the 16th ACM SIGKDD International Conference on Knowledge Discovery and Data Mining*, Washington, DC, USA, 2010, pp. 105–114
13. The MITRE Corporation, The mission of the cve program is to identify, define, and catalog publicly disclosed cybersecurity vulnerabilities. https://cve.mitre.org/, [Retrieved: Nov, 2021]
14. The MITRE Corporation, Cve - frequently asked questions (2021). https://cve.mitre.org/about/faqs.html#cve_entry_descriptions_created, [Retrieved: Nov, 2021]
15. National Institute of Standards and Technology, U.S. Department of Commerce, General information. https://nvd.nist.gov/general, [Retrieved: Nov, 2021]
16. National Institute of Standards and Technology, U.S. Department of Commerce, NVD data feeds. https://nvd.nist.gov/vuln/data-feeds, [Retrieved: Nov, 2021]
17. Offensive Security, Exploit database (2021). https://www.exploit-db.com/, [Retrieved: Nov, 2021]
18. B.L. Bullough, A.K. Yanchenko, C.L. Smith, J.R. Zipkin, Predicting exploitation of disclosed software vulnerabilities using open-source data, in *Proceedings of the 3rd ACM on International Workshop on Security and Privacy Analytics* (Scottsdale, USA, 2017), pp. 45–53
19. L. Allodi, F. Massacci, Comparing vulnerability severity and exploits using case-control studies. ACM Trans. Inf. Syst. Secur. (TISSEC) **17**(1), 1–20 (2014)
20. J. Yin, M. Tang, J. Cao, H. Wang, M. You, A real-time dynamic concept adaptive learning algorithm for exploitability prediction. Neurocomputing, 1–36 (2021)
21. J. Yin, M. Tang, J. Cao, H. Wang, M. You, Y. Lin, Vulnerability exploitation time prediction: an integrated framework for dynamic imbalanced learning. World Wide Web, 1–23 (2021)
22. J. Yin, M. Tang, J. Cao, H. Wang, M. You, Y. Lin, Adaptive online learning for vulnerability exploitation time prediction, in *Web Information Systems Engineering – WISE 2020*, Amsterdam, Netherlands, 2020, pp. 252–266
23. M. Tang, J. Yin, M. Alazab, J.C. Cao, Y. Luo, Modelling of extreme vulnerability disclosure in smart city industrial environments. IEEE Trans. Ind. Inf., 4150–4158 (2020)
24. S. Frei, M. May, U. Fiedler, B. Plattner, Large-scale vulnerability analysis, in *Proceedings of the 2006 SIGCOMM Workshop on Large-Scale Attack Defense*, 2006, pp. 131–138
25. L. Allodi, M. Cremonini, F. Massacci, W. Shim, The effect of security education and expertise on security assessments: The case of software vulnerabilities. Preprint (2018). arXiv:1808.06547
26. M. Alazab, M. Tang, *Deep Learning Applications for Cyber Security* (Springer Nature Switzerland AG, Cham, Switzerland, 2019)
27. M. Schiffman, A. Wright, D. Ahmad, G. Eschelbeck, The common vulnerability scoring system, in *National Infrastructure Advisory Council, Vulnerability Disclosure Working Group, Vulnerability Scoring Subgroup*, San Francisco, USA, 2004
28. Forum of Incident Response and Security Teams, Forum of incident response and security teams (first) (2021). https://www.cybersecurityintelligence.com/forum-of-incident-response-and-security-teams-first-5620.html, [Retrieved: Nov, 2021]
29. Forum of Incident Response and Security Teams, FIRST is the global forum of incident response and security teams (2021). https://www.first.org/, [Retrieved: Nov, 2021]
30. Oracle, Use of common vulnerability scoring system (CVSS) by oracle. https://www.oracle.com/technetwork/topics/security/cvssscoringsystem-091884.html, [Retriveved: Nov, 2021].
31. C. Sabottke, O. Suciu, T. Dumitras, Vulnerability disclosure in the age of social media: Exploiting twitter for predicting real-world exploits, in *24th {USENIX} Security Symposium ({USENIX} Security 15)*, 2015, pp. 1041–1056

32. N. Tavabi, P. Goyal, M. Almukaynizi, P. Shakarian, K. Lerman, Darkembed: Exploit prediction with neural language models, in *Thirty-Second AAAI Conference on Artificial Intelligence*, 2018, pp. 7849–7854
33. M. Edkrantz, A. Said, Predicting cyber vulnerability exploits with machine learning, in *SCAI*, 2015, pp. 48–57
34. J. Jacobs, S. Romanosky, B. Edwards, M. Roytman, I. Adjerid, Exploit prediction scoring system (epss). Preprint (2019). arXiv:1908.04856
35. O. Suciu, C. Nelson, Z. Lyu, T. Bao, T. Dumitras, Expected exploitability: Predicting the development of functional vulnerability exploits. Preprint (2021). arXiv:2102.07869

Application of Homomorphic Encryption in Machine Learning

Yulliwas Ameur, Samia Bouzefrane, and Vincent Audigier

1 Introduction to Homomorphic Encryption

This new encryption paradigm allows any entity (for example, the cloud provider) to operate on private data in encrypted form without ever decrypting it. For example, one widespread use case is outsourcing healthcare data to cloud computing services for medical studies. The goal of HE is to perform operations on the plain text while manipulating only ciphertexts (Fig. 1). Usually, we must decrypt them and then apply the desired processing to manipulate encrypted data (Fig. 2).

For some cryptosystems with algebraic structures, some operations are possible. For example, two RSA ciphertexts can be multiplied to obtain the multiplication of the two corresponding plain texts. We call this property the multiplicative homomorphic property of the "textbook RSA" cryptosystem. Another operation can also be performed on ciphertexts. For example, in the Paillier cryptosystem [1], we can add two ciphertexts to obtain the addition of the two corresponding plain texts. We call this property the additive property of the "Paillier" cryptosystem. For example, this can be useful when we are interested in e-voting applications to add encrypted votes without knowing the initial vote. Rivest, Adelman, and Dertouzos first introduced the notion of homomorphic encryption in [2]. Building a cryptosystem with both multiplicative and additive properties was a significant problem in cryptography, until the work of Gentry [3]. Gentry proposed a first fully homomorphic encryption based on ideal lattices. The HE is categorized depending on the number of mathematical operations performed on the encrypted message as following:

Y. Ameur (✉) · S. Bouzefrane · V. Audigier
CEDRIC Lab, Cnam, Paris, France
e-mail: yulliwas.ameur@lecnam.net; samia.bouzefrane@lecnam.net; vincent.audigier@lecnam.net

K. Daimi et al. (eds.), *Emerging Trends in Cybersecurity Applications*,
https://doi.org/10.1007/978-3-031-09640-2_18

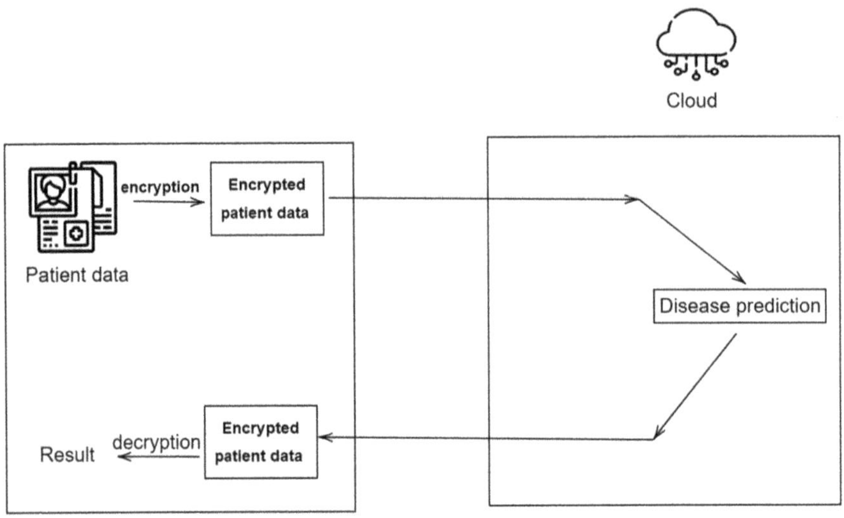

Fig. 1 Diagram showing how to manipulate encrypted data on a cloud. On the left, the user encrypts the data before sending it to the cloud, on the right, the cloud service has to decrypt the data in order to process it

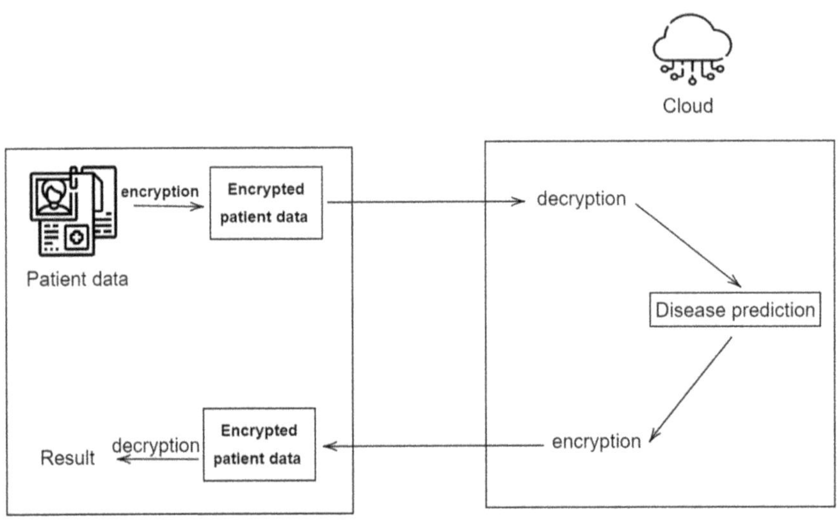

Fig. 2 Diagram showing how to manipulate encrypted data on a cloud by using homomorphic encryption. On the left, the user encrypts the data before sending it to the cloud, on the right, the cloud service can process on data by manipulating only ciphertexts

- **Partially homomorphic encryption (PHE)**: is a cryptosystem that allows a single operation (addition or multiplication) over encrypted data. When a PHE scheme allows for additions over ciphertexts, it is considered an additively homomorphic scheme [1]. When a PHE scheme allows for multiplications, it is considered multiplicatively homomorphic.

- **Somewhat homomorphic encryption (SHE)**: is a cryptosystem that allows us to perform a limited number of both additions and multiplications. SHE cryptosystems typically allow for unlimited additions but only a restricted number of multiplications.
- **Fully homomorphic encryption (FHE)**: FHE schemes are the most powerful HE schemes. They can perform both addition and multiplication, as well as circuits of any depth. The reason HE methods have a restricted circuit depth is that the encryption operation introduces noise to the data, and the decryption step removes that noise. Performing operations on ciphertexts generates more noise, which prevents correct decryption [4].

The **bootstrapping** approach is used by FHE systems to get around this as stated in [3]. Bootstrapping decreases the collected noise, allowing further computation. This procedure can be done as many times as necessary to analyze any particular circuit. However, bootstrapping is computationally costly, so many solutions do not employ it in reality. Therefore, we recommend the reader to refer to [5] for more detailed information on the different homomorphic encryption schemes. We have chosen not to describe the entire functioning of cryptosystems due to the lack of space. Also, selecting secure and efficient instantiations of the underlying cryptographic problem is hard for most of encryption and homomorphic schemes. Therefore, we have chosen to list the most studied schemes by the community of researchers and developers interested in advancing homomorphic encryption.

As usual, new cryptographic proposals need a few years before widespread adoption in the industry, as was the case of elliptic curve cryptography, post-quantum encryption and many other standardization projects. We are waiting for the standardization results and recommendations of the workshops, which include representatives from industry, government organizations and academia. This brief review aims to guide readers fast enough, even if they are not cryptography specialists, to the appropriate HE scheme by directing them to the library(ies) where HE is implementable.

1.1 HE Schemes

Research in the field of FHE may be classified into four major groups. The first family represents the difficulty based on the lattice reduction problem, which mainly comes from Gentry's seminal work [3]. The second category consists of integer-based methods [6], the hardness of which is based on the Approximate of Greatest Common Divisor (A-GCD) problem [7]. Schemes based on learning with error (LWE) [8] and ring learning with error (RLWE) [9], both reducible to lattice problems, constitute the third family. Finally, the Nth-Degree truncated polynomial ring unit (NTRU) family [10].

Table 1 Comparison of HE schemes

	Schemes				
Operation	BFV	BGV	CKKS	FHEW	TFHE
Native Add/Sub	✓	✓	✓	✗	✗
Native Mult	✓	✓	✓	✗	✗
Boolean Logic	✗	✗	✗	✓	✓
SIMD	✓	✓	✓	✓	✓

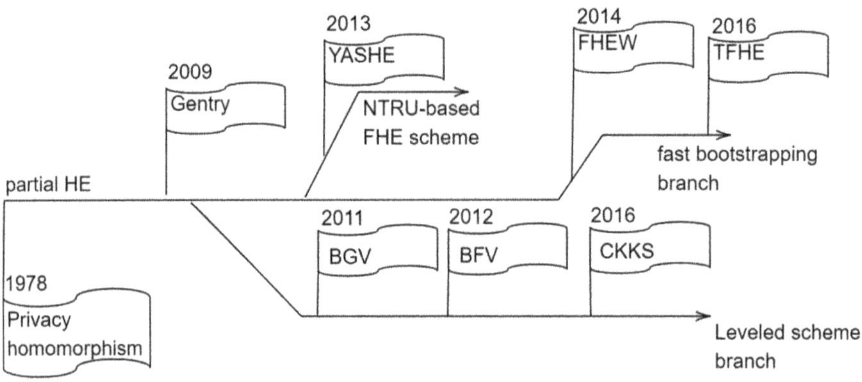

Fig. 3 Homomorphic encryption timeline

All HE schemes have common steps: key generation, encryption, decryption, and homomorphic operations on the ciphertexts. Table 1 summarizes the most implemented and studied schemes by the cryptographic community, and Fig. 3 gives an overview of the homomorphic encryption timeline.

The choice of encryption scheme has a multitude of implications:

- It specifies which operations are possible and, as a result, which types of activation and architectures may be employed.
- It can determine the plaintext space. Messages should be encoded before they may be sent in plaintext. The majority of schemes, including BGV, BFV, only support integers. CKKS can handle real numbers, but TFHE can only handle individual bits.

1.2 HE Libraries

There exist several open-source libraries for the implementation of the HE scheme. They provide key generation, encryption, decryption, and homomorphic operations for each scheme; library APIs frequently include additional features for maintaining and manipulating ciphertexts. Even though there is a lack of technical interoperability, but also a lack of conceptual interoperability; for example, even libraries that use the same scheme can provide surprisingly different results. The ongoing

Table 2 Overview of existing FHE libraries: CPU-targeting (top) and GPU-targeting (bottom)

Name	Input language	Supported schemes BFV	BGV	CKKS	FHEW	TFHE
HE-CPU-TARGETING						
Concrete	Rust	✗	✗	✗	✗	✓
FHEW	C++	✗	✗	✗	✓	✗
FV-NFlib	C++	✓	✗	✗	✗	✗
HEAAN	C++	✗	✗	✓	✗	✗
HElib	C++	✓	✓	✓	✗	✗
lattigo	Go	✓	✗	✓	✗	✗
PALISADE	C++	✓	✓	✓	✓	✓
SEAL	C++, .NET	✓	✓	✓	✗	✗
TFHE	C++	✗	✗	✗	✗	✓
HE-GPU-TARGETING						
cuFHE	C++, Python	✗	✗	✗	✗	✓
nuFHE	C++, Python	✗	✗	✗	✗	✓

standardization efforts attempt to develop a unified view of the most popular schemes (Table 2).

1.3 FHE Restrictions

Current HE methods have a significant restriction. They cannot support division operations and comparisons easily, such as the equality/inequality test. Number comparison and sign determination are critical processes for MLaaS

2 Privacy-Preserving in Machine Learning (PPML): HE Solutions

Using a third-party infrastructure reduces the problems of resources and complexity but introduces privacy issues of sensitive information. To construct a privacy-preserving framework for machine learning techniques, it must first identify the most important privacy requirements:

- Input privacy: Only the real data owner should have access to the input.
- Output privacy: The output/result of the ML methods' assessment is not permitted to be known by the cloud server.
- Model privacy: A private machine learning model that is also an asset should not be shared with anyone except its owner.

Another approach is to look if the privacy-preserving framework targets the learning phase or the inference phase of the machine learning algorithm. Depending on the framework we want to design, we have to use privacy-preserving technologies. The leading privacy-preserving machine learning techniques are

- **Multi-party computation**: These methods involve one or more trusted parties that can be used to outsource specific computations by the algorithm owner.
- **Differential privacy**: These methods rely on data randomization and perturbation. Because it affects the information, this method has the drawback of negatively influencing the model's performance.
- **Federated learning**: Federated learning is a machine learning setting where many clients collaboratively train a model under the administration of a central server while keeping the training data local.
- **Garbled circuit**: Garbled circuit, also known as Yao's garbled circuit, is an underlying technology of secure two-party computation initially proposed by Andrew Yao. GC provides an interactive protocol for two parties (a garbler and an evaluator) obliviously evaluate an arbitrary function represented as a Boolean circuit.
- **Homomorphic encryption**: An encryption that allows performing operations over encrypted data. See Sect. 1 for more detail.
- **Hybrid PPML techniques**: In addition to the above-mentioned single-protocol PPML, some commonly used frameworks typically use hybrid protocols, which combine two or more protocols by making use of the advantages and avoiding the problems of each. For example, the basic idea behind the mixed protocol that combines HE and GC is to calculate operations that have an efficient representation as Arithmetic circuits (e.g., additions and multiplications) using HE and operations that have an efficient representation as Boolean circuits (e.g., comparisons) using GC. However, converting between share systems is not simple, and the charges are very high. Furthermore, various frameworks integrate MPC with differential privacy.

We are interested therein machine learning research Using homomorphic encryption "HEML." According to this bibliometrics [11], the number of papers on HEML has constantly been rising since 2009. Each year from 2005 to 2015, fewer than 100 HEML papers were published. However, the number of publications per year increased significantly after 2015, reaching between 200 and 500 in recent years. This section summarizes recent works and gives many practical applications of homomorphic encryption for privacy-preserving purposes for each machine learning algorithm. We end the section with a summary of the work to ease the reading of this synthesis (Fig. 4).

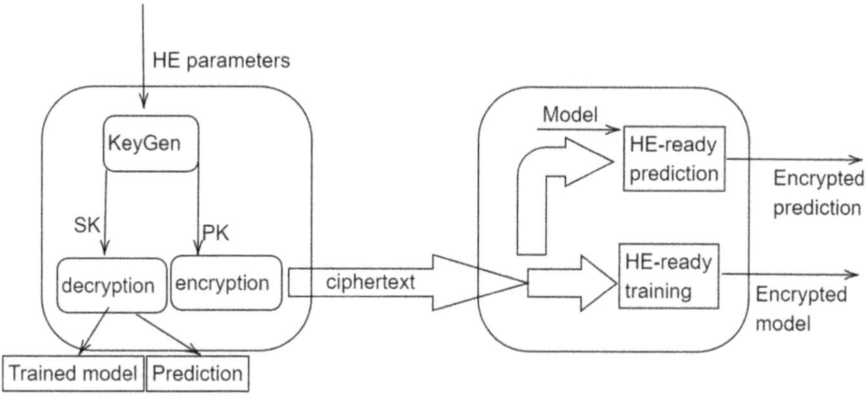

Fig. 4 HEML scenario

2.1 Logistic Regression

Logistic regression is a powerful machine learning approach that uses a logistic function to model two or more variables. Logistic models are commonly used in the medical community to predict binary outcomes, such as whether a patient requires treatment or whether a disease appears [12]. It has been utilized in applications such as evaluating diabetes patients' medications [13], and social sciences [14].

iDASH is an annual competition that attempts to deploy novel cryptographic methods in a biological environment. Since 2014, genomics and biomedical privacy have been incorporated into iDASH. Both the third track of the 2017 iDASH competition [15] and the second track of the 2018 iDASH competition driven the development of homomorphic encryption-based solutions for building a logistic regression model over encrypted data. The performance of LR training based on homomorphic encryption (HE) has improved significantly as a result of these two competitions. Homomorphic encryption has been used in much research on training logistic regression models.

Wu et al. [16] trained a privacy-preserving logistic regression model using HE; however, the time complexity of linear HE increases exponentially with the number of parameters. Aono et al. [17] used an additive HE scheme and delegated particular challenging HE computations to a trusted client, the authors in this work introduced an approximation to convert the likelihood function into a low-degree polynomial.

The issue of doing LR training in an encrypted environment was discussed by Kim et al. [18]. They used complete batch gradient descent in the training phase, using the least-squares approach to approximate the logistic regression. They also employed the CKKS scheme, which allows for a homomorphic approximation of the sigmoid function.

There is no closed-form solution to logistic regressions, so we must use non-linear optimization methods to find the maximum likelihood estimators of the regression parameters. During training, gradient descent and Newton-Raphson are the most commonly used methods. The Newton-Raphson method requires matrix inversion, and most HE schemes do not natively support division and matrix inversion. On the other hand, Gradient descent does not require the division operation and so is a better candidate for homomorphic logistic regression.

Although the gradient descent approach appears to be better suited for homomorphic evaluation than other training methods, some technical issues remain for implementation. The sigmoid function is the most challenging to evaluate since existing homomorphic encryption techniques only allow the evaluation of polynomial functions, so Taylor polynomials have been widely employed for sigmoid function approximation [19, 20]

For implementation and performance of private logistic regression (He-based solutions), see Table 3.

2.2 Naive Bayes and Decision Trees

Naive Bayesian classification is a simple probabilistic Bayesian classification based on Bayes' theorem. It uses a naive Bayesian classifier, or naive bayes classifier, belonging to the family of linear classifiers. Bost et al. [21] propose a privacy-preserving naive bayes classification algorithm. A client learns the classification of her data point X in their model without knowing the classification model or disclosing any information about her input. The model's estimated parameters are encrypted and transferred to a cloud server. The authors use two partially homomorphic encryption schemes, quadratic reciprocity [22] and Paillier [1], in the same work [21] implements a privacy-preserving strategy for three algorithms, and one of those is decision trees. This work has shown that polynomials may be utilized to express decision trees. The decision tree node values must be compared to the evaluation data, and the outputs must be used to construct the polynomial, yielding the evaluation results. For implementation and performance of private naive bayes and decision tree (He-based solutions), see Table 4.

2.3 K-Nearest Neighbors

The k-Nearest Neighbors (k-NN) a simple method that can handle continuous, categorical, and mixed data. Furthermore, as a non-parametric method, k-NN can be relevant for many data structures as long as the number of observations is sufficiently large. In addition, for a predefined number of neighbors k, the model does not require any training step since the prediction for a new observation is obtained by:

Table 3 Summary of main works on private prediction for logistic regression: "-" means that the information has not been disclosed

Ref.	HE scheme/Type	Platform	Evaluation time	Accuracy	Datasets
Logistic regression					
[16]	[1] /LHE	-	-	82.89%	Dataset SPECT—267 instances—23 features
[17]	[1] /LHE	2.60 GHz × 2 CPU, 128 GB RAM	-	73.7%	SPECTF heart dataset—267 instances—44 features)
[17]	[1] /LHE	2.60 GHz × 2 CPU, 128 GB RAM	-	80.7%	Pima diabetes dataset—768 instances—8 features
[18]	CKKS/LHE	intel Xeon 2.3 GHz processor with 16 cores and 64GB of RAM	131min	86.03%	Edinburgh—1253 instances—10 features
[18]	CKKS/LHE	intel Xeon 2.3 GHz processor with 16 cores and 64GB of RAM	101min	69.30%	Lbw—189 instances—10 features
[18]	CKKS/LHE	intel Xeon 2.3 GHz processor with 16 cores and 64GB of RAM	265min	79.23%	Nhanes3—15649 instances—16 features
[18]	CKKS/LHE	intel Xeon 2.3 GHz processor with 16 cores and 64GB of RAM	119min	68.85%	Pcs—379 instances—10 features
[18]	CKKS/LHE	intel Xeon 2.3 GHz processor with 16 cores and 64GB of RAM	109min	74.43%	Uis—575 instances,—9 features
[19]	YASHE/LHE	Intel Core i7- 3520M at 2893.484 MHz	-	-	Heart Disease Framingham—4000 instances—15 features
[20]	Linearly homomorphic encryption	Amazon EC2 c4.8xlarge machines running Linux, with 60GB of RAM each.	149.7sec	98.62%	MNIST dataset—60 000 instances—784 features

Table 4 Summary of main works on private prediction for naive bayes and decision tree "-" means that the information has not been disclosed

Ref.	HE scheme/Type	Platform	Evaluation time	Accuracy	Datasets
Naive bayes					
[21]	[1] + [22]/LHE	Two Intel Core i7 (64 bit) processors for a total 4 cores running at 2.66 GHz and 8 GB RAM	479 ms	-	Breast Cancer—2 classes—9 features
[21]	[1] + [22]/LHE	Two Intel Core i7 (64 bit) processors for a total 4 cores running at 2.66 GHz and 8 GB RAM	1415 ms	-	Nursery—9 classes—5 features
[21]	[1] + [22]/LHE	Two Intel Core i7 (64 bit) processors for a total 4 cores running at 2.66 GHz and 8 GB RAM	3810 ms	-	Audiology—14 classes—70 features
Decision tree					
[21]	[1] + [22]/LHE	Two Intel Core i7 (64 bit) processors for a total 4 cores running at 2.66 GHz and 8 GB RAM	239 ms	-	Nursery
[21]	[1] + [22]/LHE	Two Intel Core i7 (64 bit) processors for a total 4 cores running at 2.66 GHz and 8 GB RAM	899 ms	-	ECG

- Identifying the k nearest neighbors (according to a given distance)
- Computing the majority class among them (for a classification problem) or by averaging values (for a regression problem).

Homomorphic encryption has already been investigated by various authors for k-NN [23–26]. The authors of [23] suggested a homomorphic additive encryption scheme [1]. They investigated the privacy preservation in an outsourced k-NN system with various data owners. The untrusted entity securely computes the compu-

Table 5 Summary of main works on private *K*-nearest neighbors: "-" means that the information has not been disclosed

REF	HE scheme/Type	Platform	Evaluation time	Accuracy	Datasets
K-nearest neighbors					
[23]	[1]/LHE	-	-	97.85%	Cancer 1 (9 features)
[23]	[1]/LHE	-	-	96.49%	Cancer 2—569 instances,—30 features)
[23]	[1]/LHE	-	-	81.82%	Diabetes
[23]	[1]/LHE	-	-	97%	MNIST dataset—60 000 instances—784 features
[26]	[27]/FHE	Intel Core i7-6600U CPU	11.6 min	94.8	MNIST dataset—60,000 instances—784 features

tations of distances by using HE. However, the comparison and classification phases require interactions. Given that the computational and communication difficulties scale linearly, they admit that the method may not be practical for massive data volumes. The cost of communications between the entities is also a limitation in the deployment of this work [24].

Recently, [26] proposed a secure k-NN algorithm in quadratic complexity concerning the size of the database completely non-interactively by using a fully homomorphic encryption [27]. However, they assume that the majority vote is done on a clear-text domain, which is a significant security flaw that we will address here. Doing a majority vote on a clear-text domain imposes interaction between entities, which causes information leakage. For implementation and performance of private k nearest neighbors (He-based solutions), see Table 5

2.4 Neural Networks and Deep Learning

Deep learning is one of the most sophisticated techniques in machine learning, and it has received a lot of attention in recent years. It is presently employed in a variety of areas and applications, including pattern recognition, medical prediction, and speech recognition. Deep learning experiences are enhanced significantly by utilizing strong infrastructures such as clouds and implementing collaborative learning for model training. However, this compromises privacy, particularly when sensitive data is processed during the training and prediction stages, as well as when the training model is disseminated. In this section, we discuss known privacy-preserving deep learning algorithms based on homomorphic encryption, we present recent challenges concerning the intersection of HE cryptosystems and neural networks models, as well as methods to overcome limitations.

HE cannot be used naively in neural networks algorithms. There are a lot of challenges and restrictions that must be overcome. The constraints differ according to the scheme, however, several common issues emerge in most systems. The learning and inference phases of the deep learning algorithm can be distinguished.

2.4.1 Privacy-Preserving Deep Learning: Private Training

Several techniques have been proposed; they consider **collaborative training**, in which the training is performed collaboratively between different participants, or **individual training**, in which the training is performed by a single participant, such as a client who wants to use a cloud to train its model.

Aono et al. [28] proposed a solution in **collaborative learning** mode, where participants send the calculated encrypted local gradients to the cloud after each iteration of local training, starting with the initial weights obtained from the cloud. To ensure homomorphic ciphertext integrity, each participant uses a unique TLS/SSL secure channel. The cloud then updates the encrypted global weights vector, which the participants download. The approach theoretically achieves the same accuracy as standard asynchronous SGD, whereas evaluations show that MLP and CNN reach 97% and 99% accuracy, respectively. In terms of efficiency, an overhead in communication and calculation was seen; however, the authors considered it negligible. However, the accuracy/privacy trade-off might be adjusted to efficiency/Privacy, allowing the precision to be preserved while maintaining Privacy.

The privacy-preserving back-propagation technique described in [29] is based on BGV fully homomorphic encryption and Maclaurin polynomial approximation of the sigmoid activation function. The client encrypts input data and configured parameters before uploading them to the cloud, which executes one loop. The client downloads and decrypts the findings before updating its local model. It then encrypts and sends the updated parameters back to the cloud, which repeats the process. This method is repeated until the maximum error threshold or number of iterations is achieved. Although BGV encryption provides for the protection of private data throughout the learning process, it does need the approximation of the activation function. This might lead to a drop in accuracy. In terms of efficiency, while the solution achieved a two times greater efficiency, i.e., 45% of the training time of the standard model, it experienced compute and communication costs due to the encryption-related overhead.

Zhang et al. [30] employs the Taylor theorem to estimate the sigmoid activation function polynomially. The evaluation findings revealed a reduction in accuracy for both classification and prediction tasks. However, the authors proposed adding additional Taylor series terms to decrease classification loss, raising the BGV encryption level, resulting in poor performance. The method could achieve 2.5 times greater classification efficiency and two times higher overall efficiency in learning time.

Zhang et al. [31] described a more recent solution based on encryption. A client who wants to participate to the model's training transmits its data encrypted using the Paillier scheme [1] to the server, which performs all possible neural network calculations except non-linear activation functions. To continue execution, the encrypted weighted sums before each activation function are provided to the client, who will be in charge of performing the calculation. The result is then encrypted again and sent back to the server.

2.4.2 Privacy-Preserving Deep Learning: Private Inference

Fully homomorphic encryption is deployed in a line of research that performs private classification of encrypted data using a neural network that has been trained using plain data.

Gilad-Bachrach et al. [32] is the first solution for privacy-preserving deep learning for inference, developed by Microsoft Research. The approach is based on the YASHE a (leveled homomorphic encryption) LHE scheme was proposed. The user encrypts their data and sends it to the cloud, which runs the model and returns an encrypted prediction. It has since been demonstrated that the YASHE scheme is vulnerable to subfield lattice attack [33]. To make the network compatible with homomorphic encryption, max-pooling is replaced by a scaled-mean pool function, and activation functions are approximated by the square function, which is the lowest-degree non-linear polynomial function. According to the authors, these adjustments should preferably be considered during training on unencrypted data. For example, the solution could achieve 99% accuracy and 59,000 predictions per hour on a single PC for the MNIST dataset.

To overcome the heavy computation cost of HE, a dual cloud model was proposed, in which two clouds, A and B, collaborate to generate classification results in a secure environment [34]. Cloud A operates the neural network on private data encrypted by the client with Paillier cryptosystem [1], but delegates activation function computations to the cloud B since they share a key. The technique is repeated until the final layer is reached. Client A then protects the final output with a random salt from cloud B, which uses the softmax function and sends the final encrypted result to the client. A theoretical scenarios-based security and accuracy study demonstrated how the approach successfully defends against potential threats.

Chabanne et al. [35] suggested a method for classification problems over the CNN model based on BGV an FHE scheme. The combination of polynomial approximation and batch normalization is the major technological breakthrough. During the training phase, a batch normalization layer is introduced before each ReLU layer to avoid excessive accuracy deterioration, and max-pooling is replaced by average-pooling, which is more FHE-friendly and has a small overhead.

Prior to each ReLU, a batch normalization layer is introduced with a low-degree (2) polynomial approximation. When the model is complete, the user encrypts its private data and sends it to the model, carrying out the specified analysis. The evaluation findings revealed that the solution has a short running time, with comparable performance, as if there was no privacy.

Attempt to use HE for deep learning problems. They provide methods by using low-degree polynomials to approximate the most generally used neural network activation functions (ReLU, Sigmoid, and Tanh) [36]. This is a critical step in the development of effective homomorphic encryption methods. They then train convolutional neural networks using those approximation polynomial functions before implementing the models over encrypted data and evaluating its performance.

Zhu and Lv [37] suggest a recent homomorphic encryption-based approach. The user encrypts their personal information and transmits it to the server for prediction. The Paillier scheme accelerates linear, convolutional, and pooling transformations. The authors chose ReLU as the activation function and suggested, rather than utilizing polynomial approximation, an interactive protocol between the client and the server for its computation. The user gets the ReLU input data, decrypts it, and communicates the positivity or negativity of this input to the server, enabling the server to calculate the output. The evaluation findings revealed that the solution could reach near model accuracy in plain text and was similar to Cryptonet[32]. In terms of efficiency, the approach saves a significant amount of time.

Recently, authors in [38] have resulted in considerable improvements by using the scheme TFHE [27]. FHE methods permit unrestricted encrypted operations and give accurate polynomial approximations to non-polynomial activation functions using a programmable bootstrapping technique, an extension of the bootstrapping technique that allows resetting the noise in ciphertext to a fixed level while—at the same time—evaluating a function for free.

2.5 *Clustering*

Clustering is an unsupervised machine learning problem that automatically identifies natural grouping (clusters) in data. A clustering algorithm can be collaborative or individual. In both cases, a model can be based on a server, and the calculations are exclusively performed on the server or assisted by a server. In this case, some calculations are delegated to the server. Three models can be found in the literature:

1. data comes from several parties, and these parties collaborate to train a clustering model.
2. single party that holds the data but not the computational resources needed to perform the calculations. The data is outsourced to perform clustering.
3. data comes from multiple parties and is paired to build a shared database. Then the data is outsourced to perform clustering.

Cases 2 and 3 are similar; this case is called "outsourced clustering." The first case is called "distributed clustering." Plenty of clustering algorithms have been seen in the privacy-preserving framework: k-means, k-medoids, GMM, Meanshift, DBSCAN, baseline agglomerative HC BIRCH and Affinity Propagation. Among them, k-means has been intensively studied. In what follows, we focus on works that use homomorphic encryption.

2.5.1 Collaborative Clustering

In the case of collaborative clustering, several parties own data and want to collaborate to get good quality clustering without disclosing the information contained in the data. This case has been extensively studied in two parts. Liu et al. [39] interested in the case where two parties with limited computational resources would like to run k-means by outsourcing the computations to the cloud. Both parties will have a result based on both datasets. In this case, one party's data should be kept confidential from the cloud and the other party. The authors based two schemes to propose a solution: the Liu encryption and the Pallier encryption. Each party encrypts the data and sends it to the cloud. The cloud performs calculations and comparisons based on additional information about both parties. To recalculate the centers, the cloud sends the sum of all vectors to both parties, and the parties use a protocol to calculate the new centers. The authors [40] propose a protocol to perform secure k-means in the semi-honest model. In this work, the Pallier scheme has been used. The computation of the Euclidean distance requires interaction with the data owner to perform the multiplications. The comparison is performed using bit-by-bit encryption.

The authors [41] studied clustering using the k-medoids algorithm applied to intrusion detection. Multiple organizations collaborate to perform clustering and have better results without sharing the content of this information in the clear. In addition, the system relies on a semi-honest party to perform clustering using Pallier encryption. The k-medoid algorithm requires more complex operations than addition. This requires interactions between collaborators to decrypt this data at runtime and thus perform the operations.

2.5.2 Individual Clustering

A clustering is individual if only one person has data and he wants to have the results of the clustering of this data. Most of the works interested in this kind of clustering require an intermediate decryption step. The authors [42] demonstrate a solution to perform k-means using a collaboration between the client and a server. They used the BV scheme [43]. In this work, they proposed three variant solutions. Each solution takes as input a dataset of dimension nxd, an integer k which denotes a the number of clusters and a threshold of iterations. The algorithm returns a matrix of dimension kxd that indicates the cluster centers. In the first variant, the computation of the centers and the assignment are done at the client level, which implies that the client performs a lot of computations (only the distances are computed at the server level). In the second variant, the client performs the comparisons and the division. At the same time, the server calculates the distances and the assignment of the points, then the sum to calculate the new centers. This variant induces an information leakage on how the points are distributed on the clusters. Finally, a third variant tries to solve the information leakage problem by returning an encrypted assignment vector of a point instead of the clear assignment. The authors [44] propose a method for k-means that limits interaction with the data owner using the concept of "Updatable Distance Matrix (UDM)." The latter is a 3D matrix whose first two dimensions equal the number of data in the dataset, and the third dimension equals the number of attributes. Each cell in the matrix is initialized to the difference between the attributes of the data vectors. The idea is to save the encrypted data and the UDM matrix to a third party. This matrix is updated at each iteration of k-means using an offset matrix obtained by calculating the difference between the new and current centers. This method is expensive in terms of time and memory to store the UDM matrix.

The authors [45] have tried an exact implementation of k-means that requires no intermediate decryption. Instead, the method relies on building a logic circuit to perform k-means using TFHE. From a theoretical point of view, the method gives equivalent results to the plaintext version. However, this method is not feasible; with two dimensions and 400 points, the execution time has been estimated at 25 days.

The authors[46] also propose a solution that focuses on k-means. In this solution, the BGV scheme [47] is used. The authors remark that deciphering the intermediate steps at the client level is a costly operation. The proposed solution relies on using a third party as a trusted entity to decrypt the intermediate results. A private key equivalent (but different) to the owner's and a switch matrix are generated to be used by the trusted server. The proposed solution is considered secure in a semi-honest model but not in the malicious case.

3 Discussion and Challenges

Many challenges need to be tackled to apply privacy-preserving machine learning in real-world applications. Although the HE standards, platforms, and implementations described in this chapter contribute to the advancement of HEML, there are still specific remaining challenges to be tackled, including overhead, performance, interoperability, bootstrapping bottlenecks, sign determination, and common frameworks:

- **Overhead**: Compared to its unencrypted counterpart, HEML has significant overhead, making it unsuitable for many applications. However, for non-HE models, the training phase of ML comprises a computationally intensive effort. However, even with modern techniques, it becomes increasingly difficult with HE. A recent trend is to bypass the training step by employing pre-trained models to find a balance between complexity and accuracy.
- **Hardware Acceleration and parallelization:** Incorporating well-known and new leading to many algorithms is one approach to deal with the computational

overhead. High-performance computers, distributed systems, and specialized resources can all be used in HEML models. Multi-core processing units (GPUs, FPGAs, etc.) and customized chips (ASICs) provide more friendly and efficient HEML environments. Another approach to improving overall efficiency is batching and parallelizing numerous bootstrapping operations. To accelerate FHE programs, one of the research directions is to develop the ability to support multiple hardware acceleration, this is one of the projects under development in the PALISADE library[4].

- **Comparison and min/max function:** We need new methods to compare numbers which are encrypted by Homomorphic Encryption (HE). Actually, comparison and min/max functions are evaluated using Boolean functions where input numbers are encrypted bit-wise. However, the bit-wise encryption methods require relatively expensive computations for basic arithmetic operations such as addition and multiplication.
- **PPML tools:** For the deployment of these technologies, it is practically difficult to design a high-performing and secure PPML solution without a thorough HE understanding. However, PPML developers must be knowledgeable in both machine learning and security. PPML, which uses HE, has not been extensively accepted by the ML community due to HE's high barrier to entry and the absence of user-friendly tools. In terms of model accuracy, how we can ensure that the PPML Homomorphic encryption (HE), we need to develop metric for evaluating models in encrypted domain.
- **Hybrid protocols:** Adopting hybrid protocols, which combine two or more protocols to use the advantages and avoid the disadvantages of each, is a promising direction for performance improvements.
- **Homomorphic encryption (HE) with missing data:** Missing data are a significant problem, as the information available is incomplete and, therefore, less accurate. To solve this problem, we often use suppression of observations with missing data and imputation of missing data. The actual challenge is how to do these methods in the context of encrypted data by using homomorphic encryption.

References

1. P. Paillier, Public-key cryptosystems based on composite degree residuosity classes, in *International Conference on the Theory and Applications of Cryptographic Techniques* (Springer, 1999), pp. 223–238
2. R.L. Rivest, L. Adleman, M.L. Dertouzos, On data banks and privacy homomorphisms, in *Foundations of Secure Computation* (Academia Press, 1978), pp. 169–179
3. C. Gentry, Fully homomorphic encryption using ideal lattices, in *Proceedings of the 41st Annual ACM Symposium on Theory of Computing, STOC 2009, Bethesda, MD, USA, May 31 June 2, 2009*, ed. by M. Mitzenmacher (ACM, 2009), pp. 169–178. https://doi.org/10.1145/1536414.1536440
4. PALISADE Lattice Cryptography Library (release 1.11.5) (2021). https://palisade-crypto.org/

5. A. Acar et al., A survey on homomorphic encryption schemes: theory and implementation. ACM Comput. Surv. **51**(4) (2018). ISSN:0360-0300. https://doi.org/10.1145/3214303
6. M. van Dijk et al., Fully homomorphic encryption over the integers, in *Advances in Cryptology – EUROCRYPT 2010*, ed. by H. Gilbert (Springer, Berlin, Heidelberg, 2010), pp. 24–43. ISBN:978-3-642-13190-5
7. C. Gentry, Computing arbitrary functions of encrypted data. Commun. ACM **53**(3), 97–105 (2010). ISSN:0001-0782. https://doi.org/10.1145/1666420.1666444
8. O. Regev, On lattices, learning with errors, random linear codes, and cryptography, in *Proceedings of the Thirty-Seventh Annual ACM Symposium on Theory of Computing*, STOC '05 (Association for Computing Machinery, Baltimore, MD, USA, 2005), pp. 84–93. ISBN:1581139608. https://doi.org/10.1145/1060590.1060603
9. V. Lyubashevsky, C. Peikert, O. Regev, On ideal lattices and learning with errors over rings. J. ACM **60**(6) (2013). ISSN:0004-5411. https://doi.org/10.1145/2535925
10. K.R. Rohloff, D. Cousins, A scalable implementation of fully homomorphic encryption built on NTRU, in *Financial Cryptography Workshops* (2014)
11. Z. Chen et al., Bibliometrics of machine learning research using homomorphic encryption. Mathematics **9**, 2792 (2021). https://doi.org/10.3390/math9212792
12. T. Hastie, R. Tibshirani, J. Friedman, Unsupervised learning. *The Elements of Statistical Learning* (Springer, 2009), pp. 485–585
13. R. Bender, U. Grouven, Ordinal logistic regression in medical research. J. R. Coll. Physicians Lond. **31**(5), 546 (1997)
14. V. Gayle, P. Lambert, R.B. Davies, Logistic regression models in sociological research, in *University of Stirling, Technical Paper,* 1 (2009)
15. X. Wang et al., *iDASH secure genome analysis competition 2017* (2018)
16. S. Wu et al., Privacy-preservation for stochastic gradient descent application to secure logistic regression, in *The 27th Annual Conference of the Japanese Society for Artificial Intelligence,* vol. 27 (2013), pp. 1–4
17. Y. Aono et al., Scalable and secure logistic regression via homomorphic encryption, in *Proceedings of the Sixth ACM Conference on Data and Application Security and Privacy* (2016), pp. 142–144
18. M. Kim et al., Secure logistic regression based on homomorphic encryption: Design and evaluation. JMIR Med. Inf. **6**(2), e8805 (2018)
19. J.W. Bos, K. Lauter, M. Naehrig, Private predictive analysis on encrypted medical data. J. Biomed. Inf. **50**, 234–243 (2014)
20. P. Mohassel, Y. Zhang, Secureml: A system for scalable privacy-preserving machine learning, in *2017 IEEE Symposium on Security and Privacy (SP)* (IEEE, 2017), pp. 19–38
21. R. Bost et al., Machine learning classification over encrypted data. IACR Cryptol. ePrint Arch. **2014**, 331 (2015)
22. S. Goldwasser, S. Micali, Probabilistic encryption & how to play mental poker keeping secret all partial information, in *Proceedings of the Fourteenth Annual ACM Symposium on Theory of Computing*, STOC '82 (Association for Computing Machinery, San Francisco, California, USA, 1982), pp. 365–377. ISBN:0897910702. https://doi.org/10.1145/800070.802212
23. F. Li, R. Shin, V. Paxson, Exploring privacy preservation in outsourced K-nearest neighbors with multiple data owners, in *Proceedings of the 2015 ACM Workshop on Cloud Computing Security Workshop*, CCSW '15 (Association for Computing Machinery, Denver, Colorado, USA, 2015), pp. 53–64. ISBN:9781450338257. https://doi.org/10.1145/2808425.2808430
24. B.K. Samanthula, Y. Elmehdwi, W. Jiang, k-Nearest neighbor classification over semantically secure encrypted relational data. IEEE Trans. Knowl. Data Eng. **27**(5), 1261–1273 (2015). https://doi.org/10.1109/TKDE.2014.2364027
25. W.K. Wong et al., Secure KNN computation on encrypted databases, in *Proceedings of the 2009 ACM SIGMOD International Conference on Management of Data*, SIGMOD '09 (Association for Computing Machinery, Providence, Rhode Island, USA, 2009), pp. 139–152. ISBN:9781605585512. https://doi.org/10.1145/1559845.1559862

26. M. Zuber, R. Sirdey, Efficient homomorphic evaluation of k-NN classifiers. Proc. Privacy Enhanc. Technol. **2021**, 111–129 (2021)
27. I. Chillotti et al., TFHE: fast fully homomorphic encryption over the torus. J. Cryptol. **33**(1), 34–91 (2020)
28. Y. Aono et al., Privacy-preserving deep learning via additively homomorphic encryption. IEEE Trans. Inf. Forensics Secur. **13**(5), 1333–1345 (2017)
29. F. Bu et al., Privacy preserving back-propagation based on BGV on cloud, in *2015 IEEE 17th International Conference on High Performance Computing and Communications, 2015 IEEE 7th International Symposium on Cyberspace Safety and Security, and 2015 IEEE 12th International Conference on Embedded Software and Systems* (IEEE, 2015), pp. 1791–1795
30. Q. Zhang, L.T. Yang, Z. Chen, Privacy preserving deep computation model on cloud for big data feature learning. IEEE Trans. Comput. **65**(5), 1351–1362 (2016). https://doi.org/10.1109/TC.2015.2470255
31. Q. Zhang et al., GELU-Net: A globally encrypted, locally unencrypted deep neural network for privacy-preserved learning, in *IJCAI* (2018), pp. 3933–3939
32. R. Gilad-Bachrach et al., Cryptonets: Applying neural networks to encrypted data with high throughput and accuracy, in *International Conference on Machine Learning* (PMLR, 2016), pp. 201–210
33. M. Albrecht, S. Bai, L. Ducas, A subfield lattice attack on overstretched NTRU assumptions, in *Proceedings, Part I, of the 36th Annual International Cryptol- ogy Conference on Advances in Cryptology—CRYPTO 2016—Volume 9814* (Springer, Berlin, Heidelberg, 2016), pp. 153–178. ISBN:978-3-662-53017-7. https://doi.org/10.1007/978-3-662-53018-4_6
34. M. Baryalai, J. Jang-Jaccard, D. Liu, Towards privacy-preserving classification in neural networks, in *2016 14th Annual Conference on Privacy, Security and Trust (PST)* (2016), pp. 392–399. https://doi.org/10.1109/PST.2016.7906962
35. H. Chabanne et al., Privacy-preserving classification on deep neural network. Cryptology ePrint Arch. (2017)
36. E. Hesamifard, H. Takabi, M. Ghasemi, Deep neural networks classification over encrypted data, in *Proceedings of the Ninth ACM Conference on Data and Application Security and Privacy* (2019), pp. 97–108
37. Q. Zhu, X. Lv, 2P-DNN: Privacy-preserving deep neural networks based on Homomorphic cryptosystem. Preprint (2018). arXiv:1807.08459
38. I. Chillotti, M. Joye, P. Paillier, Programmable bootstrapping enables efficient homomorphic inference of deep neural networks, in *Cyber Security Cryptography and Machine Learning*, ed. by S. Dolev et al. (Springer International Publishing, Cham, 2021), pp. 1–19. ISBN:978-3-030-78086-9
39. X. Liu et al., Outsourcing two-party privacy preserving K-means clustering protocol in wireless sensor networks, in *2015 11th International Conference on Mobile Ad-hoc and Sensor Networks (MSN)* (2015), pp. 124–133. https://doi.org/1.1109/MSN.2015.42
40. Z.L. Jiang et al., Efficient two-party privacy preserving collaborative k-means clustering protocol supporting both storage and computation outsourcing. Information Sciences **518**, 168–180 (2020). ISSN:0020-0255. https://doi.org/10.1016/j.ins.2019.12.051. https://www.sciencedirect.com/science/article/pii/S0020025519311624
41. G. Spathoulas, G. Theodoridis, G.-P. Damiris, Using homomorphic encryption for privacy-preserving clustering of intrusion detection alerts. Int. J. Inf. Secur. **20**, 347–370 (2021). https://doi.org/10.1007/s10207-020-00506-7
42. A. Theodouli, K.A. Draziotis, A. Gounaris, Implementing private k-means clustering using a LWE-based cryptosystem, in *2017 IEEE Symposium on Computers and Communications (ISCC)* (2017), pp. 88–93
43. Z. Brakerski, V. Vaikuntanathan, C. Gentry, Fully homomorphic encryption without bootstrapping, in *In Innovations in Theoretical Computer Science* (2012)
44. N. Almutairi, F. Coenen, K. Dures, K-means clustering using homomorphic encryption and an updatable distance matrix: secure third party data clustering with limited data owner interaction, in *DaWaK* (2017)

45. A. Jäschke, F. Armknecht, Unsupervised machine learning on encrypted data. IACR Cryptol. ePrint Arch. **2018**, 411 (2018)
46. G. Sakellariou, A. Gounaris, Homomorphically encrypted K-means on cloud-hosted servers with low client-side load. Computing **101**(12), 1813–1836 (2019). ISSN:0010-485X. https://doi.org/10.1007/s00607-019-00711-w
47. Z. Brakerski, C. Gentry, V. Vaikuntanathan, (Leveled) Fully homomorphic encryption without bootstrapping, in *Proceedings of the 3rd Innovations in Theoretical Computer Science Conference*, ITCS '12 (Association for Computing Machinery, Cambridge, MA, 2012), pp. 309–325. ISBN:9781450311151. https://doi.org/10.1145/2090236.2090262

Part VII
Other Security Applications

The Design of Ethical Service-Level Agreements to Protect Cyber Attackers and Attackees

C. Peoples, A. Moore, and N. Georgalas

1 Introduction

Deterrence, which is the act of discouraging negative behaviours through a fear of the consequences if caught, in cyberspace is significantly different to deterrence in a non-electronic world. The challenge of anonymity, and the subsequent difficulty of attribution, identifying a person or persons associated with a crime, makes this an almost impossible situation to protect against [30]. Deterrence seeks to stop criminals in advance of taking action through an awareness that they will be punished for their crimes if caught, with the severity of the punishment influenced by the severity of the crime. Achieving deterrence effectively, however, requires ability to attribute a criminal to a crime, thus necessitating the removal of anonymous behaviour [16]. Recommendations are made by Lindsay [15] to deter others by enabling effective intrusion detection systems, as one reactive, and less preventative, example; the evidence suggests, however, that such mechanisms will ultimately, and most likely quickly, be worked around by the cyber criminals of today. The attack surface of electronic systems continues to evolve, and attackers adopt novel approaches to pursue their crimes. In an attempt to maintain anonymity in online attacks, third-party machines can become part of the end-to-end communication loop in more sophisticated attacks, thus thwarting efforts to uncover and attribute a person to a crime. It is not uncommon for cybercriminals to taunt the people that they have attacked from the background, using social media mechanisms such as Twitter [18]. As a consequence of factors such as these, Mike McConnell argued in 2010 that,

C. Peoples (✉) · A. Moore
Ulster University, Jordanstown, UK
e-mail: c.peoples@ulster.ac.uk; aa.moore@ulster.ac.uk

N. Georgalas
BT Group PLC, Martlesham, UK

K. Daimi et al. (eds.), *Emerging Trends in Cybersecurity Applications*,
https://doi.org/10.1007/978-3-031-09640-2_19

'we need to reengineer the Internet to make attribution, geolocation, intelligence analysis and impact assessment – who did it, from where, why and what was the result – more manageable'. Moving to the modern day, however, Norton Security reports more than 2200 cyberattacks every day, or one every 39 s [29]. This provides little evidence that much progress has been made on this front, and the problem remains as significant, if not more so, today. In October 2021, Forbes published, 'MORE Alarming Cybersecurity Stats for 2021!' [11]; High-profile attacks in 2021 include SolarWinds [3]; and Colonial Pipeline [28], and 2021 has been a challenging year for cybercrime: By October 2021, there were 17% more data breaches for the entire year of 2020 [2], verifying that cybercrime is in the worst position in its history.

Cybercrime has a different ethos from other forms of crime. The Minnesota House Research Department describes that economically poor areas are criminal hot spots, with the idea that a reduction in poverty can similarly reduce crime [16]. We argue that, in terms of online crimes, by way of contrast however, those who are poorer are less likely to have the resources to carry out such crimes. The Minnesota House Research Department goes further to posit that the most common method of addressing crime is enforcement and punishment. We argue that online crimes are typically low value; therefore punishment may realistically be relatively slight. Lindsay [15], in addition to recommending intrusion detection, advises making defences effective enough to deny an adversary. This is a perspective which we agree with, in that taking preventative action in advance of the act is preferable to reactive action after the event.

The National Crime Agency (NCA) describes that 61% of hackers began hacking before the age of 16 and the average age of suspects in 2015 was 17 years old [23]. It may therefore not be surprising that Norton suggests that, to protect against cyberattack, keep having dialogue with children [4]. A number of notable cybercrime attacks have been carried out by youth, such as Michael Calce (15 years old), Jack Chappell (15 years old), and Jonathan James (15 years old). We question, however, the relevance of having dialogue with children: Cameron, for example, carried out denial-of-service (DoS) attacks on gamers known to him at school. Parents got involved; however, Cameron later acknowledged that the parents did not understand what had happened [27]. He continued his life of cybercrime, which ultimately led to his arrest at 14 years old. The dialogue with his mother evidently did not help in this instance.

When considering the steps to deter cybercrime, Taddeo [31] makes reference to dealing with 'rational' beings. The NCA describes further, however, the suspected link between cybercrime and autism (although it is yet to be formally proven) [23]. We might therefore question the extent to which it is reasonable to assume that anyone who wishes to carry out cyberattacks is rational. Furthermore, evidence indicates that a number of the major notable cyberattacks have been carried out by beings who may be described as being irrational – Jack Chappell is autistic, Raphael Gray is mentally ill, and Gary McKinnon has Asperger's syndrome.

It is therefore on the basis that cybercriminals are often in their teenage years, cybercrimes are typically low gain, and it is not uncommon that there are underlying medical conditions that our research proposal is made.

2 The Challenges of Managing Online Crime

In general, it is difficult to associate a person with a crime in the online world due to the ability to remain anonymous. The significant challenge this proposes is that it makes it difficult to apply effective deterrence strategies (Fig. 1). Deterrence strategies are considered to be fundamental in inhibiting would-be attackers [16].

A more optimum approach may be to prevent criminals from attacking in advance, which can be a more possible and plausible strategy in the case of young people with mental illness than other user types. As discussed above, there are cases where a guardian has been present, such that they may be more influential in protecting against the execution of a cyberattack, as in the situation with Cameron. In another example of where an attack may have been similarly detected: Jack Chappell flooded the NatWest, National Crime Agency, the BBC, and Netflix networks [30]. If these attacks took place from home, observations on traffic activity leaving the home hub could have led to the youth being detected before criminal action was taken against them. Indeed, Chappell was arrested in April 2016 after investigators traced his Internet address to his home in the UK [1]. Quicker and more autonomous action might have been able to prevent this arrest.

Machine learning of user activity information can be applied to pre-empt where and when sensitive data may be stolen as part of a cyberattack [21]. We argue that there are opportunities to take action before this point in the criminal activity chain: The role played by the guardian in Chappell's situation could have been significant, yet it was an underused resource. The routine activity theory (RAT) [12] posits the theory that crime is more prevalent when an attacker is in close proximity to a target and a reliable guardian is absent (Fig. 2). However, in the scenarios of Cameron and Jack Chappell, guardians were present, and the attackers were far away from the attack points, yet their crimes were able to occur. The ability to carry out the crime remotely therefore removes the significance of distance between an attacker and the

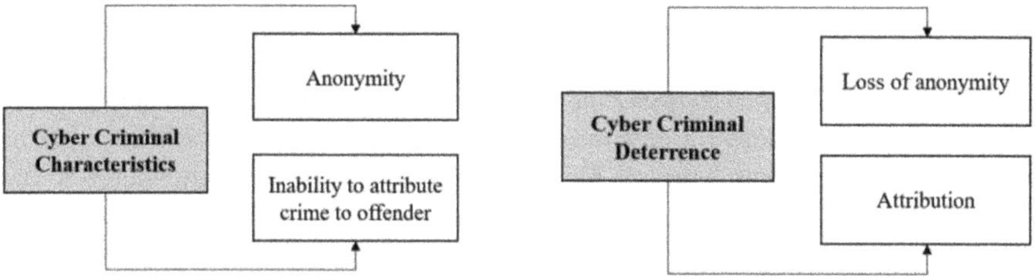

Fig. 1 One challenge of attributing crime in the online world

Fig. 2 Roles played within the routine activity theory [12]

attacked in the online world. The role of the guardian may therefore be considered to be even more significant in this scenario.

Online services, in general, are purchased through a service-level agreement (SLA). A SLA specifies the way which a network service will be provided, with quality being measured from the perspective of the person paying for it; in the case of network services, platform uptime is an attribute commonly used to indicate achievement of a SLA. A high platform uptime will facilitate greater guaranteed platform availability, with a subsequent potential higher guarantee of service availability. It is the intention that the Quality of Service (QoS) guarantees will respond directly to user Quality of Experience (QoE). We can argue that the RAT is supported through the provision of a SLA, in that the ability to accept a network service in a home is dependent on the presence of a guardian. If a member of a household has access to a network connection, it might therefore be assumed that a guardian will be present and able to act in the role of preventing crime.

However, youth today are referred to as 'Generation X' [17]; they are the members of society who do not know a life without connected technology. Guardians of this youth, on the other hand, know a different way of life and are unfamiliar with many online technologies and operations. It might therefore be considered that, in this respect, while making an online service available, there is an absence of an effective guardian, when we consider it from the perspective of the routine activity theory (Fig. 2) [12].

This fact becomes even more significant when we consider that a number of young attackers have challenges which are known to their guardians (extended from Peoples et al. [25]) (Table 1), but were unable to be stopped in their criminal behaviour, even when parents became involved.

Table 1 provides some evidence of the fact that groups of attackers can have either challenging home situations and/or a mental health condition. These unfortunate situations are unlikely to be under the direct control of the people who later become attackers: Some of them are able to turn their situation completely around, such as Cal Leeming and Kevin Mitnick. On the other hand, there are a number who are not able to and fall victim to carrying out online attacks. Several of the attackers showed visible remorse when convicted for their crimes, such as Zachary Buchta and Adrian Lamo, and a few unfortunately died at a young age, including Jonathan James and Adrian Lamo. We argue that this makes them vulnerable and that there

Table 1 Cyber attackers as vulnerable citizens

Attacker name	Why are these attackers vulnerable?	Age when attack carried out	How attack was or may have been identified
Zachary Buchta	Family tragedy and isolated at home. Suffered from anxiety and depression. Showed remorse for crimes	19 years old	Harassing phone calls
Jack Chappell	Autistic (challenges with social skills, repetitive behaviours, speech, and nonverbal communication)	15 years old	DDoS attack by flooding company networks
Edwin Robbe	Fostered, but always a troubled child. Couldn't bond with others. Suffered with anxiety	17 years old	DDoS attack by flooding company networks
Max ray Butler	Parents divorced when 14 years old; max then lived with his father	17–26 years old	Stole credit card numbers and carried out fraudulent activities
Michael Calce	Parents separated; spent week with mother and weekends with father	15 years old	Denial-of-service attacks
Raphael gray	Mentally ill	19 years old	Published credit card details from consumer websites
Jonathan James	High intelligence/mentally ill	15 years old	Intrusion into the United States Department of Defense computers
Adrian Lamo	Did not graduate from high school. Mentally ill. Showed remorse for crimes	22 years old	Broke into high-profile computer networks
Cal Leeming	Parents separated. Experienced poverty and neglect as a young child	12 years old	Responsible for stolen identities and subsequent purchasing of £750,000 worth of products
Lauri love	Asperger's syndrome	26 years old	Stolen data from US agencies, including NASA and the FBI
Gary McKinnon	Asperger's syndrome	35–36 years old	Hacked into United States military and NASA computers
Kevin Mitnick	Parents separated when he was 3 years old	16–31 years old	Computer fraud and illegally intercepting communications
Kevin Poulsen	Adopted, but had little contact with his adoptive parents. A shy, gifted child. A 24-hours-a-day hacker.	17 years old	Stealing military order
Cameron	Competitive gamer	14 years old	Flooded the IP address being used by an online gamer

are opportunities to protect similar citizens more suitably in the future – machine learning from their observed behaviours may be too late.

There is some evidence that service providers have made attempts at doing this, albeit in an ad hoc approach: Edwin Robbe was eventually attributed with the crime of flooding the network of a major telecommunication company in the Netherlands, KPN. Prior to this, KPN had made attempts to communicate with the homeowner that they had monitored suspicious network activity at the family home. Edwin was able to mislead his father by using technical wording which his father did not understand, and he took no further action [22]. This decision ultimately led to Edwin's arrest. Effort had therefore been made to take action against the signs of potential cybercriminal behaviours and activities by the service providers through alerting the guardians; however, the attacks were allowed to persist.

In this instance, the guardian failed to protect their child by not understanding the situation in its entirety. Service providers can play some role in facilitating protection by educating parents on possible relevant protective mechanisms to apply through the use of parental controls. The extent to which parental controls provide protection varies depending on the service provided – a range of protection mechanisms from a variety of service providers are summarised in Table 2.

Within the context of this work, 'protection' includes protecting other network users from the customer operating under the SLA, which might include a member of a customer's household, such as a child. This is facilitated when the parental controls can be applied and be effective in limiting malevolent behaviours. In this context, the 'potentially attacked' is being protected from an attacker, and the attacker is being protected from being able to carry out cybercrimes (Fig. 3).

The SLA goal when offered to an offender's household is therefore to prevent the offender from carrying out the attack, while in the case of a user who might be attacked, the goal of the SLA is to protect the victim. It is from this perspective that we consider network users to be vulnerable.

Considering Table 2 in more detail, the Sky Broadband Buddy gives the ability to restrict online access on a per device basis using filters. Access regulation may also occur across a home in its entirety, in addition to applying limits to the amount of time that a user can be online [9]. A log of sites visited can also be accessed. The Buddy is available as an app, in two forms – a parent app is used to manage all profiles and features, and a kid's app manages online time and activity when outside the home. The Buddy applies a rewards feature, which grants more time for favourite applications. Plusnet Safeguard is a feature of all broadband packages [8]. It enables

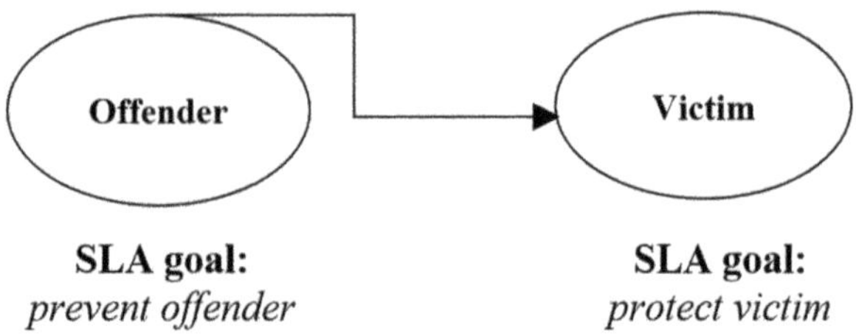

Fig. 3 The role of a SLA in protecting vulnerable online behaviours

Table 2 Parental controls on household broadband services

BT [6]	Sky [9]	Vodafone [5]	Virgin Media [7]	Plusnet [8]
Light, moderate, and strict filters, with 'strict' blocking all of the following: Pornography, obscene and tasteless, hate and self-harm, drugs, alcohol and tobacco, dating, nudity, weapons and violence, gambling, search engines, sex education, media streaming, social networking, fashion and beauty, file sharing, games Allows a list of safe sites to be created Ability to turn off filters for a period of time	Automatically blocks 18+ – rated content; additionally customise which websites are blocked Three age rating settings – PG, 13, or 18 Automatically switch to 18 age rating	Automatically blocks 18+ – rated content. Control options provided for android, iPhone, and windows 10 phone, not the responsibility of Vodafone	Parental controls configurable within each application, not the responsibility of virgin media	Ability to block adult content, gambling, social media, and violent images. Allows a list of safe sites to be created Allows a browsing time limit to be set

ability to block adult content, gambling, social media, and violent images. It also allows a list of safe sites to be created and a browsing time limit to be set.

Vodafone, EE, O2, and Three automatically block content which is rated 18+ when connected to the phone network [5]. Vodafone describes parental controls as managing the sites which children view, the apps that can be downloaded, and the purchases which can be made. The devices will not block this content, however, if the device is connected to Wi-Fi. Vodafone presents a list of options which parents have to control and manage access on their website, with the features dependent on the device being used, i.e. Android, iPhone, and Windows 10 Phone. Specific to an Android device, as one example, the recommended best advice is to create a separate user account for a child. Detailed guidance for the steps to follow to set up the controls is discussed by Martin [19]. When controlling access during browsing, the configurations made available by Chrome, IE, Firefox, and so on must be manually applied on every browser which is used across the home. This could be a relatively intensive process, dependent on the number of users and devices in the home. The Vodafone restrictions depend on the operating system used by the mobile device and the applications being used. The filters do not apply, however, when connected to Wi-Fi, either within or outside the home. Vodafone does not block content on behalf of users or provider capability to do so and instead relies on the controls made available by the hardware manufacturers and application providers.

Virgin Media does not offer an SLA for residential users. The compensation scheme is put in place by Ofcom [24]. This obviously operates with a very restricted set of attributes, compensating on the basis of loss of service, failure to instal a service, and failure to arrive for an arranged appointment. Parental controls are available for Virgin Media services on a per-application basis and are configurable within the applications themselves [7]. While Virgin provides guidance on how these controls may be applied, the provision of parental control is not an explicit feature of a Virgin Media broadband service.

When accessing the Internet via a BT account, a parent has the option to control the service accessible. With this approach, restrictions are applied to all devices connected to the service – they can be applied either on the Hub or accessed through a customer's 'My BT' account. Filter levels include 'Light', 'Moderate', and 'Strict' and filter content which include, but are not limited to 'alcohol and tobacco', 'weapons and violence', and 'gambling' (Table 2). When a user with a home SLA connects to a BT hotspot through their home account login detail, they connect subject to the performance in the region surrounding the hotspot, and not to the terms to which they have agreed for their home SLA. This may result in an inadequate service, particularly for vulnerable users.

Through this review, it can be appreciated that the opportunities of applying controls to the ways in which services are offered vary quite widely between providers – some offer capabilities to block websites, while one in our review does not offer any explicit tools or mechanisms, beyond what the application (e.g. social media or banking) offers. In addition to this, we recognise that, despite offering capabilities, all customers do not necessarily understand the significance of the configuration options or, indeed, what the residents need in response to network

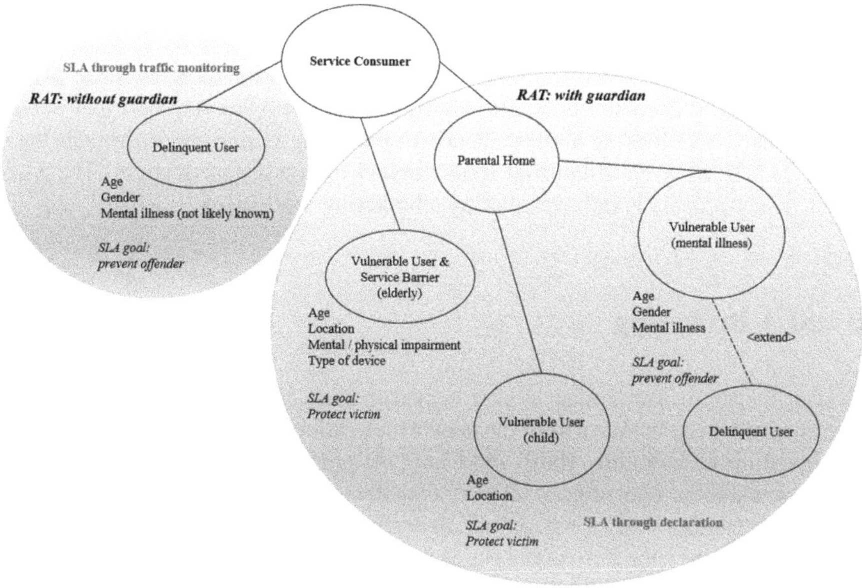

Fig. 4 Vulnerable user groups to be protected using ethical SLAs and the attributes potentially indicating their vulnerability

activities ongoing in their background. We therefore argue that there is value to be gained from automating the service protection provided. Automating the capabilities can provide additional protection through removing a layer which requires user interaction; in the context of the SLA, this involves knowing what the household needs to be protected against and asking the correct questions to ensure that it is put in place. We consider this to support the provision of 'ethical SLAs', through knowing how to respond in a manner adequate for the needs of those who can benefit from it within a household. It is therefore to this concept that our research proposal is made.

So far, the work has been positioned from the perspective of benefitting citizens who may have a potentially vulnerable member in the household. It is significant to note that our intention is that the proposal can additionally protect those who may be older and living alone. The full suite of users who we anticipate can be protected through the ethical SLA provisions are presented in Fig. 4.

We argue that all users in Fig. 4 are potentially vulnerable, in the sense of being attacked or being an attacker. Vulnerable users may have a mental illness which, in certain circumstances, can lead them to being a delinquent user, or they may be vulnerable as they are a child. This proposal does not seek to discriminate against those with autistic needs but rather to protect them and others operating within the online world. Often with disabling anxiety or high functioning characteristics with poor self-monitoring and impulse control, we recognise the benefits of limiting the extent to which they engage online, in both the volume of activities and the durations. Service consumers not in the presence of a guardian may be vulnerable through being older, or they may exploit vulnerable users through being a delinquent user. Our proposal is therefore considered to support the formation of ethical SLAs, with ethical in this context referring to protecting those who may find themselves in secondary activities as a consequence of operating online, be it through legal action or following up a financial fraud attempt on a personal account. The SLA provisioning process is influenced using a brokering mechanism.

3 SLA Brokering

Driving the ability to automatically create personalised and ethical SLAs requires the presence of a SLA broker within the network. A variety of SLA broker versions are discussed in the literature. Halili and Cico [14] propose a SLA framework, which uses a scheduling algorithm to allocate cloudlets to virtual machines (VM) based on host processing time, a scheduling algorithm which allocates cloudlets to VMs dependent on the cloudlet counts on each VM, and an algorithm to consider the VM credit to influence VM migration. This is more of an autonomous operation, ongoing in the background, as opposed to working directly with a customer in the configuration of their SLA. This mechanism works to ensure that the network will fulfil a customer's requirements. Considering the broker in more detail, it plays a facilitating role of inter-operating between cloud users and cloud service providers dependent on the user QoS requirements. The way in which this operates specifically, however, is not provided in the paper.

Haddar et al. [13] propose a multi-criteria value function or multi-attribute utility theory to allow a user to select a service from a number of choices. This operates on the basis of offering choices based on various criteria, the compensatory method, and an outranking approach, which checks if the service selected exceeds any other alternative and maintains the one with the best advantage in the case that they conflict. Metrics used to evaluate the suitability of the SLA include service type, availability, and service specification. One way it is possible to critique this approach is in relation to the technical knowledge that is required by a customer to drive their SLA selection. Without this knowledge, it is possible to assume that the SLA which the customer requests may potentially be inappropriate for their needs. Vimercati et al. [32] respond to such a criticism in their SLA provisioning proposal, proposing an approach to support SLA provision for customers who include those without particular technical competence to understand the terms involved and therefore the suitability of the SLA to which they have agreed and the scheme allocated. This is achieved by offering the various measures of a SLA, including throughput, bandwidth, performance, reliability, and reputation, to note a few. In response to each metric, customers are required to indicate their preference for a high, medium, or low service. Where users are unfamiliar with the ways in which services are offered and the meaning of the key measures, it might be questioned if they will have an understanding of whether they need a high, medium, or low service. Furthermore, 'high' is a subjective term and will most likely mean different things to different people. It may therefore be questionable if the scheme will be as useful as intended when applied in practice.

The proposal which we make is in line with the ideas of Vimercati et al. [32] that network users should not need to have significant intelligence and/or familiarity around SLA provisioning processes. We remove the need for specific knowledge on the variety of resources provisioned as part of a SLA and whether they need these to be provisioned in high, medium, or low volumes.

4 Proposed SLA Brokering Solution

Through automating the ways in which a service is provisioned to the home, we propose to adapt the extent to which a potential cybercriminal can engage with online misdemeanour using the household SLA and a potential attackee can be attacked. This recognises that cybercriminals, in a number of cases, are youth and that they carry out criminal activity from a guardian's household. It also recognises that those who are attacked can be aged at the extremes of typical user ages, in the sense of being older or younger than an otherwise average age. We recognise that a guardian can be unaware of criminal activities ongoing in their household, both in the sense of those who are attacked and those who are attackers, or a need to protect others or themselves through the configuration of parental controls on online services. We subsequently seek to provide a mechanism to implement advance preventative and protective measures. We aim to treat network users as vulnerable people as opposed to cybercriminals, in recognition of the conditions to which they might be considered to be a victim of.

The first step in this process is to gain an understanding of the household within which the potentially vulnerable user is a resident. This is achieved on the basis of location, and we have examined the mechanics behind this approach in some detail in our prior work [26]. In this chapter, we consider this from a different perspective and summarise the homes according to the metrics of income, education, technology, Internet, employment, house type, and average age. Based on the extent to which the households are considered to experience income or education or technology, they will be awarded a score between 1 and 5. An overall score will be assigned, with a lower score indicating a potentially more vulnerable household, i.e. a generally younger household, lower income, and fewer employment opportunities.

Households are considered according to two major categories – those in which the guardian may be vulnerable (Table 3) and those in which the guardian may not be vulnerable but someone in the household may be (Table 4; Figs. 5 and 6).

Homeowners can be classified automatically depending on their proximity from a town/city centre and a pre-defined characterisation of their assumed traits on this basis. Homeowners can also be classified based on their self-assignment according to the metrics of income, education, technology, Internet, employment, house type, and average age in the household – these are the core ways in which the customer

Table 3 Guardian may be vulnerable

Group	Sub-group category	Group description	Group score
Mature money	Better-off villagers Settled suburbia, older people Retired and empty nesters Upmarket downsizers	Older and retired couples, prosperous, high-income households, homeowners, less likely than average to have a smartphone, regular holidays	Income: 5 Education: 5 Technology: 3 Internet: 2 Employment: 1 House type: 5 Average age: 4
Score: 25			
Steady neighbourhoods	Suburban semis, conventional attitudes Owner occupied terraces, average income Established suburbs, older families	Home-owning families, houses are older, lower priced, three-bedroom terraced or semi-detached homes, employed in middle management, household income around average, few will go online extensively on a regular basis	Income: 3 Education: 4 Technology: 2 Internet: 2 Employment: 3 House type: 3 Average age: 3
Score: 20			
Comfortable seniors	Older people, neat and tidy neighbourhoods Elderly singles in purpose-built accommodation	Retired and empty nester couples, homes slightly below the average value for the area, modest incomes, many living off pension, unlikely to use the internet more than sporadically for practical purposes, their phone is unlikely to be able to access the internet	Income: 3 Education: 2 Technology: 1 Internet: 2 Employment: 0 House type: 2 Average age: 5
Score: 15			

classifications are observed to differ. (There is also the opportunity to offer customers the opportunity to have the automatic classification assignment made and compared against their self-characterisation to assess the accuracy of the model, although this approach is not considered as part of this work.)

Customers are scored according to their key distinguishing features. The higher the score, the greater the attention given to the metric within the household, i.e. a household with an 'income' scored 3 ('Steady Neighbourhood') will have a lower score than a household scored 5 ('Mature Money'). Customers whose scores are outliers are considered to be more vulnerable than customers whose score lies around the average. To contextualise this: A 'Mature Money' household scores 25 in Table 4, while 'Comfortable Seniors' score 15. 'Mature Money' households may be more vulnerable than others because of their financial situation and their

Table 4 Guardian may not be vulnerable but someone in home may be vulnerable

Group	Sub-group category	Group description	Group score
Executive wealth	Asset-rich families Wealthy countryside commuters Financially comfortable families Affluent professionals Prosperous suburban families Well-off edge of towners	Wealthy families in detached homes. Many own a second home. Earn more than average. Frequent internet users. Have modern technology and children with games consoles.	Income: 5 Education: 5 Technology: 5 Internet: 5 Employment: 5 House type: 5 Average age: 4
Score: 34			
Career climbers	Career-driven young families First time buyers in small, modern homes Mixed metropolitan areas	Younger people with young children. Renting, or often with a mortgage. Higher educational qualifications than usual. Some may have difficulties with debt. Confident users of new technology and frequency internet users.	Income: 3 Education: 4 Technology: 5 Internet: 5 Employment: 4 House type: 3 Average age: 2
Score: 26			
Countryside communities	Farms and cottages Larger families in rural areas Owner occupiers in small towns and villages	Lower turnover of home ownership than usual. People are older than the average. Incomes may be lower than the average but families have built-up savings. Pastimes are influenced by living in rural locations.	Income: 2 Education: 3 Technology: 3 Internet: 3 Employment: 4 House type: 4 Average age: 5
Score: 24			
Successful suburbs	Comfortably off families in modern housing Larger family homes, multi-ethnic areas Semi-professional families, owner occupied neighbourhoods	Living in homes of an average value. Homes include young children, teenagers, or young adults who haven't left home. Incomes are at least of average levels. Some savings. Occasional rather than heavy users of the internet. Phones are likely to have internet capability.	Income: 4 Education: 4 Technology: 3 Internet: 3 Employment: 4 House type: 4 Average age: 3

(continued)

Table 4 (continued)

Group	Sub-group category	Group description	Group score
Score: 25			
Starting out	Educated families in terraces, young children	Young couples in first home, starting a family. Household incomes tend to be above average. This is the internet generation, spending more time online than average. New technology includes smartphones and tablets.	Income: 4 Education: 4 Technology: 5 Internet: 5 Employment: 4 House type: 3 Average age: 3
	Smaller houses and starter homes		
Score: 28			
Student life	Student flats and halls of residence	Little in the way of incomes. Some utilising overdrafts or building up debts. Internet use likely to be extensive. Ownership of smartphones, tablets, and handheld computers well above average.	Income: 1 Education: 2 Technology: 5 Internet: 5 Employment: 1 House type: 0 Average age: 1
	Term-time terraces		
	Educated young people in flats and tenements		
Score: 15			
Striving families	Labouring semi-rural estates	Low-income families. Many rent their homes. High numbers of children are typical, high numbers of single parents. Incomes well below the national average, and unemployment is above average. A proportion may be reliant on state benefits. Majority won't have credit card. Phone is less likely to have internet capabilities. People are less likely to purchase the latest technological goods.	Income: 1 Education: 2 Technology: 1 Internet: 1 Employment: 2 House type: 1 Average age: 2
	Struggling young families in post-war terraces		
	Families in right-to-buy estates		
	Post-war estates, limited means		
Score: 10			
Young hardship	Young families in low-cost private flats	Younger people are prevalent, living in the cheapest housing in town. Educational qualifications are lower than average. There is deprivation in this group. Incomes are moderate to low, and unemployment is higher than the national average. Households with high levels of debt. Some will own smartphones, although likely to be a less expensive less fashionable model.	Income: 2 Education: 2 Technology: 2 Internet: 2 Employment: 2 House type: 2 Average age: 2
	Struggling younger people in mixed tenure		
	Young people in small, low-cost terraces		

(continued)

Table 4 (continued)

Group	Sub-group category	Group description	Group score
Score: 14			
Struggling estates	Poorer families, many houses, terraced housing Low incomes terraces Multi-ethic, purpose-built estates Deprived and ethnically diverse in flats Low-income large families in socially rented semis	Low-income families. Two-thirds rent their homes. House prices are low. High proportion of children and high number of single-parent households. Jobs reflect lower educational qualifications, of a routine nature. Incomes are low and many are claiming benefits.	Income: 1 Education: 1 Technology: 2 Internet: 2 Employment: 2 House type: 2 Average age: 3
Score: 13			
Difficult circumstances	Social rented flats, families, and single parents Singles and young families, some receiving benefits Deprived areas and high-rise flats	Higher proportion of younger people. Twice as many single parents compared to the national average. The bulk of housing is rented flats. Deprived neighbourhoods. High levels of long-term unemployment, households relying entirely on state benefits. Educational qualifications are low. Those employed are in routine semi-skilled manual jobs. Incomes are particularly low. There may be a higher than usual proportion of people with health problems. Life is a struggle.	Income: 0 Education: 1 Technology: 1 Internet: 1 Employment: 1 House type: 0 Average age: 3
Score: 7			

likelihood of owning modern technology. ‘Comfortable Seniors’ are considered to be vulnerable because of their age heavy profiles and relative financial income. ‘Steady Neighbourhoods’, by way of contrast, have average income levels and average ages, which makes them overall considered to be less vulnerable.

Once the customer classification has occurred, the vulnerability assessment is then made. All SLA requesters will be asked if they wish to be considered as a vulnerable household. In the instance that they wish to be considered as a vulnerable household, the household is first profiled at a high level (Fig. 7) to understand some demographic detail about residents in the household. This considers the age

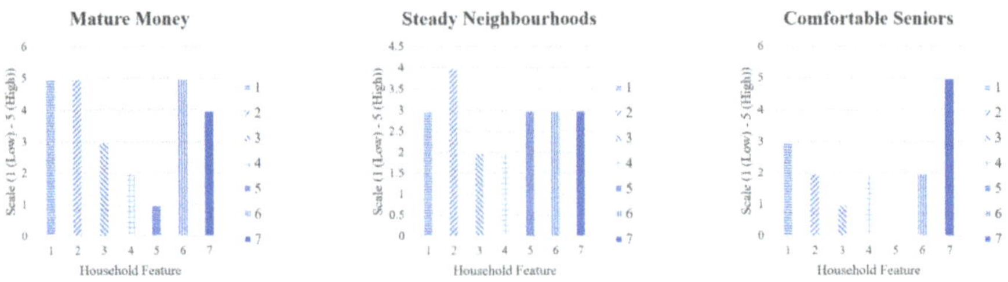

Key: 1: Income; 2: Education; 3: Technology; 4: Internet; 5: Employment; 6: House Type; 7: Average Age.

Fig. 5 Characteristics of Table 3 households according to core metrics

breakdown of residents, gender, and the presence of any physical and/or mental illness.

Further detail can then be gathered once it is ascertained that there is some degree of vulnerability in the household. Specific detail gathered in the case of a resident aged between 11 and 18 years old is captured in Fig. 8. The goal of investigating this detail is to examine the extent to which the household is vulnerable; the higher the vulnerability score, the more vulnerable the household.

In the case that male residents with mental illness are identified, a score is awarded on the basis of the severity of the mental illness. The dependency of the resident on the network connection is examined, in addition to the household owner's perspective of the vulnerability of the resident being considered – it is possible that a resident has a mental or physical condition while they are not considered to be particularly vulnerable.

Households are therefore subsequently given one of a number of vulnerability scores. A selection of user profiles is presented in Table 5 to contextualise the ways in which a household may be assigned a vulnerability score, as a function of the profiles of residents within the household. Within each field there is a taxonomy of options, and each option has been pre-assigned a score (e.g. 4 for autism, 5 for schizophrenia, and so on).

Once the household and vulnerability classifications have been made, the personalised SLA assignment can then be made.

5 SLA Assignments

The influence of the household and vulnerability classifications on the SLA assignment is captured in Fig. 9. The idea behind our approach is to throttle resource provisions when a household is identified as being vulnerable from the perspective of being an attacker. Given the multiple definitions of vulnerability from our

Table 5 User profiling and subsequent score assignments

Gender	Mental illness		Physical illness	Anything else	Reliance on Internet connection	Considered to be a vulnerable user?	Vulnerability score
Male	Yes (+5)	Autism (+4)	No	No	Yes (+2)	Yes (+5)	**16**
Male	Yes (+5)	Schizophrenia (+5)	No	No	Yes (+2)	Yes (+5)	**17**
Male	Yes (+5)	Depression (+2)	No	No	Yes (+2)	Yes (+5)	**14**
Male	Yes (+5)	Anxiety (+1)	No	No	Yes (+2)	Yes (+5)	**13**
Male	Yes (+5)	Eating disorder (+1)	No	No	Yes (+2)	Yes (+5)	**13**
Male	Yes (+5)	Addictive behaviour (+4)	No	No	Yes (+2)	Yes (+5)	**16**
Male	Yes (+5)	Autism (+4)	No	No	No	Yes (+5)	**14**
Male	Yes (+5)	Schizophrenia (+5)	No	No	No	Yes (+5)	**15**
Male	Yes (+5)	Depression (+2)	No	No	No	Yes (+5)	**12**
Male	Yes (+5)	Anxiety (+1)	No	No	No	Yes (+5)	**11**
Male	Yes (+5)	Eating disorder (+1)	No	No	No	Yes (+5)	**11**
Male	Yes (+5)	Addictive behaviour (+4)	No	No	No	Yes (+5)	**14**
Male	Yes (+5)	Autism (+4)	No	No	Yes (+2)	No	**11**
Male	Yes (+5)	Schizophrenia (+5)	No	No	Yes (+2)	No	**12**
Male	Yes (+5)	Depression (+2)	No	No	Yes (+2)	No	**9**
Male	Yes (+5)	Anxiety (+1)	No	No	Yes (+2)	No	**8**
Male	Yes (+5)	Eating disorder (+1)	No	No	Yes (+2)	No	**8**
Male	Yes (+5)	Addictive behaviour (+4)	No	No	Yes (+2)	No	**11**
Male	Yes (+5)	Autism (+4)	No	No	No	No	**9**
Male	Yes (+5)	Schizophrenia (+5)	No	No	No	No	**10**
Male	Yes (+5)	Depression (+2)	No	No	No	No	**7**
Male	Yes (+5)	Anxiety (+1)	No	No	No	No	**6**
Male	Yes (+5)	Eating disorder (+1)	No	No	No	No	**6**
Male	Yes (+5)	Addictive behaviour (+4)	No	No	No	No	9

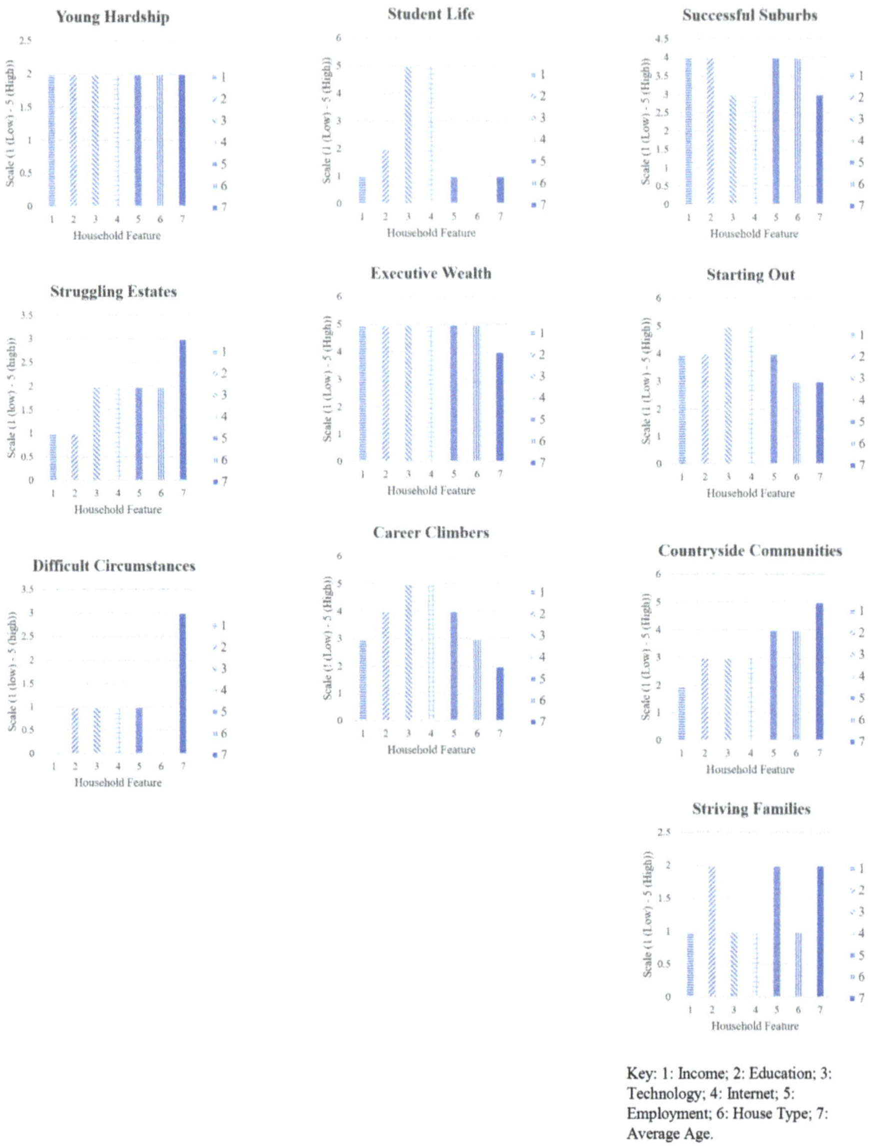

Fig. 6 Characteristics of Table 4 households according to core metrics

perspective, however, it can be appreciated how this reaction may not be the single and most appropriate reaction to take for all users dependent on their vulnerability score. It is therefore for this reason that the vulnerability score must be used in combination with the household characterisation score to ensure relevance and appropriateness of the decisions made. Combined with a high vulnerability score,

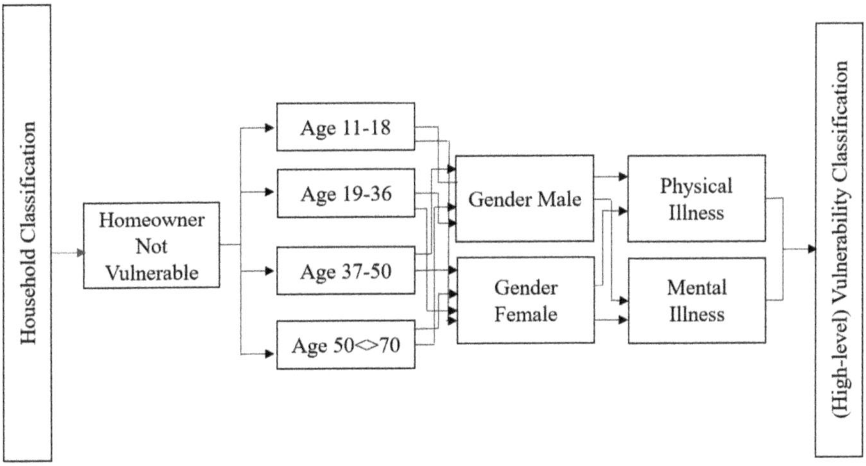

Fig. 7 Vulnerability assessment – high-level detail

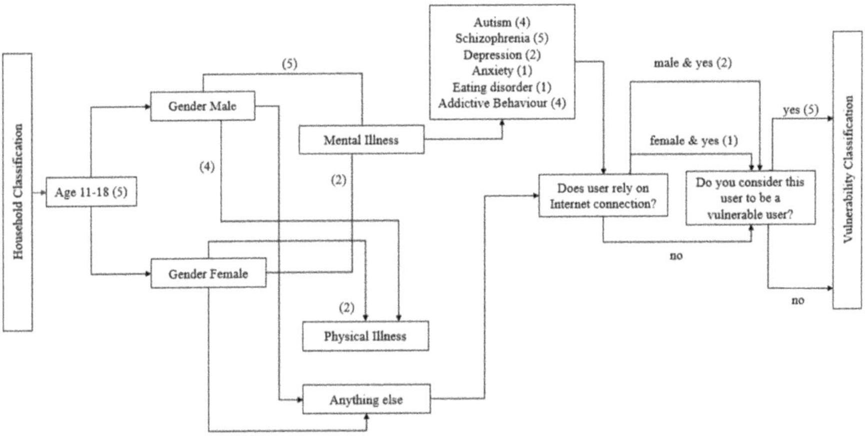

Fig. 8 Vulnerability assessment – low-level detail

it is possible to assess if the household is most likely to be vulnerable to knowing attackers or of being vulnerable to attack.

SLAs are traditionally awarded based on a restricted set of attributes, namely, platform uptime. Providing the agreed upon uptime is fulfilled, then the service provider will not be liable for customer compensation in the form of a penalty. There are opportunities, however, for more personalised SLA provisions, which can be used to distinguish between the services on offer, helping to more suitably respond to a customer's needs. Within the context of our proposal, a service plan is characterised by a set $P = \{p_1, \ldots, p_n\}$ metrics:

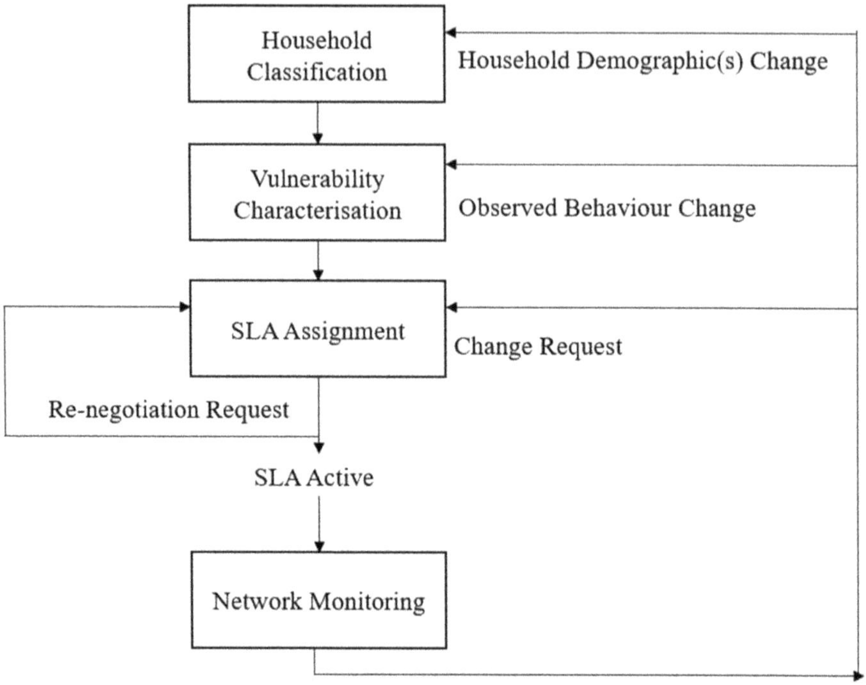

Fig. 9 SLA assignment process

p_1 uptime (%) (99% maximum)
p_2 storage space (GB) (1000 GB maximum)
p_3 minimum download speed (Mb) (65 Mb maximum)
p_4 maximum upload speed (Mb) (25 Mb maximum)
p_5 length of contract (months) (24 months maximum)
p_6 cost per month (£25 maximum)

These are attributes commonly used to define home broadband SLAs (e.g. [10]). In this setup, $p_6 \rightarrow \{p_1 \dots p_5\}$ and $\{p_1 \dots p_5\} \rightarrow p_6$. The values of $p_1 \dots p_5$ are determined by the value of p_6, and the value of p_6 is determined by the values of $p_1 \dots p_5$.

The assignments which are made for each category depend on both the household and vulnerability classification. The SLA will be provisioned according to the worst score in the household, in the instance that profiling is carried out on the basis of multiple residents present. This approach is taken on the basis that it is the worst-case scenario that we need to counteract.

A few examples of SLA allocations are examined in the following section.

6 Case Studies

Users are considered to be vulnerable in this context in their usage of online services. In the first scenario, a household is characterised with a score of 25. This indicates a relatively affluent household, where technology is relatively widely used. The following SLA assignments are made:

p_1 uptime (%): 66
p_2 storage space (GB): 666
p_3 min. Download speed (Mb): 43.3
p_4 max. Upload speed (Mb): 16.7
p_5 length of contract (months): 16
p_6 cost per month (£): 16.7.

To explain these SLA assignments in more detail: In the case of a household score of 25, and a minimum and maximum possible household scores of 34 and 7, respectively, we achieve a weight of 0.67. A normalised score is determined using $\frac{score-minScore}{maxScore-minScore}$. This is then used to weight the maximum SLA assignments which may be made. For example, for a SLA assignment where the maximum uptime assignment is 99% and maximum monthly cost is £25, the provisions made include (99)(0.67) = 66% uptime and (25)(0.67) = £16.67 monthly charge.

The household can then be considered in relation to having vulnerability. We firstly consider the impact of a vulnerability score of 20 ('Mature Money'), when the following SLA assignments are made:

p_1 uptime (%): 10.1
p_2 storage space (GB): 102.6
p_3 min. Download speed (Mb): 6.67
p_4 max. Upload speed (Mb): 2.56
p_5 length of contract (months): 2.46
p_6 cost per month (£): 2.56

We justify this vulnerability in terms of a user who needs assistance with regulating online usage – the higher the vulnerability score, the more vulnerable the user can be assumed to be. As performance needs are being significantly restricted, cost is similarly restricted.

By way of contrast, we consider a household with a vulnerability score of 10. This household is considered to be less vulnerable, and there can be fewer restrictions placed on performance and usage at this household. This justifies the higher bandwidth capacity and associated higher cost. As the household is a relatively financially stable home, with a score of 25, the monthly charge is considered to be reasonable.

p_1 uptime (%): 60.9
p_2 storage space (GB): 615.4
p_3 min. Download speed (Mb): 40
p_4 max. Upload speed (Mb): 15.38

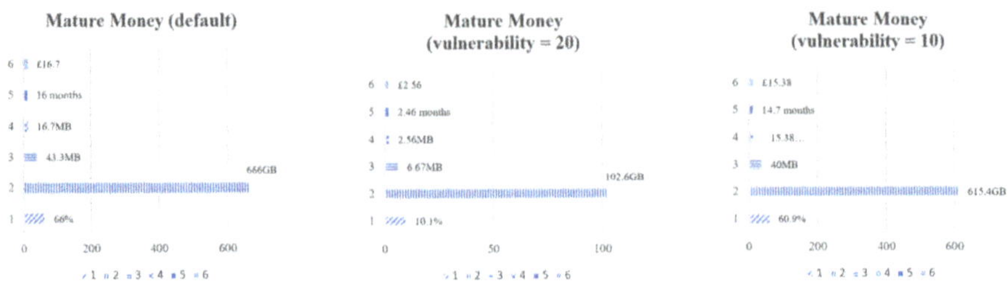

Fig. 10 SLA options offered to 'Mature Money' household

p_5 length of contract (months): 14.7
p_6 cost per month (£): 15.38

The SLA assignments made in response to the various household vulnerabilities are presented in Fig. 10.

These SLA assignments may be contrasted against those assigned to a household with a score of 10 ('Striving Families'). Without considering any vulnerabilities, the household is assigned the following SLA provisions:

p_1 uptime (%): 14.67
p_2 storage space (GB): 148.14
p_3 min. Download speed (Mb): 9.62
p_4 max. Upload speed (Mb): 3.7
p_5 length of contract (months): 3.56
p_6 cost per month (£): 3.70

With a vulnerability score of 20, the SLA becomes:

p_1 uptime (%): 2.25
p_2 storage space (GB): 22.79
p_3 min. Download speed (Mb): 1.48
p_4 max. Upload speed (Mb): 0.57
p_5 length of contract (months): 0.55
p_6 cost per month (£): 0.57

With a vulnerability score of 10, the SLA becomes:

p_1 uptime (%): 13.54
p_2 storage space (GB): 136.75
p_3 min. Download speed (Mb): 8.9
p_4 max. Upload speed (Mb): 3.4
p_5 length of contract (months): 3.3
p_6 cost per month (£): 3.4

The SLA assignments made in response to the various household vulnerabilities are presented in Fig. 11.

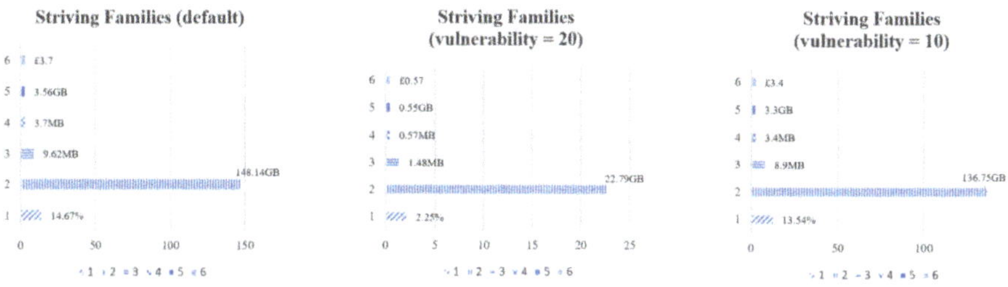

Fig. 11 SLA options offered to 'Striving Families' household

We therefore demonstrate the personalised SLAs which vary dependent on the characteristics of residents within the home, with provisions made on the basis of uptime, storage space, download speed, upload speed for an agreed contract length and cost per month.

It is possible, of course, that the customer will not be satisfied with the SLA that has been offered and will wish to re-negotiate some or all of the terms. Consider, as an example, the household with a score of 25 and a vulnerability score of 20. The SLA assignment is $[p_1, p_2, p_3, p_4, p_5, p_6] = [10.15, 102.56, 6.64, 2.56, 2.46, 6.51]$. However, the customer may be unsatisfied with $p_1 = 10.15\%$ and request that it is increased to 50%, for example, while maintaining the $p_2 \ldots p_5$ resource provisions at the same level. A weighting can therefore be determined to increase the value assigned to p_1; this will have a subsequent impact on p_6.

To demonstrate this, we can consider the household score h_1 and vulnerability score v_1, a normalised household score h_n and a normalised vulnerability score $(1 - v_n)$. The following SLA assignments are subsequently made:

$$ph_1 = (p_1)(h_n)$$

$$ph_2 = (p_2)(h_n)$$

$$ph_3 = (p_3)(h_n)$$

$$ph_4 = (p_4)(h_n)$$

$$ph_5 = (p_5)(h_n)$$

$$ph_6 = (p_6)(h_n)$$

The vulnerability score is then assigned, creating pv_n:

$$pv_1 = \left(ph_1\right)(1 - v_n)$$

$$p_{v2} = \left(ph_2\right)(1 - v_n)$$

$$pv_3 = (p_3)(1 - v_n)$$

$$pv_4 = (p_4)(1 - v_n)$$

$$pv_5 = (p_5)(1 - v_n)$$

$$p_{v6} = (p_6)(1 - v_n)$$

When a customer requests a new p_n value, p_{r1}, then

$$w_1 = \frac{pr_1}{pv_1}$$

$$pv_1 = \left(pv_1\right)(w_1)$$

$$pv_6 = \left(pv_6\right)(w_1)$$

The other pv_n values remain the same.

To contextualise this calculation in action: In the case of the customer with a household score of 25 and a vulnerability score of 20 requesting an uptime of 50%, this has a subsequent impact on their overall SLA assignment of (Fig. 12):

p_1 uptime (%): 50
p_2 storage space (GB): 102.56
p_3 min. Download speed (Mb): 6.67
p_4 max. Upload speed (Mb): 2.56
p_5 length of contract (months): 2.46
p_6 cost per month (£): 6.54

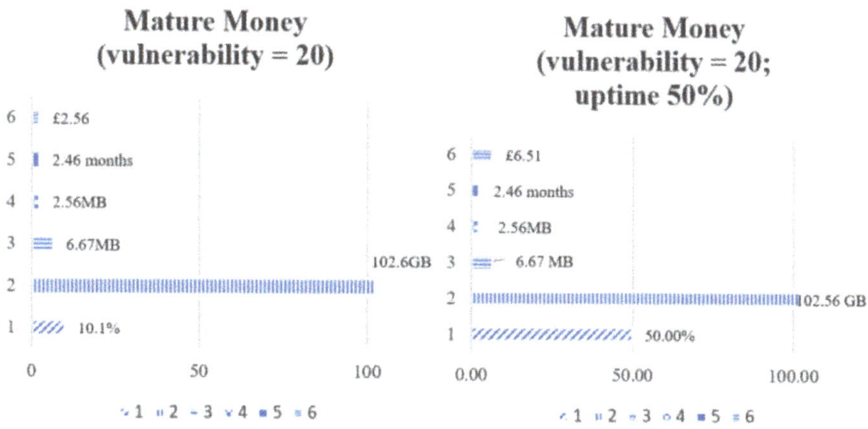

Fig. 12 SLA assignments and re-assignments for 'Mature Money' household

7 Conclusions and Further Work

When we hear the term 'vulnerable', in general, we may think about the elderly, children, the sick, or disabled members of society. While this is certainly true to some extent, we might say that we are all vulnerable when online. In the context of the online world, we can be trolled, our systems can be breached, and our transmissions can be intercepted. Coupled with the evolving ways in which attack surfaces can be breached, the anonymity with which activities can occur is among the factors leading to persistent cybersecurity attacks and vulnerabilities.

We have identified characteristics that might be used to make assumptions in relation to who is, or who may potentially become, a motivated offender in the online world. We identified that, in a number of cases, there was a background of mental health illness and/or challenging situations in the offender's home. This is knowledge that can be known if advance of such a user connecting to the network, and we believe that this information can be used to guide the ways in which user activities are monitored, or the services are offered on a user-specific basis. We believe that it is also possible to use this detail to make assumptions with regard to who may be considered to be a target – they may be identified according to age and/or gender as two metrics which have been shown to be influential in the past.

We believe it is relevant to consider these identifying characteristics within the context of users in the home – attackers may carry out attacks from here, and attackees may be attacked within this environment. Their positioning is important in relation to the scheme proposed through this work – prior research reveals that a number of both attackers and attackees are likely to be beneath the age of an adult. It is therefore natural to assume that they will be living within their parental home. As such, it is relevant to consider the extent to which the parental control can protect both attackers and attackees while they are operating at home. Even in the instance when residents in the home are not youth, there are opportunities to protect other vulnerable people, such as the elderly, through modifying the approach to service provision.

In further work, we seek to examine the ways in which a network service may be made available by allowing a customer to select the attributes of importance in their SLA. Furthermore, in our model, we recognise that the vulnerabilities of residents within the home can change, and we wish to accommodate capability within the SLA provisioning process to cope with this. Data attributes need therefore to be monitored around the home hub in a manner which is personalised to the characteristics of the household to detect any occurrence of change in behaviour. This aspect of the SLA provisioning process will be explored as part of our further work.

Acknowledgement This research is supported by the BTIIC (BT Ireland Innovation Centre) project, funded by BT and Invest Northern Ireland.

References

1. Anon, UK Man Avoids Jail Time in vDOS Case (2017). Available: https://krebsonsecurity.com/2017/12/u-k-man-avoids-jail-time-in-vdos-case/. Accessed 26 Dec 2021
2. Anon, *Data Breach Report 2020 in Review*. Identity Theft Resource Center (2021a). Available: https://www.idtheftcenter.org/post/identity-theft-resource-centers-2020-annual-data-breach-report-reveals-19-percent-decrease-in-breaches/. Accessed: 3 Feb 2022
3. Anon, SolarWinds cyberattack demands significant federal and private-sector response. U.S. Government Accountability Office (2021b). Available: https://www.gao.gov/blog/solarwinds-cyberattack-demands-significant-federal-and-private-sector-response-infographic. Accessed 26 Dec 2021
4. Anon, 8 Ways to Protect Yourself Against Cybercrime. Norton (n.d.-a). Available: https://uk.norton.com/internetsecurity-how-to-how-to-recognize-and-protect-yourself-from-cybercrime.html. Accessed 26 Dec 2021
5. Anon, Don't Worry, Be Happy ... With our Parental Controls and Filtering Advice (n.d.-b). Available: https://www.vodafone.co.uk/mobile/digital-parenting/parental-controls-and-filtering. Accessed 26 Dec 2021
6. Anon, How to Keep your Family Safe Online with BT Parental Controls and the Different Blocking Categories. BT (n.d.-c). Available: https://www.bt.com/help/security/how-to-keep-your-family-safe-online-with-bt-parental-controls-an#settingup. Accessed 26 Dec 2021
7. Anon, Parental Controls Hints and Tips (n.d.-d). Available: https://www.virginmedia.com/blog/parental-controls. Accessed 26 Dec 2021
8. Anon, Plusnet Safeguard & Plusnet Protect: Parental Control, Antivirus and Antimalware Protection (n.d.-e). Available: https://www.plus.net/broadband/extras/online-security/. Accessed 26 Dec 2021
9. Anon, Sky Broadband Buddy (n.d.-f). Available: https://www.sky.com/help/diagnostics/sky-broadband-buddy/sky-broadband-buddy. Accessed 26 Dec 2021
10. Anon, Compare Broadband Deals. MoneySuperMarket (n.d.-g). Available: https://bit.ly/3AUrkGY. Accessed: 3 Feb 2022
11. C. Brooks, MORE alarming cybersecurity stats for 2021! Forbes (2021). Available: https://www.forbes.com/sites/chuckbrooks/2021/10/24/more-alarming-cybersecurity-stats-for-2021-/?sh=7d4f32a74a36. Accessed 26 Dec 2021

12. L.E. Cohen, M. Felson, Social change and crime rate trends: A routine activity approach. Am. Sociol. Rev. **44**(4), 588–608 (1979). https://doi.org/10.2307/2094589
13. I. Haddar, Raouyane, M. Bellafkih, *Generating a Service Broker Framework for Service Selection and SLA-based Provisioning within Network Environments* (2017 Ninth International Conference on Ubiquitous and Future Networks (ICUFN), 2017), pp. 630–365. https://doi.org/10.1109/ICUFN.2017.7993868
14. M.K. Halili, B. Cico, SLA amongst users and providers in multi-cloud environment negotiation model. International Journal of Recent Contributions from Engineering Science & IT (iJES) **9**(2), 32 (2021). https://doi.org/10.3991/ijes.v9i2.22151
15. J.R. Lindsay, Tipping the scales: The attribution problem and the feasibility of deterrence against cyberattack. Journal of Cybersecurity **1**(1), 53–67 (2015). https://doi.org/10.1093/cybsec/tyv003
16. B. Johnston, *Do criminal Laws deter crime? Deterrence Theory in Criminal Justice Policy: A Primer* (Minnesota House Research Department, 2019)
17. S. Katz, Generation X: A critical sociological perspective. Generations (San Francisco, Calif.). **41**, 12–19 (2017)
18. B. Krebs, U.K. Man Avoids Jail Time in vDOS Case. Krebs on Security (2017). Available: https://krebsonsecurity.com/2017/12/u-k-man-avoids-jail-time-in-vdos-case/. Accessed 26 Dec 2021
19. J. Martin, How to use android parental controls. Tech Advisor (2019). Available: https://www.techadvisor.com/how-to/google-android/android-parental-controls-3461359/. Accessed 26 Dec 2021
20. M. McConnell, Mike McConnell on how to win the cyber-war we're losing. *The Washington Post* (2010). Available: https://cyberdialogue.ca/wp-content/uploads/2011/03/Mike-McConnell-How-to-Win-the-Cyberwar-Were-Losing.pdf. Accessed 26 Dec 2021
21. Y. Miao, C. Chen, L. Pan, Q.-L. Han, J. Zhang, Y. Xiang, Machine learning–based cyber attacks targeting on controlled information: A survey. ACM Computer Surveys **54**(7), Article 139 (2021). https://doi.org/10.1145/3465171
22. H. Modderkolk, Leave no trace: How a teenage hacker lost himself online, in *The Guardian*, (2021) Available: https://www.theguardian.com/technology/2021/oct/14/leave-no-trace-how-a-teenage-hacker-lost-himself-online. Accessed 26 Dec 2021
23. National Crime Agency, Pathways into Cyber Crime (2017). Available: https://www.nationalcrimeagency.gov.uk/who-we-are/publications/6-pathways-into-cyber-crime-1/file. Accessed 26 Dec 2021
24. Ofcom Homepage., Available: https://www.ofcom.org.uk/home. Accessed 26 Dec 2021
25. C. Peoples, J. Rafferty, A. Moore, M. Zoualfaghari, *Managing cybersecurity events using service level agreements (SLAs) by profiling the people who attack* (Springer in 'Advances in Cybersecurity Management', 2021a), pp. 221–243
26. C. Peoples, Z. Tariq, A. Moore, M. Zoualfaghari, A. Reeves, Using Process Mining to Formalise Service Level Agreement (SLA) Allocation. 2021 IEEE SmartWorld, Ubiquitous Intelligence & Computing, Advanced & Trusted Computing, Scalable Computing & Communications, Internet of People and Smart City Innovation (SmartWorld/SCALCOM/UIC/ATC/IOP/SCI) (2021b), pp. 671–676, https://doi.org/10.1109/SWC50871.2021.00100
27. C. Quevatre, The teenage hackers Who've been given a second chance. BBC (2019). Available: https://www.bbc.co.uk/news/uk-england-devon-46757849. Accessed 26 Dec 2021
28. D. Rat Nayake, The colonial pipeline attack that revolutionised ransomware landscape. British Computing Sociey (2021) . Available: https://www.bcs.org/articles-opinion-and-research/the-colonial-pipeline-attack-that-revolutionised-ransomware-landscape/. Accessed 26 Dec 2021
29. C. Stouffer, 115 Cybersecurity Statistics and Trends you Need to Know in 2021. Norton (2021). Available: https://us.norton.com/internetsecurity-emerging-threats-cyberthreat-trends-cybersecurity-threat-review.html#::text=How%20many%20cybersecurity%20attacks%20are, one%20cyberattack%20every%2039%20seconds. Accessed 26 Dec 2021

30. G. Swerling, Autistic hacker Jack Chappell 'had been exploited'. *The Sunday Times* (2017). Available: https://www.thetimes.co.uk/article/autistic-hacker-jack-chappell-had-been-exploited-kx8vwh829. Accessed 26 Dec 2021
31. M. Taddeo, The limits of deterrence theory in cyberspace. Philos. Technol. **31**, 339–355 (2018). https://doi.org/10.1007/s13347-017-0290-2. Accessed 26 Dec 2021
32. S. Vimercati, S. Foresti, G. Livraga, V. Piuri, P. Samarati, A fuzzy-based brokering service for cloud plan selection. IEEE Syst. J., 1–9 (2019). https://doi.org/10.1109/JSYST.2019.2893212

Defense Against Adversarial Attack on Knowledge Graph Embedding

Yuxiao Zhang, Qingfeng Chen, Xinkun Hao, Haiming Pan, Qian Yu, and Kexin Huang

1 Introduction

As a form of structured human knowledge, knowledge graphs have become one of the most important resources and have been widely used in real-world related applications [1], such as chat robot, big data risk control, secure investment, intelligent medical treatment, question answering system, and recommendation system.

Due to the rapid growth, a variety of knowledge graphs including Freebase [2], Dbpedia [3], YAGO [4], and NELL [8] were generated, in which knowledge facts are expressed in triples in the form of (*head entity*, *relation*, *tail entity*). For example, Fig. 1, (*Steve Jobs*, *Inventor_of*, *iPhone*) indicates that *Steve Jobs* is the inventor of *iPhone*.

Although these triples can effectively record rich knowledge, it is difficult to directly put them into machine learning models due to their underlying symbolic nature [9]. Therefore, the technology of embedding entities and relationships into vector space, called knowledge graph embedding (KGE), has been proposed and widely used to predict facts [7]. The embedding models include *translation distance-based models* (such as TransE [10]) and *semantic similarity-based models* (such as RESCAL [11]). These embedding methods can well retain the inherent characteris-

Y. Zhang · X. Hao · H. Pan · Q. Yu
School of Computer, Electronics and Information, Guangxi University, Nanning, China
e-mail: haoxinkun@st.gxu.edu.cn

Q. Chen (✉)
School of Computer, Electronics and Information, Guangxi University, Nanning, China

Department of Computer Science and Information Technology, La Trobe University, Melbourne, Victoria, Australia
e-mail: qingfeng@gxu.edu.cn

K. Huang
School of Information and Management, Guangxi Medical University, Nanning, China

K. Daimi et al. (eds.), *Emerging Trends in Cybersecurity Applications*,
https://doi.org/10.1007/978-3-031-09640-2_20

(Steve Jobs, **BornIn**, California)
(Steve Jobs, **HusbandOf**, Laurene Powell)
(Steve Jobs, **SonOf**, Joanne Carole Schieble)
(Red Paul Jobs, **SonOf**, Steve Jobs)
(Red Paul Jobs, **SonOf**, Laurene Powell)
(Steve Jobs, **InventorOf**, iPhone)
(iPhone, **ProductOf**, Apple Inc.)
(Steve Jobs, **FormerCEOOf**, Apple Inc.)
(Tim Cook, **SuccessorOf**, Steve Jobs)
(Tim Cook, **CEOOf**, Apple Inc.)

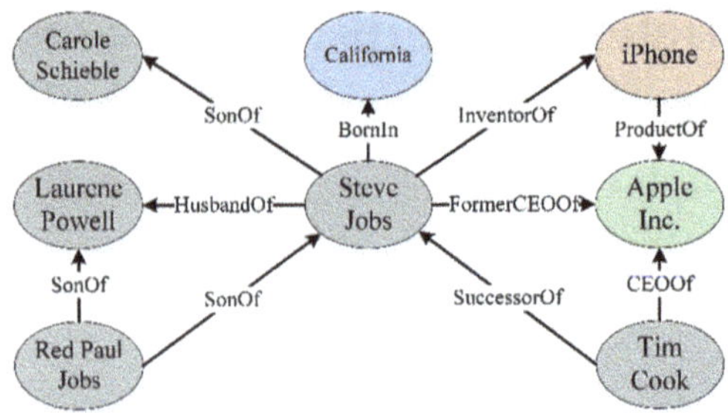

Fig. 1 An example of knowledge graph

tics of entities and relations and allow the knowledge facts to be used for various downstream tasks, such as link prediction [5, 12], question answering [13], and recommendation [14].

Despite the increasing success and popularity of knowledge graph embeddings, their robustness has not been fully studied. In fact, many knowledge graphs are based on unreliable data sources. In case of being attacked, many untrusted or even biased knowledge graph embeddings may be generated, causing serious damage or financial loss for many downstream applications [15]. For example, various recommendation algorithms (e.g., [14]) use the KGEs of products as external reference. If KGEs are manipulated, the recommended results may significantly deviate from the original expectations, which would yield a big impact on the user experience to a large extent. Therefore, we need to not only go beyond the accuracy of link prediction but also pay attention to whether these representations are robust and stable and which facts they use to predict.

Unfortunately, defending such adversarial attacks against KGE involves several major challenges. First, such attacks usually focus on local knowledge graph properties, and the manipulation of adding/deleting a small number of triples is often unobvious to be detected [15]. Further, it is difficult for existing defense methods against adversarial behaviors on machine learning models to be directly applied to defense adversarial attacks on KGEs.

In this paper, we focus on analyzing the vulnerability of KGEs and improving the robustness of KGEs against adversarial attack. Unlike traditional machine learning methods, the adversarial defense schema we proposed consists of an offline stage (adversarial training) that helps the KGEs to minimize the influence of perturbations on attacking targets during the training step and an online stage (adversarial perturbation detection) that detects and filters the perturbations during preprocessing step. These two stages are intertwined, and the application of perturbation detection before adversarial training is supposed to have better performance.

The remaining content of this paper is organized as follows. In Sect. 2, the background and related works of adversarial attacks are briefly introduced. In Sect. 3, the proposed adversarial defense method is explained in more detail. The experiment results and related analysis are shown in Sect. 4, and the conclusion is offered in Sect. 5.

2 Related Works

This section introduces the research background and relevant works, including *knowledge graph embedding*, *adversarial attack on knowledge graph embedding*, and *adversarial defense.*

2.1 Knowledge Graph Embedding

Knowledge graph is a widely applied structured representation of facts, which is composed of entity, relation, and semantic description. Generally, a knowledge graph is viewed as a set of triples (h, r, t), where h and t denote the head entity and tail entity, respectively, and r denotes the relation between h and t. Knowledge graph embedding (KGE) transforms the entities and relations of knowledge graph into low-dimensional continuous vector space, which not only facilitates the operation of knowledge graph but also reserves the original structure and semantics of knowledge graph. The research on KGE mainly focuses on the following four aspects: *representation space, scoring function, encoding model, and auxiliary information of knowledge graph* [1].

The key issue of representation learning is to learn the low-dimensional distributed embedding of entities and relations. Existing representation spaces mainly include real-valued point-wise space (including vector, matrix, and tensor space), complex vector space, Gaussian space, and manifold space [6]. Among them, point-wise Euclidean space is widely used to represent entities and relations and to project relations and entities embeddings into vector or matrix space. For example, TransE [10] represents entities and relations in d-dimensional vector space, i.e., $\boldsymbol{h},\boldsymbol{r},\boldsymbol{t} \in \mathbb{R}^d$, and makes the embedding follow the translation principle $\boldsymbol{h} + \boldsymbol{r} \approx \boldsymbol{t}$. In order to solve the problem of insufficient single space for entities and relations (limitations in dealing with one-to-many, many-to-one, and many-to-many complex relations), TransR [16] further introduced the separation space of entities and relations, using projection matrix $\boldsymbol{M_r} \in \mathbb{R}^{k \times d}$ that projects an entity ($\boldsymbol{h},\boldsymbol{t} \in \mathbb{R}^k$) into a relation ($\boldsymbol{r} \in \mathbb{R}^d$)space.

Scoring function is applied to measure the plausibility of fact triple (h, r, t), i.e., namely, $f(\boldsymbol{h}, \boldsymbol{r}, \boldsymbol{t})$. There are two typical scoring functions, including *distance-based* function and *similarity-based* function. The former measures the plausibility of facts by calculating the distance between entities, in which the translation model with $\boldsymbol{h} + \boldsymbol{t} \approx \boldsymbol{r}$ relation is widely used [17]. The scoring function based on semantic similarity measures the plausibility of facts through semantic matching [18]. Semantic matching usually adopts multiplication formula, i.e., $\boldsymbol{h}^{\mathrm{T}}\boldsymbol{M_r} \approx \boldsymbol{t}^{\mathrm{T}}$. The representation space parameters and scoring functions of KGE models used in this paper are shown in Table 1.

The encoding model includes *linear/bilinear model, factorization model*, and *neural network model*. The linear model represents the relations as a linear/bilinear mapping by projecting the head entity into the representation space close to the

Table 1 Representation space parameters and scoring functions $f(\boldsymbol{h}, \boldsymbol{r}, \boldsymbol{t})$ of KGEs used in this paper

Model	Representation space parameters	Scoring functions
TransE	$\boldsymbol{h},\boldsymbol{r},\boldsymbol{t} \in \mathbb{R}^k$	$-\|\boldsymbol{h}+\boldsymbol{r}-\boldsymbol{t}\|_{1/2}$
DistMult	$\boldsymbol{h},\boldsymbol{r},\boldsymbol{t} \in \mathbb{R}^k$	$\boldsymbol{h}^{\mathrm{T}}$ diag $(\boldsymbol{r})\boldsymbol{t}$
ComplEx	$\boldsymbol{h},\boldsymbol{r},\boldsymbol{t} \in \mathbb{C}^k$	$\mathrm{Re}\left(\boldsymbol{h}^{\mathrm{T}} \operatorname{diag}(\boldsymbol{r})\, \overline{\boldsymbol{t}}\right)$

Note that $\overline{\boldsymbol{t}}$ denotes conjugate for complex vector $\boldsymbol{t}$

tail entity [19]. Factorization model aims to decompose relation data into low-rank matrices for representation learning [11]. The neural network model encodes relation data with nonlinear neural activation and more complex network structure [20]. In order to facilitate more effective knowledge representation, some embedding models combine auxiliary information, such as *text description* [21], *type constraint* [22], *relational path* [24], *visual information* [23], and *knowledge graph* [25] itself.

2.2 *Adversarial Attack Against Knowledge Graph Embedding*

Given a target fact triple ($e^{h,\text{ target}}$, r^{target}, $e^{t,\text{ target}}$) that does not exist in the training set, the $e^{h,\text{ target}}$ and r^{target} denote the embedding of head entity and the embedding of relation, respectively. Based on the description in Sect. 1, a slight perturbation on knowledge graph can make a well-trained KGE generate wrong embeddings and further lead to wrong prediction. We can define a general form of the objective for adversarial attacks, which aims to minimize the plausibility of ($e^{h,\text{ target}}$, r^{target}, $e^{t,\text{ target}}$) calculated by scoring function f, i.e., $f(e^{h,\text{ target}}, r^{\text{target}}, e^{t,\text{ target}})$. Figure 2 shows an example of the adversarial attack against knowledge graph embedding on anti-money laundering application. Original KGE model predicted the target triple (*Person2*, *Associated_with*, *Perssion1_the_suspect*) as true. However, after adding a perturbation triple (*Acc2*, *Transfers*, *Acc3*), the target triple is predicted as false by the retrained KGE model. We omit the situation of graph-level adversarial attacks since the magnitude of the whole knowledge graph is relatively large and almost no adversarial attacks are implemented to degrade the overall KGE performance.

Most adversarial attacks [15, 26–28] proposed in recent years can be categorized into white-box attack and poisoning attack based on adversarial attacker's knowledge and capacity. Attacker's knowledge means how much information an attacker knows about the victim model [29–31]. Usually, there are two settings: *white-box attack* (i.e., all information about the model parameters, training input, and the labels are given to the attacker) and *black-box attack* (i.e., the attacker knows nothing about the model parameters or training input and can only do black-box query for output scores or labels). Attacker's capacity means the stage (i.e., the model training and model testing) at which the adversarial attacks occurred according to the attacker's capacity to insert adversarial perturbation. Poisoning attack means attacking happens before

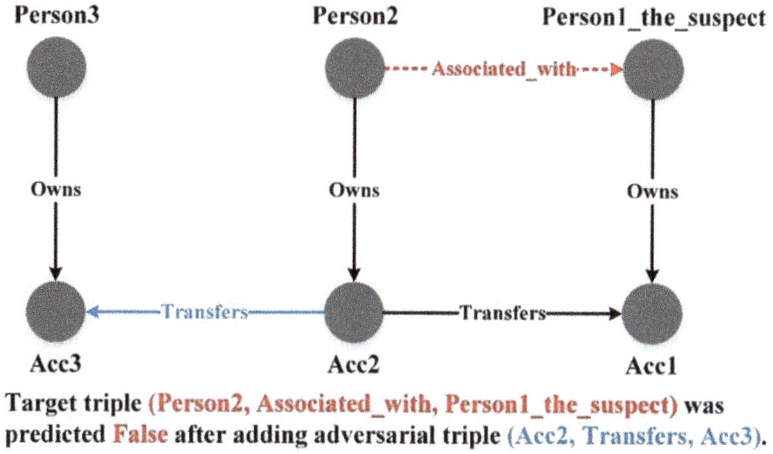

Fig. 2 Adversarial attack against knowledge graph embedding on anti-money laundering application which consists of two types of entities: person and bank account (*Acc*)

the model was trained, and the attacker can add "poisons" into the model training data, letting trained model make mistakes [32]. In contrast, evasion attack means attacking happens after the model was trained or in the test phase, and the model is fixed and the attacker cannot change the model parameters [33].

Due to the discrete and combinatorial nature of knowledge graph, we cannot directly use the gradient-based approach to back-propagate the wrong prediction results on the victim triples. Also, due to the heterogeneity of knowledge graph, existing attack methods on graph data cannot be directly applied to attack the KGE [34]. Solving the above problems is challenging. However, several methods have been proposed recently [15, 26–28]. Zhang et al. [15] proposed perturbation benefit score to calculate the score for all possible candidate triples and select triples with the highest perturbation score. Pezeshkpour et al. [26] used a vanilla auto-encoder to reconstruct discrete entities and relations which made the generation of perturbation triples tractable by performing efficient gradient-based continuous optimization. Bhardwaj et al. [27, 28] proposed an attack method aiming to improve the prediction confidence of KGE model on decoy facts through different inference patterns. Bhardwaj et al. [27, 28] heuristically constructed influence score and generated adversarial perturbation triples according to the influence score. These experiment results have shown that even a small number of perturbations can give rise to a great impact on the KGE models.

2.3 Adversarial Defense Against Adversarial Attack

In previous sections, several instances have been used to show that knowledge graph embeddings are quite vulnerable to well-designed attacks. The vulnerability of knowledge graph embedding results in great risk while applying them in security sensitive applications. To defend against graph adversarial attacks, different countermeasure strategies have been proposed [35, 36]. The first adversarial defense work, conducted by Feng et al. [35], used projected gradient descent to generate perturbations and injected them into the training graph data such that the trained model could correctly classify the future adversarial perturbations (Adversarial training). Xu et al. [36] proposed detection approaches to find adversarial perturbations on graph data (Perturbation detection). To our knowledge, there is no existing research on adversarial defense on knowledge graph. Inspired by the above adversarial defense methods, we thus start from the definitions of adversarial defenses against attack especially on knowledge graph.

Given a victim KGE, a victim fact triple set, and an adversarial attack model, the adversarial defense aims to improve the victim KGE's security and applicability against attack as far as possible at the same time. The KGE's security denotes the degree of its safety and ability to solve the adversarial attack perturbation. The applicability denotes the ability to repair performance when facing attack. Practically, adversarial defense needs to keep the balance between security and applicability according to the corresponding type of adversarial attack. Therefore, most adversarial defenses on graph data only target at resisting evasion attacks, which means that the attack happens during the test period [37, 38]. Nevertheless, it is challenging for adversarial defense on knowledge graph data, since recently proposed corresponding attacks belong to poisoning attack, which means experimentally effective poisons are already injected into the training data before victim KGE is trained. Thus, adversarial defenses on knowledge graph data should focus on reducing the influence of perturbations on victim fact triple set.

In the following section, we proposed to combine adversarial training (aiming to improve the applicability of KGE) and perturbation detection (aiming to improve the security of KGE) on knowledge graph data for the first time. These methods are described in more detail in Sect. 3.

3 Two-Stage Adversarial Defense Approaches

In this section, we present our two-stage adversarial defense approaches, which include the offline stage (adversarial training) and online stage (perturbation detection). The principle and mechanism of adversarial training and perturbation detection are first illustrated. Then the mathematical and algorithmic details of these approaches are explained.

3.1 Adversarial Training

Inspired by the successful application of generative adversarial networks (GANs) [39] in deep learning area, we use a GAN-based framework during the adversarial training stage. Generative adversarial networks (GANs) were originally proposed for generating samples in a continuous space such as image. Generally, a GAN model consists of *generator* and *discriminator*. As its name implies, the generator part receives a random noise z and outputs an image according to z, recorded as $G(z)$. The discriminator is a distinguishing network to decide whether an image is "true" (from the ground truth dataset) or "fake" (generated by the generator). The discriminator takes x (i.e., an image) as input and output $D(x)$. $D(x) = 1$ means that the probability of image x being a true image is 100%, while $D(x) = 0$ means that the probability of image x being a true image is 0%. When training a GAN, the generator and the discriminator conduct a minimax game, in which the generator tries to generate "true" image and the discriminator tries to distinguish the image obtained by the generator.

Due to the discrete and combinatorial nature of knowledge graph, the generator cannot directly generate samples through gradients. Cai and Wang [40] used policy gradient to train the generator on knowledge graph (KBGAN) for the first time. According to KBGAN, a KGE model with *similarity-based* scoring function can be used as the generator G, and a KGE model with *distance-based* scoring function can be used as the discriminator. Specifically, the objective of the generator is formulated as maximizing the following expectation of negative distances:

$$R_G = \sum_{(h,r,t)\in T} \mathbb{E}\left[-f_D\left(h', r, t'\right)\right] \tag{1}$$

where T is the training triple set, $f_D(h, r, t)$ is the scoring function of the discriminator, $\mathbb{E}\,[.]$ calculates the expectation of negative distances, and $(h', r, t')P_G(h', r, t' \mid h, r, t)$ denotes the probability distribution on negative triples (h', r, t') given a positive triple (h, r, t):

$$P_G\left(h', r, t'|h, r, t\right) = \frac{\exp f_G\left(h', r, t'\right)}{\sum \exp f_G\left(h^*, r, t^*\right)} \tag{2}$$

where $f_G(h, r, t)$ is the scoring function of the generator and $(h^*, r, t^*) \in \text{Neg}(h, r, t)$ means the set of candidate negative triples. Note that KBGAN generates $\text{Neg}(h, r, t)$ by uniformly sampling. And, the policy gradient theorem [41] was used to obtain the gradient of R_G.

Meanwhile, the objective of the discriminator is to minimize the marginal loss between the target triple and adversarial triple. The marginal loss function is formulated as the following formula, where γ is the margin and $[.]_+$ is the hinge function.

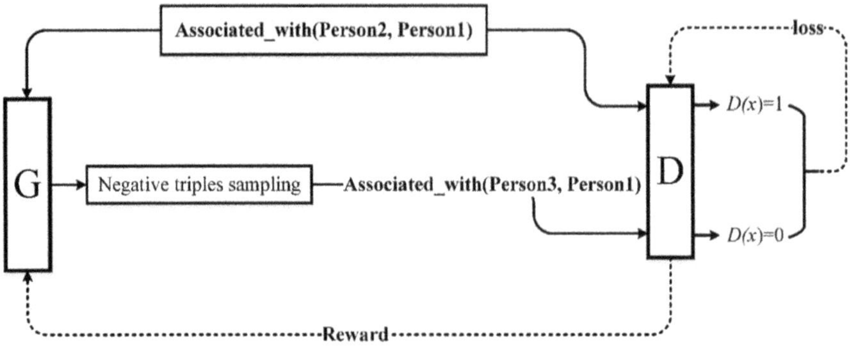

Fig. 3 An overview of the adversarial training framework

$$L_D = \sum_{(h,r,t)\in T} \left[f_D(h,r,t) - f_D\left(h',r,t'\right) + \gamma\right]_+ \quad (3)$$

Although Cai and Wang [40] utilized the GAN framework to generate a good discriminator and improve its performance under various settings. It was limited to the sampling step in generator which might sample true positive facts instead of real effective negatives. And the policy gradient it used needs to calculate *baseline* (a constant used to assist the training of generator) to reduce the variance. Here, we propose an improved method by using the gradient-based attack scoring function, which directly calculates the derivate of head or tail entity. The framework of our adversarial training approach is shown in Fig. 3. For the target triple (*Person2, Associated_with, Perssion1*), the generator G calculates a probability distribution of adversarial perturbation triples and then samples one negative triple (e.g., (*Person3, Associated_with, Person1*)) as output. The discriminator D receives the generated perturbation triple as well as the true target triple and computes the score. The generator and the discriminator are alternatively trained.

Algorithm 1 summarizes the adversarial training process. According to the GAN training setting, both the generator G and discriminator D require pre-training. G and D can use the same KGE model for convenience. As shown in Algorithm 1, we first calculate probability distribution $P_G(h', r, t' \mid h, r, t)$ for negative triples. And then, we obtain the gradient of generator and discriminator in Line 6 and Line 8, respectively. Parameter r in Line 7 denotes the reward for generator. The procedure runs until it reaches the maximum epoch or converges.

Algorithm 1: Adversarial Training

Input: generator G with parameter θ_G and the scoring function $f_G^{(h,r,t)}$, and discriminator D with parameters θ_D and scoring function $f_D^{(h,r,t)}$, and target triple (h, r, t).

Output: adversarially trained discriminator D

loop

  $G_G \leftarrow 0$, $G_D \leftarrow 0$; // gradient of parameters of G and D;

  Sample adversarial triples as Pertubation(h, r, t);

  Allocate probability $P_G(h^{'}, r, t^{'} \mid h, r, t)$ for Pertubation(h, r, t);

  Sample one adversarial triple according to $P_G(h^{'}, r, t^{'} \mid h, r, t)$;

  $G_D \leftarrow G_D + \nabla_{\theta_D} L_D$; // accumulate gradients for D

  $r \leftarrow f_D(h^{'}, r, t^{'})$;

  $G_G \leftarrow G_G + r\nabla_{\theta_G} R_G$; // accumulate gradients for G

  $\theta_G \leftarrow \theta_G + \eta_G G_G$, $\theta_D \leftarrow \theta_D + \eta_D G_D$ // update parameters

end loop

3.2 *Perturbation Detection*

Unlike adversarial learning strategy in Sect. 3.1, perturbation detection aims to find adversarial examples injected into dataset. Thus, in this section, we use perturbation detection strategy to detect the malicious triples added by a potential attacker.

Since the malicious examples are relatively rare, Xu et al. [36] proposed to sub-sample several small graphs to perform defense on graph data. Meanwhile, according to the adversarial attack against KGE, triples in the sub-graph which involve the target triple are more likely to be attacked. Therefore, adversarial attack generally happens in the sub-graph of target triple. The sub-graph and an example of perturbation detection are shown in Fig. 4.

LinkPred Given a perturbed knowledge graph $G^{'}$, the most concise and straightforward idea is to perform a link prediction algorithm on $G^{'}$. Given a triple (h, r, t), replacing the h or t entity with other entity and calculating the score according to the scoring function f is a conventional method for link prediction. In practice, we use $\text{rank}_{(h,r,t)}$ (the rank of $f(h, r, t)$ among replacing triples) to measure the plausibility of (h, r, t). If the $\text{rank}_{(h,r,t)}$ is high, it is probable that (h, r, t) could be a maliciously added triple.

SubGraphLinkPred The *LinkPred* method is simple. However, compared with the large amount of the knowledge graph triples, the perturbations are insignificant. Meanwhile, it will require intensive time and computing resources to use *LinkPred* method on all triples on the knowledge graph. To solve the problem, we focus on the sub-graph that involves the target triple. The intuition behind this method is that perturbation budget is limited, and according to the adversarial attack methods

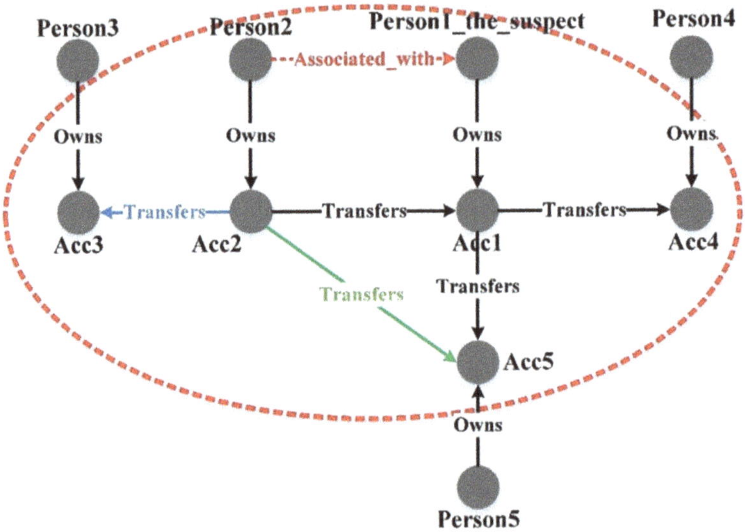

Fig. 4 Sub-graph of target triple (*Person2*, *Associated_with*, *Peron1_the_suspect*) is shown in the ellipse. The perturbation triple (*Acc2*, *Transfers*, *Acc3*) is added into the sub-graph that contains target triple. The perturbation triple and triple (*Acc2*, *Transfers*, *Acc5*) are filtered by *SubGraphGen* method. Then, the perturbation triple is detected by *LinkPred*

introduced above, the more the perturbations related to the target triple, the better the attack effect. Thus, we defined the sub-graph of target triple as:

$$\text{Subgraph}\,(h, r, t) = \left\{ \left(h', r, t\right) \cap \left(h, r, t'\right) | h' \in G \text{ and } t' \in G \right\} \tag{4}$$

where G denotes the knowledge graph. For a given target triple, we first find the sub_graph of target triple and apply *LinkPred* on the sub_graph triple set as well.

Algorithm 2: Perturbation Detection

Input: manipulated knowledge graph G' and the scoring funtion f and target triple (h, r, t).

Output: maliciously added perturbation triple

calculate $\text{Sub\_graph}(h, r, t)$

divide $\text{Sub\_graph}(h, r, t)$ into two subsets: $\text{Sub\_graph}_1(h, r, t)$ and $\text{Sub\_graph}_2(h, r, t)$

delete $\text{Sub\_graph}_1(h, r, t)$ and $\text{Sub\_graph}_2(h, r, t)$ in G' and train model separately

for two trained model:

    for every entity pair (h,t) in $\text{Sub\_graph}(h, r, t)$:

        complete the triple with relation (h, r^*, t)

    obtain new sub_graph $\text{Sub\_graph}'(h, r, t)$

8: compare the different parts of two $\text{Sub}\_\text{graph}'(h, r, t)$
9: apply *LinkPred* on the different parts
10: choose the triple with highest probability

SubGraphGen The *SubGraphLinkPred* is easy to compute and apply, but there still exists a problem. Although the adversarial additional triples are very likely appeared in the sub-graph of target triple, the model learns the wrong information as well. It is possible that using *LinkPred* method cannot find out the adversarial additional triples at all. To alleviate the above problem, we first delete the triples in $\text{Sub}\_\text{graph}(h, r, t)$ and train the model on the remaining *true* triples. Then we apply a graph generation method (knowledge graph completion) using the trained true model. Considering that if we delete the triple set $\text{Sub}\_\text{graph}(h, r, t)$ at a time, the target triple (h, r, t) will be isolated, and we also cannot apply graph generation method. Thus, we divide $\text{Sub}\_\text{graph}(h, r, t)$ into two parts, $\text{Sub}\_\text{graph}_1(h, r, t)$ and $\text{Sub}\_\text{graph}_2(h, r, t)$, and train two models on them separately. Generally, adversarial adding attacks select one of the most influential perturbation facts at a time due to the attack budget. If $\text{Sub}\_\text{graph}_1(h, r, t)$ contains the perturbation fact, then $\text{Sub}\_\text{graph}_2(h, r, t)$ does not contain, and vice versa. Thus, one of the two trained models is a manipulated model and the other is a clean model. We apply graph generation method on two models and compare the different parts generated by the two models. As such, the adversarial additional triple is in the different parts and can be found by *LinkPred*. The pseudo code of *SubGraphGen* is shown in Algorithm 2.

4 Experiment Results

4.1 Datasets

In this paper, we use two common KGE benchmark datasets in our experiment, including FB15K-237 and WN18RR. FK15K-237 is derived from Freebase, which is a large knowledge graph consisting of a large number of real-world facts about movies, actors, awards, sports, and sports team. WN18RR belongs to WordNet, which is a large lexical knowledge graph. The training set and the test set of these two datasets are already fixed. According to Dettmers et al. [42], we use FB15K-237/WN18RR instead of FB15K/WN18 to ensure the datasets do not have inverse relation test leakage. Following the filter setting [10], we remove all the negative triples that already exist in the training, validating, and testing sets. The statistics of datasets used in this paper are shown in Table 2.

Table 2 The statistics of WN18RR and FB15K-237

	WN18RR	FB15K-237
Entities	40,559	14,505
Relations	11	237
Training	86,835	272,115
Validation	2824	17,526
Test	2924	20,438

4.2 Baseline and Target Models

Since there are no existing adversarial defense methods proposed at present against adversarial attack on knowledge graph embedding, we investigate the effectiveness of the adversarial training stage by comparing the performance before and after applying our defense methods. For the perturbation detection stage, we specially design *Random_select* (random filter out one triple) and *SubGraphLinkPred* (simply filter out triple in sub-graph using LinkPred method) as baselines for comparison with our proposed *SubGraphGen* detection method. *SubGraphGen_2* denotes that we can filter two most suspicious triples which may contain the adversarial attack triple. According to the protocol of adversarial attacks [15], we first train a victim KGE model on the original dataset. Then, we inject the adversarial perturbations generated by the following attack methods into the training set and re-train a new KGE model on the manipulated training set. Last, given target facts and the perturbed training sets, we implement our defense methods and evaluate the effectiveness under different settings.

We evaluate our defense methods on three most representative KGE models: TransE, DistMult, and ComplEx whose representation space parameters and scoring functions are shown in Table 1. We apply our defense methods against random edits and the state-of-the-art attack methods. Specifically, *Random_a* adds a random triple into the training set, while *Random_d* deletes a random triple from the training set. *Direct_Add* and *Direct_Del* are the attack methods from Zhang et al. [15]. *CRIAGE* is an attack method especially against DistMult model in Pezeshkpour et al. [26]. Bhardwaj et al. [27, 28] use different *RIP* (*Relation Inference Patterns*) to generate perturbations including symmetry pattern, inversion pattern, and composition pattern. Note that the *RIP* focuses on adversarial additional attack. For convenience, we calculate the average results of these patterns in experiment. *IA* (*Instance-Attribution*) uses different metrics including Dot Metric, l_2 Metric and Cosine Metric to implement attack by calculating similarity between perturbation and target triple (Bhardwaj et al. [27, 28]). Also, we compute the average results of these metrics in experiment. Due to the different target facts selected by the attack methods, we randomly select 100 samples in the test set as target triples and keep the same to ensure fair evaluation.

4.3 Metrics and Experiment Settings

Following previous works like Feng et al. [35] and Xu et al. [36], we use the MRR and Hit@10 as our evaluation metrics. The MRR is the mean reciprocal rank of all the ground truth triples, and the Hit@10 is the proportion of correct triples ranked in top 10. The larger the MRR and the Hit@10 are, the better the performance of models on testing set is. We also use success rate (SR, the higher the better) as the evaluation metric to evaluate the ability of our perturbation detection method to detect perturbation triples.

Due to limited computation resource, we set the dimensions of embeddings of KGE models to $k = 50$ and the training epochs to 500 times. For uniform standard, we use the standard implementation provided by THUNLP/OpenKE[1] [43], and other parameters are set according to default values. For target attack methods, we set the attack budget (the number of perturbations) to 1 in most cases. The defense methods are implemented based on pytorch and python 3. The code runs on a server with RTX 2080 Ti GPUs.

4.4 Results and Analysis

In this section, we display the experimental results and discuss the effectiveness of our proposed two-stage adversarial defense approaches. The results of adversarial training stage against attack under different settings are reported in Tables 3, 4, 5, 6, 7, and 8. Tables 3, 4, 5, and 6 reveal the improvement in MRR and Hit@10 by adversarial training stage against different attacks on the FB15K-237 and WN18RR datasets. Figure 5 shows the performance of perturbation detection stage against adding attack. Tables 7 and 8 display the effectiveness of the combination of two-stage defense approaches.

Adversarial Training Performance Here, we analyze the effectiveness of our adversarial training method by comparing the improvement of KGE performance. The results of the adversarial training method against deleting attack on FB15K-237 and WN18RR are shown in Tables 3 and 4; the results of the adversarial training method against adding attack on FB15K-237 and WN18RR are shown in Tables 5 and 6. In these tables, the first row is the original performances of KGE models. The second block of rows is the reduction in MRR and Hit@10 due to different attacks. The bottom block of rows are results improved by our adversarial training defense method against attacks. For each block, the underlined results are the best ones among our implementations. And we report the *best* improvement in percentage relative to the attacked MRR, which is computed as *(defensed−attacked)/attacked* * 100%.

[1] https://github.com/thunlp/OpenKE

Table 3 Adversarial training stage against adversarial deleting attack on FB15K-237

		TransE		ComplEx		DistMult	
		MRR	Hit@10	MRR	Hit@10	MRR	Hit@10
Original		0.26	0.49	0.24*	0.42*	0.24*	0.42*
Attack Method	Random_a	0.25	0.47	0.22	0.38	0.21	0.39
	Random_d	0.24	0.46	0.2	0.36	0.2	0.38
	Direct_Del	0.19	0.37	0.19	0.35	0.18	0.39
	CRIAGE	-	-	-	-	0.18	0.4
	IA	0.22	0.41	0.11	0.17	0.12	0.31
Defense Method	AT_Random_a	**0.27**	**0.5**	0.23	0.4	0.22	0.4
	AT_Random_d	0.25	0.47	0.24	0.41	**0.26**	**0.48**
	AT_Direct_Del	**0.24**+26%	**0.44**	0.22	0.4	0.2	0.4
	AT_CRIAGE	-	-	-	-	**0.21**+17%	**0.42**
	AT_IA	0.24	0.42	**0.18**+64%	**0.34**	0.18	0.39

Numerical values marked with * are derived from Dettmers et al. [42]

Table 4 Adversarial training stage against adversarial deleting attack on WN18RR

		TransE		ComplEx		DistMult	
		MRR	Hit@10	MRR	Hit@10	MRR	Hit@10
Original		0.19	0.51	0.44*	0.51*	0.43*	0.49*
Attack Method	Random_a	0.18	0.5	0.42	0.49	0.4	0.45
	Random_d	0.17	0.49	0.4	0.44	0.38	0.39
	Direct_Del	0.11	0.26	0.36	0.4	0.3	0.27
	CRIAGE	-	-	-	-	0.34	0.37
	IA	0.13	0.31	0.28	0.31	0.2	0.26
Defense Method	AT_Random_a	0.18	0.49	0.41	0.5	0.42	0.48
	AT_Random_d	0.16	0.5	0.41	0.48	0.4	0.47
	AT_Direct_Del	**0.14**+27%	**0.39**	0.39	0.45	0.36	0.4
	AT_CRIAGE	-	-	-	-	0.36	0.41
	AT_IA	0.13	0.33	0.36	0.4	0.31	0.36

Numerical values marked with * are derived from Dettmers et al. [42]

Obviously, our proposed adversarial training method can effectively defend most state-of-the-art attack models, to a certain extent. Note that, against random adding attack and deleting attack, some results of adversarial training defense are better than the original results. For example, the MRR of TransE on FB15K-237 is improved to 0.27 by adversarial training against *Random_a attack*, while the original MRR is 0.26. It is not strange because of the characteristic of adversarial training. As we have known, most KGE models generate negative training triples by replacing the head or tail entity of a triple with randomly selected entity, which can be easily distinguished from positive triples. While our adversarial training method specially uses a generator from GAN framework to assist training. Meanwhile, we can find that the effectiveness of adversarial training against deleting attack is better than against adding attack. It is likely because the adversarial adding triples are already injected

Table 5 Adversarial training stage against adversarial adding attack on FB15K-237

		TransE		ComplEx		DistMult	
		MRR	Hit@10	MRR	Hit@10	MRR	Hit@10
Original		0.26	0.49	0.24*	0.42*	0.24*	0.42*
Attack method	Direct_Add	0.24	0.42	0.21	0.38	0.2	0.39
	CRIAGE	–	–	–	–	0.19	0.36
	RIP	0.24	0.42	0.19	0.36	0.18	0.36
	IA	0.23	0.4	0.16	0.34	0.18	0.34
Defense method	AT_Direct_Add	0.24	0.44	0.23	0.4	0.21	0.4
	AT_CRIAGE	–	–	–	–	0.21	0.39
	AT_RIP	0.24	0.42	0.2	0.37	0.2	0.38
	AT_IA	0.24	0.42	0.18	0.36	0.2	0.38

Numerical values marked with * are derived from Dettmers et al. [42]

Table 6 Adversarial training stage against adversarial adding attack on WN18RR

		TransE		ComplEx		DistMult	
		MRR	**Hit@10**	**MRR**	**Hit@10**	**MRR**	**Hit@10**
Original		0.19	0.51	0.44*	0.51*	0.43*	0.49*
Attack Method	Direct_Add	0.16	0.48	0.32	0.36	0.36	0.4
	CRIAGE	-	-	-	-	0.36	0.39
	RIP	0.16	0.46	0.3	0.34	0.32	0.35
	IA	0.15	0.42	0.35	0.4	0.25	0.3
Defense Method	AT_Direct_Add	0.18	0.49	**0.38**+19%	**0.45**	0.39	0.41
	AT_CRIAGE	-	-	-	-	0.37	0.4
	AT_RIP	0.17	0.48	0.34	0.42	**0.38**+19%	**0.4**
	AT_IA	0.16	0.46	0.37	0.41	**0.32**+28%	**0.36**

Numerical values marked with * are derived from Dettmers et al. [42]

into the training set and have a certain impact on the pre-trained models, although adversarial training can generate adversarial negative triples.

Perturbation Detection Performance The effectiveness of perturbation detection stage is revealed in Fig. 4 and Tables 7 and 8. According to Fig. 5, we can find that our *SubGraphGen* method performs best against *Direct_Add* attack with SR of 0.41 and it also performs well against the other three attacks with SRs over 0.3. It is easy to understand as *Direct_Add* selects and adds triples that are directly related to the target triples. The *SubGraphGen* focuses on the sub-graph of the target triple as well. However, *SubGraphGen_2* with more fault-tolerant does not enhance the performance significantly compared to *SubGraphGen* as we expected. For example, the SRs against *Instance-Attribution* of *SubGraphGen_2* and *SubGraphGen* are 0.36 and 0.32, respectively. It may be because the sub-graph is divided into two parts which make KGE fail to obtain complete information of the sub-graph. Generally, our perturbation detection method has a certain effect against adding attacks with average SR of 0.35. On the other hand, as shown in Tables 7 and 8, perturbation

Table 7 The combination of adversarial training stage and perturbation detection stage against adversarial adding attack on FB15K-237. AT+PD in the table denotes the combination of two stages

		TransE		ComplEx		DistMult	
		MRR	Hit@10	MRR	Hit@10	MRR	Hit@10
Original		0.26	0.49	0.24*	0.42*	0.24*	0.42*
Attack Method	Direct_Add	0.24	0.42	0.21	0.38	0.2	0.39
	CRIAGE	-	-	-	-	0.19	0.36
	RIP	0.24	0.42	0.19	0.36	0.18	0.36
	IA	0.23	0.4	0.16	0.34	0.18	0.34
Defense Method	AT+PD_Direct_Add	0.26	0.47	0.24	0.41	0.23	0.41
	AT+PD_CRIAGE	-	-	-	-	0.22	0.41
	AT+PD_RIP	0.24	0.43	0.23	0.4	0.21	0.4
	AT+PD_IA	0.25	0.44	**0.22**+38%	**0.4**	0.22	0.39

Numerical values marked with * are derived from Dettmers et al. [42]

Table 8 The combination of adversarial training stage and perturbation detection stage against adversarial adding attack on WN18RR

		TransE		ComplEx		DistMult	
		MRR	Hit@10	MRR	Hit@10	MRR	Hit@10
Original		0.19	0.51	0.44*	0.51*	0.43*	0.49*
Attack Method	Direct_Add	0.16	0.48	0.32	0.36	0.36	0.4
	CRIAGE	-	-	-	-	0.36	0.39
	RIP	0.16	0.46	0.3	0.34	0.32	0.35
	IA	0.15	0.42	0.35	0.4	0.25	0.3
Defense Method	AT+PD_Direct_Add	**0.2**+25%	**0.52**	**0.42**+31%	**0.49**	0.41	0.45
	AT+PD_CRIAGE	-	-	-	-	0.4	0.44
	AT+PD_RIP	0.19	0.5	0.4	0.47	**0.4**+25%	**0.3**
	AT+PD_IA	0.18	0.49	0.41	0.47	**0.38**+52%	**0.4**

Numerical values marked with * are derived from Dettmers et al. [42]

detection combined with adversarial training performs good against attacks. For example, on FB15K-237, the MRR of ComplEx is improved from 0.16 to 0.22 by the combination of two stages against *Instance-Attribution* attack.

Defense Performance of Two-Stage Combination Perturbation detection method just focuses on adding attacks; the effectiveness of the combination of adversarial training stage and perturbation detection stage is described in Tables 7 and 8, respectively. For example, the MRR of DistMult on WN18RR is improved from 0.25 to 0.38 by the combination against *Instance-Attribution* attack. In experiment, given an attacked knowledge graph, we first use perturbation detection to filter out possible additions. Then we apply adversarial training on the filtered knowledge graph. The

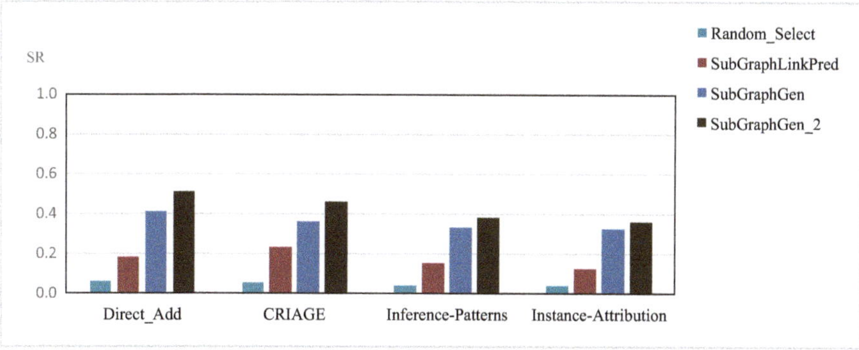

Fig. 5 Performance of our SubGraphGen perturbation detection method which achieves an average SR of about 0.35 for adversarial adding attacks

results on Tables 7 and 8 show that the combination can effectively defend against adding attacks. Comparing the results between Table 5 with Table 6 and Table 7 with Table 8, we can find that perturbation detection has a good filtering effect and is helpful for the improvement of adversarial training.

4.5 *Case Study*

To further verify the effect of our two-stage defense method, we offer an example in Table 9. We randomly select a target fact triple (The Last King of Scotland, film_genre, War film) from the training set of FB15K-237, and the link prediction task of the KGE is to predict the missing tail entity (The Last King of Scotland, film_genre, War film). Because the relation of target triple "film_genre" belongs to a one-to-many relationship, multiple tail entities may be predicted. As shown in Table 9, the first column is the target fact triple, and the second column shows the link prediction results of the original KGE model. We can find that the true tail entity "war film" ranks second in the prediction results which means that the MRR of the original KGE model on the target triple is 0.5. The third column shows that, under a well-designed attack, the KGE model fails to correctly predict the true tail entity ("war film"). The fourth column demonstrates that our defense method can effectively defend the well-designed adversarial attack, with the MRR on the target triple being 1.0 and the true tail entity "war film" ranking first in the prediction results again.

Table 9 Example of defense of adversarial attack on FB15K-237

Target fact triple	Original predict	Predict under attack	Predict after defense
(The Last King of Scotland, film_genre, **War film**)	Forest Whitaker **War film** Biographical film Searchlight Pictures Academy Award for Best Actor	Romance film Titanic Forest Whitaker Searchlight Pictures Drama film	**War film** Searchlight Pictures Forest Whitaker Academy Award for Best Actor Peter Morgan

5 Conclusions

We propose a novel two-stage method for defensing the adversarial attack on knowledge graph embedding. We designed adversarial training and perturbation detection approaches and applied them to defend the state-of-the-art attacks. For adversarial training stage, we use GAN framework to improve the capacity of the trained model to distinguish perturbations. For perturbation detection stage, we use the sub-graph of target triple to filter out perturbations. We investigate the effectiveness of proposed method against attacks under different settings, and the results showed that our two-stage method could defend against state-of-the-art attacks effectively.

Our proposed adversarial training approach improves the performances of victim KGE models, especially on FB15K-237, with MRR of TransE under adversarial deleting attack rising from 0.11 to 0.18; our proposed perturbation detection approach achieves average SR of about 0.35 against adversarial adding attacks.

There are some works we will address in the future to improve the adversarial defense. We will design a more simplified and effective method to generate and utilize adversarial negative triples. We will also consider similarity to detect perturbations.

Acknowledgment The work reported in this paper was partially supported by a National Natural Science Foundation of China project 61963004, two key projects of Natural Science Foundation of Guangxi 2017GXNSFDA198033, and a key research and development plan of Guangxi AB17195055.

References

1. S. Ji, S. Pan, E. Cambria, P. Marttinen, S.Y. Philip, A survey on knowledge graphs: Representation, acquisition, and applications. IEEE Trans. Neural Netw. Learn. Syst., 1–21 (2021)
2. K. Bollacker, C. Evans, P. Paritosh, T. Sturge, J. Taylor, in *Proceedings of the ACM SIGMOD International Conference on Management of Data*. Freebase: A Collaboratively Created Graph Database for Structuring Human Knowledge (Vancouver, 2008), pp. 1247–1250

3. J. Lehmann, R. Isele, M. Jakob, A. Jentzsch, D. Kontokostas, P.N. Mendes, C. Bizer, Dbpedia: A large-scale, multilingual knowledge base extracted from wikipedia. Semantic Web **6**, 167–195 (2015)
4. F. Suchanek, G. Kasneci, G. Weikum, in *Proceedings of the International Conference on World Wide Web*. Yago: A Core of Semantic Knowledge (Banff, 2007), pp. 697–706
5. Q. Chen, Y. Li, K. Tan, Y. Qiao, S. Pan, T. Jiang, Y. Chen, Network-based methods for gene function prediction. Brief. Funct. Genomic **20**, 249–257 (2021)
6. Q. Chen, Y. Qiao, F. Hu, Y. Li, K. Tan, M. Zhu, C. Zhang, Community detection in complex network based on APT method. Pattern Recogn. Lett. **138**, 193–200 (2020)
7. W. Lan, Y. Dong, Q. Chen, R. Zheng, J. Liu, Y. Pan, Y. Chen, KGANCDA: Predicting circRNA-disease associations based on knowledge graph attention network. Brief. Bioinform. **23**, bbab494 (2022)
8. A. Carlson, J. Betteridge, B. Kisiel, B. Settles, E. Hruschka, T.M. Mitchell, in *Proceedings of the AAAI Conference on Artificial Intelligence*. Toward an Architecture for Never-Ending Language Learning (Atlanta, 2010), pp. 1306–1313
9. Q. Wang, Z. Mao, B. Wang, L. Guo, Knowledge graph embedding: A survey of approaches and applications. IEEE Trans. Knowl. Data Eng. **29**, 2724–2743 (2017)
10. A. Bordes, N. Usunier, A. Garcia-Duran, J. Weston, O. Yakhnenko, in *Proceedings of the Advances in Neural Information Processing Systems*. Translating Embeddings for Modeling Multi-Relational Data (Lake Tahoe, 2013), pp. 2787–2795
11. M. Nickel, V. Tresp, H. Kriegel, in *Proceedings of the International Conference on Machine Learning*. A Three-Way Model for Collective Learning on Multi-Relational Data (Bellevue, 2011), pp. 809–816
12. A. Rossi, D. Barbosa, D. Firmani, A. Matinata, P. Merialdo, Knowledge graph embedding for link prediction: A comparative analysis. ACM Trans. Knowl. Disc. Data **15**, 1–49 (2021)
13. X. Huang, J. Zhang, D. Li, P. Li, in *Proceedings of the ACM International Conference on Web Search and Data Mining*. Knowledge Graph Embedding Based Question Answering (Melbourne, 2019), pp. 105–113
14. Y. Xian, Z. Fu, S. Muthukrishnan, G. de Melo, Y. Zhang, in *Proceedings of the International ACM SIGIR Conference on Research and Development in Information Retrieval*. Reinforcement Knowledge Graph Reasoning for Explainable Recommendation (Paris, 2019), pp. 285–294
15. H. Zhang, T. Zheng, J. Gao, C. Miao, L. Su, Y. Li, K. Ren, in *Proceedings of the International Joint Conference on Artificial Intelligence*. Data Poisoning Attack Against Knowledge Graph Embedding (Macao, 2019), pp. 4853–4859
16. Y. Lin, Z. Liu, M. Sun, Y. Liu, X. Zhu, in *Proceedings the AAAI Conference on Artificial Intelligence*. Learning Entity and Relation Embeddings for Knowledge Graph Completion (Austin, 2015), pp. 2181–2187
17. G. Ji, S. He, L. Xu, K. Liu, J. Zhao, in *Proceedings of the Annual Meeting of the Association for Computational Linguistics and International Joint Conference on Natural Language Processing (Volume 1: Long Papers)*. Knowledge Graph Embedding via Dynamic Mapping Matrix (Beijing, 2015), pp. 687–696
18. B. Yang, W.T. Yih, X. He, J. Gao, L. Deng, in *Proceedings of the International Conference on Learning Representations*. Embedding Entities and Relations for Learning and Inference in Knowledge Bases (San Diego, 2015), pp. 1–13
19. T. Trouillon, J. Welbl, S. Riedel, É. Gaussier, G. Bouchard, in *Proceedings of the International Conference on Machine Learning*. Complex Embeddings for Simple Link Prediction (New York City, 2016), pp. 2071–2080
20. A. Bordes, X. Glorot, J. Weston, Y. Bengio, A semantic matching energy function for learning with multi-relational data. Mach. Learn. **94**, 233–259 (2014)
21. Z. Wang, J. Zhang, J. Feng, Z. Chen, in *Proceedings the Conference on Empirical Methods in Natural Language Processing (EMNLP)*. Knowledge Graph and Text Jointly Embedding (Doha, 2014), pp. 1591–1601

22. S.Guo, Q. Wang, B. Wang, L. Wang, L. Guo, in *Proceedings of the Annual Meeting of the Association for Computational Linguistics and International Joint Conference on Natural Language Processing (Volume 1: Long Papers)*. Semantically Smooth Knowledge Graph Embedding. (Beijing, 2015), pp. 84–94
23. R. Xie, Z. Liu, H. Luan, M. Sun, in *Proceedings of the International Joint Conference on Artificial Intelligence*. Image-Embodied Knowledge Representation Learning (Melbourne, 2016), pp. 3140–3146
24. N. Lao, W.W. Cohen, Relational retrieval using a combination of path-constrained random walks. Mach. Learn. **81**, 53–67 (2010)
25. H. Paulheim, Knowledge graph refinement: A survey of approaches and evaluation methods. Semantic Web **8**, 489–508 (2017)
26. P. Pezeshkpour, Y. Tian, S. Singh, in *Proceedings of the Conference on Automated Knowledge Base Construction and Conference of the North American Chapter of the Association for Computational Linguistics: Human Language Technologies (Volume 1: Long and Short Papers)*. Investigating Robustness and Interpretability of Link Prediction via Adversarial Modifications (Amherst, 2019), pp. 3336–3347
27. P. Bhardwaj, J. Kelleher, L. Costabello, D. O'Sullivan, in *Proceedings of the Annual Meeting of the Association for Computational Linguistics and International Joint Conference on Natural Language Processing (Volume 1: Long Papers), Virtual Event*. Poisoning Knowledge Graph Embeddings via Relation Inference Patterns (2021a), pp. 1875–1888
28. P. Bhardwaj, J. Kelleher, L. Costabello, D. O'Sullivan, in *Proceedings of the Conference on Empirical Methods in Natural Language Processing*. Adversarial Attacks on Knowledge Graph Embeddings via Instance Attribution Methods (Punta Cana, 2021b), pp. 8225–8239
29. K. Xu, H. Chen, S. Liu, P.Y. Chen, T.W. Weng, M. Hong, X. Liu, in *Proceedings of the International Joint Conference on Artificial Intelligence*. Topology Attack and Defense for Graph Neural Networks: An Optimization Perspective (Macao, 2019), pp. 3961–3967
30. Y. Sun, S. Wang, X. Tang, T.Y. Hsieh, V.G. Honavar, Node injection attacks on graphs via reinforcement learning (2019), arXiv preprint, arXiv:1909.06543
31. H. Chang, Y. Rong, T. Xu, W. Huang, H. Zhang, P. Cui, W. Zhu, J. Huang, The general black-box attack method for graph neural networks (2019), arXiv preprint, arXiv:1908.01297
32. B. Wang, N.Z. Gong, in *Proceedings of the ACM SIGSAC Conference on Computer and Communications Security*. Attacking Graph-Based Classification via Manipulating the Graph Structure (London, 2019), pp. 2023–2040
33. H. Wu, C. Wang, Y. Tyshetskiy, A. Docherty, K. Lu, L. Zhu, in *Proceedings of the International Joint Conference on Artificial Intelligence*. Adversarial Examples on Graph Data: Deep Insights into Attack and Defense (Macao, 2019), pp. 4816–4823
34. W. Jin, Y. Li, H. Xu, Y. Wang, S. Ji, C. Aggarwal, J. Tang, Adversarial attacks and defenses on graphs. ACM SIGKDD Explorations **22**, 19–34 (2021)
35. F. Feng, X. He, J. Tang, T.S. Chua, Graph adversarial training: Dynamically regularizing based on graph structure. IEEE Trans. Knowl. Data Eng. **33**, 2493–2504 (2021)
36. X. Xu, Y. Yu, B. Li, L. Song, C. Liu, C. Gunter, in *Proceedings of the International Conference on Learning Representations*. Characterizing Malicious Edges Targeting on Graph Neural Networks, (2018), (under review)
37. Z. Deng, Y. Dong, J. Zhu, Dong Batch virtual adversarial training for graph convolutional networks (2019), arXiv preprint, arXiv:1902.09192
38. D. Zügner, S. Günnemann, in *Proceedings of the ACM SIGKDD International Conference on Knowledge Discovery & Data Mining*. Certifiable Robustness and Robust Training for Graph Convolutional Networks (Anchorage, 2019), pp. 246–256
39. I. Goodfellow, J. Pouget-Abadie, M. Mirza, B. Xu, D. Warde-Farley, S. Ozair, A.C. Courville, Y. Bengio, in *Proceedings of the Advances in Neural Information Processing Systems*. Generative Adversarial Nets (Montreal, 2014), pp. 2672–2680

40. L. Cai, W.Y. Wang, in *Proceedings of the Conference of the North American Chapter of the Association for Computational Linguistics: Human Language Technologies (Volume 1: Long Papers)*. KBGAN: Adversarial Learning for Knowledge Graph Embeddings (New Orleans, 2018), pp. 1470–1480
41. R. Sutton, D. McAllester, S. Singh, Y. Mansour, in *Proceedings of the Advances in Neural Information Processing Systems*. Policy Gradient Methods for Reinforcement Learning with Function Approximation (Denver, 2000), pp. 1057–1063
42. T. Dettmers, P. Minervini, P. Stenetrop, S. Riedel, in *Proceedings of the AAAI Conference on Artificial Intelligence (AAAI-18)*. Convolutional 2D Knowledge Graph Embeddings (New Orleans, 2018), pp. 1811–1818
43. X. Han, S. Cao, X. Lv, Y. Lin, Z. Liu, M. Sun, J. Li, in *Proceedings of the Conference on Empirical Methods in Natural Language Processing: System Demonstrations*. Openke: An Open Toolkit for Knowledge Embedding (Brussels, 2018), pp. 139–144

Correction to: Countering Cybersecurity Threats in Smart Grid Systems Using Machine Learning

Mais Nijim, Hisham Albataineh, Viswas Kanumuri, Ayush Goyal, Avdesh Mishra, and David Hicks

Correction to:
Chapter 14 in: K. Daimi et al. (eds.), *Emerging Trends in Cybersecurity Applications*, https://doi.org/10.1007/978-3-031-09640-2_14

The chapter was inadvertently published with incorrect online content. It has been corrected so that the chapter DOI gets re-directed properly.

The updated original version for this chapter can be found at
https://doi.org/10.1007/978-3-031-09640-2_14

K. Daimi et al. (eds.), *Emerging Trends in Cybersecurity Applications*,
https://doi.org/10.1007/978-3-031-09640-2_21

Index

K. Daimi et al. (eds.), *Emerging Trends in Cybersecurity Applications*,
https://doi.org/10.1007/978-3-031-09640-2

Milton Keynes UK
Ingram Content Group UK Ltd.
UKHW020202211024
449810UK00001B/8

9 783031 096396